Foreign Direct Investment and Technological Change
Volume II

The Globalization of the World Economy

Series Editor: Mark Casson
Professor of Economics
University of Reading, UK

1. Transforming International Organizations
   William G. Egelhoff
2. Strategic Alliances
   Paul W. Beamish
3. Structural Change, Industrial Location and Competitiveness
   Joanne E. Oxley and Bernard Yeung
4. Developing and Newly Industrializing Countries (Volumes I and II)
   Chris Milner
5. Historical Foundations of Globalization
   James Foreman-Peck
6. Trade and Investment Policy (Volumes I and II)
   Thomas L. Brewer
7. The Globalization of Financial Services
   Mervyn K. Lewis
8. Foreign Direct Investment and Technological Change (Volumes I and II)
   John Cantwell

Future titles will include:

The Globalization of Business Firms from Emerging Economies
Henry Wai-Chung Yeung

Privatization, Deregulation and the Transfer to Markets
David Teece, Leonard Waverman and Pablo T. Spiller

Financial Market Integration and International Capital Flows
David Vines

Globalization and Labour Markets
David Greenaway and Robert C. Hine

Wherever possible, the articles in these volumes have been reproduced as originally published using facsimile reproduction, inclusive of footnotes and pagination to facilitate ease of reference.

Foreign Direct Investment and Technological Change

Volume II: Technology Creation and its Economic Impact

Edited by

John Cantwell

Professor of International Economics
University of Reading, UK

THE GLOBALIZATION OF THE WORLD ECONOMY 8

An Elgar Reference Collection
Cheltenham, UK • Northampton, MA, USA

Published by
Edward Elgar Publishing Limited
Glensanda House
Montpellier Parade
Cheltenham
Glos GL50 1UA
UK

Edward Elgar Publishing, Inc.
136 West Street
Suite 202
Northampton
Massachusetts 01060
USA

A catalogue record for this book is available from the British Library.

Library of Congress Cataloguing in Publication Data

Foreign direct investment and technologial change / edited by John Cantwell.
(The globalization of the world economy: 8)
(An Elgar reference collection)
Includes bibliographical references.
Contents: v.1. Theories of technological change—v.2. Technology creation and its economic impact.
1. Investments, Foreign—European Union countries.
2. International business enterprises. 3. Technology transfer—Economic aspects—European Union countries. 4. Technological innovations—Economic aspects. I. Cantwell, John. II. Series. III. Series: An Elgar reference collection.
HG5422.F67 1999
332.67'3'094—dc21 99–21905
CIP

ISBN 1 85898 929 9 (2 volume set)

Printed and bound in Great Britain by MPG Books Ltd, Bodmin, Cornwall

Contents

Acknowledgements

The editor and publishers wish to thank the authors and the following publishers who have kindly given permission for the use of copyright material.

Blackwell Publishers Ltd for articles: Edwin Mansfield, David Teece and Anthony Romeo (1979), 'Overseas Research and Development by US-Based Firms', *Economica*, **46**, 187–96; Sanjaya Lall (1979), 'The International Allocation of Research Activity by US Multinationals', *Oxford Bulletin of Economics and Statistics*, **41** (4), November, 313–31; Lars Håkanson and Udo Zander (1988), 'International Management of R&D: The Swedish Experience', *R&D Management*, **18** (3), July, 217–26.

Cambridge University Press for excerpt: David C. Mowery and Nathan Rosenberg (1989), 'International and Domestic Collaboration in Research and Development', in *Technology and the Pursuit of Economic Growth*, Chapter 9, 238–73.

Elsevier Science Ltd for articles: John Hagedoorn and Jos Schakenraad (1992), 'Leading Companies and Networks of Strategic Alliances in Information Technologies', *Research Policy*, **21**, 163–90; Ari Kokko (1994), 'Technology, Market Characteristics, and Spillovers', *Journal of Development Economics*, **43** (2), April, 279–93; Frédérique Sachwald (1998), 'Cooperative Agreements and the Theory of the Firm: Focusing on Barriers to Change', *Journal of Economic Behavior and Organization*, **35** (2), April, 203–25.

Ove Granstrand for his own excerpt: (1979), 'R&D and Corporate Internationalization', in *Technology, Management and Markets: An Investigation of R&D and Innovation in Industrial Organizations*, Chapter Six, viii, 107–29, references.

Helbing & Lichtenhahn Verlag AG Basel for article: Gary Hewitt (1980), 'Research and Development Performed Abroad by U.S. Manufacturing Multinationals', *Kyklos*, **33** (2), 308–27.

Journal of International Business Studies for article: John Hagedoorn and Rajneesh Narula (1996), 'Choosing Organizational Modes of Strategic Technology Partnering: International and Sectoral Differences', *Journal of International Business Studies*, **27** (2), 265–84.

Kluwer Law International for article: Terutomo Ozawa (1991), 'Japan in a New Phase of Multinationalism and Industrial Upgrading: Functional Integration of Trade, Growth and FDI', *Journal of World Trade Law*, **25** (1), February, 43–60.

Macmillan Press Ltd for excerpt: Sanjaya Lall (1987), 'Multinationals and Technology Development in Host LDCs', in John H. Dunning and Mikoto Usui (eds), *Structural Change, Economic Interdependence and World Development: Proceedings of the Seventh World Congress of the International Economic Association, Madrid, Spain*, Volume 4: Economic Interdependence, Chapter 13, 193–209.

Management International Review for article: Marina Papanastassiou and Robert Pearce (1997), 'Technology Sourcing and the Strategic Roles of Manufacturing Subsidiaries in the U.K.: Local Competences and Global Competitiveness', *Management International Review*, **37**, 5–25.

MIT Press Journals for article: Bruce Kogut and Sea Jin Chang (1991), 'Technological Capabilities and Japanese Foreign Direct Investment in the United States', *Review of Economics and Statistics*, **LXXIII**, 401–13.

Oxford University Press for articles: Pari Patel (1995), 'Localised Production of Technology for Global Markets', *Cambridge Journal of Economics*, **19** (1), February, 141–53; John Cantwell (1995), 'The Globalisation of Technology: What Remains of the Product Cycle Model?', *Cambridge Journal of Economics*, **19** (1), February, 155–74.

Pinter Publishers, an imprint of Cassell for excerpt: François Chesnais (1988), 'Multinational Enterprises and the International Diffusion of Technology', in Giovanni Dosi, Christopher Freeman, Richard Nelson, Gerald Silverberg and Luc Soete (eds), *Technical Change and Economic Theory*, Chapter 23, 496–527.

Routledge for excerpt: John H. Dunning and Rajneesh Narula (1996), 'The Investment Development Path Revisited: Some Emerging Issues', in John H. Dunning and Rajneesh Narula (eds), *Foreign Direct Investment and Governments: Catalysts for Economic Restructuring*, Chapter 1, 1–41.

Springer Verlag GmbH & Co. KG for article: Tommaso Perez (1997), 'Multinational Enterprises and Technological Spillovers: An Evolutionary Model', *Journal of Evolutionary Economics*, **7** (2), June, 169–92.

United Nations Conference on Trade and Development (UNCTAD) for articles: Terutomo Ozawa (1992), 'Foreign Direct Investment and Economic Development', *Transnational Corporations*, **1** (1), February, 27–54; John H. Dunning (1996), 'The Geographical Sources of the Competitiveness of Firms: Some Results of a New Survey', *Transnational Corporations*, **5** (3), December, 1–29.

John Wiley & Sons Ltd for article: Paul Almeida (1996), 'Knowledge Sourcing by Foreign Multinationals: Patent Citation Analysis in the U.S. Semiconductor Industry', *Strategic Management Journal*, **17**, Special Issue, Winter, 155–65.

Every effort has been made to trace all the copyright holders, but if any have been inadvertently overlooked the publishers will be pleased to make the necessary arrangement at the first opportunity.

In addition the publishers wish to thank the Library of the London School of Economics and Political Science, the Marshall Library of Economics, Cambridge University and B & N Microfilm, London for their assistance in obtaining these articles.

Part I
The Internationalization of Technology Creation

[1]

Economica **46,** 187–196

Overseas Research and Development by US-Based Firms

By Edwin Mansfield, David Teece and Anthony Romeo

University of Pennsylvania, Stanford University, University of Connecticut

Introduction

In recent years, the overseas research and development activities of US-based firms have become the focus of controversy. Some observers view such activities with suspicion, since they regard them as a device to "export" R and D jobs, or as a channel through which American technology may be transmitted to actual or potential foreign competitors.[1] Others, particularly the governments of many developing (and some developed) countries, view them as highly desirable activities that will help to stimulate indigenous R and D in these countries. Indeed, the United Nations Group of Eminent Persons recommended that host countries require multinational corporations to contribute towards innovation of appropriate kinds, and to encourage them to do such R and D in their overseas affiliates.[2] Although the amount of controversy in this area might lead one to believe that the nature of existing overseas R and D activities of US-based firms has been studied quite thoroughly, this is far from the case. The unfortunate truth is that economists have devoted little or no attention to even the most basic questions concerning these activities.[3] In this paper, we try to fill part of this gap.

I. Overseas R and D Expenditures: 1960–1980

How big are the overseas R and D expenditures of US-based firms, now and in the past, and how big do firms expect them to become by 1980? To help answer this question, we constructed a sample of 55 major manufacturing firms, this sample being divided into two parts. The first subsample, composed of 35 firms, included major US-based firms in the chemical, petroleum, electrical equipment, metals and machinery, drugs, glass, food and paper industries. The second subsample, composed of 20 firms, included major manufacturing firms in the southern New England and Middle Atlantic states. Table 1 shows the percentage of R and D done overseas by these firms, during 1960–1974 (for the first subsample) or 1970–1974 (for the second subsample), as well as the estimated value of this percentage in 1980. In each subsample, about 10 per cent of the total amount spent on R and D by these firms was carried out overseas in 1974. Based on the 35-firm subsample, it appears that this percentage grew substantially during the 1960s and early 1970s.[4] Based on the estimates provided by the firms in the sample, this growth will continue, but at a reduced rate, during the rest of the 1970s; by 1980 they estimated that about 12 per cent of their R and D expenditures will be made overseas.

Because of the importance in the innovation process of close communication and cooperation among R and D, marketing, production and top management, Vernon (1974) and others have argued that a firm's R and D activities will tend to be centralized near its headquarters. Why then do these US-based firms spend

TABLE 1

PERCENTAGE OF COMPANY-FINANCED R AND D EXPENDITURES CARRIED OUT OVERSEAS, 1960–1980: 55 FIRMS

	1960*	1965†	1970‡	1972†	1974	1980§
35-firm subsample:						
Weighted mean	2	6	6	8	10	10
Unweighted mean	2	4	5	7	8	8
Standard deviation	3	7	7	8	10	8
20-firm subsample:						
Weighted mean	—	—	4	—	9	14
Unweighted mean	—	—	5	—	8	11
Standard deviation	—	—	7	—	10	14

* Data were not available for 4 firms in the 35-firm subsample.
† Data were not available for 1 firm in the 35 firm subsample.
‡ Data were not available for 1 firm in the 35 firm subsample and 1 firm in the 20 firm subsample.
§ Data were not available for 9 firms in the 35 firm subsample.
Source: see Section I.

about 10 per cent of their R and D dollars overseas? There are a variety of possible reasons, including the presence of environmental conditions abroad that cannot easily be matched at home, the desirability of doing R and D aimed at the special design needs of overseas markets, the availability and lower cost of skills and talents that are less readily available or more expensive at home, and the greater opportunity to monitor what is going on in relevant scientific and technical fields abroad. In our sample, practically all of the firms doing R and D overseas say that the principal reason is to respond to special design needs of overseas markets. In their view, there are great advantages in doing R and D of this sort in close contact with the relevant overseas markets and manufacturing units of the firm.

II. FACTORS INFLUENCING THE PERCENTAGE OF A FIRM'S R AND D EXPENDITURES CARRIED OUT OVERSEAS

What determines the percentage of its R and D that a firm conducts overseas? Given the fact that overseas laboratories seem to be so closely geared to the special design needs of foreign markets (and the firm's overseas plants), we would expect that the percentage of a firm's R and D expenditures carried out overseas would be directly related to the percentage of the firm's sales that is derived from abroad. Firms with relatively small foreign markets would be expected to spend relatively little on overseas R and D. Further, we would expect that the percentage of a firm's R and D expenditures carried out overseas would be more closely related to the percentage of its sales from foreign subsidiaries than to its percentage of sales from exports. This is because much overseas R and D is in support of foreign manufacturing operations.[5]

Holding constant the percentage of a firm's sales that come from abroad, we would expect that the percentage of a firm's R and D expenditures that is carried out overseas would be directly related to the firm's size. Economies of scale require that R and D laboratories be a certain minimum size if they are to be relatively efficient. If it is going to establish an overseas laboratory, the firm must have a big enough prospective market (in the area served by this laboratory) to support a laboratory of minimum economic scale. If the percentage of a firm's sales that comes from abroad is held constant, the probability that this prospective market will be of the requisite size is an increasing function of the absolute size of the firm.

Further, we would expect that, holding constant both the firm's sales and its percentage of sales coming from abroad, there would be inter-industry differences in the percentage of a firm's R and D expenditures carried out overseas. For example, we would expect this percentage to be relatively high in the pharmaceutical industry because some firms, according to industry sources at least, have moved a substantial amount of R and D abroad to avoid Food and Drug Administration regulations. Also, foreign regulations sometimes require that R and D be done locally. Because of these regulatory considerations, as well as other factors discussed below, we might expect the drug firms in our sample to carry out a relatively high percentage of R and D overseas.

In addition, we would expect that, if the firm's sales, its percentage of sales coming from abroad and its industry are held constant, there will be differences over time in the percentage of a firm's R and D carried out overseas, owing to changes in the profitability of locating R and D overseas rather than in the United States (as well as bandwagon effects).[6] In general, during the period covered here, we would expect these effects of time to be positive, since cost differentials and other factors favoured the expansion of overseas R and D.

To test these hypotheses, we carried out two sets of computations. First, we pooled the 1970 and 1974 data for the 35-firm subsample, and regressed each firm's percentage of R and D expenditures carried out overseas on its percentage of sales from abroad, its sales, an industry dummy variable, and a time dummy variable. The results are

$$\text{(1)} \quad \underset{}{A_{it}} = \underset{(0.44)}{-1.13} + \underset{(0.34)}{0.73T_t} + \underset{(1.81)}{0.15Q_{it}} + \underset{(3.09)}{0.004S_{it}} + \underset{(5.41)}{16.10D_i}, \quad (\bar{R}^2 = 0.50;\ n = 51)$$

where A_{it} is the percentage of the ith firm's R and D expenditures carried out overseas in year t (1970 or 1974), T_t is a dummy variable that equals 1 if t is 1974 and 0 if t is 1970, Q_{it} is the percentage of the ith firm's sales derived from abroad in year t, S_{it} is the ith firm's sales (in millions of dollars) in year t, and D_i is a dummy variable that equals 1 if the ith firm is in the drug industry and 0 otherwise.[7] Each regression coefficient's t-ratio is given in parentheses.

Second, we pooled the 1970 and 1974 data for the 20-firm subsample, and regressed each firm's percentage of R and D expenditures carried out overseas on the same variables as in equation (1), except that we split the percentage of sales from abroad into two parts—the percentage of sales from foreign subsidiaries, and the percentage of sales from exports—and we redefine the industry dummy to include both chemicals and drugs, not drugs alone.[8] The results are:

$$\text{(2)} \quad A_{it} = \underset{(1.16)}{2.79} + \underset{(1.91)}{4.40T_t} + \underset{(4.47)}{0.322F_{it}} - \underset{(2.32)}{0.539E_{it}} - \underset{(1.13)}{0.00162S_{it}} + \underset{(1.76)}{4.52C_i}, \quad (\bar{R}^2 = 0.54;\ n = 39)$$

where F_{it} is the percentage of the ith firm's sales from foreign subsidiaries in year t, E_{it} is the percentage of the ith firm's sales from exports in year t, and C_i is a dummy variable that equals 1 if the ith firm is in the drug or chemical industries and 0 otherwise.

The econometric results generally are in accord with our hypotheses. As expected, equation (1) shows that there is a direct and statistically significant relationship between a firm's percentage of sales derived from abroad and its percentage of R and D expenditures carried out overseas. And when sales derived from abroad are disaggregated in equation (2), a firm's percentage of sales from foreign subsidiaries has a highly significant positive effect on its percentage of R and D expenditures carried out overseas, while its percentage of sales from exports has a significant negative effect, which suggests that these firms' exports may be more R and D-intensive than their domestic sales.[9] With regard to our hypothesis that A_{it} would be directly related to S_{it}, the results of equation (1) bear this out, but in equation (2), S_{it} does not have a statistically significant effect (and its regression coefficient has the wrong sign). As expected, most of the industry and time dummies are statistically significant.[10]

III. Overseas R and D: Minimum Economic Scale and Relative Cost

As noted above, many governments, particularly of developing countries, favour the establishment in their nations of overseas R and D laboratories by US-based firms. One factor influencing the practicality of establishing a laboratory of a certain type in a particular overseas location is the extent of economies of scale in such laboratories. If the minimum economic scale for a laboratory of this type is quite large, a firm must be prepared to shift considerable R and D resources abroad if the laboratory is to be competitive.[11] Despite the fact that data concerning the minimum economic scale of R and D laboratories of various types would be of value to many kinds of microeconomic studies, practically no information is available on this score. In this section, we present the estimates (obtained from 27 members of the 35-firm subsample) of the annual R and D expenditures for an overseas laboratory of minimum economic scale. Although these estimates should be treated with caution, they are of considerable interest, since they seem to be the first systematic evidence on this topic.[12]

The results, shown in Table 2, indicate that the minimum economic scale tends to be quite substantial in most industries. On the average, for a single product line it was estimated that the expenditures per year for an R and D facility of minimum economic scale would be about $1 million in pharmaceuticals and glass, about $2 million in electrical equipment and petroleum, and about $5 million in chemicals. However, the minimum economic scale seems to vary considerably, depending on the responsibilities of the laboratory. It is less for a laboratory that performs either research or development than for one that performs both, and less for a laboratory that deals with a single product line than for one that deals with several product lines. For a laboratory that is concerned entirely with minor product changes, the average estimated expenditure per year for an R and D facility of minimum economic scale is only about $500,000 per year—and in some industries it is substantially less. In interpreting the results in Table 2, the dispersion among the estimates is almost as interesting as the averages. The estimates in each industry vary enormously, reflecting the fact that the minimum economic scale of an R and D laboratory depends on the specific type of work to be done, as well as the fact that opinions differ on this score even among experts.[13]

TABLE 2

ESTIMATED ANNUAL R AND D EXPENDITURE FOR OVERSEAS LABORATORY OF MINIMUM ECONOMIC SCALE, 1975: 27 FIRMS ($million)

	Single product line				Several product lines			
Industry*	Research	Development	Minor product changes	Research and development	Research	Development	Minor product changes	Research and development
Chemicals ($n = 7$)								
Mean	2·42	3·27	0·72	4·57	2·50	2·46	1·39	3·31
Range	1·0–5·0	0·1–14·0	0·03–2·0	0·46–20·0	1·5–3·5	0·26–6·0	0·24–3·5	0·16–6·7
Petroleum ($n = 6$)								
Mean	1·64	1·46	0·28	2·30	2·60	2·23	1·18	3·40
Range	0·25–3·0	0·25–3·0	0·15–0·50	0·40–5·0	0·40–5·0	0·40–4·5	0·25–3·0	0·6–7·5
Drugs ($n = 2$)								
Mean	3·25	0·50	0·12	1·00	6·00	1·30	0·35	†
Range	1·5–5·0	0·5–0·5	0·10–0·13	0·50–1·5	5·0–7·0	0·60–2·0	0·19–0·50	†
Electronics and electrical equipment ($n = 5$)								
Mean	1·00	1·15	0·40	2·00	2·42	3·95	0·77	6·75
Range	0·8–1·2	0·36–2·0	0·20–0·50	1·25–2·75	1·0–5·0	0·54–8·0	0·51–1·0	1·25–10·0
Glass ($n = 2$)								
Mean	0·70	0·85	0·42	1·38	0·73	1·53	0·71	1·88
Range	0·1–1·3	0·4–1·3	0·08–0·75	0·50–2·25	0·20–1·25	0·80–2·25	0·16–1·25	1·0–2·75
Total ($n = 27$)								
Mean	1·58	1·82	0·47	3·00	2·62	2·83	1·23	4·10
Standard deviation	1·34	2·78	0·45	4·10	1·87	2·76	1·30	3·68

* n is the number of firms that provided estimates.
† An estimate was obtained from only one firm.
Source: see Section I.

According to many observers, one major reason why US-based firms have carried out R and D overseas is that costs have tended to be lower there. However, very little information has been published concerning the extent of this cost differential, and how it has varied over time. To help fill this gap, we obtained data from the 35-firm subsample concerning the ratio of the cost of R and D inputs in Europe, Japan and Canada to those in the United States in 1965, 1970 and 1975.[14] The results, shown in Table 3, indicate that there was a very substantial cost differential in 1965: on the average, the cost of R and D inputs seemed to be about 30 per cent lower in Europe, 20 per cent lower in Canada and 40 per cent lower in Japan than in the United States. And although there was some increase in R and D costs relative to those in the United States during 1965–1970, the cost differential remained quite substantial in 1970.

TABLE 3

MEAN RATIO OF COST OF R AND D INPUTS IN SELECTED OVERSEAS LOCATIONS TO THAT IN THE UNITED STATES, 1965, 1970 AND 1975.* 35 FIRM SUBSAMPLE†

Year	Location		
	Europe‡	Japan	Canada
1965	0·68	0·56	0·82
1970	0·74	0·60	0·86
1975	0·93	0·90	0·96

* Note that there are many costs of communication and coordination in a multinational network of laboratories. See Mansfield (1974).

† Usable data were obtained from 19 firms. Many of the rest had no overseas R and D experience.

‡ There are considerable differences within Europe in the level of R and D costs. According to a number of firms in our sample, costs tend to be relatively low in the United Kingdom and relatively high in West Germany.

However, between 1970 and 1975 the situation changed drastically. Owing in part to the depreciation of the dollar relative to other currencies between 1970 and 1975, the cost differential was largely eliminated for many firms. On the average, the cost of R and D inputs was estimated to be about 10 per cent lower in Japan, and about 5 per cent lower in Europe and Canada, than in the United States in 1975. Of course, this helps to explain the fact (noted in Section I) that the percentage of R and D carried out overseas was expected to increase less rapidly between 1974 and 1980 than in the period prior to 1974. Since the cost differential between overseas and domestic R and D was smaller, it is quite understandable that firms would expect this percentage to grow less rapidly than in earlier years.[15]

IV. OVERSEAS R AND D: NATURE OF WORK AND RELATION TO DOMESTIC R AND D

Some observers, as we have seen, are suspicious of overseas R and D because they fear that it may be a channel through which American technology may "leak out" to foreign competitors. The extent to which such a leakage is likely to occur depends in part on the nature of the work being carried out in the overseas laboratories of US-based firms. For example, if such work is focused largely on the modification and adaptation of products and processes for the local market, there is less need to transfer much of the firm's most sophisticated technology overseas than if the work is focused on major product or process developments intended for a worldwide market. Based on information obtained from 23 firms in our sample, it appears that these firms' overseas R and D activities tend to focus on development rather than research, on product and process improvements rather than on new products and processes, and on relatively short-term, technically safe work.

Specifically, on the average the percentage of overseas R and D going for basic research is about 6 percentage points less than the percentage of domestic R and D going for basic research; the percentage of overseas R and D going for applied research is 10 percentage points less than the percentage of domestic R and D going for applied research; while the percentage of overseas R and D going for development is 16 percentage points greater than the percentage of domestic R and D going for development. Moreover, about three-fourths of these firms' overseas R and D expenditures are aimed at product or process *improvements* and *modifications*, not at entirely new processes or products. This percentage is much higher than for all domestic R and D.

Firms seem to differ considerably in the extent to which they have integrated their overseas R and D with their domestic R and D.[16] Worldwide integration of overseas and domestic R and D exists in almost one-half of the firms (with overseas R and D) in our sample, according to the firms. On the other hand, about one-sixth say that they attempt no such integration, and the rest say that some limited integration is attempted.

Finally, of how much value is overseas R and D to a firm's US operations? Policy-makers are interested in this question because it must be considered in any full evaluation of the effects of overseas R and D (and foreign direct investment) on America's technological position *vis-à-vis* other countries. Unfortunately, practically no evidence exists on this score. To shed a modest amount of light on this question, we obtained estimates from 27 firms in our sample concerning the percentage of their 1975 overseas R and D expenditures with no commercial applicability to their US operations. The results indicate that, on the average, about one-third of these firms' overseas R and D expenditures have no such applicability. Also, we asked each firm to estimate the amount that it would have to spend on R and D in the United States to get results of *equivalent value to its US operations* as a dollar spent overseas. The results, which are only rough, indicate that, on the average, a dollar's worth of overseas R and D seems to result in benefits to these firms' US operations that are equivalent to about 50 cents' worth of R and D carried out in the United States.

V. CONCLUSIONS

Overseas R and D expenditures by US-based firms topped the billion-dollar mark in the early 1970s. In 1974 they amounted to about one-tenth of total domestic company-funded R and D expenditures, and the firms in our sample reported that they expected them to amount to a larger proportion of their domestic R and D expenditures by 1980.[17] When compared with the total R and D expenditures in various host countries, their size is perhaps even more striking. In the early 1970s about one-half of the industrial R and D performed in Canada and about one-seventh of the industrial R and D performed in the United Kingdom and West Germany was done by US-based firms.[18]

Yet despite the magnitude and importance of these overseas R and D activities, little is known about their purpose, nature or effects. Indeed, the very existence of such activities is ignored in all current econometric studies carried out to estimate the effects of R and D on US productivity growth. Our purpose in this paper has been to present some basic information concerning the size, nature, minimum economic scale and relative cost of overseas R and D. The limitations of these findings should be recognized. In particular, our results pertain to a sample of 55 firms, and some of the data obtained from the firms were necessarily rough. None the less, we believe that these results, although only a first step, shed substantial new light on this topic.

ACKNOWLEDGMENTS

The work on which this paper is based was supported by a grant to Edwin Mansfield from the Division of Policy Research and Analysis of the National Science Foundation. Of course, the views expressed here are not necessarily those of the Foundation. We are grateful to the more than 50 firms that provided us with data concerning their overseas R and D activities. Some of the results were presented in papers given by Mansfield at Johns Hopkins, Vanderbilt and Yale Universities, and at the National Science Foundation, as well as the 1978 annual meeting of the American Economic Association.

NOTES

1 For discussion of this point of view, see David (1974) and Conference Board (1976).

2 United Nations (1974). *The Impact of Multinational Corporations on Development and on International Relations*. New York, p. 70.

3 Caves (1974), Hufbauer (1974), Mansfield (1974) and Stobaugh (1974) have pointed out the need for work concerning this and related aspects of international technology transfer. For some interesting case studies, see Ronstadt (1975). Also, see Mansfield, Romeo and Wagner (1979) for some related findings regarding international technology transfer.

4 The Conference Board (1976) has estimated the total overseas R and D expenditure of US-based multinational firms in 1971–1973. According to its estimates, overseas R and D constituted about 9–10 per cent of total R and D expenditures carried out by US firms during these years. This agrees quite well with our results for 1972 and 1974. The proportion of firms in our sample with no overseas R and D is somewhat lower than that reported by the Conference Board for firms of comparable size, but this may be due to different industry mix, the later year, or sampling error. The US Department of Commerce (undated) has estimated the total overseas R and D expenditure of US firms in 1966. According to its figures, overseas R and D constituted about 7 per cent of all R and D expenditures carried out by US firms in 1966 (see Conference Board, 1976). This agrees reasonably well with our result for 1965. In 1978 the National Science Foundation published data for 1976 which indicated that overseas R and D constituted about 7 per cent of all R and D expenditures by US manufacturing firms.

5 Suppose that a firm's desired R and D expenditures in a given year equal

$$R = a_u S_u + a_f S_f + a_e S_e$$

where S_e is its export sales during the relevant year, S_f is its sales through foreign subsidiaries, S_u is its sales from domestic plants to domestic customers, a_u is the proportion of sales to domestic customers that it wants to devote to R and D, a_f is the proportion of sales through foreign subsidiaries that it wants to devote to R and D, and a_e is the proportion of export sales that it wants to devote to R and D. If only the R and D in support of foreign subsidiaries is done overseas, it follows that the proportion of its R and D expenditures carried out overseas equals

$$P = \frac{a_f F}{a_u U + a_f F + a_e E}$$

where F is the proportion of its sales from foreign subsidiaries, U is the proportion of its sales to domestic customers, and E is the proportion of its sales that are exports. Under these circumstances, it can be shown that $\partial p/\partial F$ is always positive, but whether or not $\partial p/\partial E$ is positive depends on whether or not $a_u > a_e$. Of course, this model is a polar case, but it illustrates the point in the text.

[6] In terms of the highly simplified model in n. 5, the a's are a function of time. (Also, as indicated previously in the text, they are a function of the firm's size.)

[7] Other industry dummies were tried in equation (1), but D_i was the only one that was statistically significant. The reason why n is less than 70 is that data could not be obtained concerning the percentage of sales from abroad for all firms in both years.

[8] The reason for this redefinition is that none of the firms in this subsample is really an ethical drug fim. (There are several such firms in the other subsample.) The closest we could come to ethical drugs is the chemical firms, some of which do some work in the drug area (broadly defined).

[9] This result concerning E_{it} would be expected if $a_e > a_u$ in n. 5, and if the extremely simple model given there were valid. However, although it may be a reasonable approximation to regard some firms' overseas R and D as being entirely in support of foreign subsidiaries, this is far from the case in other firms.

[10] The industry dummy is much larger and more highly significant in equation (1) than in equation (2) because, as pointed out in n. 8, none of the firms in equation (2) is really an ethical drug firm. Because of sampling variation, the estimate of D_i in equation (1) is probably too large. One of the drug firms in our sample carried out an unusually large percentage of its R and D overseas.

[11] By minimum economic scale we mean the smallest scale that realizes all, or practically all, of the relevant economies of scale.

[12] Freeman *et al.* (1965) have presented some relevant data concerning the electrical equipment industry. Eight members of the 35-firm subsample could not provide estimates, sometimes because they had no experience on which to base such estimates.

[13] These figures help to explain why, holding other factors constant, smaller firms in equation (1) tend to carry on a smaller percentage of their R and D overseas than bigger firms. But they should not be interpreted as saying that smaller firms are squeezed out completely. The estimates in each industry vary enormously. In most industries, at least some of the respondents felt that research and development could be carried out effectively with an annual budget of $500,000, and that minor product changes could be carried out with one of about $100,000. Although these levels of expenditure are hardly trivial, they are within the reach of many firms other than the billion-dollar giants.

Needless to say, these results in no way contradict the finding by many economists that small firms and independent inventors continue to play an important role in the inventive process. Their contribution is frequently in the earlier stages of the inventive process, where the costs are relatively low. Further, according to some observers, costs tend to be lower in smaller organizations, and the figures in Table 2 reflect the perceptions of large firms. For some relevant discussion, see Mansfield *et al.* (1977).

[14] The relative cost of R and D inputs is the ratio of the annual cost of hiring an R and D scientist or engineer (together with the complementary amount of other inputs) in various overseas locations to do the sort of work carried out there to the annual cost of hiring a comparable R and D scientist or engineer (together with the complementary amount of other inputs) to do the same sort of work in the United States. Each firm was asked to estimate this ratio for each year. Many of the estimates were based on studies the firms seem to have carried out in recent years on this topic.

[15] If very significant differences exist between the productivity of US and overseas R and D personnel, they may offset the observed differences in the relative costs of inputs. About 80 per cent of the firms in our sample regarded the productivity of their R and D personnel in Canada, Europe and Japan to be no lower than those in the United States. Thus, this factor cannot offset the observed difference in the relative cost of R and D inputs in the great majority of firms in our sample.

[16] By integration, we mean that the firm's worldwide R and D is viewed as a whole, and laboratories are given worldwide missions, if this seems desirable.

[17] In early 1977, the United States Treasury put into effect a new regulation (1.861-8) that, according to some observers, may increase the amount of R and D done overseas by US-based firms. Since the forecasts in Table 1 were made before this new regulation was announced, they do not take this factor into account. Also, they do not take account of recent changes in exchange rates, which may have tended to discourage overseas R and D.

[18] See Conference Board (1976, p. 86).

REFERENCES

CAVES, R. (1974). Effect of international technology transfers on the U.S. economy. *The Effects of International Technology Transfers on U.S. Economy*, Washington, DC: National Science Foundation.

CONFERENCE BOARD (1976). *Overseas Research and Development by U.S. Multinationals, 1966–1975*. New York: Conference Board.

DAVID, E. (1974). Technology export and national goals. *Research Management*, **17**, 12–16.

FREEMAN, C., HARLOW, C. and FULLER, J. (1965). Research and development in electronic capital goods. *National Institute Economic Review*, **34**, 40–91.

HUFBAUER, G. (1974). Technology transfers and the American economy. In *The Effects of International Technology Transfers on U.S. Economy*. Washington, DC: National Science Foundation.

MANSFIELD, E. (1974). Technology and technological change. In *Economic Analysis and the Multinational Enterprise*, (J. Dunning, ed.). London: George Allen and Unwin.

—— *et al.* (1977). *The Production and Application of New Industrial Technology*. New York: W. W. Norton.

——, ROMEO, A. and WAGNER, S. (1979). Foreign trade and U.S. research and development. *Review of Economics and Statistics*, forthcoming.

RONSTADT, R. (1975). R and D abroad: the creation and evolution of foreign R and D activities of U.S. based multinational enterprises. Unpublished DBA thesis, Harvard University.

STOBAUGH, R. (1974). A summary and assessment of research findings on U.S. international transactions involving technology transfers. *The Effects of International Technology Transfers on U.S. Economy*. Washington, DC: National Science Foundation.

US DEPARTMENT OF COMMERCE (undated). *U. S. Direct Investments Abroad, 1966*, Part II. Washington, DC: US Government Printing Office.

VERNON, R. (1974). The location of economic activity. In *Economic Analysis and the Multinational Enterprise* (J. Dunning, ed.). London: George Allen and Unwin.

[2]

Extract from 'R&D and Corporate Internationalization'

ABBREVIATIONS

For corporations

Ch–Corporation	= Chemical Corporation
El–Corporation	= Electronic Corporation
E&F–Corporation	= Engineering & Food Corporation
E&S–Corporation	= Engineering & Steel Corporation
M&M–Corporation	= Mining & Metallurgy Corporation
P&P–Corporation	= Pulp & Paper Corporation
Ph–Corporation	= Pharmaceutical Corporation
Tp–Corporation	= Transportation Corporation

El–Subsidiary, Ph–Subsidiary and Tp–Subsidiary (within aerospace) refer to specific subsidiaries in the corresponding corporations.

Prefix to positions

D	= Divisional level

Other abbreviations used are:

MNC	= Multinational corporation
R&D	= Research and development
S&T	= Science and technology

R&D AND CORPORATE INTERNATIONALIZATION

6.1. INTRODUCTION

6.1.1. Aims and Limits

The aim of this chapter is to explore the relationship of on one hand R&D operations in a corporation and on the other hand the process of internationalization of a corporation and the rationales for this process will be penetrated. A description and normative discussion of current practice in managing R&D operations in a multinational setting will, however, be limited.

6.1.2. Concepts

The term *internationalization* in this study is taken to mean the expansion of certain operations to other nations. Corporate internationalization then refers to corporate operations in general and internationalization of R&D then refers to the R&D operations of a corporation. Moreover, internationalization may refer to both the process of expansion and a state in this process at a specific point in time. In the latter case a degree of internationalization may be assessed. By a multinational corporation (MNC), one usually means a corporation which has acquired a certain degree of internationalization. There are several ways to assess a degree of internationalization or a degree of multinationality. The process of internationalization of an industry, a corporation or a unit or function in a corporation and the states in this process may be described by many variables. These variables are, moreover, often correlated. Thus, if one uses MNC's as a category of corporations, analyses and conclusions will be dependent on the choice of variables used in defining this category.

Reviews of concepts and variables used and proposed in studies of MNC's are found, for example in Aharoni (1971), Robinson (1967), Perlmutter (1969), and Vahlne (1974). In short, Aharoni distinguishes between definitions of multinationality based on:

a) Structural variables. (For example the number of nations in which the corporation has located operations of different kinds or the number of nations represented among owners or corporate management).

b) Performance variables. (For example, absolute or relative measures of sales, assets, or employees related to operations in different nations).

c) Behavioral variables. (For example, national or multinational orientation in top management).

Attempts have been made to classify MNC's into different types depending on how connections to and considerations of different nationalities influence the conduct of operations. Robinson (1967) distinguishes between international, multinational, transnational, and supranational corporations. The latter two types cannot be assigned a nationality and supranational corporations in addition are controlled by international legislation and authorities. In a similar vein of thought, Perlmutter (1969) distinguishes between etnocentric, polycentric, and geocentric corporations, the latter type operating without any considerations of nationality and for which the distinction between parent nations and host nations does not apply. However, almost all corporations, which are multinational by some standard, still have a single or double nationality in a clear sense and much xenophobia around MNC's derives from this fact.

In this study it has been considered more relevant to use degrees of internationalization than to specify a concept of multinationality. Certainly all corporations in the sample have a nationality, and all of them have a substantial character of multinationality with respect to foreign location of production or foreign location of sales. Half of them also have multinational R&D in the sense that R&D operations are located in three or more nations. (See Table 10:3.)

As already mentioned, degrees of internationalization and multinationality may be assessed in different ways. The relative measure of internationalization commonly used here will be the ratio between non--domestic employees to total employees. Sometimes also simple counts of nations and nationalities will be used. A pure increase in export intensity will not be considered as an increase in internationalization. However, the presentation and discussion will mainly be made in qualitative terms.

6.1.3. Problem Background

It is almost superfluous to point out the concern over large, multinational corporations and their relations to nation states, labor organizations, and society as a whole. Corporations and nations become connected in a matrix network and it may be asked to what extent such a structure contribute to international stability and distribution of wealth. Almost, all MNC's still have a clear nationality at the same time as they have a certain amount of discretion in international affairs. Additional basic questions to be asked then is why certain corporations become multinational and why certain nations become hosts for parts of MNC's.

Scientific and to a certain extent also technological knowledge have in a sense historically been internationalized public goods. A growing internalization of R&D in industry and the rise of MNC's may alter this picture in the long run. At the same time it has been claimed that MNC's contribute to international diffusion of technology. In fact, the possibilities of cross-national transfer of non-physical resources may characterize the operations of a MNC to larger extent than the multinational location of physical resources.

6.2. METHODOLOGY

The empirical findings have to be judged in the light of the exploratory nature of the study. Some occurencies, tendencies and counterexamples may be described but little can be said about the extent to which something is true. Not only that the sample is small, the level of precision varies among the corporations in the sample. For one thing it is difficult to get an accurate picture in a large, diversified MNC. As diversification increases, the product level becomes a more appropriate level relative to the corporate level for studying technological innovation and diffusion. Research efforts could be distributed proportional to the complexity of the objects in the study in this sense and with the exception of the El-Corporation this has actually been the case, roughly speaking. Still, however, the main focus has remained on the corporation and its R&D function.

Another complication derives from the varying points in time when the corporations start to internationalize. One may find expressed motives and rationales in documents why certain corporations internationalized early, while it is harder to say why such a step was not taken far back in time in other corporations.

Finally, there are several ways to quantify a degree of internationalization. Although accurate as single measures, they cannot by themselves capture the multidimensional and qualitative content of internationalization.

6.3 LITERATURE SUMMARY

International economics has been a recognized subject since long. In fact, many topics wich are relevant today were raised an discussed in the early 19th century, not only by classical economists but also by practitioners as well as by investigators like Babbage (1832). Mostly the focus has been on international trade and productive factors such as capital and labor, while the technology factor has been neglected (see Kindleberger 1968). This focus is also reflected in earlier theory formation, and in the 1960's and 1970's much criticism of different theories has been raised (see, for example, Vernon 1966 and Leroy 1976).

Although MNC's have existed a century or so, the literature about MNC's is of recent date. (Note 6:1.) Parts of literature originate from labor organizations, from government organizations, and from educational institutions. Many nations have a nationality-oriented literature on MNC's concerning both the movement abroad of domestic industry and the domestic establishments of foreign industry, see, for example, SOU (1975:50), Iger (1976), and Johansson (1967). Nations like USA and Japan with viable MNC's have gained a special interest, see, for example, Vernon (1971) and Yoshino (1976). Assessments of current international competitive strength of domestic industries are frequently made (Ohlsson 1976, IVA 1979) and studies of the history of national MNC's are emerging (Carlsson 1979, Johansson and Wiedersheim-Paul 1975).

A large part of the literature on MNC's is in the business management tradition (Robinson 1978, Ghertman and Leontiades 1978). There are several foci, such as corporate or strategic planning in MNC's (Steiner 1969, Channon 1973, Ringbakk 1976), international marketing, international finance, relationships between headquarters and subsidiaries, organizational structure and strategy in MNC's (Fouraker and Stopford 1968), international patent planning (Shipman 1967), international tax planning, and transfer of managers in MNC's (Edström and Galbraith 1977).

Literature on R&D, technology, and innovation in relation to international trade and MNC's is growing. Gruber et al. (1967) found close relations on the macro level between R&D and international trade. Leroy (1976) has surveyed product strategies of some MNC's in innovation and diffusion of new products. The role of MNC's in technology transfer has been much discussed, (see Quinn 1969, SOU 1975:50.) The role of patents in innovation and diffusion has been studied (Silberston 1975) as well as the role of international licensing (Wilson 1977).

Literature which more specifically focuses on internationalization of R&D and the conduct of multinational R&D operations in MNC's is much more scarce. Ronstadt (1978) reports on R&D investments made abroad by seven United States MNC's and Dalborg (1974) reports on considerations in locating R&D in some Swedish MNC's. In addition there are some case studies (Hanson 1971, Papo 1971, Potter 1971, Van Rumker 1971).

6.4. EMPIRICAL FINDINGS

6.4.1. History of R&D and Corporate Internationalization

In this section the empirical observations will be presented in summarized form. A fuller account of historic patterns and present strategic thinking and behavior in the corporations under study is presented in Chapters 4, 5, and 7. The present organization and location of R&D are given in Chapter 10. Figure 6:1 illustrates the nation-wide expansion of operations. Much more may be said about what operations have been located when, where, and why. There are two main factors, however, which are important in the early process of corporate internationalization, namely product inventions and raw materials. These factors interact with other factors, primarily initial attitudes towards internationalization, the small size of the Swedish market, and the need for heavy investments in exploiting raw materials. However, it is hardly possible to assess the relative importance of different factors. The history of corporate and R&D internationalization will be treated below for two groups of corporations according to the main factors product inventions and raw materials. It should be noted, however, that this it but a rough grouping. (See Chapter 4.)

The corporations that are primarily product-invention-based are the El-, E&F-, E&S-, and Ph-Corporations. For these corporations, R&D in the form of a product invention with a world-wide sales potential obviously was most decisive in their initial internationalization. All four corporations had small domestic markets, and in the cases of the E&F- and the E&S-Corporations there was a conscious striving by the inventor from the beginning to "capture the world market." These corporations also early made foreign direct investments both in marketing and production. The high level of product technology motivated the use of their own subsidiaries rather than agents in international marketing. Also, the large and growing advanced markets in industrialized nations like England, France, Germany, and USA were primary targets. (Note 6:2.) Production operations were at an early stage located in some of these nations as well. Among the reasons were a desire to overcome protective trade barriers. (Note 6:3.)

This kind of entrepreneurial internationalization during the decades around the turn of the century is a conspicuous feature. One may, for instance, note that in a small nation there were incentives to go abroad because of market size, but on the other hand, there was among many people a fear of the technical superiority and large-scale advantages of the industry in large industrialized countries. Also a kind of reverse to internationalization occurred in that some Swedish companies, including the Ph-Corporation and the Tp-Corporation, were formed with foreign ones as a protoype, for instance, the German chemical industry, the German and Swiss pharmaceutical industry, and the American car industry.

112

FIGURE 6:1

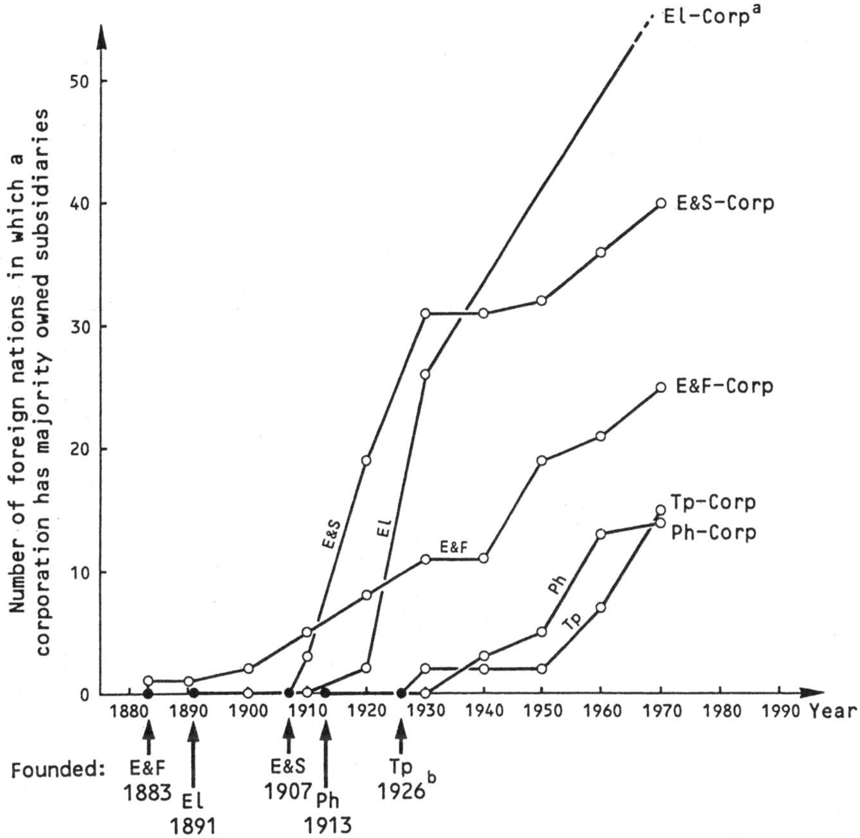

Note:

a) 64 foreign nations in 1978.

b) Legally, the Tp-Corporation was founded in 1915. However, operations were initiated in 1926.

Sources: Corporate documents, SOU 1975:50.

The subsequent internationalization in the E&F- and E&S-Corporation is hard to assess in general terms. One observes a mixture of acquisitions abroad, formation of new subsidiaries, periods of market expansion, and periods of economic and technological consolidation, investments in new plants abroad, etc. What might be observed for the E&F- and the E&S-Corporations in the 1930's and 1940's is a stage of more gradual internationalization following the initial rapid one. R&D seem to have played a minor role in this second phase of further internationalization, at least in the E&S- and E&F-Corporations. Also, rapid invention-based internationalization was just not a possibility in the early stages of industrialization as the Ph-Corporation exemplifies. The Ph-Corporation had existed domestically for four decades before a rapid invention-based internationalization took place after World War II.

The internationalization of the R&D function in terms of foreign location was mainly a feature during and after World War II. While the establishment of sales subsidiaries and production plants abroad and, to some extent, also foreign acquisitions seem to have resulted from rather conscious strategies, the same does not hold for R&D. Two factors seem to have been of primary importance in initial internationalization of R&D, namely local ambitions in subsidiaries and acquisitions "containing" R&D. The El-Corporation provides a special case in that R&D personnel went to foreign sibsidiaries to escape the Nazis. The small size of the domestic supply market for R&D personnel has then contributed to continued internationalization of R&D after World War II.

World War II brought about a breakdown of corporate control of foreign operations, and international coordination of R&D was not a primary issue immediately after World War II. Due to ambitions among technicians and management in subsidiaries with production facilities, some R&D developed. It is also interesting to note how some subsidiaries in the El-Corporation and the E&S-Corporation have started R&D in connection with contract work for the local national defense. National secrecy is here useful for local management in gaining more independence vis-à-vis corporate headquarters. The same goes for the anti-trust laws of the United States, which together with interests of local management, to some extent have impeded coordination with subsidiaries in that country. Traditionally, the United States subsidiaries have gained a high degree of independence. Of course, there are many more factors intervening, for example, the need for adapting products to local markets and the need to have local quality control and technical services, but often local management and engineers have been a primary source of influence in initiating R&D work abroad. Although a local objective, it has not initially been a part of corporate strategy. Moreover, acquisitions of foreign companies "containing" some R&D have, in general, been made for other reasons than to internationalize R&D. Thus, one may conclude that R&D has not been initially internationalized due to a conscious corporate strategy.

As far as subsequent internationalization of R&D is concerned, there is a component of gradual behavior as in the case of total corporate internationalization, but there are also examples of strategic decisions. The E&S-Corporation, which had internationalized early, had diversification and multinational coordination as emerging strategic issues in the 1960's. Multinational R&D coordination was part of these issues, and it was (rightly) thought that a substantial investment in a central R&D laboratory would promote coordination. The rationale behind location abroad of this laboratory was, of course, complex, but a more distinct reason was that a non-domestic, geographically "neutral", and central location would facilitate communication and coordination. The laboratory was also intended to be truly multinational in the sense of having a mizture of nationalities among the regular personnel in contrast to some other internationalized R&D organizations consisting of several national R&D-laboratories. It might also be noted that the chosen foreign location of the central R&D laboratory was neither close to corporate headquarters, nor to local production or marketing facilities.

This example of a foreign location of an R&D-laboratory was not the only one in Swedish engineering industry in the first half of the 1970's. R&D personnel economy, specialized fields of knowledge, recruitment problems, and closeness to sources of knowledge and markets were (and are) influencing R&D managers to consider foreign R&D locations. So far, however, there has not been a "wave" of R&D locations abroad, and most corporations want to have a large part of their R&D "at home", at least as far as the more advanced parts of R&D are concerned.

The second group of corporations consists of those which are primarily raw-material-based and clearly includes the M&M-Corporation and the P&P-Corporation and somewhat less clearly, the Ch-Corporation. These corporations have been oriented around domestic resources and capital investments to exploit these resources, and they formerly perceived no specific incentive to internationalize. It was not until the 1960's that integration forwards, diversification, and foreign supply of raw materials emerged as strategic issues. There also were internally developed unwritten policies or management attitudes that limited internationalization. Thus, for instance, the M&M-Corporation is said to have respected each nation's sovereignity over natural resources and also to have had the policy to "stop at the refined metal." In the mining business there has been much secrecy around process technology, which until recently has impeded technology trade. From this brief description it then seems as if internationalization of the M&M-Corporation has previously been blocked in several ways. However, one may note that some international interaction among metal producers also has existed concerning, for example, plant capacities, and also some R&D for widening the application base of metals and by-products.

Without going too far into reasons behind the late start of an internationalization process of this group of corporations, one may just add the influence of production and transportation economies on management concern in these industries. Marketing and to some extent also diversification have traditionally had a lower "share of mind."

6.4.2. Present Situation

Corporate Internationalization

Figure 6:2 shows recent corporate development with respect to diversification and internationalization. (See also Chapter 7 for present corporate policies in these respects.) The spread is apparent, but a common trend towards a higher degree of corporate internationalization may be observed. Although data are largely missing, degrees of internationalization with respect to product divisions and different functions could be similarly assessed. In general, marketing is more internationalized than production which is more internationalized than R&D for each corporation. For the Ph-Corporation the degrees of internationalization for marketing, production, and R&D were in 1975, 0.72, 0.48, and 0.17 respectively. For the E&F-Corporation R&D is internationalized to a degree of roughly 0.2. Thus, in both cases R&D is internationalized to roughly one third of the degree of corporate internationalization.

FIGURE 6:2

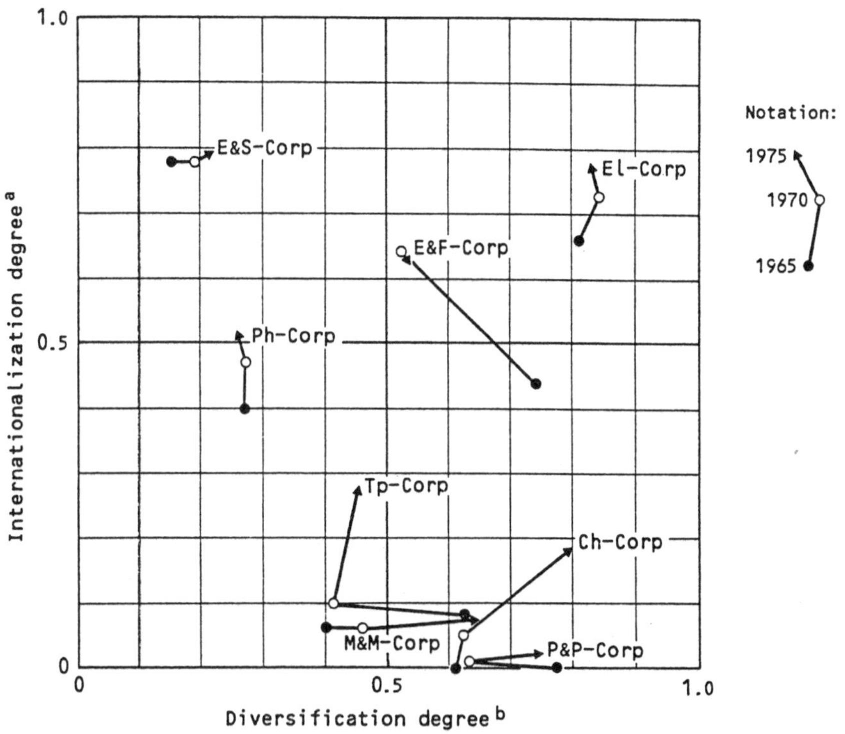

Note:

a) Employees abroad/total numbers of employees.

b) Sales outside largest product area/total sales.

The highly internationalized corporations also have a high ratio of foreign sales to total sales but the converse does not hold as seen from Table 6:1. The raw-material-based corporations are particularly export intensive and together with the early internationalized corporations they have stayed fairly stable so far in the 1970's with respect to both internationalization degree and foreign sales intensity. The group of corporations presently in transition in these respects are the Ch-Corporation, the Ph-Corporation, and the Tp-Corporation. Of these, only Tp--Corporation has a higher pace of change in internationalization degree than in foreign sales intensity. It must, however, also be noted that the early internationalizers, the E&F-Corporation, the El-Corporation, and the E&S-Corporation, are in transition in the sense that coordination of multinational operations has been increasing. For example, this gives an increase in cross-national deliveries and world trade, which also may give an increase in export intensity (that is, exports from many nations to many nations rather than from one nation to many nations.)

TABLE 6:1

Internationalization and Foreign Sales Intensity[a)]

Variable		Corporation							
		Ch	El	E&F	E&S	M&M	P&P	Ph	Tp
Foreign sales/ total sales	1970	0.21	0.9	0.84	0.92	0.45	0.66	0.43	0.65
	1975	0.39	0.9	0.83	0.91	0.47	0.60	0.54	0.71
Employees abroad/total employees	1970	0.05	0.73	0.64	0.78	0.06	0.01	0.47	0.10
	1975	0.18	0.77	0.63	0.79	0.07	0.02	0.51	0.27

a) Internal deliveries excluded.

In dealing with reasons for or objectives of internationalization of a corporation, one has to view them in relation to other strategic objectives such as growth, profitability, and diversification. (See Chapters 5 and 7.) A list of expressed objectives linked to internationalization includes:

- o Growth through larger markets
- o Diversification of markets with different features of maturity, business cycles, economic conditions, etc.
- o Decreased overall dependence on local input and output markets and nation state conditions
- o Increased possibilities to benefit from transfer and allocation of resources
- o Increased access to local markets
- o Increased access to international capital and labor markets
- o Increased access to sources of raw materials
- o Increased possibilities to recruit managers and R&D personnel
- o Increased access to international S&T

One may note the complexity of a strategy of further internationalization, a strategy which is encountered in almost all the corporations studied. Many objectives are also self-reinforcing, for example, growth requiring investments which require large sales on international markets to pay off or internationally marketable products requiring heavy R&D, efforts which again require large sales to pay off. Also, exploitation of arising business opportunities as well as a willingness to learn from international ventures may be simple but general reasons for an international outlook.

The ways to achieve internationalization are many, but for late internationalizers, such as the Ch-, M&M-, P&P-, and also the Tp-Corporation, acquisitions or cooperation are the most common. The need to secure markets has led the M&M-Corporation to integrate forward and make acquisitions abroad, and one acquisition in the early 1970's was made very much because of the international marketing resources of the acquired company. This acquisition also led to a more radical diversification. One further reason for the striving for internationalization in the M&M-Corporation is that the domestic raw materials are complex ores in low concentrations, so the corporation has developed an unique know-how which is now looked upon as internationally marketable. Moreover, M&M has started to explore foreign raw material sources.

One could go on like this and describe the complex interrelatedness of corporate objectives. (See also Chapter 7.) However, it is enough here to point out the diversity of cooperative options and give a few examples:

- o The Ch-Corporation has recently tried to merge with another large Swedish corporation with a large network of foreign subsidiaries, but in the nearest future, the plans will probably not go further than a certain cooperation.

- o A Swedish subsidiary of the Ch-Corporation capitalizes on its highly developed and specialized technology by selling it to third world countries. In order to cope with demands for national ownership, one makes so-called management agreements and provides the management of the facilities, while the corporations are nationally majority owned. Extension of these agreements in the future is supposed to be achieved by developing attractive proprietory technology in the meantime.

- The Tp-Subsidiary performs further development of jet engines on a contract basis for a large American manufacturer that has done original development work. R&D in this area is highly expensive, which makes joint ventures attractive both for small companies that cannot enter the market by themselves and for large corporations that can increase their R&D pay-off by licensing to companies that will not constitute competitive threats.
- The Ph-Subsidiary has had marketing cooperation with a large pharmaceutical corporation in Switzerland. When the Ph-Subsidiary built up its internal R&D, it partly benefited from cooperation with this corporation. Later on, when the Ph-Subsidiary was marketing a product based on its R&D, the product was licensed to the Swiss corporation for the United States market, despite the Ph-Corporation having a United States subsidiary.
- The general shortage or raw materials and energy sources has generated a wave of joint ventures since large risks and capital requirements are involved. Traditionally, however, the Swedish industry has been strongly oriented around domestic sources of raw materials.

The corporations with a developed network of sales subsidiaries naturally want to use it also for the marketing of new products on an international scale. A repellant effect may then occur if the subsidiaries are oriented around a traditional product. This has happened to the Ph-, E&F-, and E&S-Corporation. The effect is similar to the one encountered in development work, where traditions tend to dominate over new development work. Instead of trying to push the new product trough the existing marketing organization, responses may be to build a new marketing organization, to license the product, or to market it on a joint-venture basis, possibly giving the existing marketing organization a "first refusal right."

Thus, a tendency to utilize existing marketing resources does by no means exclude the use of acquisitions, interfirm cooperation, joint ventures with nation states, or development of new marketing resources abroad. In fact, there is a broad variety of options employed within and between the corporations. While there was formerly a concordance between corporate and product level regarding internationalization, there is now in a full--blown, diversified MNC a diversity of international operations that makes it hard to extract simple patterns.

One common and important feature of especially the El-, E&S-, and E&F-Corporations is the striving for multinational coordination. As described above, internationalization has proceeded through a stage that especially because of World War II was not concomitant with coordination. Local ambitions among managers and engineers, adaption to local conditions, nationalism, local opportunities, protectionism, barriers to communication, etc. are all contributing factors. Corporate control by mainly financial indicators such as growth, profitability, and market shares has given independence to successful subsidiaries. "As long as you grow profitably you are left alone" is often heard in subsidiaries. Tight corporate policies, guidelines, and standards have often been absent, sometimes consciously so since it is thought on the corporate level that one has to be cautious about centralization in multinational operations. Without going very deeply into the actual way these corporations coordinate, a few features, may be pointed out.

First, a matrix organization of product divisions and foreign subsidiaries is common. (See Figure 10:1.) The matrix may be more or less full blown with respect to horizontal and vertical degrees of coverage, and many variations of a matrix organization occur. (Note 6:4.)

Second, it is common to use the center concept as an integrating device. Such centers are profit centers, strategy centers, development centers, production centers, or supply centers. The formation of centers sometimes coincides with or is contained within divisional boundaries or national borders, but development, production, and marketing responsibilities of foreign subsidiaries may be relocated. This is a kind of centrally coordinated internationalization of subsidiaries, which represents a new stage in the internationalization process of product-invention-based corporations. In the sample here it is only the E&S- and the El-Corporation that definitely have reached this stage.

The formation and location of centers of different kinds are sometimes complex, at least as far as development and production centers are concerned. Local management ambitions clearly are important, and one encounters conscious strategies on the subsidiary level such as striving for:

- Suitable structure without conglomerate growth
- Products with advanced technology, low labor intensity, and markets with advanced industrial customers. Systems development is preferable in this respect over development of components
- Product concentration, for instance limited engagement in product divisions
- No production without accompanying development

The infrastructure of different nations is also important, of course, and industrial nations with advanced technology and advanced customers are attractive for development work but often not for non-advanced mass production due to high wages. Development and production responsibilities may then be transferred according to how products and markets develop, although such transfers may involve a great deal of discord.

Third, coordination requires information and communcation along with power for implementation. A structure of committees, boards, temporary working groups, consultants, agents, scanners, coordination units, liason people, and international commuters is present with various degrees of "richness" in the corporations. This is not only a matter of large staff organizations for intrafirm coordination but also a matter of environment scanning and finding opportunities for external cooperation as well. Case 6:1 gives an example of coordination of multinational R&D.

Case 6:1. The E&S-Corporation

During the 1960's the E&S-Corporation sold ball bushings under license. In 1971 corporate management decided that the E&S-Corporation should develop ball bushings of its own, since the prospects of continued licensing agreements were uncertain. The small innovation company at the corporate level started the development work, but the German subsidiary was dissatisfied with the results and started development work along alternative routes. The first evaluation of the two internal R&D competitors was made in April 1974 at a corporate management meeting, which approved of continued development both in Sweden and Germany. In early 1975 a thorough review was made of the nine alternatives developed so far. Through a sequence of meeting at the project level, at the central R&D-laboratory, in the product R&D management committee at the corporate level and finally in corporate management, a corporate R&D project based on an alternative emanating from Germany, was approved. The work on this project was done in the central R&D-laboratory in Holland. A new review in late 1975 assessed substantial technical problems. At the same time the pressure for rapid market introduction of a bushing of their own increased due to the licensing situation. Development work based on another alternative, emanating from the innovation company, was therefore decided upon by the product R&D management committee in early 1976, and a new corporate R&D project with a new project leader was approved. This alternative involved essentially known manufacturing techniques, and with an energetic project leader the project resulted in market introduction of a new bushing in August 1976. However, work continues on the project, which is classified as a project of corporate interest aiming at a product new to the E&S-Corporation within non--conventional bearings, which is an identified diversification class apart from conventional rolling bearings and a class within which growth has been found particularly desirable and natural.

Some important features of this case are:

- Initiation of self-development sparked by an uncertain licensing situation.
- Fruitful internal R&D competition and subsequent evaluation and direction of work made possible by a system for R&D coordination.
- The many 'turns' of the development work with respect to project objectives, location of work, origins of alternatives, market aspects, project leadership and status.
- The long time from initiation to market introduction in spite of the fact that similar products have long been on the market and the experience of the E&F-Corporation in related technical fields.

In relation to the multinational context of this case of product development, one may observe that it does not conform to the 'ordered' patterns of innovation within and diffusion between nations

Internationalization of R&D

In general, there is a more or less expressed policy and established behavior to "keep" R&D domestic in the M&M-, P&P-, Ch-, and Tp-Corporations. These corporations are late internationalizers and it has so far been considered rather natural to have R&D together with other domestic production and marketing operations, especially since process R&D is a large part of R&D in these corporations. Foreign R&D is only marginally existing and then in the form of cooperative R&D and as a byproduct of acquisitions.

In the product-invention-based corporations the E&S-, E&F-, El-, and Ph-Corporations, which were early internationalizers, except for the Ph-Corporation, there is a tendency to keep the more basic and far--reaching R&D domestic and/or close to corporate headquarters. (A parallel holds for production in several cases in the total sample in that advanced production is done domestically and assembly and supporting production may be located abroad.) The geographical location is, however, not necessarily decisive, and the central laboratory of the E&S-Corporation that is located abroad, is organizationally "closer" to headquarters than, for instance, the domestic laboratory in the Ph-Subsidiary is to the headquarters of the Ph-Corporation.

Reasons for the tendency towards domestic research and advanced development in the highly internationalized corporations (E&S, El, E&F, Ph) are more complex, and one should note that all these corporations, with the possible exception of E&F, actually perform some non-domestic research. Some reasons for the tendency are top management's perception of the need to have vital R&D close to its supervision, to have access to the R&D potential in Sweden, for instance, in pharmaceutical and metallurgical research, and also tradition, nationalism, and cautiousness. The cost/benefit ratio of Swedish scientists and engineers is said to have risen in relation to for example nations such as USA. The issue of R&D location is, however, more difficult to treat on such quantitative economic grounds than production. The Ph-Corporation, which has a strong interdependence between R&D, internationalization, and growth, also considers R&D scale advantages, risk dispersion at the subsidiary and divisional level, and closeness to medical universities. There are also ambitions and requirements emerging in different nations to locate R&D together with production within the nation, which is also enhancing market contact and adaption. Clinical research, for example, is appropriate to internationalize as well as quality control and R&D for market adaption.

This kind of decentralization of those parts of R&D that are more tied to products and markets is also considered to be natural and in some cases a deliberate policy. Clearly such decentralization tend to imply internationalization of parts of R&D in already internationalized corporations. But also the more genuine new product development is internationalized in the E&S- and the El-Corporation through the establishment of different development centers among the foreign subsidiaries. A deliberate policy change from local development for local markets to mobile local development for global markets is taking place in these corporations. This is, however, difficult to accomplish not only because of uncertainty and market differences, but also because new products and variants influenced by local conditions develop in subsidiaries. Much R&D is duplicated, and corporate coordination may come in late, especially if technology moves fast but internationally homogeneously or if international market differentiation is high. Thus R&D and internationalization have another mutually reinforcing element in this sense.

There is finally an important notion of internationalization of R&D that is more connected to the flow of knowledge than to the location of R&D work. This is related to technology transfer or diffusion. Technology transfer refers to both national and organizational boundaries and may be accomplished through communication, licensing, espionage, mobility of people, mobility of responsibility, and cooperation, among other means. In a special sense knowledge also flows together with a product.

The state of a technology tends to be internationally more homogeneous the more effective the cross-national flow of knowledge is. There are also several examples of R&D being internationalized in this sense. The increasing cost of industrial R&D in certain technologies stimulates both supply of and demand for technology. In general, most R&D is also performed by others, and scanning, access, cooperative, and absorption capabilities are developed. External technological developments may imply a wider opportunity as well as a threat to MNC's, which further enhances the incentives to keep abreast or become attractive for cooperation.

Technological leads may diffuse rapidly, which the E&S-Corporation and some other Swedish corporations have experienced, and technological lags may be hard to catch up, which the El-Corporation has experienced in the field of general-purpose computers. Not seldom, the necessity at least to keep abreast has to supersede internal coordination when technology moves fast as for the El-Corporation. The situation is similar in the E&F- and the E&S-Corporation, which operate in mature industries with a "differential cost/benefit warfare" that may require timely responses also to small advances in a conglomerate of technologies.

Alternatives to keeping abreast in international S&T are to develop technological leads in interstices or to develop competitive advantages through systems development or to develop an ability to serve certain categories of customers, for example governments. Large orders of whole systems often require unique development work, temporary organizations, joint ventures, and a spectrum of managerial skills. The whole complex of international competition and marketing in technology-based industries must, however, be left aside here.

6.5 DISCUSSION

6.5.1 Empirical Summary

Historically, two groups of corporations are discernible, the raw-material--based ones and the product-invention-based ones. (See Table 6:2 for a simplified picture.) The grouping is rough and, for example, the Tp-Corporation does not fit in. Also the progression of stages is not always clear.

The product-invention-based corporations initially internationalized rapidly based on the product invention. After a second stage of mainly gradual internationalization, a third stage has emerged with multinational coordination as a main feature. The third stage is, however, more diverse and unordered in the options employed. Corporations in this stage are also large and diversified, and patterns of internationalization are differentiated at the divisional, subsidiary, and product level. Generally, there is a trend towards temporalism and pluralism, a break-up of traditional patterns of internationalization, and increased emphasis on technological and managerial knowledge.

Internationalization of the R&D function has formerly followed the internationalization of production, and often acquisitions and local ambitions among managers and engineers have been strongly influential. Initial internationalization of R&D was hardly part of a corporate strategy. At present there is a tendency to keep research and advanced development domestic and/or organizationally close to corporate headquarters and decentralize and internationalize development work. Permanent location of R&D operations has decreased in importance, and temporary organizations and flows of knowledge are gaining emphasis. The link between R&D and internationalization appears to grow stronger.

The raw-material-based group of corporations has only recently begun to strive for internationalization and generally tries to accomplish this by joint ventures and acquisitions rather than by internal R&D. Also, these corporations have started to exploit foreign raw material sources and market their technology internationally. So far R&D in this group is mainly kept domestic.

TABLE 6:2

Stages of Corporate Internationalization

Corporate Characteristics	Stage I	Stage II	Stage III
Product-invention-based	Rapid internationalization with mainly majority-owned sales subsidiaries	Decreased relative pace of internationalization with a variety of corporate behaviors	Multination coordination of R&D, production and marketing an additional feature in internationalization
Raw-material-based	Domestic operations with varying degrees of export	Internationalization mainly through acquisitions and joint ventures and to some extent through exploiting internal technological knowledge	(Not yet observed)

6.5.2. Corporate Internationalization

Concerning general patterns of corporate internationalization it is hard to be conclusive on the limited basis of this study. There is in this sample of corporations, all of which have small domestic markets, a correspondence between rapid internationalization and product invention implying a global technological advance. Internationalization of production has further been stimulated by national trade barriers.

That the pattern of corporate internationalization in general is characterized by a gradual component, has been emphasized by Johansson and Wiedersheim-Paul (1975) and Carlsson (1979). It is naturally hard to agree on distinctions between gradual and non-gradual. The pace of internationalization may be assessed in absolute or relative terms, qualitative as well as quantitative ones. A simple count of nations in which a corporation has subsidiaries does not give more than a first approximation of expansion patterns.

Provided there are shifts over time in the pace of internationalization, although not necessarily giving rise to distinct stages, one may ask about the causes to these shifts. Certainly the pace of global industrialization, wars, and periods of protectionism are influential as well as the advent of inventions but it is also conceivable that corporate internationalization interplays in a periodic fashion with other dimensions of corporate development, such as diversification or consolidation. At least in the early phases of corporate development, this interplay may be observed in that no corporation initially internationalized in parallel with substantial diversification. Thus, the pace of internationalization, whether gradual or not, is influenced by shifting external conditions in interplay with the internal dynamics of corporate development. Corporate development, in turn, has a feature of sequential development in different dimensions.

6.5.3 Internationalization of R&D

Concerning internationalization of different functions, Johansson and Wiedersheim-Paul (1975) in their study of internationalization of four Swedish corporations, propose the concept of the establishment chain as a specific sequence of functional types of operations a given corporation is likely to follow in locating operations in a given nation over time. They conclude: "The establishment chain - no regular export, independent representative (an agent), sales subsidiary, manufacturing - seems to be a correct description of the order of the development of operations of the firms in individual countries" (Johansson and Wiedersheim-Paul 1975, p. 321). This pattern also finds support to some extent in this study. In fact, the establishment chain may be extended in the sense that manufacturing operations are likely to be followed by R&D or at least D operations. Such an extension also finds support in Steele (1975), p. 212: "Once some technical capability has been established, domestic scientists and engineers are certain to see additional opportunities for improvement. There is an almost irresistible creepage from production engineering upstream into design and development."

On the other hand, the establishment chain is not valid in all respects. (Cf. the concept of the innovation chain.) Progression through the sequence may be rapid, partial or disrupted. This is especially so if separate products are in addition considered and the corporation is in stage III with multinational coordination of different functions. This leads to the consideration of multinational product strategies as well as to the consideration of decisions to locate R&D abroad.

Leroy (1976) constructs a typology for multinational product strategies in a "two-country world" by:

1. Defining 8 basic states S1-S8 which refer to the eight combinations of parent or host country locations of product development, production, and markets.
2. Defining a multinational product strategy as a sequence of basic states occurring over time. (That is in principle a trajectory in a state space.)
3. Assuming that product know-how is once and for all developed domestically or abroad.

This gives a set of product strategies, some of which are not feasible. In all, Leroy's typology encompasses 58 feasible multinational product strategies. An example of these is:

> "The product is developed, manufactured, and marketed in the parent country. Then it is exported to the host country. Later manufacturing operations start in the host country for that market. Finally, the product is exported to the parent country".

Using this typology, Leroy studies five large North-American multinational corporations and the occurrence of the different strategies for different products in these corporations. Some of Leroy's empirical conclusions deserves to be quoted in extenso here:

> "The first conclusion is that in all cases, except one, the firms initially became involved internationally with a unique product that had been a success in their domestic market. They initially exported the product and as sales in international markets increased this led them to investment abroad. This corresponds to the classic product strategy.
>
> The second conclusion is that after having achieved a dominant position with their unique product internationally, the firms diversified their activities into other product lines. This diversification was achieved by internal development of products and/or acquisition of firms with different technologies.
>
> ...
>
> The fourth conclusion is that these firms developed a worldwide out-look and consequently they are able to see the potential of new technology on a world scale. It follows that the firm's products will eventtually diffuse on a worldwide scale."
> (Leroy 1976, p.128)

Leroy's findings in relation to theory indicated that the international product cycle theory had very limited relevance. Moreover, many of the presently existing theories were found to be only broadly consistent with his findings, a reason being that the theories are oriented around aggregated foreign investment behavior of firms, which makes the theories less specific in terms of decision making at the corporate level.

This study does not focus primarily on multinational product strategies, and frequency statements cannot be made in relation to Leroy's findings. What may be said is that his typology of strategies (or behaviors) at the product level has to be expanded in order to include some behaviors observed in this study. Normally product innovations have multiple sources and are subjected to subsequent development in the course of diffusion. (See Case 6:1.) Generally one may consider:

a) Nation of origin of components or raw materials
b) Nation of origin of R&D contributions
c) Nation of production location
d) Nation of market location

Also, there are different ways to make provisions for supply, R&D, production and marketing resources, primarily by internal generation, external cooperation, or acquisitions. Furthermore, origins and flows of capital and personnel may be added but this is a possibility of secondary importance here. With just a two-country world, there are then 16 basic states with respect to a)-d). If, moreover, the three principal ways to make provisions for each of the four resources corresponding to a)-d) are taken into account, there are theoretically six raised to the fourth power = 1296 basic states. If then the set of nations is expanded and sequences of basic states are considered as strategies, the combinatorial explosion of options is a fact. This is not said as an exercise but to emphasize the diversity of possible and actually employed behaviors and the impossibility of basing managerial decisions on a comprehensive evaluation of alternatives in advance. Even Leroy's comprehensive list of strategies is too limited for describing possible behaviors of, for example, a MNC developing and marketing whole systems internationally. For corporations in stage III, far-reaching analysis and conclusions are impeded by lack of data. One has also to consider the difficulties in coordinating complex operations. People in MNC's complain themselves that they cannot overview the operations or gain insight into product histories and patterns of diffusion.

Concerning internationalization of R&D and decisions to locate R&D facilities abroad, several factors have been found in this study to be influential, such as:

- Local ambitions among management and engineers
- Acquisitions "containing" R&D
- Scale advantages, both small-and large-scale advantages
- Integration with production and/or marketing functions
- Differentiation with respect to national market conditions, for example need for market adaption, local product testing, and quality control
- Cost/benefit ratio for scientists and engineers
- Access to national R&D potentials and infrastructure
- Communication costs and difficulties to transfer technology
- Recruitment considerations
- Subsidiaries are psychologically encouraged by having R&D
- Creating good-will locally
- Tradition and attitudinal preferences for different nations

Usually internationalization of R&D has followed the pattern of corporate internationalization and has not been strongly connected to a corporate strategy, as mentioned. Exceptions exist, however, such as the central R&D laboratory in the E&S-Corporation. It may also be noted that tax advantages in locating R&D abroad in this sample do not appear to have been influential, although there is a growing awareness of these not only regarding R&D costs but also regarding tax evasion through patenting in low-tax nations.

Concerning multinational coordination of R&D for corporations in stage III, it is difficult to assess advantages of scale, synergy, and multinationality, which may be achieved. Certainly multinational coordination presents difficulties, which may outweigh coordination benefits. Steele (1976) essentially adopts this standpoint regarding R&D. Several factors are important in an assessment, such as communication possibilities, the extent to which the corporation has a long-standing experience as an MNC, corporate history and culture, and type of technologies and markets. The highly internationalized corporations generally operate in oligopolies, and both markets and technologies often have low international differentiation.

A high pace of advances in technology as for instance in electronic components, may require R&D flexibility and responsiveness, at the expense of immediate coordination with other technologies and functions.

There is also the problem of secrecy in coordination by communication. In the case of Eastman Kodak the international homogeneity of its technologies among other things makes coordination between different R&D laboratories worthwhile (Hanson, 1971). On the other hand, economic and technological differences among nations call for regionalized and loosely coordinated agricultural R&D in Farbenfabriken Bayer of West Germany (van Rumker 1971).

A situation similar to that of Eastmann Kodak is reported by Papo (1971) for IBM with a uniform and compatible product line and R&D performed in multinational laboratories. These are, however, reported to be multinationally located, not multinationally composed to any large degree. This is also essentially true for the multinational research organization of the El-Corporation with a dominating domestic laboratory. IBM, the El-Corporation, and the E&S-Corporation share the feature of continental R&D coordination with looser coordination between USA and Europe. The one-day return travelling distance actually seems to supersede market and/or technology homogeneities.

Finally, concerning the relationship between R&D and internationalization at the corporate level, there are a variety of connections. R&D initiated internationalization in invention-based corporations, especially in small nations. In fact there are almost no examples of successful internationalization in which a technological advantage has not played a crucial role. It is also important to realize that the product inventions behind initial internationalization of the corporations in this sample were based on R&D and not on a flash of genius out of nowhere. In the cases of the E&F-, E&S-, El-, and Ph-Corporation the inventors were well-educated men performing development work, albeit on an experimental rather than a theoretical basis. Moreover, there was a second important product invention made or acquired within about 15 years after the first one was made in all these corporations. It might also be added that the international product cycle theory does not apply to almost any of these inventions or at least the initial progression through the stages in the theory was very rapid in that the products were developed for world markets and also production was soon located abroad (except for the El-Corporation).

The coupling between R&D and international operations has been mutually strengthening. Although looked upon as a means in an internationalization process, the causality is intertwined in MNC's. Without R&D they would mostly be unable to stay on international markets, and without these markets they hardly could pay for their R&D. This is especially true in pharmaceuticals, electronics, and advanced engineering. The correlation between R&D intensity and degree of internationalization is 0.48, between R&D intensity and total sales 0.42, and between R&D intensity and degree of diversification -0.20 in this sample. (See Chapter 5.) Thus there is reason to believe in a strong coupling between R&D, size, and multinationality.

Technological knowledge is, however, economized in more ways than just through direct pay-off in sales of products. There are examples of how such knowledge may be licensed, cross-licensed, exchanged for raw materials, utilized in sales of systems, or as a means to cope with national requirements or attract foreign partners for cooperation. One trend seems to be an increased frequency of such behaviors which for the corporation involve temporary international relations in which knowledge of management and technology plays a vital part. Thus the conception of multinationality based on a more permanent location of operations is becoming somewhat diffuse. It is probable that the internationalization of corporations in the ordinary sense will be more and more counterbalanced by nationalization, at least in the form of increased nation state influence over physical operations rather than formal ownership of physical resources. It is, however, also probable that the traditional conception of an MNC has to be geared more towards the ability and performance in managing international relations. As pointed out above, knowledge of management and technology is vital in this respect and the global information processing capacity of an MNC is gaining increased importance.

6.6. CONCLUSIONS

Regarding R&D and corporate internationalization, two groups of corporations may be distinguished, those based on product inventions allowing an early and rapid internationalization and those based on raw materials tying their locations to domestic supply markets until recently, albeit with high export intensity.

There is a common trend towards increased internationalization. Late internationalizers have a preference for acquisitions and joint ventures. The pace of internationalization has historically been influenced by shifting external conditions in interplay with the internal dynamics of corporate development. A lead in product technology in combination with internationally oriented entrepreneurship have been crucial factors in internationalization, while advanced process technology has been associated with domestication.

Corporations also internationalize their R&D, although there is a tendency to domesticate the R-part. The establishment chain model may thus be extended to R&D in some respects. Initial internationalization of R&D was not, in general, part of corporate strategy but often resulted from acquisitions and local ambitions.

At present a wide variety of strategies are employed by MNC's. The highly internationalized product-invention-based corporations have entered a stage of multinational coordination, which reinforces a strong coupling between R&D, size, and multinationality. Small domestic markets, a high degree of internationalization, multinational coordination, and increasing emphasis on technological and managerial knowledge give a limited relevance to models of orderly patterns of internationalization, innovation, and diffusion.

LIST OF REFERENCES

AHARONI, Y. 1971. 'On the Definition of a Multinational Corporation.' Quarterly Review of Economics and Business, Autumn 1971.

BABBAGE, C. 1832. On the Economy of Machinery and Manufactures. Charles Knight, London, 1832.

CARLSON, S. 1979. Swedish Industry Goes Abroad. An Essay on Industrialization and Internationalization. Studentlitteratur, Lund, 1979.

CARLSSON, B. and DAHMEN, E. et al. 1979. Teknik och industristruktur – 70-talets ekonomiska kris i historisk belysning. Industrins Utredningsinstitut och Ingenjörsvetenskapsakademien, Almqvist & Wiksell, Stockholm, 1979. (In Swedish.)

CHANNON, D. F. 1973. The Strategy and Structure of British Enterprise. The Macmillan Press Ltd., London, 1973.

DALBORG, H. 1974. Research and Development – Organization and Location. EFI, Stockholm, 1974.

EDSTRÖM, A. and GALBRAITH, J. 1977. 'Transfer of Managers as a Coordination and Control Strategy in Multinational Organization.' Administrative Science Quarterly, Vol. 22, June 1977, pp. 248–263.

FOURAKER, L. and STOPFORD, J. 1968. 'Organizational structure and the multinational strategy.' Administrative Science Quarterly, Vol. 13, 1968, pp. 47–64.

GHERTMAN, M. and LEONTIADES, J. (Eds.) 1978. European Research in International Business. North-Holland Publ., Amsterdam, 1978.

GRUBER, W., MEHTA, D. and VERNON, R. 1967. 'The R&D Factor in International Trade and International Investment of United States Industries.' Journal of Political Economy, Febr. 1967, pp. 20–37.

HANSON, W. T. 1971. 'Multinational R&D in practice. Eastman Kodak Corporation.' Research Management, Jan. 1971, pp. 47–50.

IGER, S. 1976. Den svenska ekonomins internationalisering. Drivkrafter och utvecklingsmönster. Sekretariatet för Framtidsstudier, Rapport nr 206, Stockholm, 1976. (In Swedish.)

IVA, 1979. Kunskap och Konkurrenskraft. Tekniskt industriella utvecklingsproblem i internationellt perspektiv. Ingenjörsvetenskapsakademien Meddelande Nr 223, Stockholm, 1979. (In Swedish.)

JOHANSSON, H. 1967. Utländsk företagsetablering i Sverige. SNS, Stockholm, 1967. (In Swedish.)

JOHANSSON, J. and WIEDERSHEIM-PAUL, F. 1975. 'The Internationalization of the Firm – Four Swedish Cases.' The Journal of Management Studies, Oct. 1975, pp. 305–322.

KINDLEBERGER, C. P. 1968. International Economics. 4th ed., Richard D. Irwin, Homewood, Illinois, 1968.

LEROY, G. 1976. Multinational Product Strategy. A Typology for Analysis of Worldwide Product Innovation and Diffusion. Praeger Publ., New York, 1976.

OHLSSON, L. 1976. Svensk verkstadsindustris internationella specialisering. Industriens Utredningsinstitut, Almqvist & Wiksell, Stockholm, 1976. (In Swedish.)

PAPO, M. 1971. 'How to Establish and Operate Multinational Labs.' Research Management, Jan. 1971, pp. 12–19.

PERLMUTTER, H. 1969. 'The Tortuous Evolution of the Multinational Corporation.' Columbia Journal of World Business, Vol. 4, No. 1, Jan.–Febr. 1969.

POTTER, B. V. 1971. 'Effective Information and Technology Transfer in Multinational R&D.' Research Management, Jan. 1971, pp. 20–27.

QUINN, J. B. 1969. 'Technology Transfer by Multinational Companies.' Harvard Business Review, Nov.–Dec. 1969, pp. 147–161.

RINGBAKK, K.-A. 1976. 'Strategic Planning in a Turbulent International Environment.' Long Range Planning, June 1976, pp. 2–10.

ROBINSON, R. 1967. International Management. Holt, Rinehart & Winston, New York, 1967.

ROBINSON, R. 1978. International Business Management. A Guide to Decision Making. 2nd Ed., The Dryden Press, Hinsdale, Illinois, 1978.

RONSTADT, R. 1978. Research and Development Abroad by U.S. Multinationals. Praeger Publishers, New York, 1978.

van RUMKER, R. 1971. 'Multinational R&D in practice. Chemagro Corporation.' Research Management, Jan. 1971, pp. 50–54.

SHIPMAN, J. R. 1967. 'International Patent Planning.' Harvard Business Review, March–April 1967, pp. 56–72.

SILBERSTON, A. 1975. 'Impact of the Patent System on the Creation and Diffusion of New Technology.' Omega, Vol. 3, No. 1, 1975. pp. 9–22.

SOU 1975:50. Internationella koncerner i industriländer. Samhällsekonomiska aspekter. Betänkande av Koncentrationsutredningen, Industridepartementet, Stockholm, 1975. (In Swedish.)

STEELE, L. W. 1975. Innovation in Big Business, Elsevier North-Holland Publishing Company, New York, 1975.

STEINER, G. A. 1969. Top Management Planning. Macmillan, New York, 1969.

VAHLNE, J.-E. 1974. 'Svenska Multinationella Företag. Prövning av ett par definitioner.' Published in Vahlne (ed.): Internationellt Företagande. Norstedts, Stockholm, 1974. (In Swedish.)

VERNON, R. 1966. 'International Investment and International Trade in the Product Cycle.' Quarterly Journal of Economics, May 1966.

VERNON, R. 1971. Sovereignty at Bay. Basic Books, New York, 1971.

WILSON, R. 1977. 'International licencing of technology: empirical evidence.' Research Policy, 6, 1977, pp. 114–126.

YOSHINO, M. Y. 1976. Japan's Multinational Enterprises. Harvard Univ. Press, Cambridge, Mass., 1976.

[3]

THE INTERNATIONAL ALLOCATION OF RESEARCH ACTIVITY BY US MULTINATIONALS*

By SANJAYA LALL

I INTRODUCTION

It now seems to be widely accepted in the empirical literature on trade and investment that enterprises need some form of 'monopolistic advantage' to compete in the highly imperfect markets for products and factors that characterize modern industry. Studies of the patterns of comparative advantage and of direct investment flows, starting from different premises, have come increasingly to converge on a set of explanatory factors that are practically identical with the 'barriers to entry of new competition' which have been identified by the industrial organization literature to lead to the growth of large firms, and to the emergence of concentrated market structures, within the advanced countries.[1] Thus, the forces that lead to successful growth at home also seem to provide the leading firms with the edge that they need to sell their products abroad or to set up foreign affiliates and serve overseas markets by producing abroad.

The choice between exporting from home and producing abroad (the choice, that is, of the location where the monopolistic advantages are to be exploited) has been analysed at length, and found to depend on the evolution of the product or technology cycle,[2] on the ease with which the various advantages can be transferred abroad,[3] on various cost and market conditions as suggested by location theory[4] and the exigencies of oligopolistic interaction.[5]

One of the most important of monopolistic advantages is technological superiority, based upon expenditures on the broad set of activities termed 'research and development' (R & D). The creation and maintenance of a technological lead provides enterprises with a powerful competitive edge which is exploited, according to the product cycle hypothesis, first by exporting from home, when certain conditions render its transfer abroad impossible, and later by investing abroad, when competition from imitators and the standardization of the new technology renders its transfer both profitable and technically feasible. A central assumption of the traditional product cycle hypothesis, and indeed of all explanation of trade and investment based on 'technology gaps', is that the creation of new technology

* I am grateful to Donald Hay and Teddy Jackson for useful comments on an earlier draft.

[1] The links between industrial economics, trade theory and direct investment theory are noted by Dunning (1977) and empirically explored by Lall (1980). The same links are analysed, in the context of the theory of the firm, by Hirsch (1976). While the traditional analysis of trade does not distinguish 'monopolistic advantages' that accrue to specific firms from competitive advantages belonging to countries in general, it seems sensible (see Dunning's contribution to this issue) to draw such a distinction in world markets in which the bulk of exporting and direct investment activity is undertaken by a relatively few large firms in each industry. On the role of US multinationals in that country's trade and investment see the US Tariff Commission (1973) as well as Helleiner's paper in this issue.

[2] Vernon (1966), (1977) and Magee (1977).

[3] Hirsch (1976), Lall (1980).

[4] Dunning (1973).

[5] Knickerbocker (1973).

occurs at home. The country of origin of the multinational corporation (MNC) which is enjoying the fruits of its R & D internationally remains the main base of new investments in R & D. It provides a stream of fresh innovations that replenishes its stock of exportable products while maturing products and technologies are being shifted abroad to be exploited in low cost locations. The MNC is, therefore, the agent of dynamic comparative advantage, but in a particular sense. It allocates production activity internationally according to comparative cost advantages, but only for mature technologies; it keeps its new technologies at home where its main R & D is located.

As Vernon notes (in this issue), the original product cycle hypothesis has been qualified by several new developments. One of these is the tendency of US MNCs to undertake an increasing proportion of their R & D abroad (we shall present some evidence on the foreign R & D propensity of these companies in the following section). This may be interpreted as a further twist to the cycle process, but one with potentially drastic implications to the entire sequential process. Consider the extreme case. With the shifting abroad of mature technologies, MNCs gain increasing experience of operating in foreign locations. This leads them to set up R & D support activities abroad which themselves increase in significance over time. Given the location advantages of overseas R & D (assuming, not unrealistically, that skilled manpower is generally cheaper abroad than in the US), this ultimately leads MNCs to 'source' this activity internationally in exactly the same way as they 'source' certain products and components from cheap labour areas. They become, in this sense, the agents of comparative advantage *for new technology generation as well as for mature technology diffusion*: ultimately, the home country may serve simply as the final market for products developed and made elsewhere.

This representation of MNC behaviour—as true international citizens locating *all* their activities according to global comparative cost (an advanced version of Vernon's 'global scanner')—is a recurrent one in the literature. But is it really valid? Does the evidence really support the contention that R & D activity will become increasingly 'untied' from the home country? And what determines the extent to which different industries undertake foreign as compared to domestic R & D?

This paper attempts to provide a preliminary answer to these questions. It addresses itself, in particular, to the last one. By undertaking a statistical examination of inter-industry differences in the propensity to undertake foreign R & D by US MNCs (defined as foreign R & D as a percentage of total R & D), it explores the factors which influence the location of R & D as between the home country and foreign countries, and enables us to draw some tentative conclusions about the future evolution of the phenomenon.

The conclusions must, however, remain very tentative. Cross-section analysis of data from one year is not the best way to analyse a dynamic phenomenon. The only year for which detailed data for the US are available, 1966, is already well in the past, and may not reflect the present situation correctly. Even for this year the detail is not sufficient, and foreign R & D data are given at a fairly high level of aggregation. And, finally, there is little received theory on the determinants of

R & D location, so that we are probing new territory where we may miss some important landmarks. Nevertheless, the available information is the best we have (no other country besides the US publishes any data on this), and this attempt may, given due caution, cast some light on a possibly important trend in international economics.

Section II briefly presents the evidence on foreign R & D by US MNCs. Section III analyses the influences on R & D location. Section IV describes the variables used in the statistical tests and the results obtained. Section V draws the main conclusions.

II Evidence on Foreign R & D by US Multinationals

The US Department of Commerce carried out a comprehensive survey of the activities of all US firms with overseas operations for the year 1966. This survey collected, among other things, data on the MNCs' foreign and domestic R & D activity; these data were subsequently published in the US Tariff Commission's (1973) study of US multinationals. An Appendix to this paper presents the Tariff Commission's data, showing the dollar values of R & D in total, in the US and overseas, as well as the 'propensity to conduct R & D overseas', defined simply as foreign R & D as a percentage of total R & D. All these figures pertain only to R & D undertaken by US MNCs (i.e. not by US firms without foreign investments and not by foreign firms in the US). It is assumed that these data are the correct indicator of the multinationals' decision on the location of the R & D function, though the US R & D spending includes finance provided by the US government (we cannot correct for this).

The appendix table shows that in 1966 US MNCs conducted 6.5 per cent of their total R & D overseas. The propensity to conduct foreign R & D differed markedly across industries. Such industries as beverages, aluminium, other metals, textiles and printing conducted negligible (less than $500,000) R & D abroad, while others, like soap and cosmetics, plastics, other chemicals, industrial machinery, other electrical machinery and lumber, wood and furniture, conducted between 15 and 77 per cent abroad. In terms of the value of expenditures, only a few industries accounted for most of overseas R & D: thus, electrical and non-electrical machinery and transport equipment by themselves spent 62 per cent of the total, while transport equipment alone accounted for 25 per cent. But these are not necessarily the industries which spend a high proportion of their total R & D abroad. Being high R & D spenders in the US (the three accounted for 67 per cent, and transport equipment by itself for 33 per cent, of the total US figure), their propensity to shift R & D overseas was not always higher than the average. Several low R & D spenders in the US, on the other hand, had a high propensity to undertake research overseas. Given the research intensity of operations in the different sectors, therefore, there are clearly diverse factors at work affecting this propensity: it is these which we shall investigate below.

While the 1966 data remain the most comprehensive and the most detailed, two other studies have appeared recently which examine foreign R & D by smaller samples of US MNCs. Creamer (1976), reporting on a survey conducted by the Conference Board, presents data on overseas R & D for 1966–75 for a sample of 444 US firms, of which 265 had no foreign R & D and 179 had some. These 444 companies accounted for 55 per cent of all industrial R & D in US, and their total R & D spending was distributed over sectors in roughly the same pattern as R & D for all US firms, with the (expected) difference that space and defence-related companies had a higher weight in the latter since they had refused to participate in the Conference Board survey. Creamer discussed in detail the possible sources of bias in his sample and the methods employed to extrapolate from the sample to the universe of US MNCs. The details of these procedures need not concern us here. What is of more general interest is the propensities shown by the firms: between 1966 and 1972 R & D spending by foreign affiliates rose much faster than such spending by their parent companies and by US industry more generally. According to Creamer's estimates (and these are not exactly comparable with figures given in our appendix), overseas R & D spending in 1966 came to 4.6 per cent of US parent companies' spending and 3.4 per cent of total US industrial R & D.[6] By 1972 these percentages rose to 7.9 and 6.2 respectively, a 70–80 per cent increase in the foreign share. In 1973–75 the shares dropped marginally, presumably due to the general recession and the cutting down of foreign activity generally, but the long-term trend seemed to be to allocate an increasing portion of total R & D to foreign affiliates.[7]

Both Creamer and the Tariff Commission note that overseas R & D is highly concentrated geographically in other developed countries. In fact, Canada, the UK and West Germany, the main recipients of US direct investment by themselves account for two-thirds of overseas R & D by the US.[8] There is also a high degree of concentration by company: a relatively few large firms account for most of R & D spending both in the US and abroad.[9]

Creamer finds that both parent firms and affiliates have similar patterns of research work: research-intensive parents have research-intensive affiliates, and both concentrate on similar fields of scientific or engineering endeavour. Both

[6] Creamer (1976), p. 35. In turn US MNCs contributed, in 1971, 51 per cent of total industrial R & D in Canada and 14 per cent each in the UK and W. Germany (p. 86).

[7] This is confirmed by a report just published in *Nature* (Vol. 279, 24 May, 1979, p. 278), according to which a National Science Foundation survey for 1974–77 found overseas R & D to rise by 41 per cent, and domestic R & D by 32 per cent, over the three year period.

[8] Creamer (p. 5) reports a positive rank correlation coefficient of 0.92 between the country-wide location of direct investment and R & D expenditure in chemicals, machinery and transport equipment, and concludes that 'overseas expenditures on R & D follow the investment dollar in the science-based industries'.

[9] The degree of concentration in R & D spending within the US can be illustrated for 1973 with the following data extracted from the NSF (1975). Of total R & D, including non-manufacturing, of $21 billion, the top 4 R & D performing firms accounted for 19 per cent, the top 40 for 66 per cent and the top 100 for 79 per cent. The degree of concentration was much higher within the main industrial sectors. Thus, the 4-firm R & D concentration ratio for each sector spending over $500 million on R & D in 1975 was as follows: industrial chemicals ($1133 m.) 58 per cent; pharmaceuticals ($618 m.) 43 per cent; petroleum ($504 m.) 53 per cent; non-electrical machinery ($2144 m.) 46 per cent, of which office and computing machinery ($1325 m.) 90 per cent; communication equipment ($3064 m.) 64 per cent; 'other' electrical equipment ($1861 m.) 87 per cent; motor vehicles ($2411 m.) 94 per cent; aircraft and missiles ($5051 m.) 57 per cent; and instruments ($914 m.) 65 per cent. Creamer notes that there is a significant positive correlation between size of domestic and overseas R & D spending in a sub-sample of 60 MNCs. For an analysis of the influence of firm size on R & D effort see Kay (1979).

spend relatively little on basic research[10] though when only company-funded R & D is taken into account, the parents spend some 7 times more on this than their affiliates. Both spend most of their R & D on 'development' work, but the lines between the different kinds of research are difficult to draw and we cannot infer too much from the precise figures.[11]

Finally, Creamer finds that US parent companies provide considerable direct funding for overseas R & D. In 1972 nearly 30 per cent of overseas research activity was financed from the US. Correspondingly, affiliates carried out some R & D to support directly the activities of their US parent; 26 per cent of their total R & D was deemed to be for this purpose. Thus, US MNCs not only passively permitted their affiliates to engage in R & D for purposes related to the latter's own production, but actively supported and benefited from it. To the extent that there was overseas R & D into problems of parent MNC interest, we may take the evidence to indicate that MNCs are undertaking a conscious (and, presumably, economically rational) reallocation of some research activity internationally. Creamer does not, unfortunately, pursue this line of enquiry, and the evidence does not permit a more definite conclusion.

The other study of R & D MNCs is by Ronstadt (1977), who studied a sample of seven firms.[12] Despite the small number of observations Ronstadt's findings are of interest because he explores in detail *why* overseas R & D is undertaken. He classifies such R & D into four kinds by purpose: (a) 'to help transfer US technology to a foreign market'; (b) to conduct 'indigenous technology work' expressly for the foreign market; (c) to 'develop new and improved products for simultaneous application' in the US and elsewhere, or 'global product work'; and (d) to undertake 'corporate technology work' specifically for the parent company.[13] After examining 55 foreign R & D units he concludes that function (a), transfer of technology, was the most common one, being the primary function of nearly 75 per cent of the number of foreign R & D units. Function (b) accounted for the smallest number (2 units), while functions (c) and (d) were about equal at 5 and 4 units respectively.

While these numbers are difficult to interpret, and comparative values of R & D are not given, Ronstadt provides a plausible explanation of the process of foreign research based on his interviews and impressions. Technology transfer work, reasonably small in scale and relatively inexpensive, is needed to implement all

[10] On the official US classification, 'basic' research is original investigation without a specific commercial objective, 'applied' research is new scientific work with specific commercial objectives regarding new products or processes, and 'development' research is non-routine technical work in translating research findings into actual products and processes. Taking company-funded R & D (excluding government financed) for 1972, parent companies spent 7.3 per cent, 33.9 per cent and 58.7 per cent of their R & D on the three respectively, while affiliates spent 1.1 per cent, 29.9 per cent and 69.0 per cent; the corresponding figures for total R & D are 1.9 per cent, 15.4 per cent, 82.7 per cent for parent companies, and 1.4 per cent, 29.9 per cent and 68.9 per cent for affiliates.

[11] The US Tariff Commission (1973) suggests, but without providing details, that 'practically all of the basic research of US industry is done in the central research headquarters in the US. The few companies which have established overseas laboratories do more development work there than research ... Duplication of efforts by the parent and the foreign affiliates is shunned because of the high cost of research' (p. 585). Note that this does not correspond with Creamer's analysis, who argues that there is a fairly high incidence of 'applied' work by the affiliates.

[12] Union Carbide (chemicals and plastics), Corning Glass Works, Otis Elevators, CPC International (food products), Exxon Corporation Energy Businesses, Exxon Chemical Company and IBM.

[13] Ronstadt (1977), p. xx.

sorts of 'unstandardized' technology abroad, and is thus the most widespread form of foreign R & D activity.[14] Once established, such transfer units often evolve into indigenous technology units, especially in larger, older production centres. Finally, these sometimes evolve further into global product units when the relevance of indigenous technology work to foreign markets as a whole, and the potential productivity of foreign units, is recognized by the parent. At this stage the foreign R & D unit is quite large and expensive, and is brought under the direct control of the MNC head office.[15]

The first three types of overseas research work (which correspond roughly to development, applied and basic R & D in the usual sense), thus show a natural progression based on the age, growth and relative size of the foreign *manufacturing* subsidiary. The fourth type of unit evolves, according to Ronstadt, independently of foreign manufacturing, and engages mainly in basic scientific work which is needed to sustain competitiveness. It is undertaken abroad because of the availability of the appropriate researchers abroad who are unwilling to move to the US. Though Ronstadt denies this motive, it is plausible that such R & D units are also maintained in foreign countries which are technologically advanced in certain fields, not only in order to engage high-quality but immobile scientists, but also to share in the 'fall out' effects of being close to a successful research community (it is likely, for instance, that the British government's decision to set up an electronics research centre in California's 'Silicon Valley' was partly motivated by the desire to be close to the centre of technical change).

Ronstadt also states that taking advantage of cheaper R & D labour abroad did not influence the decision of his sample of seven MNCs to set up foreign R & D units. Since there is no doubt that foreign researchers were considerably cheaper (Creamer notes, on p. 72 of his study, that in 1972 foreign scientists and engineers were 34 per cent cheaper, and R & D support personnel 54 per cent cheaper, for foreign affiliates than for US parents), Ronstadt's finding runs counter to what we would expect on grounds of economic rationality. If MNCs were rational in allocating their resources, and if foreign researchers were as productive as their US counterparts, his finding could be due to one or more of three factors: first, his interviewees (mainly concerned with R & D activity) were not conscious of all the economic aspects of R & D allocation; second, that the cost advantage of foreign R & D, while present in the allocation decision, was heavily outweighed by other factors; and, third, that his particular sample of MNCs was economically irrational. The first two explanations seem to be fairly reasonable, and we shall ignore the third. In the following sections, we shall assume that cost pressures did provide a motive to re-allocate R & D to foreign locations, but that they may have been counteracted by other forces which affect this location decision.

[14] This is supported by remarks made in Creamer's and the US Tariff Commission's studies.

[15] In Lake's (1979) terminology, autonomous R & D units primarily serving local needs have a 'parallel structure' of R & D organization while those controlled from the centre have a 'hierarchical structure'. Lake does not explore their relationship and significance empirically, but his analysis (pp. 150–2) seems to suggest that MNCs in technologies at the mature stage of the life-cycle would prefer parallel structures while those with new and valuable technology would prefer the hierarchical one. This is somewhat at variance with Ronstadt's characterization of the *evolution* of foreign R & D from a decentralized to a centralized system, regardless of the stage of the technology cycle.

The only other studies of R & D abroad by MNCs have been made in particular host countries (mainly Canada),[16] and have been directed to exploring whether foreign firms do as much R & D as domestic firms. They have not gone into the subject of our present enquiry, of what determines the propensity of MNCs to allocate R & D between their home and host countries, and we shall not discuss them here.

None of the existing studies provides an explanation of the determinants of the propensity to conduct foreign R & D across different industries. While the findings described above are certainly suggestive, they do not add up to even a preliminary hypothesis about why different sectors have the markedly different propensities which our appendix table reveals. Let us now consider our attempt to provide such an explanation.

III Influences on R & D Location

It has always proved difficult to provide a satisfactory economic explanation of R & D behaviour as such, of how firms decided expenditures on research and how they allocate their funds between different types of research activity. While the usual profit-maximization hypothesis goes some way in explaining R & D spending (with due consideration taken of such factors as 'technological opportunity', firm size, inter-firm rivalry, market structure, learning and so on), it fails to come to grips with one important fact about innovative activity—its inherent uncertainty and unpredictability.[17] We avoid this problem to some extent by taking the amount of total R & D by MNCs as given and simply considering its allocation between the home country and abroad, but we do not escape it entirely: to the extent that R & D decisions in general take place in an uncertain environment, and are the result of learning and adaptation of corporate rules-of-thumb,[18] we may be wrong to treat the location decision as a rational one from the economic point of view. However, the economic approach is the only one open to us and, despite its limitations, we must rely on it.

The economic considerations which can help us explain the international allocation of R & D activity of MNCs derive more from theories of industrial location than those of MNC behaviour. While the literature on MNC expansion offers a fairly full explanation of the industrial characteristics that give rise to the location of productive facilities abroad, it assumes, as we noted at the start, that the monopolistic advantages which underlie foreign expansion are based in activities in the home country. In fact, the product cycle hypothesis can provide an explanation of why MNCs may *not* shift their basic and applied (as opposed to development) research abroad. As Vernon explains,[19] the need, in the early stages

[16] The Canadian government has been concerned about the possibility that foreign investors do insufficient R & D in Canada and that they keep their basic technological functions centralized in their home country (what has been termed the 'truncation' of R & D). For a careful evaluation of the truncation argument, and relevant references, see McFetridge (1977), pp. 29–36. His findings support the reasoning of the Canadian government.

[17] For a comprehensive review of the literature and new empirical analysis see Kay (1979).

[18] *Ibid.*

[19] Vernon (1977), pp. 41–5.

of major technological work, to coordinate scientific, engineering, planning, financial and marketing activities, and to keep a close watch on expensive and high-risk investments, will induce both the R & D function and its initial implementation (in production) to be kept at home. Thus, by implication, only relatively minor adaptive and development work will be relegated abroad.

There is a great deal of truth in this argument, but it has two deficiencies. First, as the evidence shows, a certain amount of basic and applied research *is* undertaken by foreign affiliates, to serve their own needs, the needs of the parents, as well as the needs of the MNC as a whole. Second, it is too general, and does not take into account possible inter-industry difference in the pressures that induce R & D of different types to be kept at home rather than located abroad.

To address ourselves to such possible differences, we need to draw, but selectively, upon the economics of industrial location.[20] There are obvious differences between the location of a particular form of activity within a larger complex of manufacturing operations by a firm and the location of a manufacturing plant, but there are enough similarities to render this effort worthwhile. In order to make an investigation empirically manageable within the very limited data available, however, we shall need to make various simplifying assumptions. Let us consider these assumptions and set out our hypotheses.

First, we assume that R & D personnel abroad are significantly less costly than in the US (a reasonable assumption for 1966, the year to which our data pertain). On pure labour cost considerations, therefore, there is an unequivocal pressure on all industries to undertake more R & D outside the US. The pressure will be greater the more an industry spends on research activity: thus, we expect to find a positive correlation between research intensity of industries and their propensity to undertake foreign R & D.

Second, R & D is an activity with considerable economies of scale, which exert a strong pressure against its location abroad. These economies arise from two factors: first, a research effort has to be of a certain minimum size to contain the different sorts of scientific specializations that need to work together. We assume that this factor is equal between industries. Second, while particular teams may be able to conduct research into well-defined problems in separate locations, the successful commercialization of a major innovation may require close and frequent communication between them. Thus, the 'transport cost' of having to maintain necessary communications between R & D teams in different locations may be high, and may rise higher if such communication involves different languages, cultures and educational backgrounds. However, as scale economies in R & D are exhausted, and as MNCs grow in size abroad and gain experience of controlling foreign operations, these 'transport costs' may be expected to diminish. MNCs will develop closely integrated systems of information exchange across countries, impose standardized patterns of management and operation and may even require a common language at the higher levels of the organization. Thus, given a uniform pressure to centralize R & D between industries, the more internationally spread an

[20] For a succinct exposition of the very large literature on industrial location see Miller (1977).

industry's manufacturing operations, the more it will be able to undertake foreign R & D economically, i.e. multinationality will be positively correlated with foreign R & D propensity.

Third, given the very highly developed technological infrastructure of the US, it is possible that it provides a larger variety and quantity of highly skilled research labour than foreign countries. Furthermore, if research skills developed 'in house' are significant in the conduct of R & D, the original base of research (in the US) will enjoy an advantage over foreign locations simply by virtue of its age. We assume that skill requirements are reflected in the average wages paid by each industry, and postulate that such wages will be inversely related to the propensity to conduct R & D overseas. The assumption that skill requirements for research activity are adequately represented by the average wage is, of course, an important simplification, but we do not possess data on the wages paid to R & D personnel in sufficient detail to employ this more direct measure. As with 'transport costs', however, we expect that a greater degree of multinationality will lead to more foreign R & D because of greater accumulated skills abroad.

Fourth, there may be certain benefits of 'agglomeration' in placing R & D in an area with advanced technology in the relevant activity. Such benefits may result from spillovers of information from other R & D units, access to trained personnel (though this may also be a potential cost), established links with universities or government institutions, and the existence of an appropriate infrastructure for specific kinds of research. For certain industries (electronics) agglomeration benefits may favour R & D in the US, for others (chemicals, food processing) they may also favour it abroad (say, in Europe). Since we cannot take account of this factor in our calculations, we shall assume that its influence is equal between industries, again a deliberate simplification.

Fifth, government regulation, subsidies and tax policies may influence the location of R & D units in the US and abroad. Many governments place a premium on attracting R & D activity and offer various incentives to MNCs as well as local firms to stimulate research expenditure.[21] In certain strategic industries, the home government may prevent the undertaking of any foreign R & D at all (there are good examples of this in the US). On the other hand, in some industries stringent regulations at home on the commercialization of new technologies (for health or environmental reasons) may induce firms to shift more innovative activity abroad than they otherwise would.[22] Policy measures may thus be expected to exercise different effects on different industries in their R & D location decision. We cannot take this possibility into account for lack of sufficient information, and have to assume it absent. The most extreme cases of government influence—in strategic industries—are, however, excluded from our sample, since they do not feature as multinational producers at all; and even in the case of pharmaceuticals it is probable that the effects of stringent regulations were not evident as early as 1966.

[21] See the US Tariff Commission (1973), p. 587.

[22] This is argued for the US pharmaceutical industry by Grabowski and Vernon (1977) and Schnee (1978), and supported by the report in *Nature*, *op. cit.*, according to which the industry more than doubled its foreign R & D in 1974–77 while increasing it by 34 per cent in the US.

Sixth, one way of exploiting the possession of advanced technology is by licensing it to other enterprises rather than by setting up or expanding foreign affiliates. There are several conditions under which licensing will be preferred to direct investment[23]—where the innovating firm lacks the financial or managerial resources to expand abroad, where the licence is for a technology which does not yield exceptional profits, where the technology is swapped for a similar licence from another firm, or where the market in question is too small or risky for direct investment. The literature suggests that in general large firms in high-technology sectors will tend to exploit commercially valuable technology by direct investment, and that licensing will be more commonly used by MNCs for older and/or more stable technologies. Since there does exist an active market in the sale of licences by US MNCs, it is plausible that to make the relevant technologies more saleable some foreign R & D is undertaken to adapt them to non-US conditions. The licensing of technology to other countries may, in other words, be entrusted by US MNCs to their affiliates abroad, who are, by virtue of their own experience of adapting US technologies, able to make the transfer more efficiently. Thus, we expect that the more an industry tends to use licensing as a means of exploiting its technology (measuring this by overseas royalty receipts as a percentage of R & D spending in the US), the more will it promote foreign R & D to make it more saleable in third markets. The causal connection may, of course, also run the other way. A larger R & D establishment abroad may, by providing a greater stock of readily transferable technology, promote a greater use of licensing, but we assume this away for present purposes.

There is, however, another, more serious, problem with this hypothesis of higher royalty receipts causing higher foreign R & D: it is possible that both are themselves caused by a third factor which we have not accounted for—the age of the technology in question. We have suggested that licensing by MNCs is more common for older and stabler technologies. It is also likely that it is easier to expand R & D abroad in older technologies which have less 'linkages' (considered next) with head office functions than new ones. Thus, both licensing and foreign R & D may grow together and not be causally related. This cannot be satisfactorily investigated with present data, since we cannot separate new and old technologies within industries. But it is an interesting possibility and we shall return to it below.

Finally, and most importantly, the extent to which R & D can be shifted abroad depends on the 'linkages' between R & D and other activities in the MNC. Given that the US is the main research base of its firms, and that it contains the main markets for its industries[24] and the main decision-making centres of its MNCs, the relative strength of these linkages determines the extent to which a shift to foreign R & D can be economical. It is hypothesized here that the main factor which will

[23] See Dunning's paper in this issue for a discussion of some of these conditions.

[24] For summary data on the extent of 'foreign involvement'—by exports and production by affiliates—of US industries in 1970, see the appendix to Lall (1980). In no case does affiliate production exceed 55 per cent of domestic production at the industry level; in 22 out of 25 industries it is under 30 per cent of US production.

influence the strength of the linkages is the extent to which *major applied research* can be separated from the managerial and marketing functions in the MNC.

Basic scientific research is not particularly closely linked to the other operational parts of the firm (though perhaps the strategic financial planners may wish to be near investments involving very high degrees of uncertainty);[25] in any case, there are no *a priori* grounds for expecting its linkages to differ significantly between industries. At the other end, minor development work is inextricably linked with the location of production, since it consists mainly of translating given technologies into practical use, and of tailoring them to specific material and marketing needs in each production unit. It is not, however, closely linked with strategic planning decisions in the head office of the firm. Different industries clearly have different needs for such development work. The mechanical engineering sectors, where detailed design is a crucial part of the production process, seem to need it much more than process or other industries. Such R & D will be undertaken in each production unit, local and foreign, and will lead to high research-intensity for the industry in every location. Thus, it may result in a high absolute value of foreign R & D (as we saw earlier), as well as high R & D in relation to sales, but not in a high *propensity* to undertake foreign R & D. So this particular linkage cannot account for inter-industry differences in this propensity.

As far as applied research into new products and processes is concerned, we suggest that certain R & D functions can be more easily 'de-linked' in process industries than in engineering ones. In a process industry like chemicals, research into new products can be conducted separately from research into new processes. Process technology in several such industries is sometimes held by independent engineering companies, suggesting that the two forms of R & D can be separated efficiently. Furthermore, the product does not have to be 'tailored' to individual customer requirements, so that a close monitoring of, and immediate reaction to, the main markets is not required by the R & D department of the innovating firm. In engineering industries like machinery, on the other hand, the core of the innovative process is the improvement, 'tailoring' and testing of new designs and it is more difficult to separate such functions as exploring the needs of major customers, bringing scientific and technical skills to bear on these needs, getting rid of bottlenecks in successive stages of production, ensuring an adequate supply of new components of various kinds, marketing the product and reacting to 'feed back' from the users.[26] In other words, there is a closer need for continuous interaction between all the major innovative functions and the procurement, production, management and marketing functions in engineering industries than in the process industries, and the former will find it more difficult to shift their major R & D abroad than the latter.

It is possible, of course, that both sorts of industries find it necessary to keep certain types of major innovation near the head office or the major production

[25] Lake (1979) also notes that firms may like to keep important innovative work at home because the risk of 'leaking' research findings may increase in foreign countries.

[26] For an historical analysis of the capital goods industries in the US which bears out the existence of these close 'linkages' see Rosenberg (1976).

centres. What is being suggested here is that it is *relatively* easier to separate parts of the R & D function for process than for engineering based industries. The 'linkages' do, of course, get weaker over time in all cases and for older technologies even engineering industries may (as Ronstadt's study suggests) be able to upgrade minor adaptive work abroad to meet certain basic innovative needs, especially when the affiliate has reached a relatively large size.

This suggests the following testable hypothesis: in process industries, a greater degree of research intensity should lead to a higher propensity to undertake R & D abroad because of the cost advantages of foreign research; in engineering industries, a greater degree of research intensity (representing a heavier emphasis on new products) should lead to a greater centralization of R & D in the US. In both sets of industries, however, the propensity to undertake foreign R & D is expected to rise with the relative size and age of affiliate production: the more experienced a firm is in foreign production, the larger the size of its 'minor' R & D abroad, and the greater the significance of foreign markets, the easier it will be to bear the costs of 'de-linking' R & D from the home base, and the greater will the benefits be of exploiting cheap and experienced manpower in the affiliates.

We have now identified a set of factors which are expected to explain the observed inter-industry variations in the propensity of US MNCs to undertake foreign R & D under a set of simplifying assumptions. Let us see how well the evidence supports our arguments.

IV The Empirical Tests

All the data on domestic and foreign R & D were taken from the US Tariff Commission (1973), supplemented by Creamer (1976). The propensity to undertake foreign R & D was the dependent variable in all the regressions. Two- and three-digit industries were combined so as to obtain 28 observations (see the appendix). Where a three-digit industry (e.g. farm machinery) was used, the relevant two-digit industry (non-electrical machinery) was excluded from the regression: an unexceptionable procedure where the levels of aggregation do not differ greatly between the observations. The statistical procedure was OLS multiple regression, with all the variables expressed in natural logarithms. Semilog and linear formulations were also tried, and yielded similar but less significant results.

The independent variables, all calculated from the Tariff Commission's study, are as follows:

FS: The foreign spread of each industry, measured by the sales of overseas affiliates of US firms as a percentage of total domestic sales of each industry. This variable was chosen to represent the relative foreign experience and involvement of MNCs in each sector, and was expected to be positively related to the dependent.

RD: The technological intensity of each industry in the US, measured by R & D spending by MNCs in the US as a percentage of total US industry sales within the US. It may be noted that we are relating MNC R & D to sales by all firms, but appropriate data at the detailed level are not available for either group (MNC or all machinery). Since the MNCs account for the bulk of R & D in each industry, however, this is not likely to cause a major bias. It is expected that this variable will be negatively related to foreign R & D propensity in engineering industries and positively in others.

WAGE: The skill requirements of each industry, measured by the average level of employee remuneration. As noted earlier, this is an unsatisfactory measure of the factor we wish to measure—skill requirements in R & D—but is the best available. It is expected that it is negatively correlated with foreign R & D propensity.

ROYALTY: Royalties and fees earned abroad by US MNCs as a percentage of their R & D in the US, a measure of their propensity to rely on licensing as a means of exploiting their technology. There is an important drawback to this measure, which unfortunately we cannot overcome with available data, that it includes royalties paid both by affiliated firms and by third parties. We are interested only in the latter since the former is an intra-firm transaction at least partly influenced by taxation and financial strategy.[27] However, the Tariff Commission data do not allow us to separate the two forms of royalties, and we must exercise due caution in interpreting the results.

The empirical literature on MNCs suggests that some of these independent variables will be strongly inter-correlated. Since technological superiority has been found in a large number of studies to lead to overseas investment, we would expect FS and RD to be collinear. Technological superiority also generally calls for a higher level of skills in the managerial (and possibly production) personnel, so that RD may be expected to be positively correlated with WAGE. Finally, we would expect a negative correlation between RD and ROYALTY, on the ground that research-intensive industries would tend to have a relatively high proportion of new products and processes, which would tend to be exploited by foreign production rather than licensing. Many of these relationships did obtain for US industry as a whole and for the sub-groups, and will be noted below. We try to avoid the worst collinearity problems in our presentation of the results.

The regressions were run for all US industries together as well as for three sub-groups: process industries, engineering industries and 'other' industries. There were 12 observations for process industries: grain mill products, beverages, other foods, paper, drugs, soaps, industrial chemicals, plastics, other chemicals, rubber, textiles and printing and publishing. There were nine observations for engineering industries: farm machinery, industrial machinery, office machines, electronic and other non-electrical machinery, electrical (household and industrial) equipment, radio and TV, other electrical machinery, transportation equipment and instruments. And there was a rag-bag of seven other industries: primary metals, fabricated metals, aluminium, other metals, lumber, wood and furniture, stone, clay and glass, and 'other' manufacturing. We had no particular expectation of the influence on foreign R & D propensity in this last group.

[27] See Kopits (1976) for an analysis of this aspect of MNC technical payments.

TABLE 1

Determinants of the Propensity to Conduct R & D Overseas

Equation No.	*Industry Group*	*FS*	*WAGE*	*RD*	*ROYALTY*	*Constant*	$\bar{R}^2$	*F*
1	All 28 Industries	1.01^{b} (2.01)	−1.99 (−0.66)			10.54	0.08	2.21
2	″			0.478^{c} (1.44)	0.678^{c} (1.66)	−0.65	0.05	1.66
3	″	0.913^{b} (2.02)				−0.63	0.10	4.08
4	12 Process	1.76^{a} (3.58)				−4.86	0.52	12.83
5	″			0.615^{c} (1.45)		2.32	0.09	2.09
6	″	2.07^{a} (3.63)	−3.53 (−1.05)			14.34	0.52	7.03
7	″			1.08^{b} (2.29)	1.26^{c} (1.74)	−5.32	0.24	2.78
8	″	1.77^{a} (3.53)			0.369 (0.76)	−6.93	0.50	6.43
9	9 Engineering			-1.59^{a} (−7.16)		9.29	0.86	51.26
10	″			-1.41^{b} (−2.88)	0.10^{b} (2.89)	8.43	0.84	22.32
11	″	-0.82^{c} (−1.60)			1.04^{a} (4.78)	3.99	0.74	12.13
12	″				1.01^{a} (4.21)	10.10	0.68	17.76
13	″		−1.70 (−0.57)		0.98^{a} (3.81)	10.04	0.64	8.18
14	7 Other			-3.70^{b} (−3.09)	1.43 (1.40)	1.64	0.56	4.82
15	″	−2.09 (−1.47)	-12.08^{c} (−1.62)			81.60	0.21	1.80
16	″			-3.10^{b} (−2.54)		7.02	0.48	6.46
17	″		−2.30 (−0.32)	-2.87^{c} (−1.91)		20.08	0.36	2.70

Notes: R^2 corrected for degrees of freedom; *t* value in parentheses. Degrees of significance shown as follows: a: 99 per cent; b: 95 per cent; c: 90 per cent (one-tail test).

Table 1 sets out the main results of our investigation. Equations 1–3 show the results for the sample as a whole, 4–8 for process industries, 9–13 for engineering industries and 14–17 for other industries. Let us describe them in turn.

All Industries: None of the regressions was very successful in terms of explaining the inter-industry variation in the dependent, but most of the independent variables had the expected signs. The sign of RD in equation 2—positive and significant at 90 per cent—indicates that rising research intensity leads in general to greater R & D abroad; however, the effect is rather weak and may not (see below) be meaningful. FS by itself 'explains' the maximum amount of variation in the dependent; clearly, greater foreign experience and spread help US MNCs to overcome the centralizing tendencies that normally govern research activity. FS and RD have a high correlation coefficient of 0.67, confirming the expected link between technological superiority and foreign investment.

WAGE has a negative sign but does not reach significance; as expected, it is significantly related to RD (simple correlation coefficient of 0.57). ROYALTY is positively and significantly (at 90 per cent) related to the dependent, but whether this is due to licensing causing greater foreign R & D, or to both licensing and foreign R & D being related to 'mature' technology, is difficult to say. The latter interpretation may be supported by the fact that RD and ROYALTY have a negative correlation of −0.47. RD by itself has no noticeable effect on the dependent; in equation 2, its significance may be due to its collinearity with ROYALTY.

Process Industries: In this group the simple correlation coefficients between RD and FS (0.71) and RD and WAGE (0.67) are stronger than for all industries together, suggesting that research activity in the US, based on high levels of skill, are exceptionally strong in pushing US firms to invest abroad. The influence of FS on foreign R & D propensity is also much stronger, with the former (significant at 99 per cent) by itself explaining over half the variation in the dependent. RD, with a (slightly) stronger effect than for all industries, also has a significant independent effect on foreign R & D propensity: however, by itself RD only explains 9 per cent of the variation in the dependent. It is possible that its effect is due to its strong collinearity with FS, and that overseas R & D is primarily caused by greater foreign experience rather than its lower cost.[28] The evidence is compatible with the argument that certain kinds of R & D can be 'de-linked' in process industries, but we cannot establish this clearly because of multi-collinearity problems.

WAGE has the anticipated negative sign but does not achieve significance. ROYALTY has a strong negative relationship with RD (−0.57), suggesting, as before, that it is the less technologically advanced ones which choose licensing as the route to exploit their advantage. It has a positive influence on the dependent, especially in equation 7 when tried with RD. The addition of ROYALTY adds greatly to the explanatory power of RD, with the $\bar{R}^2$ rising from 0.09 to 0.24, and the level of significance rising from 90 to 95 per cent. This may tentatively be interpreted as supporting the argument that higher R & D costs in the US lead MNCs to seek cheaper locations abroad: when the effect of licensing (and, by implication, of more mature technologies) is taken into account, the effect of higher research intensity is to increase overseas R & D.

Engineering Industries: The results obtained for this group are, despite the small number of observations, the most significant in statistical terms and the most interesting in the context of the present enquiry. They strongly confirm that higher technological intensity in engineering industries leads to a centralization of research activity: RD by itself explains 86 per cent of the variation in foreign R & D propensity, and the coefficient is negative and highly significant. RD also

[28] It may also be due to the fact that in certain process industries (chemicals and food processing), the technological prowess of European countries is equal to or exceeds, that of the US. Thus, US MNCs may locate some R & D there to enjoy 'agglomeration' benefits.

has a very high negative correlation with ROYALTY (−0.88) and the latter appears as the only factor which promotes R & D overseas. ROYALTY by itself has an $\bar{R}^2$ of 0.68, and is significant at 99 per cent; whether this is due to the need to undertake foreign R & D to adapt licensed technology or simply to the age of the technology licensed we cannot say.

Curiously enough, FS has a rather low correlation with RD (0.16), and it seems to exercise a negative (but barely significant) effect on foreign R & D propensity. This suggests that greater foreign spread and experience is, contrary to expectations, unable to counterbalance the centralizing tendencies that affect R & D location in these industries. This finding casts doubt on any simple projection that MNCs will become the agents of dynamic comparative advantage in allocating research activity as they have tended to become in allocating certain kinds of production activity, certainly where some of the most important 'high technology' industries are concerned. Engineering MNCs do undertake substantial amounts of R & D in every place they place a plant, but most of this seems to be devoted to minor technological work and to be concentrated in older technologies.

As before, WAGE has the correct sign but fails to reach significance.

Other Industries: The regressions for this group yield results similar to those for engineering industries: RD has a significant (at 95 per cent) negative effect on foreign R & D propensity. FS also has a negative effect but just fails to reach significance. WAGE has a negative effect and, in combination with FS (equation 15) manages to reach significance at 90 per cent. ROYALTY has, as before, a positive effect, but is not statistically significant. The simple correlation coefficient between RD and ROYALTY is of interest here: it is positive and moderately strong (0.36). Thus, in this group greater technological intensity at home also leads to a greater use of licensing abroad.

To sum up our findings: the propensity of US MNCs to conduct foreign R & D is subject to different influences according to the nature of the industry. In process industries, greater foreign spread and experience are strongly related to this propensity, and higher technological intensities seem to promote more research abroad; here, therefore, there are some grounds for supposing that an internationalization of research activity according to comparative cost is taking place. In engineering industries, on the other hand, greater technological intensity seems to call for more centralization of R & D in the US, and the effect of foreign spread is insufficient to counter this powerful centripetal force. This seems to support Ronstadt's impression that cost factors do not figure large in the R & D location decision, and that most foreign R & D is in minor adaptive technological work. It also strengthens the traditional product cycle analysis to the extent that in most of the high technology sectors important new products tend to be innovated and launched in the home market, regardless of the 'internationalization' of the innovation firms. The cycle may be shortened over time, but the sequence of events may still remain.

The availability of highly-skilled personnel in the US seems to strengthen its attraction as a research base to all sorts of MNCs, but the statistical evidence on this is weak. The need to tailor technologies for the licensing market serves, on the other hand, to promote foreign R & D in all the industry groups, but the calculations are difficult to interpret clearly because of the strong likely influence of the age of technology on both foreign R & D and licensing. If it is the latter which is the important factor, we may infer that foreign research work will concentrate on adapting and improving relatively well diffused technologies.[29]

V Conclusions

This paper has tried to explain inter-industry differences in the propensity of US MNCs to undertake foreign R & D. Despite using rather limited and out-of-date information, and making various simplifying assumptions, it has thrown up some interesting suggestions about how this propensity is differently influenced in different groups of industries.

In process industries, where major technological efforts on products and processes are not crucially linked to each other or to the main production centres and markets, it appears that such factors as international experience and cost advantages tend to promote greater reliance on foreign R & D. In engineering industries, on the other hand, where innovation centres around product development and testing, it seems much more difficult to separate any part of major R & D activity from the main market and the centres of decision making (both in the US). In these industries, greater technological intensity seems to create a greater centralization of research activity. In terms of absolute value of R & D expenditures, it is these engineering industries which predominate in overseas research; however, most of this research seems to be concentrated on relatively minor technological activity and on techniques which are relatively mature.

Our findings are subject to a number of qualifications because of the nature of the data and the levels of aggregation. They do, however, suggest that there is a need, on the part of economists, to look more deeply into the complex processes of industrial innovation, and in particular into inter-industry differences in the way that the innovative process fits in with other important functions. They also suggest that the product cycle hypothesis retains a strong base in many high-technology industries (though, of course, evidence on more recent trends may lead us to modify this[30]); the 'internationalization' of production need not lead to a similar internationalization of all the activities of manufacturing firms.

Institute of Economics and Statistics,
Oxford.

[29] This provides indirect support, at least in engineering industries, for the Canadian fears that US MNCs 'truncate' their R & D abroad (see McFetridge 1977), and keep the most basic and important innovative work at home.

[30] See Vernon's paper in this issue for evidence that MNCs may be speeding up the process of transferring innovations (if not actual R & D) abroad.

APPENDIX

Domestic and Foreign R & D Expenditure by US MNCs, 1966

	In US ($ m.)	Abroad ($ m.)	Total ($ m.)	Foreign R & D[b] Propensity (Percentage)
All manufacturing	7598	526	8124	(6.5)
Food products	136	18	154	(11.7)
Grain mill products	41	2	43	4.7
Beverages } Other	95	16	111	{ 0.0[d] 14.4
Paper and allied products	64	3	67	4.5
Chemicals	1258	74	1332	(5.6)
Drugs	303	25	328	7.6
Soaps and cosmetics	66	13	79	16.5
Industrial chemicals	777	8	785	1.0
Plastics	31	12	43	27.9
Other	81	16	97	16.5
Rubber products	127	4	131	3.1
Primary and fabricated metals	312	10	322	(3.1)
Primary (excl. aluminium)	130	5	135	3.7
Fabricated (excl. aluminium, copper & brass)	138	5	143	3.5
Primary and fabricated aluminium } Other	44	0[c]	44	{ 0.0 0.0
Non-electrical machinery	743	90	833	(10.8)
Farm machinery and equipment	119	13	132	9.8
Industrial machinery and equipment	184	44	228	19.3
Office machines	108	5	113	4.4
Electronic computing equipment & other	332	28	360	7.8
Electrical machinery	1814	103	1917	(5.4)
Electrical machinery and equipment[a]	1100	13	1113	1.2
Radio, TV, electronic components	685	28	713	3.9
Other	29	62	81	76.5
Transportation equipment	2537	134	2671	5.0
Textiles and apparel	29	0[c]	29	0.0
Lumber, wood & furniture	25	61	86	70.9
Printing & publishing	17	0[c]	17	0.0
Stone, clay and glass	103	4	107	3.7
Instruments	372	21	393	5.3
Other	61	4	65	6.2

Notes:
[a] Includes household appliances.
[b] Expenditure on R & D abroad as a percentage of total R & D. Figures in parentheses are excluded in the statistical tests.
[c] 0 signifies sum is positive but less than $500,000.
[d] Beverages conduct negligible R & D abroad.
Source: US Tariff Commission (1973), p. 582, and Creamer (1976), pp. 123–6.

REFERENCES

Creamer, D. (1976). *Overseas Research and Development by United States Multinationals 1966–1975*, N. York: The Conference Board.

Dunning, J. H. (1973). 'The Determinants of International Production.' *Oxford Economic Papers*, pp. 289–336.

Dunning, J. H. (1977). 'Trade, Location of Economic Activity and the MNE: A Search for an Eclectic Approach,' in B. Ohlin *et al.* (eds.), *The International Allocation of Economic Activity*, London: Macmillan.

Grabowski, H. G. and Vernon, J. M. (1977). 'Consumer Protection Regulation in Ethical Drugs.' *American Economic Review, Papers and Proceedings*, pp. 359–64.

Hirsch, S. (1976). 'An International Trade and Investment Theory of the Firm.' *Oxford Economic Papers*, pp. 258–70.

Kay, N. M. (1979). *The Innovating Firm*, London: Macmillan.

Knickerbocker, F. T. (1973). *Oligopolistic Reaction and Multinational Enterprise*, Boston: Harvard University.

Kopits, G. F. (1976). 'Intra-firm Royalties Crossing Frontiers and Transfer-pricing Behaviour,' *Economic Journal*, pp. 791–805.

Lake, A. W. (1979). 'Technology Creation and Technology Transfer by Multinational Firms,' in R. G. Hawkins (ed.), *Research in International Business & Finance, The Economic Effects of Multinational Corporations*, Vol. I, Greenwich (Conn.): JAI Press.

Lall, S. (1980). 'Monopolistic Advantages and Foreign Involvement by US Manufacturing Industry,' *Oxford Economic Papers* (March).

McFetridge, D. C. (1977). *Government Support of Scientific Research and Development: An Economic Analysis*, Toronto: University of Toronto Press, for the Ontario Economic Council.

Magee, S. P. (1977). 'Information and the Multinational Corporation: An Appropriability Theory of Direct Foreign Investment,' in J. N. Bhagwati (ed.), *The New International Economic Order*, Cambridge (Mass.); MIT Press.

Miller, E. W. (1977). *Manufacturing: A Study of Industrial Location*, University Park (Penn.): Pennsylvania State University Press.

NSF (1975). *Research and Development in Industry, 1973*, Washington, D.C.: National Science Foundation.

Ronstadt, R. (1977). *Research and Development Abroad by US Multinationals*, New York: Praeger.

Rosenberg, N. (1976). *Perspectives on Technology*, London: Cambridge University Press.

Schnee, J. E. (1978). 'International Shifts in Innovative Activity: The Case of Pharmaceuticals,' *Columbia Journal of World Business*, Spring, pp. 112–22.

US Tariff Commission (1973). *Implications of Multinational Firms for World Trade and Investment and for US Trade and Labor*, Washington, D.C.: G.P.O.

Vernon, R. (1966). 'International Investment and International Trade in the Product Cycle,' *Quarterly Journal of Economics*, pp. 190–207.

Vernon, R. (1977). *Storm over the Multinationals: The Real Issues*, London: Macmillan.

[4]

KYKLOS, Vol. 33 – 1980 – Fasc. 2, 308–327

RESEARCH AND DEVELOPMENT PERFORMED ABROAD BY U.S. MANUFACTURING MULTINATIONALS

GARY HEWITT*

I. INTRODUCTION

'... a regime of North Atlantic Multinational Corporations would tend to produce a hierarchical division of labor between geographical regions corresponding to the vertical division of labor within the firm. It would tend to centralize high-level decision-making occupations in a few key cities in the advanced countries, surrounded by a number of regional sub-capitals, and confine the rest of the world to lower levels of activity and income, *i.e.*, to the status of towns and villages in a new Imperial system ...'

'The viability of the multinational corporate system depends upon the degree to which people will tolerate the uneveness it creates.' S. HYMER, 'The Multinational Corporation and the Law of Uneven Development'[1].

The research and development (*R & D*) activities of American multinational manufacturing companies (*AMMCs*) is an area where HYMER's description and implicit warning seem particularly apt. This is because *R & D* is an excellent example of a high level, high income corporate activity, and because available evidence clearly establishes that *AMMCs* concentrate their *R & D* within the U.S.[2].

* Department of Economics, Concordia University (Montreal). The author wishes to thank Professors SALIM RASHID (Dartmouth College) and WILLIAM SIMS (Concordia University) for their comments and encouragement.

1. HYMER (1972, pp. 114 and 130).

2. For survey evidence, see DUERR (1970, pp. 2–19) and BEHRMAN (1969, p. 82).

In 1966, the foreign assets of U.S. majority owned foreign affiliates in manufacturing were *19.7%* of the domestic assets of U.S. foreign investing manufacturing firms. The corresponding percentage for *R & D* performed abroad was only *6.9%*. These data were taken from U.S. Department of Commerce (1976, pp. 152, 224, 226, and 229).

RESEARCH AND DEVELOPMENT

The present study focusses on the reverse side of that centralization. It seeks to explain how and why the *AMMC* comes to locate *R & D* outside the U.S. Such research should prove helpful to host country policy makers charged with obtaining a larger local share of multinational corporate *R & D*[3]. If HYMER is right, the success or failure of their efforts might well contribute to the survival or demise of the 'multinational corporate system'.

The literature directly relating to this topic is composed of two surveys. The results of the first were released by M. C. DUERR in 1970 and those of the second by R. C. RONSTADT in 1977[4]. RONSTADT's work is particularly relevant owing to its greater analytic emphasis. Based on close examination of the foreign *R & D* activities of seven *AMMC*s, RONSTADT concluded that four specific kinds of such work exist and '... that the vast majority of foreign *R & D* investments seems to follow a particular evolutionary pattern'[5].

What follows below is a scenario describing the typical *AMMC*'s foreign *R & D* evolution placed in the context of its general development from exporter to global firm. Much detail is suppressed in order to highlight the importance of changes in the firm's foreign business strategy and to extend RONSTADT's work through establishing the significance of the *AMMC*'s *R & D* versus marketing orientation. The scenario is later summarized in such a way that its insights are brought to bear on explaining cross firm variation in the percent of corporate *R & D* located outside the U.S. Such variation is indirectly explored in the multivariate regression analysis presented in the last part of the paper. The exploration is 'indirect' because, as is often the case in industrial organization work, data availability requires the use of industry aggregates as substitutes for firm level data. It will be seen that the empirical test supports both the paper's scenario and the general approach underlying it. A few words about that general approach are necessary before proceeding to the scenario itself.

3. At least one host country government, Canada, has already included local *R & D* employment as a factor of interest in deciding whether or not to permit a particular foreign direct investment to take place. See *Foreign Investment Review* (1977, p. 18).

4. DUERR (1970) and RONSTADT (1977).

5. RONSTADT (1978, p. 22).

GARY HEWITT

An explicit or implicit assumption of much economic theorizing is that information costs nothing to obtain and to use. Firm decision making is seldom discussed because it amounts to a trivial procedure of choosing from among *all* possible alternatives the one maximizing some particular objective function. If this standard paradigm were employed here, the scenario would become a formal model revolving around two variables – relative *R & D* factor costs across countries and differing factor intensities among the various types of *R & D*. There would be no need to even mention the variables comprising the very heart of this paper's scenario and bearing the weight of its empirical test.

A real world *AMMC* must deal with the difficulties imposed on it by the high cost of finding *and* using pertinent information. How it tackles these problems has a direct effect on the decisions it makes. HERBERT SIMON and others have extensively explored the information search and decision processes actually used in business enterprises and have developed what are now called 'satisficing' or 'behavioural' models of the firm[6]. Such models have proven themselves in empirical investigation, including some related to foreign direct investment[7].

The essence of the 'satisficing' or 'behavioural' models lies in their identifying the factors responsible for provoking, steering and terminating a firm's search for decision alternatives. The firm comes into focus not as an all knowing, profit maximizer, but rather as an incompletely informed, problem solving, multi-goal 'satisficer'[8]. The 'satisficer' label is used because, once provoked into an information search by unsatisfactory performance on one or more of its goals, the firm will accept the first 'satisfactory' alternative it finds. What

6. For examples of this work, see SIMON (1957) and CYERT and MARCH (1963), especially ch. 6.

7. See for example JOHANSON and VAHLNE (1971, pp. 23–32); also AHARONI (1966).

8. Readers familiar with game theory or with the literature on decision making under uncertainty might wonder why such models have been rejected here. The answer, paraphrasing SIMON (1957b, p. 202), is that such approaches are 'fundamentally wrongheaded' in not fully confronting the issues of 'bounded rationality' and information costs. For a similar view, see CYERT and MARCH (1963, p. 119).

is 'satisfactory' depends on the firm's past experience and its initial search results.

The 'satisficing' approach provides the basic foundation for the scenario discussed below. It will be particularly visible when the *AMMC* is described as taking 'defensive' measures, or when importance is attributed to 'lobbyists' (information sources) internal to the firm, or when changes in the firm's basic strategy or policy is being discussed. The scenario is built around the successive organization stages that extensive survey work revealed to be good reflections of the *AMMC*'s evolving foreign business strategy[9].

II. THE SCENARIO

1. Exports

The typical *AMMC* begins its international experience through the export of products enjoying some monopoly advantage[10]. Unless foreign laws or market differences make it imperative, there will be no *R & D* undertaken in connection with these initial foreign sales. There is simply no perceived challenge to do so at this stage, and the export sales volume is probably insufficient to warrant such *R & D* spending in any case.

2. Autonomous Subsidiaries

As the firm's foreign sales increase, its export related technology matures, and the potential for foreign competition grows, the erstwhile exporter will eventually, according to the product cycle tradition, be forced into *defensive* foreign direct investment[11]. Elaborating on this reaction, C. A. Michalet has said: 'Essentially ... we

9. See Stopford and Wells (1972). – We are not alone in assigning great importance to the effect of organization structures linking parents and subsidiaries. Cordell (1971, p. 42), working from a Canadian survey, stated that: 'The role of *R & D* and the capacity for innovation are determined in almost every case by the organizational model chosen for the total firm and the extent to which North American operations are rationalized.'

10. See Vernon (1971, pp. 66–77); and Wells (1972, especially pp. 11–15).

11. See Vernon (1971, p. 71); Stopford and Wells (1972, pp. 19–20); and Aharoni (1966, pp. 64 and 176).

are dealing with an oligopolistic strategy whose aim is to maintain a threatened technological position by other than innovative means[12].' This stage is labelled 'autonomous subsidiaries' because, as STOPFORD and WELLS discovered: 'The men who are sent out to manage these fledging subsidiaries are generally allowed virtually unlimited powers of decision and action, at least at first[13].' The basic concerns of such managers center on producing and marketing abroad the parent's major, previously exported, product lines[14].

Having opted for foreign direct investment rather than innovation, and having chosen to concentrate on a product line previously established abroad, the typical 'autonomous' stage *AMMC* will show little interest in *R & D* related to adapting parent *product* technology to better suit foreign market conditions – henceforth referred to as 'product adaptive' *R & D*. The firm will be even less inclined towards *R & D* intended to create new products expressly for the foreign market – henceforth labelled 'local (*i.e.* related to a subpart of the firm's global market) original' *R & D*[15]. In contrast, there will be an acute awareness of the need for *R & D* related to foreign purchase of production inputs and to scaling down production runs for a smaller market. At least some of this 'process adaptive' *R & D* appears to be done in foreign subsidiaries[16]. How much is located abroad depends on:

(a) how easily the parent company's *R & D* facilities could handle the extra work;

(b) how much underemployment exists among the firm's *R & D* capable foreign production workers;

(c) how geographically separable is 'process adaptive' *R & D* from its production application; and

12. MICHALET (1973, p. 17).

13. STOPFORD and WELLS (1972, p. 20).

14. MICHALET (1973, pp. 35–36) provided some pertinent insights on this score.

15. Otis Elevator Co. provided a good example of 'local original' *R & D* focussed on a foreign market when it designed a small elevator specifically for low rise European buildings. Since Otis, at the time, had nothing similar for sale in the U.S., the small elevator project was qualitatively different from 'product adaptive' *R & D*. See RONSTADT (1978, p. 12).

16. See CORDELL (1971, p. 43); and DUERR (1970, pp. 9–11).

(d) how much host government pressure is applied to win some of the *AMMC*'s *R & D* employment.

This list is motivated by the *AMMC*'s 'satisficing' outlook (*i.e.* foreign *R & D* is much more likely to be considered if the parent does not already have the facilities needed to do it), by PENROSE's findings that firms tend to grow so as to fully employ resources already hired and integrated into the firm[17], and by evidence that firms are indeed influenced by host government pressure to boost local *R & D* employment[18].

Before proceeding to the next stage, let us emphasize the importance of early establishment of foreign *R & D* facilities. Once these are set up, for whatever reason, an active lobby group arguing in favour of foreign *R & D* is created within the firm.

3. *The International Division* (*ID*)

As foreign sales continue to expand and foreign subsidiaries multiply, parent company attention will eventually focus on the need to co-ordinate foreign operations, especially as concerns intercompany sales, exports, and the raising of capital. Such co-ordination will be the task assigned to a newly created international division (*ID*) whose head reports directly to the parent's chief executive officer[19].

It is at the *ID* stage that a fundamental orientation difference, identified by STOPFORD and WELLS, first becomes significant[20]. A minority of the *AMMCs* they surveyed appear to rely on the exploitation of a narrow product line composed of what the product cycle writers refer to as 'mature' products. These are the 'marketing oriented' (*i.e.* advertising intensive) *AMMCs*. In contrast, the majority counterpart group is made up of heavy *R & D* spenders offering a very diverse product line.

The minority group's foreign subsidiaries will normally be manufacturing their parents' full product lines before or soon after the

17. See PENROSE (1959), ch. 5.

18. See BEHRMAN (1970, pp. 61–62).

19. See STOPFORD and WELLS (1972, pp. 21–22).

20. This paragraph's ideas are rooted in evidence presented by STOPFORD and WELLS (1972, pp. 35–62). Supporting comments can be found in BEHRMAN (1970, pp. 65–68).

ID stage is reached. Since it is international production transfer that creates a need for 'process adaptive' *R & D*, and since such work is the mainstay of foreign performed *R & D* during the *AMMC*'s 'autonomous' stage, the *ID* period would seem to spell decline for a marketing oriented *AMMC*'s foreign *R & D* effort. This no doubt would be the case were it not for the fact that an end to product transfer abroad also threatens to slow the growth of foreign production. RONSTADT's survey revealed that a foreign subsidiary faced with this situation might succeed in altering its parent's policy of locating all new product development in the U.S. To do this it has to identify new investment opportunities different from existing U.S. businesses as well as demonstrate the managerial and technical capability to develop appropriate products to meet those opportunities[21]. The probability then of 'local original' *R & D* being performed abroad during a marketing oriented *AMMC*'s *ID* stage is related to: the diversity of its parent's basically narrow product line; the degree of international taste or use differences pertinent to the firm's products; and the scope of the subsidiary's 'autonomous' stage *R & D* programs.

Blessed as they are with a large and growing set of innovative products, the *R & D* intensive *AMMC* does not experience any difficulty in continuing to stimulate foreign growth during the *ID* stage with products transferred from the U.S. Their foreign subsidiaries will feel no great pressure to challenge parent policy of innovating in and for the U.S. market and later spreading 'sunk' *R & D* costs through exports and foreign production[22]. On the foreign *R & D* front, 'process adaptive' *R & D* will remain as much demanded as ever, while 'local original' work will not even be actively considered.

4. Global Structure

As firms expand their foreign sales and/or foreign product diversity, they increasingly come to see the U.S. – rest of the world dichotomy, institutionalized by the international division, as an obstacle to fur-

21. See RONSTADT (1978, p. 12).

22. Evidence of just such a parent policy in many *AMMC*s can be gleaned from revealing quotes scattered throughout DUERR's report. See DUERR (1970, pp. 2–7 and 15).

ther growth. When this happens, the *ID* will be dissolved into one of two basic global structures discussed below. Either set-up will encourage the development of worldwide perspective, co-ordination and control.

(a) Global Area Divisions

This structure is generally chosen by the marketing oriented, narrow product line group of firms[23]. Owing to the maturity of their products, such enterprises can be expected to eagerly pursue production rationalization and marketing standardization *within each new area division*[24]. Area centered tactics will allow them to lower their costs while maintaining an appropriate marketing diversity. The firm's *R & D* effort will typically be determined by its area orientation, meaning that the 'local original' *R & D* begun previously will be reorganized to take on an area rather than country specific focus.

(b) Global Product Divisions

R & D intensive, product diverse firms will normally find that expanding foreign product diversity so increases communication between foreign and domestic product divisions that it becomes expedient to unite them under common roofs[25]. Once this is done, the firm will be well on its way to adopting a 'one world' market and innovation perspective. No longer will the firm be content to innovate in and for the U.S. treating the rest of the world as an afterthought. Corporate *R & D* will become 'global original' in character – *i.e.*, innovations will be conceived and production planned with the entire world market in mind. Recent examples of such innovations are the Fiesta and Chevette automobiles – the U.S. entries in what *Business Week* has termed: '... a marked trend toward so-called "world cars" – standardized autos designed to be built and sold in any major market with few changes[26].'

As 'global original' *R & D* takes hold within the *AMMC*, other forms of *R & D* are pushed out. 'Process adaptive' work is rendered obsolete by worldwide production rationalization, while 'product

23. See Stopford and Wells (1972, p. 48); and Cordell (1971, p. 32).
24. See Stopford and Wells (1972, pp. 56, 58 and 61).
25. See Stopford and Wells (1972, p. 30), as well as Cordell (1971, p. 32).
26. *Business Week* (1978, p. 102).

adaptive' and 'local original' *R & D* find themselves out of place in a corporation bent on exploiting worldwide market similarities. It is precisely in servicing such similarities that the *AMMC* will realize a marketing and *R & D* based comparative advantage over its competitors restricted to one national market.

The foregoing argument is not meant to question RONSTADT's findings of a general trend from 'adaptive' to 'local original' *R & D* among the mostly *R & D* intensive *AMMCs* he surveyed[27]. Instead it is intended to support two contentions compatible with RONSTADT's work:

1. *compared to their marketing oriented cousins*, *R & D* intensive, globally structured *AMMCs* will emphasize 'global original' *R & D* and leave 'local original' work *underdeveloped;* and

2. there will be a very dim future for foreign *R & D* within the *R & D* intensive, globally organized *AMMCs* unless their foreign subsidiaries are able to force decentralization of 'global original' *R & D* spending. Such a decentralization would represent a clear break from previous corporate policy of locating all 'original' *R & D* within the U.S.

Assuming that *AMMCs* act as 'problem solving satisficers', the most important forces aiding foreign subsidiaries in their efforts to decentralize 'global original' *R & D* are likely to be:

(a) full employment of the firm's domestic *R & D* facilities coupled with underemployment of those abroad; and/or

(b) a pressing competitive need to tap *R & D* resources unavailable in the U.S.[28].

For the first time in the scenario, some role, even if a minor one, must be allowed to international differences in *R & D* factor costs and availability. Only at the global stage will a typical *AMMC* have sufficient foreign experience to be effectively aware of such differences. It will be exceedingly rare however, that relative *R & D* costs and capabilities will assume so great an importance that the *AMMC* will decide to move *all* its *R & D* to one foreign location. Even if

27. See RONSTADT (1978, pp. 15–19). Note that his '*TTUs*' and '*ITUs*' are engaged in what we have termed 'adaptive' and 'local original' *R & D* respectively.

28. For supporting evidence, see RONSTADT (1978, pp. 13–15); and DUERR (1970, pp. 12–15).

heterogeneous *R & D* factor intensities did not rule that out, concern to keep abreast of developments in the U.S. would[29]. The large, wealthy, highly competitive U.S. market will normally remain the *AMMC*'s most important pioneer market. In consequence, the firm's foreign *R & D* proponents will find their efforts hindered by various decentralization costs – foregone economies of scale plus extra coordination and communication expenses – that favour keeping *R & D* centralized within the U.S.

III. SCENARIO SUMMARY

Ronstadt has found that a change in the purpose of foreign *R & D* appears to be a necessary condition for its rapid expansion[30]. This is quite consistent with what the present study's scenario would lead one to expect. Such changes of purpose in favour of either 'local original' or 'global original' *R & D* reflect conscious policy decisions to *decentralize*, or to decentralize further, an *AMMC*'s total *R & D* effort. It seems that rapid absolute growth of foreign *R & D* is the chief means by which that decentralization is normally effected. This brings us to the following generalizations:

1. When the marketing intensive *AMMC* moves from 'autonomous' to '*ID*' organization and simultaneously shifts from 'process adaptive' to 'local original' work as its principal type of foreign *R & D*, the percent of corporate *R & D* located outside the U.S. will typically grow. Such growth will usually continue when the international division is in turn replaced by global area divisions, and decentralized 'local original' *R & D* acquires an area division focus. The U.S. market from then on will be treated as one among many.

2. The *R & D* intensive *AMMC* will normally stick to 'process adaptive' *R & D* abroad during both the 'autonomous' and '*ID*' stages. One would expect little change in the percent of its *R & D* located abroad until global product divisions are set up. At that time, foreign *R & D* may be drastically reduced as a share of total

29. One could say that A.J. Cordell has directly established this point by showing that the *R & D* efforts of some Canadian multinationals have tended to gravitate to the U.S. – their largest single market. See Cordell (1971, p. 74).

30. See Ronstadt (1978, p. 15).

corporate *R & D* or it may be expanded. Which way things go depends on whether or not the *AMMC* decides to decentralize its 'global original' *R & D* work.

3. Comparing the percent of total corporate *R & D* performed abroad by a pair of otherwise similar marketing and *R & D* oriented firms, it will typically be the case that:

(a) at the autonomous stage – the marketing oriented *AMMC* will be ahead owing to its roughly similar propensity to do 'process adaptive' *R & D* abroad, coupled with its smaller overall corporate *R & D* budget;

(b) at the *ID* stage – the marketing oriented firm will retain its earlier lead because it begins to decentralize 'original' *R & D* work while the *R & D* intensive firm does not;

(c) at the global stage – a still further widening of the gap in favour of the marketing oriented *AMMC* whose 'local original' *R & D* specialization lends itself more to decentralization than the 'global original' variety emphasized by the *R & D* intensive firm.

IV. EMPIRICAL WORK

A thorough testing of this study's model would require enough *firm* level data to adequately populate six separate classes (three reporting stages times two orientation groups). What is in fact available are 1966 vintage data representing *twenty*, approximately two digit manufacturing *industries*. Any idea of separate tests for each reporting/orientation class must be abandoned and industries must be treated as if they were firms. Despite these problems, it remains true that the empirical test was inspired by the study's scenario and offers some indication of the latter's plausibility.

The test itself consists of attempting, by multivariate regression analysis, to explain cross industry variation in *PFRD* – the percent of an American industry's worldwide *R & D* performed abroad in majority owned foreign affiliates[31]. There are eight explanatory, or independent, variables:

31. Anyone concerned about running a restricted dependent variable should know that the ratio of an industry's foreign divided by its domestic *R & D* was also tried in place of *PFRD*. The results as to signs, t statistics, and F test were practically identical.

RESEARCH AND DEVELOPMENT

1. The percent of an industry's subsidiaries belonging to parents having neither international divisions nor global organization structure (*i.e.*, percent made up of 'autonomous' subsidiaries – *PAUT*).

2. The percent of an industry's foreign subsidiaries belonging to globally organized systems (*PGLOBE*).

Running the above two dummy variables and leaving out a third covering firms with *ID* structures, means choosing to highlight the differential impact of non-*ID* structures on *PRFD*. Specific coefficient signs for *PAUT* and *PGLOBE* cannot be predicted because, as discussed in the scenario summary, one must expect different things according to whether industries with different *PAUT* or *PGLOBE*, *but otherwise exactly similar*, fall into the marketing or *R & D* oriented class. For a sample of marketing oriented industries, one would foresee *PAUT* having a negative coefficient, and *PGLOBE* a positive one. Insignificant coefficients for both dummies would seem to be the safest bet for *R & D* oriented industries. For a mixed sample of marketing and *R & D* industries, all that can be safely predicted is that *PAUT*'s coefficient should not be positive, nor *PGLOBE*'s negative.

3. The average age of U.S. foreign subsidiaries in each manufacturing industry (*AGE*). *Ceteris paribus*, the longer *AMMCs* have operated abroad the more likely are they to have learned about the potential advantages of moving away from their initial policy of centralizing all *R & D* within the U.S. Since all 'real world' learning takes times, *AGE* should be positively signed. This prediction is consistent with RONSTADT's finding that, no matter what the purpose of foreign *R & D*, it tends to grow once established[32].

4./5. Domestic advertising expense divided by domestic sales (*i.e.* *a*dvertising *i*ntensity – *AI*), calculated over all firms in the U.S. industry, and worldwide *R & D* expenditure divided by estimated worldwide sales (*i.e.* *w*orldwide *R & D* *i*ntensity – *WRDI*), calculated *just* for firms undertaking foreign direct investment[33].

These two variables are intended to account for the *R & D* versus marketing orientation difference so much highlighted in the study's

32. See RONSTADT (1978, p. 18).

33. The data necessary to calculate *LAI* on the same basis as *WRDI* were unobtainable.

scenario. For reasons elaborated in the scenario summary, a positive coefficient for *AI* and a negative one for *WRDI* are predicted.

6. Two estimates, separately employed, of broadly defined *R & D* economies of scale:

(a) *ARDC* – the *R & D* cost per *R & D* scientist or engineer in companies having 5000 or more employees (*a*verage *R & D c*ost)[34].

(b) *PRDSP* – the *p*ercentage of an industry's *R & D* performed in *s*mall *p*rograms (*i.e.*, under $50,000 – the smallest program size disclosed).

Though the research scenario mentioned *R & D* economies of scale solely in connection with firms considering a decentralization of 'global original' *R & D*, it is safe to presume that such economies of scale hinder, to some extent, all *R & D* decentralization. A negative sign for *ARDC* and a positive one for *PRDSP* are therefore expected. *ARDC* and *PRDSP* are separately employed in the regressions following since there is no *a priori* means of identifying which best represents *R & D* scale economies[35].

7. Domestic professional, technical and kindred (*PTK*) workers, *exclusive* of those probably employed in *R & D* pursuits, divided by domestic employee totals (*i.e. d*omestic *t*echnical *int*ensity – *DTINT*).

The greater is *DTINT*, the greater is the probability of production associated *PTK* workers being located abroad even from the earliest days of foreign production. The presence and possible underemployment of such *R & D* capable workers will, as earlier argued, favour the performance of *R & D* abroad. Consequently, a positive sign for *DTINT* is predicted.

It is suspected that the higher is *DTINT* the more inseparable are *R & D* and production activities. If this is in fact true, it provides another reason for *DTINT*'s positive sign prediction.

8. Majority owned foreign affiliate exports to the U.S. divided by total foreign sales – *i.e. ex*port to the *U.S. i*ntensity (*XUSI*).

34. We could have run an *ARDC* unrestricted as to firm size but chose not to on the grounds that U.S. multinationals are typically very large firms. See Vernon (1971, p. 12). It is *R & D* economies of scale *as perceived by U.S. multinationals* that are pertinent to our work.

35. As would be expected, there is a significant negative simple correlation between *ARDC* and *PRDSP*.

RESEARCH AND DEVELOPMENT

The higher is *XUSI*, the more likely are an industry's firms to have granted a worldwide product mandate to their foreign subsidiaries. Such a production mandate may or may not include doing the associated *R & D* work depending on whether *R & D* is best located near production or near the largest target market (usually the U.S.). *XUSI*'s coefficient must be left unpredicted because of a lack of information about the extent of *R & D* and production complementarity.

Many questions could be posed concerning possible non-linear relationships between *PFRD* and each of the independent variables. Unfortunately, this study's scenario is not equipped to answer such questions. There are, however, good *a priori* grounds for expecting non-linear relationships between *PFRD* and each of *AGE*, *WRDI*, *AI*, and *ARDC*. Being a percentage, *PFRD* has a strict upper bound. This is not the case for the four non-percentage independent variables just cited. There must be threshold values above which unit increases in these four variables cease to have a linear partial effect on *PFRD*. Just in case some of the actual data used exceed those thresholds, the special four are re-specified as logarithms before entering the regressions – hence the '*L*' preceding their variables names. Though logarithmic specification alters a regression coefficient's interpretation, it has no effect on its expected sign. All predictions therefore remain unchanged.

Referring to the regression results, it is clear that *PAUT*, *PGLOBE*, *LAGE*, *LWRDI* and *DTINT* have behaved as expected[36]. *LAI* and the two economies of scale variables (*ARDC* and *PRDSP*) have failed to obtain statistically significant predicted signs. While no ready excuse can be offered for *LAI*'s performance, it can be concluded that, just as this paper's scenario would suggest, *R & D* economies of scale are not very important deterents to *R & D* decentralization within *AMMC*s.

XUSI's statistically significant positive coefficient gives tentative support to the idea that *R & D* and production are locationally complementary activities. This interpretation rests on assuming that *XUSI* indeed measures the degree to which foreign subsidiaries have been granted worldwide product mandates.

36. Recall that non-positive and non-negative signs are predicted for *PAUT* and *PGLOBE* respectively.

GARY HEWITT

Regression results (dependent variable – *PFRD*)

Independent	Coefficient	('t' statistic)	Coefficient	('t' statistic)
C	-1.0726	(-2.636)**	-1.0374	(-5.330)***
PAUT	0.0455	(0.376)	0.0468	(0.388)
PGLOBE	0.1384	(2.837)**	0.1383	(2.813)**
LAGE	0.2376	(5.041)***	0.2377	(4.870)***
LAI	0.0104	(0.929)	0.0099	(0.965)
LWRDI	-0.0706	(-4.801)***	-0.0704	(-4.361)***
LARDC	0.0060	(0.105)		
PRDSP			-0.0302	(-0.045)
DTINT	0.9018	(3.051)**	0.9134	(2.962)**
XUSI	0.4674	(2.248)**	0.4667	(2.180)*
	$R^2 = 0.82$; F = 6.26***		$R^2 = 0.82$; F = 6.26***	

Number of observations = 20
* Significant at the 90% level, two tailed test. ** 95% level. *** 99% level.

Data sources:

PFRD – Based on unpublished data, collected as part of the 1966, U.S. Direct Investment Abroad Survey, kindly supplied by G. R. Kruer, International Investment Division, Bureau of Economic Analysis, U.S. Department of Commerce. – These data were supplemented where necessary, from U.S. Tariff Commission (1973), Table 6 – '*R&D* spending by multinational firms in manufacturing', page 582.

PAUT and *PGLOBE* – Both rest on unpublished data purchased from Harvard's Graduate School of Business Administration. Similar data, though not disaggregated by industry, are contained in tables found in Vaupel and Curhan (1969, pp. 175–177).

AGE – Taken from Vaupel and Curhan (1969, pp. 134–155).

AIC – Based on data contained in *Advertising Age* (1968, pp. 77–78).

WRDI – Taken from unpublished data, again supplied by G. R. Kruer.

ARDC and *PRDSP* – Found in: National Science Foundation (1969), Tables 7 and 51 respectively.

DTINT – Domestic *PTK* workers and total employment data were taken from U.S. Department of Commerce (1972), Tables 1 and 8. Domestic *R&D* employee totals were found in the National Science Foundation (1971), Table B-25, January 1970 column.

XUSI – Based on unpublished data supplied by A. Gilbert, International Investment Division, Bureau of Economic Analysis, U.S. Department of Commerce.

RESEARCH AND DEVELOPMENT

Multicollinearity in the two regressions has been checked by regressing each independent variable in turn on all the others. Only four of the resulting sixteen regressions had 'significant' (*i.e.* 95% level, two tailed test) F statistics[37]. Two of those cases involved *LWRDI* and two, *DTINT* as dependent variable. Not surprisingly, *LWRDI* and *DTINT* are themselves positively related.

V. CONCLUSIONS

It is worth noting that 82% of the cross industry variation in *PFRD* has been explained without using any explanatory variable measuring the cost advantages of an American versus a 'rest of the world' *R & D* location. While such a variable certainly belongs in the regressions run, it is and will remain prohibitively difficult to obtain unless the OECD or some other international body publishes suitable comparative data[38]. Significant multicollinearity between this omitted independent variable and the eight others used in the regressions is, with the possible exception of *LWRDI*, a very remote possibility. This implies that there would be a near orthogonality between the omitted and included independent variables[39]. It seems safe therefore to conclude that relative *R & D* costs could explain a maximum of just 18% of the *R & D* decentralization behaviour of *AMMCs*. That finding plus the good empirical showing of independent variables resting on the importance of changes in firm strategy (*PAUT*, *PGLOBE*) and of organizational learning time (*LAGE*) tend to recommend this study's 'satisficing' approach. Theorists preferring to base their work on assuming that information costs nothing to obtain and to use are free to argue that this paper's analysis has a short run relevance only. It is correct that in a long enough run, *assuming an unchanging world*, *AMMCs* will locate their

37. None of the four had an R^2 greater than the parent regression's R^2.

38. What is needed is an expansion of the work contained in OECD (1975).

39. To the extent that an omitted independent variable lies in a subspace orthogonal to that of all the included independent variables, the regression coefficients are consistent and unbiased, but their estimated variance is increased. This study's tests probably err, therefore, on the conservative side. See MADDALA (1977, p. 156).

R & D guided primarily by relative *R & D* factor costs. Such knowledge is not much help though to policy makers operating in a rapidly changing world. To host country policy designers desiring to obtain more *R & D* from the *AMMC*, this study offers two insights:

1. Some comfort can be taken from *LAGE*'s 'significant' positive coefficient. Even if nothing much is done to encourage it, *AMMC*s will tend in time to decentralize their *R & D*. That process could probably be accelerated however by host countries adopting entry regulations insisting that *AMMC*s make an early start on foreign *R & D*. The sooner such activities begin, the sooner will the *AMMC* become aware of the cost advantages of doing *R & D* abroad. Perhaps host government subsidy programs will be required to ensure that such cost advantages do in fact exist.

2. Pressuring U.S. multinationals to grant product mandates to their foreign subsidiaries, as the Canadian Province of Ontario is now doing, could well lead to a large *R & D* employment payoff for host countries[40]. This is the practical message that probably lies hidden behind *XUSI*'s positive coefficient.

This study's scenario summary argued that marketing intensive *AMMC*s will probably locate a larger percent of their corporate *R & D* abroad than will their *R & D* intensive cousins. Because of *LAI*'s poor empirical performance, this view must now be regarded as questionable. This is particularly necessary given the fact that when the study's regression pair was re-run with *LRDAD* replacing both *LWRDI* and *LAI*, the regression F statistics became barely instead of very comfortably 'significant'. *LRDAD* is the log of the ratio of American *R & D* expense divided by American advertising costs.

It must be admitted that *LWRDI*'s 'significant' negative regression coefficient can be explained without reference to marketing versus *R & D* orientation. Such an explanation rests on treating *LWRDI* as an index of an industry's 'original' as opposed to 'adaptive' *R & D* emphasis. To the extent that *LWRDI* is such an index

40. See Dowling (1979, p.5). – If the Ministry of State for Science and Technology gets its way, Canadian Federal Government policy might soon second Ontario's. See Wex (1979, p. 6).

and to the extent that all firms have a greater propensity to centralise 'original' as compared with 'adaptive' *R & D*, one would expect *LWRDI* to be negatively related with *PFRD*.

Further research into the whole issue of 'local original' *R & D* performed abroad is needed before utterly rejecting the hypothesis that marketing intensive *AMMCs* are more likely to engage in it than are their *R & D* intensive counterparts. It would be an important step forward if any such work were tested using company rather than industry data. The present study suggests that the effort required to assemble such data is well warranted.

REFERENCES

Advertising Age, 'Percentage of Sales Invested in Advertising in 1965–1966 for 262 Separate Product Classifications' (1968), November 4.

AHARONI Y.: *The Foreign Investment Decision Process*, Boston, Harvard Business School, 1966.

BEHRMAN J. N.: *National Interests and the Multinational Enterprise*, Englewood Cliffs, N.J., Prentice-Hall, 1970.

BEHRMAN J. N.: 'Some Patterns in the Rise of the Multinational Enterprise', *Research Paper 18* Chapel Hill, Graduate School of Business Administration, University of North Carolina at Chapel Hill, 1969.

Business Week: 'To a Global Car' (1978), November 20.

CORDELL A. J.: 'The Multinational firm, Foreign Direct Investment, and Canadian Science Policy', Science Council of Canada, *Special Study* No. 22, Ottawa, Information Canada, 1971.

CYERT R. M. and MARCH J. G.: *A Behavioural Theory of the Firm*, Englewood Cliffs, Prentice-Hall, 1963.

DOWLING D.: 'Grossman spells out Ontario strategy – Living with multinationals', *Financial Post* (1979), March 31.

DUERR M. G.: '*R & D* in the Multinational Company – A Survey'. *Managing International Business*, 8 (1970), New York, National Industrial Conference Board, 1970.

Foreign Investment Review, Vol. 1 (1977), No. 1.

HYMER S.: 'The Multinational Corporation and the Law of Uneven Development' in: J. N. BHAGWATI (Ed.), *Economics and World Order from the 1970's to the 1990's*, New York, Macmillan, 1972.

JOHANSON J. and VAHLNE J. E.: 'The Internationalization Process of the Firm – A Model of Knowledge Development and Increasing Foreign Market Commitments', *Journal of International Business Studies*, Vol. 8 (1977), No. 1, pp. 23–32.

MADDALA G. S.: *Econometrics*, New York, McGraw-Hill, 1977.

GARY HEWITT

MICHALET C.A.: 'Transfer of Technology and the Multinational Firm', annexed to Organization for Economic Co-operation and Development, Directorate for Scientific Affairs, Ad Hoc Group on Industrial Innovation, *Multinational Enterprises and the Transfer of Technology*, Paris, OECD, 1973.

National Science Foundation, *Research and Development in Industry, 1966*, Washington: U.S. Government Printing Office, 1968.

National Science Foundation, *Research and Development in Industry 1971*, Washington: U.S. Government Printing Office, 1973.

Organization for Economic Co-operation and Development, Directorate for Science, Technology and Industry: *Survey of the resources devoted to R-D by OECD Members Countries, International Statistical Year, 1971*, Business Enterprise Sector Statistical Tables and Notes, Vol. 1 (1975), Paris, OECD.

PENROSE E.T.: *The Theory of the Growth of the Firm*, Oxford, Basil Blackwell, 1959.

RONSTADT R.C.: *Research and Development Abroad by U.S. Multinationals*, New York, Praeger Books, 1977.

RONSTADT R.C.: 'International *R&D*: the Establishment and Evolution of Research and Development Abroad by Seven U.S. Multinationals', *Journal of International Business Studies*, Vol. 9 (1978), No. 1, pp. 7–24.

SIMON H.A.: *Administrative Behaviour*, New York, Macmillan, 1957a.

SIMON H.A.: *Models of Man*, New York, John Wiley & Sons, 1957b.

STOPFORD J.M. and WELLS L.T. (Jr.): *Managing the Multinational Enterprise*, New York, Basic Books, 1972.

U.S. Department of Commerce, Bureau of the Census, *1970, Census of Population Subject Reports PC(2)-7c*, Occupation by Industry, Washington: U.S. Government Printing Office, 1972.

U.S. Tariff Commission, *Implications of Multinational Firms for World Trade and Investment and for U.S. Trade and Labour* – a report submitted to the Committee on Finance, U.S. Senate, Washington: U.S. Government Printing Office, 1973.

U.S. Department of Commerce: *U.S. Direct Investment Abroad, 1966, Final Data*, Washington, U.S. Government Printing Office, 1976.

VAUPEL J.W. and CURHAN J.P.: *The Making of Multinational Enterprise – A Sourcebook of Tables Based on a Study of 187 Major U.S. Manufacturing Corporations*, Boston: Graduate School of Business Administration, Harvard University, 1969.

VERNON R.: *Sovereignty at Bay*, New York, Basic Books, 1971.

WELLS L.T. (Jr.): 'International Trade: The Product Life Cycle Approach', in: L.T. WELLS (Jr.), *The Product Life Cycle and International Trade*, Boston, Harvard Business School, 1972.

WEX S.: 'World product mandate – A first step toward an industrial strategy', *Financial Post* (1979), July 7.

RESEARCH AND DEVELOPMENT

SUMMARY

This study develops and empirically tests a 'scenario' describing the evolution of a typical American multinational manufacturing corporation's (*AMMC*'s) foreign research and development (*R & D*) activity. It finds that international differences in *R & D* factor costs probably have very little effect on *R & D* location decisions. The critical variables appear instead to be less traditional things like the time elapsed since the *AMMC* first engaged in foreign production. That such variables were discovered and emphasized in the study is a result of the 'satisficing' approach adopted in its 'scenario'. In vindicating that approach, the paper's findings also show the need for economic theorizing to take explicit account of the fact that information costs something to obtain and to use.

[5]

International management of R&D: The Swedish Experience

Lars Håkanson and Udo Zander

Institute of International Business, Stockholm School of Economics, Box 6501, S—113 83 Stockholm, Sweden.

Abstract

This paper summarizes the results of in-depth case studies of international R&D management in four Swedish multinational corporations. The results suggest that foreign volumes and shares of R&D expenditures are rapidly increasing. Through a rapidly increasing number of foreign acquisitions, foreign R&D laboratories have been added. As part of international rationalization of production, foreign units have been given global product mandates with responsibility both for manufacturing and R&D. In maturing foreign subsidiaries, routine technical activities have often evolved into proper R&D. Tight labour market conditions for engineers in Sweden have made it increasingly necessary to exploit existing technical capacity, regardless of location.

Concurrently, companies face new pressures for international coordination and control of R&D. Coordination is required to reduce product differentiation, to facilitate technology transfer and to ensure the technical and market compatibility of products and components developed at different locations but sold as part of total systems. Drawing on the experience in the four investigated companies, the paper discusses how systems and procedures can be developed to meet the need for coordinated international R&D management.
Keywords: R&D in multinational corporations, Location of R&D, Decentralization of R&D, Coordination and control.

INTRODUCTION

Multinational corporations (MNCs) tend to centralize research and development to the home country. Foreign R&D, where it occurs, is marginal to the development of corporate technology. Moreover, there is almost unanimous consensus, as expressed by one prominent observer, that '. . . the last activity of the firm to be organized on an international basis — if it is at all — is R&D'. (Terpstra 1977, p. 25)

This article questions these assumptions. It reports the results of an exploratory research project based on intensive case studies of four Swedish multinationals.[1] The results suggest that, in an increasing number of companies, systems and procedures must be devised to achieve coordination and exploitation of R&D on an international scale. The article argues that implementation of such systems and procedures tends to affect strategic dimensions of corporate organization and behaviour. An attempt is made to identify some of these consequences.

The research project was instigated by two related sets of observations, both of considerable interest — and cause for concern — for the management of Swedish industry:

First, casual observation, ranging from press articles to interviews with managers of major Swedish corporations, suggests that, for a number of reasons, increasing amounts of industrial R&D are being located abroad.

Second, Swedish international enterprises are today facing new environmental and internal conditions, different from those that historically shaped the evolution of that — singularly successful — set of managerial styles and administrative practices that Hedlund & Åman (1984) dubbed 'the Swedish Model'.

[1] The research project was carried out at the Institute of International Business at the Stockholm School of Economics in cooperation with the Swedish Association of the Mechanical and Electrical Engineering Industries (Sveriges Mekanförbund). A full report of the project is available in Håkanson & Zander (1986).

R&D IN SWEDISH MNCs

Many Swedish MNCs were founded to pursue and exploit unique technical capabilities based on domestic inventions and development of imported technology. Due to the limited size of the Swedish market, international market penetration was often very rapid.

Today, a small number of large MNCs perform a dominating proportion of Swedish R&D (Håkanson 1980, 1983; Swedenborg 1982). In 1978, MNCs accounted for some 70 per cent of total R&D in Swedish industry. This percentage can be compared to their share of Swedish manufacturing employment, 47 per cent, and their share of total Swedish exports, 58 per cent.

In 1978, the latest year for which data are available, the dominant proportion (86%) of R&D was located at home, in the neighbourhood of the corporate head office and major producing units. However, in that year, some 40 Swedish MNCs also carried out R&D outside Sweden. Moreover, the number of companies performing R&D at foreign locations had increased considerably during the preceding decade.

THE SWEDISH MODEL OF INTERNATIONAL MANAGEMENT

Hedlund's & Aman's (1984) study of the management of foreign subsidiaries in Swedish MNCs focussed on that significant portion of Swedish industry which is organized by highly internationalized companies with a moderate degree of product diversification, selling technologically advanced systems and products, primarily on industrial markets.

Their findings suggested that Swedish international companies are gradually moving away from the informal and personalized managerial styles and administrative practices that traditionally prevailed. Concurrent with the introduction of 'global product divisions' — with explicit responsibility for foreign subsidiary performance — new emphasis is placed on formal coordination and control systems, explicit goals and evaluation criteria.

In contrast, the inherited 'Swedish Model' of management was characterized by highly autonomous subsidiaries, coordinated through informal networks of personal contacts between corporate top management and the general managers of major subsidiaries. Foreign subsidiary managers were generally given much freedom to design their operations to suit local conditions. Control of subsidiary strategy was maintained largely through informal means. The small recruitment basis and policies of internal promotion helped to create strong corporate cultures and common value systems, often reinforced by years of personal friendships and acquaintance.

In spite of the evolution of complex, highly differentiated communication and control systems, Swedish MNCs continued to rely heavily on the pervasiveness of corporate culture, socialization patterns and common value systems to influence subsidiary strategy. This also remained true, at least initially, in companies that abandoned the traditional 'mother-daughter' organizational structure. However, divisionalization along product lines — a form of organization that became common in the 1960s — already marked the first step towards more 'global organizational structures' (Hedlund 1984).

The organizational adjustments described by Hedlund & Åman as 'the transition of the Swedish Model' suggest that informal mechanisms and reliance on common value systems to support rather loosely structured coordination and control systems no longer suffice to ensure the necessary level of headquarters control over subsidiary operations and strategy.

Several tendencies jointly explain the changes observed: with growing volume and complexity of operations, centralized control by the CEO is becoming increasingly difficult. Increasing competitive pressures on maturing product markets have forced corporations to improve performance by reducing slack and increasing efficiency, inducing tighter control of subsidiary performance. Growing size and new generations of managers, increasingly of non-Swedish origin, erode the efficiency of socialization to maintain corporate cultures and common value systems.

Thus, both internal pressures and changes in the competitive environment help to account for the introduction of new organizational forms and administrative practices. The reasons for and the effects of these changes are of course manifold. *Here, however, we shall concentrate on the role of technology and the management of R&D.*

THE EMPIRICAL STUDY

The aim of the research project was to study the experience of a number of Swedish MNCs in the international management of R&D. Specifically, the purpose was to analyze the structures and systems developed to perform R&D and technology transfer in an international context. Focus was set on two sets of questions:

(1) What is the historical rationale for present structures and systems? How do they relate to overall organization and strategy? How do environmental factors (changes in the level and type of technological change in the firm's product markets, changes in communication techniques, etc.) influence the R&D organization?
(2) How do structures and systems affect the R&D process (idea generation, selection, control and implementation of projects, transfer and utilization of new technology)?

Our research was based on in-depth study of four Swedish multinational companies. In addition to printed, publicly available material, the principal source of information was interviews with managers at corporate headquarters in Sweden and at selected subsidiaries abroad. In all, some 25 interviews were conducted and subsidiary managers visited in the U.S., Canada, the U.K., Holland, Italy, France and Finland.

The companies studied — AGA, Dynapac, LKB-produkter, and the SAB NIFE division of the Sonesson Group — are firms with high percentages of sales and employees outside Sweden. They are large enough to have several important subsidiaries and substantial experience of international business.

The industries in which the four companies operate differ considerably: LKB is a very successful company in a 'high-technology business', basing its competitive position on its research capabilities. SAB NIFE and Dynapac are both active in mature industries, where they have both held leading positions. AGA is active in a process industry, where R&D is directed at finding new applications for given products. The industry could perhaps be described as a rejuvenated mature business, with very high growth rates during recent years.

Although suited to an explorative study of this kind, the disadvantage of the case study method is, of course, the limited generalizability of the results. Our conclusions should therefore be seen as empirically and theoretically founded hypotheses which need to be tested on larger samples of firms.

FOUR CASE STUDIES

SAB NIFE AB

SAB NIFE is the result of a merger between SAB — a multinational supplier of advanced products for the rail transportation industry — and Nife Jungner, world-wide supplier of alkaline batteries and stand-by power systems. A common denominator for the two businesses is their orientation towards the railbound traffic market, which accounts for about 60 per cent of the turnover.

More than 90 per cent of sales are generated abroad and foreign subsidiaries account for 75 per cent of the 3,500 employees. In addition to production in Sweden, SAB NIFE has plants in 13 other countries. Through sales companies and agents, products are marketed in more than 100 countries.

Organizationally, foreign subsidiaries are grouped into two areas. Area 1 consists of Sweden, North and South America, and Area 2 of Western Europe, Australia, and agent sales. Separate Product Managers have global responsibility for SAB and NIFE products, respectively.

SAB NIFE's experience of R&D in the power electronics area provides an interesting illustration of the problems of R&D management in a multinational company:

NIFE (originally Nife Jungner) has been the world market leader in alkaline nickel-cadmium battery cells for some considerable time. Power systems were first developed around the expertise in the battery field. A growing proportion of battery output is now sold as an integrated part of the company's power systems.

Electronic equipment was originally exported from Sweden, but local market needs/wants often demanded product modifications. Several subsidiaries now manufacture electronic equipment locally.

In NIFE, before the take-over, the Swedish R&D organization appears to have undergone an extended period of stagnation — largely due to the difficult financial situation of the company. R&D resources were cut back and for some time no new products were launched.

In order to remain competitive, several foreign subsidiaries extended their engineering function into development work of their own. This process was largely unplanned and no formal systems were developed to ensure coordination and control of R&D carried out abroad. Following Sonesson's acquisition of NIFE Jungner, it became clear that such systems had to be developed in order to avoid duplication of effort and to ensure the efficient world-wide exploitation of R&D results.

In a first attempt, R&D in the power electronics area were concentrated in three Development Centres, in Brazil, Canada and Italy. A possible fourth Centre would be located in Sweden. Each unit would employ 5-7 engineers and concentrate on a particular technical area. To ensure cohesion, the overall R&D programme would be guided and coordinated by the Product Manager in Sweden.

When implementing this new organization, several problems became evident. Subsidiaries felt that their right to take R&D initiatives had been removed. Instead, they turned to the Product Manager for directives, specifications for new products, etc. Of course, the Product Manager in Sweden was in a very poor position to provide guidance without extensive consultations with the foreign market organization, a burdensome and time consuming task. Moreover, foreign subsidiaries cut their R&D budgets, in expectation of central funding.

The new organization was soon abandoned. Responsibility for R&D was explicitly returned to the subsidiaries. At present, no project is allowed to start unless a foreign subsidiary manager assumes direct responsibility and is willing to provide the finance needed. If the local market is too small to justify the investment, sister companies in potential markets may be engaged to form a Reference Group responsible for the formulation of appropriate product specifications. The Swedish Product Manager participates in these Reference Groups, providing a mechanism for overall coordination.

A subsidiary which performs R&D is guaranteed exclusive rights to the product developed, i.e. top management explicitly forbids other subsidiaries to engage in overlapping projects.

Concurrently, R&D at divisional headquarters at Landskrona, Sweden, was strengthened and roles were redefined. Central R&D now concentrates on long-term developments of general interest to the group. R&D projects may be carried forward to the prototype stage but should then be transferred to subsidiaries for detailed design and implementation in accordance with specific market needs. Transferring projects in this manner is considerably facilitated by the recently introduced Computer Aided Design (CAD) system.

DYNAPAC AB

In 1984, Dynapac had a turnover of 1,619 MSEK and employed some 2,800 persons. Foreign operations accounted for more than two thirds of employment and sales outside Sweden represented 96 per cent of the turnover.

The Dynapac product range consists of two sectors: heavy construction equipment — vibratory and static rollers, asphalt finishers, cold planers, etc. — and light equipment — small rollers, vibrating tampers, poker vibrators, etc. The following account focusses on the heavy equipment sector.

The products, initially developed and produced by the Swedish company, were marketed in Brazil and the U.S. at an early stage. When introducing the new products — which embodied an entirely new technology for compaction — adaptation to local market characteristics was essential. The establishment of local capacity for technical adaptation formed the nucleus of a geographically decentralized R&D organization, with major R&D units in the U.S., Brazil and Sweden. Through acquisitions, R&D was later established also in Japan and Germany.

Due to local differences in customer markets and the availability of components, each of the manufacturing units initially developed its own versions of the major products. Since the end of the 1970s, a major effort has been made to reduce the number of product variants and to develop and agree on a set of identical product specifications.

In the heavy equipment sector, four divisions — in Sweden, the U.S., Brazil and Japan — are responsible for sales within a specific geographical area. Responsibility for product development, documentation and maintenance of product standards has been divided between the four divisions: The Swedish R&D organization is responsible for the development of vibrating rollers. Asphalt removers are developed in the U.S., asphalt distributors in West Germany and static rollers in Japan. Rubber-wheeled rollers are developed both in Brazil and Japan. The division of tasks reflects both local market requirements and local technical competence.

Product Councils control and authorize all product development and decide on the distribution of product responsibility. The Product Councils consist of the CEO of Dynapac, staff representatives, divisional heads, marketing and construction managers, but no sales subsidiary managers.

Each of the six divisions prepares its own R&D budget and there is no coordination except through the Product Councils. At regular intervals, a total corporate R&D budget is computed by adding the six separate budgets. However, due to different organizational structures, reflected in different accounting practices, it is very difficult to prepare a relevant total estimate.

A special function at corporate staff, Production & Engineering Control, maintains and revises rules, standards and procedures for R&D in the group.

LKB-PRODUKTER AB

The operations of LKB include the development, manufacture and marketing of instruments, reagents and methods for biochemistry, analytical chemistry, biotechnology and clinical chemistry. The main customer groups are research institutes, hospitals, and industrial enterprises in the pharmaceutical, food and chemical industries.

In 1984, LKB's sales were 817 MSEK, of which approximately 96 per cent were generated in foreign markets. The number of employees was 1,414, two thirds of which were outside Sweden.

LKB has manufacturing units (Product Companies) in Sweden, Finland and Great Britain and sales companies and agents in almost 80 countries.

Corporate headquarters are located in Bromma, Sweden, where one of the Product Companies is also situated. The Finnish and British Product Companies, Wallac and Biochrom, were both acquired because of their technical capacities. In both, R&D was expanded following the acquisition.

The three Product Companies operate fairly independently of each other when it comes to R&D matters. All Product Companies work in basically the same areas of technology, e.g. electronics, chemistry and optics, but, historically, there have been rather few overlaps in products and applications. However, technical developments — particularly in the electronics field — tend to increase the need for coordination and cooperation to avoid duplication of effort, to ensure compatibility of products, etc. An additional reason is the increased emphasis on systems sales, with components developed and manufactured in the three Product Companies.

In 1982, a so-called R&D Forum was established, to meet three times a year. The location of the meetings switches between the home-countries of the Product Companies. In addition, a series of regular and *ad hoc* committees have been formed to deal with specific coordination projects, standardization of systems, process techniques for the production of chemicals etc. As a general rule, LKB is small enough to be able to rely quite successfully on informal information channels and the *ad hoc* formation of committees to deal with specific issues — there is not much need for formal procedures.

R&D strategy and planning issues within each division are handled at monthly meetings of Programme Councils, consisting of the functional heads. In order to coordinate strategies, joint committees for marketing and R&D have been formed, with representatives of all three Product Companies. These committees are responsible for the formulation of overall R&D and marketing strategies and consist of top management within R&D and marketing, respectively, but without representatives from the sales subsidiaries.

AGA AB

Aktiebolaget Gasackumulator (AGA) came into operation in 1904 based on innovations in the field of lighthouse technology. The construction of gas plants to supply the navigational systems installed by AGA laid the foundation for world-wide expansion. Today, AGA is one of the world's leading producers of industrial and medical gases.

In 1984, gas accounted for three-quarters of group turnover of 5,632 MSEK. Foreign markets accounted for about 80 per cent of sales, and three quarters of the 10,422 employees, were engaged abroad. With more than 200 plants, AGA is active in 22 countries.

The operations of the AGA group are organized regionally. The local Gas Companies belong to one of the six Gas Regions: the Nordic countries; Europe 1 (BeNeLux and West Germany); Europe 2 (Austria, France, Spain and Switzerland); North America; South America 1 (Colombia, Ecuador, Peru and Venezuela); and South America 2 (Argentina, Brazil, Chile and Uruguay).

Central staff for administration, control, and finance is located at the Lidingö headquarters, near Stockholm. The so-called Gas Staff, responsible for marketing, production technology and R&D is also found at Lidingö. The task of the Gas Staff is to develop group policies, new business opportunities and new technical solutions, and to give marketing and technical support to the Gas Companies.

In general, the Gas Companies have quite an independent position, reporting primarily to the respective regional managers. Each Gas Company is a separate operative unit responsible for production, distribution, and marketing.

High transportation costs preclude long shipments of gas and virtually no gas products are transported across national borders. The primary competitive weapons in the industry are price, systems and applications know-how, service, delivery reliability and market confidence. Volume growth lies in new gas applications, and strong customer orientation is vital.

During recent years, AGA has concentrated its resources on the development of gas technology and gas applications. New applications areas are rapidly emerging — half of all gas applications areas today were unheard of ten years ago. An important task is to transfer new technology to customers and to customize solutions in accordance with their needs.

AGA Innovation is the corporate department responsible for research, development and technical support for gas technology and gas applications. This work requires a broad understanding of the technologies and conditions of the customer industries. Projects are therefore often carried out in cooperation with customers and with universities and other research establishments.

R&D at AGA is highly centralized — some 85—90 per cent of all R&D is performed at Lidingö. One reason for centralization is that development of new gas applications requires considerable resources, qualified expertise, testing and other equipment, etc. The costs of these resources cannot easily be defrayed by the individual gas companies. Moreover, know-how in the industry is internationally valid and often has a global market. The market structure — with few customers in each country — is a further reason for centralization.

While the central R&D organization seems to be able to operate in relatively close contact with the Nordic subsidiary organization, keeping in touch with market developments in more distant markets is somewhat more problematic. In the past, there has been a tendency to focus development primarily on applications directed towards Scandinavian customers.

Partially in response to this problem, AGA has lately located some R&D abroad, especially in the U.S. However, further geographical and organizational decentralization of R&D will raise new problems in terms of cooordination and information exchange. Thus, while decentralization would facilitate market contact, foreign subsidiaries currently have limited resources to spare and would find it difficult to participate in technology transfers to third countries.

The ability to swiftly transfer technology to new markets is a major competitive weapon in the gas industry. However, although AGA Innovation has successfully developed a series of new applications, foreign subsidiaries are often reluctant to assume the investments necessary to introduce new applications in the marketplace. For subsidiary management, who are largely evaluated on the basis of short term returns, hiring a new applications engineer in the hope of a new customer contract is often not a very attractive proposition.

PATTERNS IN THE EVOLUTION OF INTERNATIONAL R&D MANAGEMENT

Forces for Decentralization

Traditionally, MNCs have conducted the bulk of their R&D at central laboratories in the vicinity of the corporate head office. Although this still remains the predominant pattern, many Swedish MNCs have in the last decade — inadvertently or by design — located increasing amounts of R&D in foreign units.

In Dynapac, the first major subsidiaries — in Brazil and the U.S. — both established their own R&D capacity at a very early stage. A rather typical example of the Swedish Model, these early subsidiaries were only very loosely integrated with the group — each was allowed to develop quite independently to meet local conditions. Later, production and R&D were allocated to specialized units as part of Dynapac's international rationalization.

In SAB-NIFE, R&D — especially in the area of power electronics — became increasingly decentralized partly due to the inability of the parent organization to furnish subsidiaries with new products. Faced with this problem, some entrepreneurial subsidiary managers themselves initiated R&D to meet acute needs. In consequence of rather loose control mechanisms and the fairly autonomous position of subsidiaries, the technical competence in the power electronics area gradually gravitated from corporate headquarters to subsidiaries, without effective action on the part of top management.

In LKB, the two foreign product companies were both acquired on account of their technical capacity. The R&D units at the acquired companies were allowed to develop without much guidance from headquarters. Since each specialized in a different product area, there was no need for very strong coordination.

In AGA, foreign R&D accounts for a much smaller share of the total than in the other three companies. However, during the last few years, a tendency to locate more R&D abroad can be noticed. The primary reason is to achieve closer contact with potential customer industries that are not well developed in Sweden. A contributory factor has been that the R&D unit in Sweden has been growing so fast that difficulties in absorbing new personnel have been encountered.

Thus, a number of factors help to account for the increasing geographical decentralization of R&D:

- The evolution of technical support functions at foreign subsidiaries to encompass more advanced product development tasks.
- Increasing reliance on acquisitions as a strategy for international growth.
- International rationalization of production, i.e. the establishment of specialized manufacturing units with groupwide supply responsibility.
- Difficulties in recruiting qualified technical expertise on the tight labour market for technicians in Sweden.
- The exploitation of foreign entrepreneurial and technical talent, sometimes in compensation for stagnating R&D at home.
- Advantages of proximity to customers and foreign research establishments.

Increasing Need for Coordination

In the past, R&D at foreign subsidiaries were only very loosely integrated with those of headquarters and the rest of the group. This reflected both the tradition of the 'Swedish Model' — i.e. strong foreign subsidiary managers and very loose control systems — and the fact that foreign R&D volumes tended to be small and directed solely at the local market.

One of our main conclusions is that traditional practices do not suffice to achieve the required level of international coordination and integration of R&D. Organizational adjustments lag behind and new structures and procedures will have to be developed. The companies in our sample have all started to implement new coordinative instruments, but the experience of Swedish industry in this area is still limited.

In Dynapac, the division of production and R&D responsibility makes it imperative that R&D performed by product responsible divisions meets not only the needs of the local market but is coordinated with the needs of other divisions and markets as well.

In SAB-NIFE, the task of coordination is to ensure that wasteful duplication of efforts be avoided and that the technical capacity of subsidiaries be exploited to benefit the technological development of the entire group.

In LKB, increasing coordination needs primarily reflect technical requirements in the electronics field: the need to coordinate the development of software and purchasing of components. These needs were strengthened by the increased emphasis on systems sales and the need to use product design to establish and maintain a coherent 'corporate image'.

In AGA, coordinative instruments are primarily needed to ensure fast and efficient exploitation of new applications know-how in international markets.

Thus, for a variety of reasons, companies today face a much stronger need than in the past to coordinate and exploit R&D on an international scale:

- The cost escalation of R&D reinforces the need to avoid duplication of effort and to ensure that R&D is performed on an adequate scale.
- Increasing competition on mature product markets makes it imperative to cut production and inventory costs through limiting product differentiation and increasing standardization.
- Greater importance of lead times when launching new products places new demands on efficient technology transfer and cooperation between R&D units.
- Increasing reliance on strategies of 'systems selling', implying a need for common scheduling and coordination of R&D projects in order to ensure technical compatibility and preparedness for launching system components on specified common dates.
- Intensified international competition and recruitment problems in Sweden reinforce the need to fully exploit the technical capacity of the group, regardless of location.

New Strategic Balance

One central consequence of the developments outlined above is to shift the strategic mode of operations of the group. In terms suggested by Bartlett (1984, cf. also Hedlund 1986), the traditional 'Swedish Model' can be described as a 'decentralized federation', characterized by loose, simple controls and decentralization of strategic decisions, the main coordinative instrument being financial flows between subsidiaries and the parent.

Increasingly, Swedish firms are now moving towards an 'integrated network model', characterized by tight and complex controls and high subsidiary involvement in the formulation and implementation of strategies. In consequence of the strategic allocation of tasks between specialized units, heavy flows of technology, finance, people and materials tie subsidiaries to each other and to the parent.

These tendencies have important impacts on the creation and dissemination of corporate technology:

> Traditionally, central managers had been responsible for taking the accumulated wisdom of the parent company . . . and transferring it abroad. Today, sophisticated MNCs recognize that no one country is the source of all technological innovation, no one market is continually more advanced than others, and no one group of managers has a corner on innovative ideas. In fact, one of the advantages in

being an MNC is having managers around the world able to spot trends in one market that later may spill over into others, to observe technical developments with global implications, or to develop successful strategies or approaches that can be tried in other parts of the company. Rather than acting as a one-way valve, division level central managers can act as the monitor of worldwide developments, the repository of useful knowledge and experience developed throughout the corporation, and transfer agent of that accumulated wisdom to other parts of the organization as appropriate. (Bartlett 1984, p.37)

These changes imply a new strategic balance, with high subsidiary involvement in strategy formulation and much more complex relationships between units. These changes, in turn, imply new roles and capacities for headquarters, divisions and foreign subsidiaries.

The central task of corporate level R&D staff is to act as a liaison between the R&D organization and corporate management, i.e. monitoring the R&D portfolio of the group to ensure its conformity with overall corporate strategy and to explicate technological threats and opportunities in the formulation of corporate strategy. Corporate staff must also facilitate communication within the group, develop and maintain common standards for R&D work and support long-range research of common corporate interest. However, central R&D rarely has the competence to exercise detailed control of R&D carried out abroad.

Divisional R&D staff carries the responsibility for the bulk of the R&D effort and must ensure world-wide coordination of R&D within its product areas. A central task in this context is to allocate R&D responsibilities in a way that both matches the technical capacity of different units and minimizes the need for close and costly control. However, coordinative instruments must not override the authority of foreign subsidiary managers. Coordination should instead encourage active participation in the strategy formulation process and ensure that local initiatives are efficiently exploited.

Line responsibility for foreign R&D units should rest with local subsidiary management with a great amount of discretion as to the allocation of R&D funds. Only in this way will decision processes be flexible enough to permit the rapid exploitation of new business opportunities. However, projects must be monitored by divisional staff in order to ensure their conformity to overall strategy and the efficient world-wide exploitation of R&D results. Active participation during strategy formulation requires foreign subsidiary managers to assume active scanning of markets, competitors and new technological developments.

Systems and Procedures

In order to support this division of responsibility and the required level of interaction between units, systems and procedures must be developed to facilitate world-wide coordination and information exchange:

Whereas in traditional hierarchical organizations, communications are generally channelled via the head office, 'integrated networks' are characterized by much more complex communication patterns, involving a great deal of lateral contact between subsidiaries. Facilitating such flows — by means of organizational directives and technical communication facilities — is a vital task for corporate management.

The complex information flows characteristic of an 'integrated network' can only be partly formalized. Especially in the area of R&D, formal communication channels must be supported by extensive informal networks based on personal contacts between officials in different parts of the organization. Such informal contacts are vital during several stages of the innovation cycle, ranging from idea handling to implementation and technology transfer. Rotational programmes, conferences and activities in Product Councils are examples of mechanisms that support the creation of informal networks.

In most companies, Product Councils constitute a major mechanism for coordination and cooperation. Achieving a wide representation of foreign subsidiaries in Product Councils is often necessary in order to fully exploit their competence and increase their understanding of, and commitment to, corporate strategies. However, this must be balanced by the need to maintain flexibility and rapid decision processes. Hence, wherever possible, it is often an advantage to create several smaller 'reference groups', each responsible for a particular product area.

In order to ensure that technical development conforms with corporate strategy and does not become the inadvertent result of uncoordinated decisions, it is vital to compile a corporate R&D budget by adding the budgets of all R&D units. This, in turn, requires that R&D budgets are prepared according to common accounting principles, as defined by corporate headquarters. As a rule, the corporate R&D budget will not be a strong instrument for the allocation of funds, but rather a mechanism for monitoring the long term direction of R&D.

A central mechanism for R&D coordination is the establishment of corporate funding of long range research and projects of common corporate interest. The central fund can be financed through license payments from subsidiaries which should all have access to corporately financed technology. However, in addition to corporate funds, subsidiaries should have the possibility to finance R&D out of their own operations and be given incentives to also exploit new products on foreign markets.

REFERENCES

Bartlett, C. A., (1984), 'Organization and Control of Global Enterprise: Influences, Characteristics and Guidelines'. Paper presented at Harvard Business School 75th Anniversary Colloqium on Competition in Global Industries.

Hedlund, G., (1984), 'Organization In-Between: The Evolution of the Mother-Daughter Structure of Managing Foreign Subsidiaries in Swedish Multinational Corporations'. *Journal of International Business Studies*, Vol. 15, Fall, pp. 109–123.

Hedlund, G., (1986), 'The Hypermodern MNC — A Heterarchy?'. *Human Resource Management*, Vol. 25, No. 1, pp. 9–35.

Hedlund, G. & Åman, P., (1984), *Managing Relationships With Foreign Subsidiaries*. Stockholm: Sveriges mekanförbund.

Håkanson, L., (1980), *Multinationella företag: FoU-verksamhet, tekniköverföring och företagstillväxt* Stockholm: SIND 1980: 4 Liber.

Håkanson, L., (1981), 'Organization and Evolution of Foreign R&D in Swedish Multinationals'. *Geografiska annaler*. Ser B., Vol.63, pp.47–56.

Håkanson, L., & Zander, U., (1986), *Managing International Research and Development*. Stockholm: Sveriges mekanförbund.

Swedenborg, B., (1982), Svensk industri i utlandet. En analys av drivkrafter och effekter. Expertrapport från direktinvesteringskommittén. *SOU* 1982: 27.

Terpstra, V., (1977), 'International Product Policy: The Role of Foreign R&D'. *Columbia Journal of World Business*, Vol. 12, pp. 24–32.

[6]

Cambridge Journal of Economics 1995, **19**, 141–153

Localised production of technology for global markets

Pari Patel*

Given the trends towards increasing globalisation of markets and of production, the globalisation of technology remains a subject of considerable interest to analysts and policy makers in the 1990s. The main conclusion of this paper is that there is no systematic evidence to suggest that widespread globalisation of technological activities occurred in the 1980s. The evidence, based on the US patenting activities of 569 of the world's largest firms (based in 13 countries, and covering 17 product groups), shows that for an overwhelming majority of them technology production remains close to the home base. The analysis also shows the dangers of generalising on the basis of anecdotal evidence from a small sample of firms from a particular country or sector.

1. Introduction

1.1. Background

The aim of this paper is to examine the volume and trends in the extent of internationalisation of technological activities of the world's largest firms based on data on their US patenting. There are two sets of reasons for addressing this issue.

The first is that debate continues about the implications of the globalisation of markets for the location of innovative activities. A number of recent analyses (Cantwell, 1992; Chesnais,1992; Dunning, 1992; Granstrand and Soljander, 1992; Hakanson, 1992; and Howells, 1990) have argued that large multinational firms, who are the main actors in the globalisation of markets, are now globalising their technological activities. The two main contentions of these studies are first, that this process has accelerated since the 1980s, and second, that the reasons underlying firms' decisions to locate R&D outside their home countries have changed. In the earlier literature on the internationalisation of business the emphasis was on the role of home markets in determining firms' technological advantage. In particular, Vernon (1966) argued that, having established a new product or a new production process in the home market, firms would subsequently export and/or locate production facilities in overseas locations. This process would inevitably involve some overseas R&D activity mainly concerned with adapting the products (e.g. to account for differences in consumer tastes) and the production processes (e.g. to account for differences in the labour market) to suit the local market

Manuscript received 15 February 1993; final version received 11 June 1993.

*University of Sussex. This paper is based on research done at the Centre for Science, Technology and Energy Policy (STEEP), established by the UK Economic and Social Research Council (ESRC) in the Science Policy Research Unit (SPRU) at the University of Sussex. It has benefited from comments from Keith Pavitt, Daniele Archibugi, Ove Granstrand and an anonymous referee of this journal.

0309 166X/95/010141+13 $08.00/0

conditions. However, in his more recent work, Vernon (1979) suggests that in some (high technology) industries the product cycle has become highly compressed with firms engaged in programmes of almost simultaneous innovations in several major markets. In addition, in the most recent analyses (quoted above) there is a strong emphasis on the role of 'supply-side' factors in the firms' decisions concerning the location of R&D. Thus firms are now assumed to assess a wide range of different geographic locations purely in terms of the strength of their science and technology base and the availability of adequately qualified scientists and engineers. In such a scenario the home country of the firm is assessed on the same basis as all the other potential locations. Another element in such analyses is the role played by the latest advances in information and communications technology in solving the problems of co-ordination and control posed by having technological activities scattered in a number of different locations.

The second reason for addressing this subject is that there is still a debate concerning the available empirical evidence. In an earlier study (Patel and Pavitt, 1991) we argued, on the basis of US patenting data for the world's largest manufacturing firms, that the importance of the foreign technological activities of large firms had been exaggerated. We also showed, using correlation analysis, that home country conditions have a major impact on the levels and trends of the technological activities of nationally-based large firms. This analysis has been criticised for three sets of reasons: (i) patenting is a particularly poor measure and different measures, e.g. those based on R&D expenditures, would have produced different results; (ii) the evidence from case studies of individual firms or sectors cannot be reconciled with these results; and (iii) by providing only a static picture of the early 1980s, the widespread increases in the extent of globalisation in the late 1980s were neglected. In Section 3 we address each of these issues in turn, but before doing so we summarise briefly the main results of the most recent studies that have looked at the nature and extent of globalisation of technology.

1.2. Other recent studies

Archibugi and Michie (1995, this issue) analyse quantitative evidence by making a distinction between three different (but complimentary) meanings of globalisation of technology (or 'techno-globalism') as used in the literature. They are: (i) the global exploitation of technology; (ii) the global production of technology; and (iii) global technological collaboration. In relation to the first of these there is very little controversy, as there is strong evidence (quoted in their paper) showing that firms are increasingly exploiting their innovations on a global scale. Most of the controversy surrounds the empirical evidence related to the other two meanings.

The studies concerned with analysing the global production of technology can be categorised into two groups. In the first group are those,[1] based on private surveys or case studies in selected companies, industries, or countries, concerned with understanding the nature and the motivation for firms locating technological activities in foreign locations. The main results of these studies are: (i) most overseas R&D is still concerned with adapting existing products and processes to host country market conditions (Pearce and Singh, 1992); (ii) 'political' factors have been an important influence on firms in making decisions about locating their technological activities outside the home base (Hakanson, 1992); and (iii) most firms envisage an increase in the extent of activities

[1] See for example, Hakanson, 1992; Pearce and Singh, 1992.

undertaken abroad and also a change in the nature of these activities[1] (Hakanson, 1992; Pearce and Singh, 1992).

The second group of studies comprises those concerned with a quantitative assessment of the extent of globalisation based on different measures of technological activities. Within this group are studies using data on the number of R&D personnel (Howells, 1990; Wortmann, 1990; Dorrenbacher and Wortman, 1991); on R&D expenditures (US NSF Science Indicators Reports); and on patenting activities (Etemad and Seguin-Delude, 1987; Patel and Pavitt, 1991; Cantwell, 1992). None of the results reported in these studies shows that the foreign technological activities of the large firms are more important than those in their home base (most put the proportion of foreign activities between 10 and 30%). Most of these studies also show that there has been an increase in the extent to which firms are locating technological activities outside the home country.

Quantitative studies based on the third definition of 'techno-globalism' analyse the trends and patterns in international collaborative agreements between firms involving a substantial technology element (Chesnais, 1989; Freeman and Hagedoorn, 1992). The main results of these studies are:[2] (i) over 90% of all the agreements in the 1980s were between firms based in the Triad countries;[3] (ii) around 20% were between firms in Europe and the USA, 14% between those in Japan and the USA, and around 6% between Japanese and European firms, with the remaining 60% being inter-country agreements.

It is in the context of these studies that we address two sets of questions in Section 3 of this paper. First, what is the extent to which firms are locating the production of their technology outside the home base, as shown by data on their US patenting activities, and have there been changes since the early 1980s? Second, what are the main locations of large firms' technological activities in the second half of the 1980s and have these changed since the early 1980s? Before turning to the results, Section 2 outlines the main elements of the database used in looking at these questions.

2. Dataset

The dataset has been compiled from information, supplied by the US Patent Office, on the name of the company, the technical class, and country of origin of the inventor, for each patent granted in the USA from 1969 to 1990. The main difficulty in using these data at the corporate level is that many patents are granted under the names of subsidiaries and divisions that are different from those of the parent companies and are therefore listed separately. An additional difficulty is that the names of companies are not unified, in the sense that the same company may appear several times in the data, with a slightly different name in each case.

Consolidating patenting under the names of parent companies can only be done manually on the basis of publications like *Who Owns Whom*. In our earlier study we unified the names and consolidated the data for 636 of the world's largest firms, in terms of sales, for the year 1984. Beginning with that sample, in the present study we have performed that same exercise on the basis of *Who Owns Whom* for 1988. After taking account of the firms that ceased trading and those that were taken over by other firms in

[1] For example Pearce and Singh (1992) report that for some firms an increasing proportion of the work of their foreign laboratories is concerned with development work and not just adaptation.

[2] See Freeman and Hagedoorn (1992).

[3] Europe, Japan and the United States.

Table 1. *The distribution of the 569 large firms in the sample by nationality and principal product group*

Product group	Europe[a]	Japan	North America[b]	Total
Chemicals	18	25	28	71
Pharmaceuticals	7	4	14	25
Mining and petroleum	14	7	23	44
Textiles etc.	4	5	9	18
Rubber and plastics	3	2	5	10
Paper and wood	8	6	17	31
Food	11	11	19	41
Drink and tobacco	7	1	9	17
Building materials	10	6	10	26
Metals	24	13	19	56
Machinery	23	13	32	68
Electrical	13	17	28	58
Computers	4	2	11	17
Instruments	1	8	11	20
Motor-vehicles	15	18	10	43
Aircraft	7	0	12	19
Other transport	2	1	2	5
Total	171	139	259	569

Notes: [a]For the total number of firms from each European country, see Table 5.
[b]243 US firms and 16 Canadian firms.

the sample, we have 569 firms in our present sample. For each firm we also have information, from published sources (*Fortune, Disclosure*), on sales, employment, country of headquarters and principal product group.

In an earlier paper (Patel and Pavitt, 1991) we described the main advantages and disadvantages of using patent statistics as a measure of technological activities. Briefly, their main advantages over other indicators such as R&D expenditures are that they are available over a long period of time and can be broken down in great statistical detail according to geographic location and technical area. Their main disadvantage is that they do not satisfactorily measure two important areas of technology, namely, software and biotechnology.

The main added advantage of our current dataset is that, for the first time, it enables us to examine trends in the internationalisation of technological activities. Other analyses using similar data (our own previous work and that of Cantwell, 1992) have been based on a set of firms consolidated for one particular year, and hence have not been able to analyse changes. With the present dataset we can examine these changes by comparing the geographic distribution of the firms' patenting activities between the two sets of consolidations: 1984 and 1988. Such a comparison of the two cross-sections also enables us, in Section 3, to look at the extent to which changes in internationalisation have been achieved through mergers and acquisitions or through organic growth.

Before describing the results, we list in Table 1 the numbers of large firms in our database according to their home country and principal product group.[1] American firms accounted for 45% of the sample, European firms 30% and Japanese firms 25%. Within

[1] There are 16 Canadian firms that have been aggregated under the US total for convenience.

Table 2. *A comparison of the overseas R&D and patenting activities of US firms*: proportion of the total undertaken outside the US*

	R&D 1979–1984	Patenting 1979–1984	R&D 1985–1989	Patenting 1985–1990
Chemicals	6·9	6·9	9·3	8·7
Pharmaceuticals	10·4	11·0	12·7	12·5
Electrical	6 8	6·3	5·3	8·1
Machinery	8·4	8·5	9·3	9·1
Metals	5·3	3·6	6·1	3·1
Transport	9·9	5·0	11·0	6·6
Instruments	6·7	9·2	6·1	7·0
Total (of all product groups)	8·0	6·3	7·9	7·8

Notes: [a]The patent data are for the 243 US large firms in our database and the R&D data refer to all US firms covered by the NSF R&D Survey.
Source: Science Indicators, 1991 and SPRU Large Firms Database.

Europe, the largest contributors were the UK (54), followed by Germany (42) and France (25). In terms of the industrial distribution, 21% had machinery and metal goods as their principal product group, 16% had chemicals and pharmaceuticals, and 13% electrical (including electronics) and computing machinery.

3. International location of technological activities: results of the analysis

3.1. Comparison of US patenting and R&D expenditures

Apart from the data on US patenting, the only other systematic evidence of the extent of internationalisation of technological activities of firms is that on self-financed R&D expenditures of US firms and their overseas subsidiaries collected by the National Science Foundation in the US. In Table 2 we compare, for the 1980s, the proportion of R&D spent by US firms outside the US with the proportion of patenting activities undertaken by the US-based firms in our database from outside the US.[1] The two measures are almost identical at the aggregate level and for chemicals, pharmaceuticals, machinery and instruments.[2] For firms in transport and metals the proportion of R&D spent abroad is consistently higher than the proportion of their patenting abroad. Nevertheless, both these datasets clearly show that on average US firms undertook less than 10% of their technological activities outside the United States in the 1980s. The similarity between them also allows us to place some confidence in the results reported below which are based on patenting activities alone.

3.2. Evidence at the firm level

One of the main advantages of our dataset described in Section 2 is that it allows us to examine the patterns of internationalisation of technological activities at the level of the individual firm. In Table 3 we present the cumulative distribution of firms according to

[1] It should be noted that the two datasets involve different sets of firms: the NSF R&D survey almost certainly covered more firms than those covered in our large firm patenting data base (243). Nonetheless, the basis for comparison between R&D and patenting is two strongly overlapping populations of large firms, defined according to their principal product group.

[2] The correlation coefficient between the two measures (using a pooled sample of 14 observations) is 0·59 which is significantly different from zero at the 5% level.

Table 3. *Distribution of firms according to their foreign technological activities, 1985–1990*

Percentage of US patents from outside the home country ≥ (%)	Cumulative distribution	
	No. of firms	Percentage
90	0	0·0
80	9	1·6
70	21	3·7
60	30	5·3
50	43	7·6
40	56	9·8
30	75	13·2
20	99	17·4
10	157	27·6
1	341	59·9
0	569	100·0

the proportion of their US patenting undertaken outside the home country. Of the 569 firms in the sample, 43 located more than half their technological activities outside the home country. Any case studies or anecdotal evidence based on these 43 firms would be seriously biased, as Table 3 also shows that for more than 40% of the sample (228 firms) the proportion executed abroad was less than 1%, and for more than 70% (341 firms) it was less than 10%.

Table 4 contains the name, nationality,[1] and product group for each of the top 50 firms (amongst those with more than 20 patents granted in 1985–90) sorted according to the degree of internationalisation of their technological activities. The list contains the names of some firms, such as Ciba Geigy, Philips and Shell, that have been used in the past as examples of 'global firms', but it also omits others such as IBM, which undertakes less than 15% of its total patenting from outside the United States.

In terms of nationality, by far the largest contributors to this list are firms from the UK (21 out of 50), followed by those from the smaller European countries (17 out of 50). On the other hand, there is only a single firms from each of France, Germany and Japan, and only five from the US. Thus any analysis that just concentrated on UK firms would produce results which would not be applicable to firms based elsewhere. The table also shows that for 32 of the 50 firms the United States is the most important foreign location of technological activities, and within Europe, it shows the importance of Germany as a location (for 10 out of the 50 firms).

The evidence presented in this section shows that US patenting data are a reasonable measure for analysing the extent of internationalisation of technological activities, and that such analysis needs to be based on systematic evidence, i.e. evidence covering a wide range of firms from different countries and product groups. It is to such an analysis that we now turn.

3.3. Recent trends in internationalisation

3.3.1. Aggregate level. Table 5 shows, for the 569 firms aggregated according to their nationality, the proportion of their total US patenting in 1985–90 that originated from

[1] There are three firms that may be described as having dual nationality and we have arbitrarily assigned their home country: ABB (Switzerland), Shell (Netherlands) and Unilever (UK).

Table 4. *Top 50 firms with largest share of US patenting outside the home base, 1985–1990*

Company	Nationality	Product group	% Abroad	Main foreign location	%
Hanson	UK	Building materials	88·8	USA	87·9
Molson	Canada	Drink and tobacco	87·0	USA	73·9
Grand Metropolitan	UK	Drink and tobacco	87·0	USA	87·0
Nova	Canada	Mining and petroleum	86·7	USA	83·3
Coats Viyella	UK	Textiles	82·9	France	54·3
Esselte	Sweden	Paper and wood	82·3	Germany	40·3
GKN Engineering	UK	Vehicles and parts	82·1	Germany	62·2
RTZ	UK	Mining and petroleum	79·2	USA	67·5
FKI Babcock	UK	Machinery	75·6	USA	51·2
Akzo	Netherlands	Chemicals	74·1	Germany	44·8
Imperial Metal Industries	UK	Machinery	72·7	USA	72·7
Unilever	UK	Food	72·6	USA	52·2
Shell	Netherlands	Mining and petroleum	71·6	USA	61·4
BOC	UK	Chemicals	70·8	USA	65·1
Inco	Canada	Metals	70·6	USA	67·0
SKF	Sweden	Machinery	70·2	Germany	45·6
Racal Electronics	UK	Electrical	69·9	USA	67·0
ITT	USA	Electrical	68·4	Germany	54·0
Labofina	Belgium	Mining and petroleum	67·3	USA	63·3
Wellcome	UK	Pharmaceuticals	67·3	USA	62·5
Nestle	Switzerland	Food	66·7	USA	43·3
Solvay	Belgium	Chemicals	64·7	Germany	33·3
Pilkington	UK	Building materials	61·1	USA	47·3
Redland	UK	Building materials	60·7	USA	53·6
Gillette	USA	Machinery	60·1	Germany	40·4
ABB	Switzerland	Electrical	59·3	Sweden	31·6
TI Group	UK	Machinery	59·2	USA	49·0
Electrolux	Sweden	Electrical	57·9	USA	37·9
Hoffman La Roche	Switzerland	Pharmaceuticals	55·0	USA	36·2
Philips	Netherlands	Electrical	53·8	USA	15·1
BAT	UK	Drink and tobacco	53·6	USA	40·4
Universal Match	Sweden	Paper and wood	50·9	UK	24·6
Sandoz	Switzerland	Pharmaceuticals	50·4	USA	30·2
Northern Engineering	UK	Machinery	50·0	Canada	30·0
ICI	UK	Chemicals	49·6	USA	38·7
Hawker Siddeley	UK	Machinery	48·5	USA	46·5
Alcan Aluminium	Canada	Metals	48·2	UK	29·4
BTR	UK	Machinery	47·5	USA	37·7
Reckitt & Colman	UK	Food	46·7	France	26·7
American Standard	USA	Building materials	45·3	Germany	36·5
BP	UK	Mining and petroleum	44·6	USA	25·9
Ciba-Geigy	Switzerland	Chemicals	42·3	USA	23·2
Elkem	Norway	Metals	41·9	USA	32·6
Metallurgie Hoboken-Overpelt	Belgium	Metals	41·4	USA	27·6
Hoechst	Germany	Chemicals	41·2	USA	31·9
Sara Lee	USA	Food	40·7	Netherlands	40·7
Black & Decker	USA	Machinery	39·9	Germany	15·7
Saint-Gobain	France	Building materials	39·8	Germany	26·7
Norsk Hydro	Norway	Chemicals	39·1	Sweden	21·9
Dainippon Ink & Chemical	Japan	Chemicals	38·2	USA	37·7

Table 5. *Geographic location of large firms' US patenting activities, according to nationality, 1985–1990 (percentage shares)*

			Of which			
Firms' nationality	Home	Abroad	USA	Europe	Japan	Other
Japan (139)	99·0	1·0	0·8	0·2	—	0·0
USA (243)	92·2	7·8	—	6·0	0·5	1·3
Italy (7)	88·2	11·8	5·3	6·2	0·0	0·3
France (25)	85·7	14·3	4·8	8·7	0·3	0·6
Germany (42)	85·1	14·9	10·4	3·9	0·2	0·4
Finland (7)	82·0	18·0	1·6	11·5	0·0	4·9
Norway (3)	67·9	32·1	12·7	19·4	0·0	0·0
Canada (16)	67·0	33·0	24·9	7·3	0·3	0·5
Sweden (13)	60·8	39·2	12·6	25·6	0·2	0·8
UK (54)	57·9	42·1	31·9	7·1	0·2	3·0
Switzerland (8)	53·3	46·7	19·6	26·0	0·6	0·5
Netherlands (8)	42·2	57·8	26·1	30·6	0·5	0·6
Belgium (4)	37·2	62·8	22·2	39·9	0·0	0·6
All firms (569)	89·1	10·9	4·1	5·6	0·3	0·8

Note: The parenthesis contains the number of firms based in each country.

different geographic locations, as identified by inventor addresses. According to this measure, firms from Japan, United States, Italy, France and Germany undertook 85% or more of their technological activities from within their home country. Only Belgian and Dutch firms executed more of their technological activities outside their home country than within. In terms of locations, the table also shows that very little of these firms' activities were located outside the 'Triad' countries, reinforcing the view that what occurred in the late 1980s can, at best, only be described as 'triadisation' and not 'globalisation'. Within the 'Triad' countries, Japan was the least favourable foreign location for these large firms, and Europe was slightly more favourable than the US. What our detailed data (not reported here) show is that within Europe, Germany was the most important foreign location for firms of all nationalities.

There is some evidence in Table 5 to suggest that geographic proximity has an influence on location decisions: for firms from the smaller European countries (Belgium, Finland, Netherlands, Norway, Sweden and Switzerland) Europe was more important than the United States, and the Canadian firms preferred the US to Europe. The main exceptions to this were the UK large firms, who undertook around 32% of their total technological activities in the United States and less than 7% within Europe.

3.3.2. Trends. As mentioned in Section 2, one of the main advantages of our dataset is that it allows us to examine changes in the importance of the various locations over the 1980s. In order to do this we compare the two geographic distributions of US patenting: (i) for firms as they were consolidated for 1984, and their total US patenting for the period 1979–1984; and (ii) for firms consolidated for 1988 and their total patenting for 1985–1990 (as reported in Table 5).

Figure 1 shows changes in the proportion of firms' total activities (again aggregated according to their nationality) located outside their home countries, and the extent to which these changes were due to increases (or decreases) in the importance of Europe or

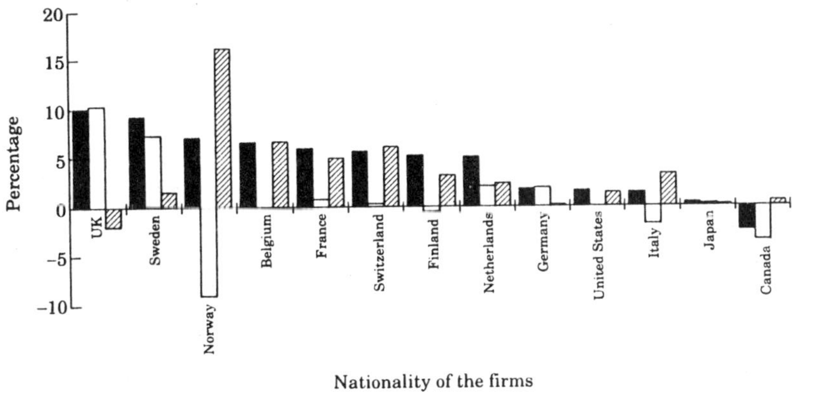

Fig. 1. *Changes in the location of US patenting activities of the world's largest firms, 1979–1984 to 1985–1990.* *Notes:* (■) Abroad; (□) USA; (▨) Europe.

the United States as locations. By way of illustration, the figure shows that UK large firms increased their technological activities abroad by around 10%, comprising a 10% increase in the importance of United States, and a 2% decrease in the importance of Europe, with all the other locations increasing by 2% (not shown in the figure). All firms, except those from Canada, have increased the proportion executed outside the home country in the second half of the 1980s. Furthermore, the results of the trend analysis confirm and reinforce those reported above in terms of levels: Europe increased in importance as a location for firms based in Norway, Belgium, France, Switzerland, Finland and Italy; there was little change in the activities of the Japanese firms; and UK large firms increased their activities in the USA.

3.3.3. Causes of increased internationalisation: acquisitions vs organic growth. The total change in the number of patents granted to a firm from one period to the next can be decomposed into that due to mergers and acquisitions and that due to a secular trend. The latter can be interpreted as being the result of organic growth, i.e., the change resulting from the firm's decision to increase technology production either from existing R&D facilities or by creating new facilities in a particular location. In terms of our data, the trend element is identified by taking the firm (or a set of firms) as consolidated for 1984 and looking at the change in its patenting from 1979–1984 to 1985–1990. The change due to mergers and acquisitions is identified by calculating, for the period 1985–1990, the difference between the patenting of the firm as it was consolidated for 1984 and for 1988.

In Figure 2 we have taken the total US patenting of firms executed from outside the home base and considered the proportion due to mergers and acquisitions and that due to a secular trend. By way of illustration, of the increase in the total number of patents granted to UK large firms from outside the UK, around 60% can be attributed to acquisitions and around 40% to increasing activities due to organic growth. A comparison of Figures 1 and 2 shows that, in general, the firms with the largest increases in activities outside their home base did so more by a process of mergers and acquisitions than by organic growth. On the other hand, for firms based in the US, Netherlands,

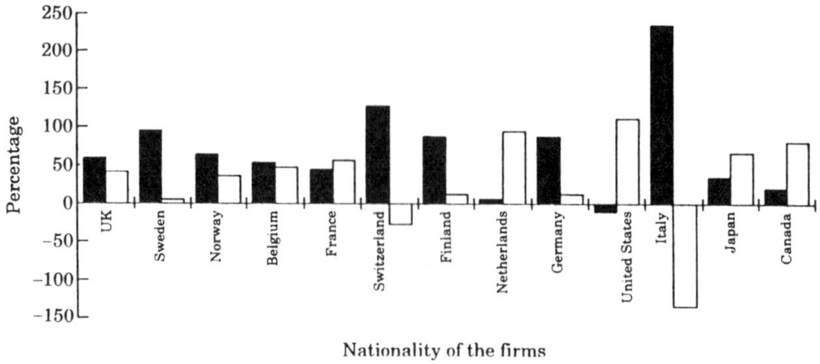

Fig. 2. *Decomposition of the total change in the number of US patents granted to large firms from outside the home country, 1979–1984 to 1985–1990.*
Notes: (■) Mergers and acquisitions; (□) organic growth.

Table 6. *Geographic location of large firms' US patenting activities, according to product group, 1985–1990 (percentage shares)*

Product group	Abroad (change[a])	Of which USA	Europe	Japan	Other
Drink and tobacco (17)	30·7 (− 6·6)	17·5	10·9	0·4	1·8
Food (41)	26·3 (6·1)	15·7	8·6	0·2	1·7
Building materials (26)	20·8 (9·4)	9·2	9·9	0·1	1·6
Other transport (5)	19·7 (9·5)	2·0	6·8	0·0	10·9
Pharmaceuticals (25)	16·7 (1·0)	5·5	8·3	1·1	1·7
Mining and petroleum (44)	14·7 (3·0)	9·1	3·5	0·1	1·6
Chemicals (71)	14·5 (3·1)	8·0	5·1	0·4	1·0
Machinery (68)	13·7 (1·2)	3·4	9·0	0·1	1·1
Paper & wood (31)	11·9 (3·9)	2·6	8·4	0·1	0·8
Metals (56)	11·4 (0·4)	5·1	5·9	0·1	0·3
Electrical (58)	10·2 (1·9)	2·6	6·9	0·3	0·4
Computers (17)	9·0 (0·7)	0·1	6·6	1·1	1·1
Rubber and plastics (10)	6·1 (2·4)	0·9	2·4	0·4	2·4
Textiles etc. (18)	4·7 (0·4)	1·5	1·8	0·8	0·6
Motor vehicles (43)	4·4 (− 0·9)	0·9	3·3	0·1	0·2
Instruments (20)	4·3 (− 0·8)	0·4	2·8	0·3	0·8
Aircraft (19)	2·9 (0·7)	0·3	1·8	0·1	0·7
All firms (569)	10·9 (1·1)	4·1	5·6	0·3	0·8

Note: [a]See text for the definition of change.

Canada, France and Japan organic growth was a relatively more important means of increasing their foreign technological activities.

3.3.4. Sectoral breakdown. Table 6 describes the main geographic locations of technological activities for the 569 firms aggregated according to their principal product group. It shows that the most internationalised firms (both in terms of levels and in terms of changes in the 80s) are those with their principal activities in drink and tobacco, food, building materials, other transport, pharmaceuticals, and mining and petroleum, with between 15 and 30% of their activities located outside the home base. What is of interest is that, with one exception (pharmaceuticals) these are not the 'high-tech' product groups or ones that are normally associated with having a 'world mandate' (automobiles and computers). In fact they are precisely the product groups where adaptive R&D for the purposes of serving local markets is important. Technological activities for such localised adaptation occur due to differences in consumer tastes (food, drink and tobacco), differences in government regulations (building materials and pharmaceuticals), and in order to exploit local natural resources (food, mining and petroleum). Indeed, the evidence here suggests that firms involved in the product groups with higher than average R&D intensity—computers, aircraft, instruments and electrical equipment—are below average in terms of the internationalisation of their technological activities. One reason for this could be that in these 'high tech' products, links between R&D and design, on the one hand, and production, on the other, are particularly important in the launching of major new products and benefit from geographical proximity.

4. Main findings and conclusions

The main conclusion of this paper is that there is no systematic evidence (i.e. covering firms from different countries and product groups) to suggest that widespread globalisation of the production of technology has occurred in the 1980s. The evidence presented above, based on the US patenting activities of 569 firms (based in 13 countries, and in 17 product groups), shows that for an overwhelming majority of them technological activities are located close to the home base. The analysis also shows the dangers of generalising on the basis of anecdotal evidence from a small sample of firms from a particular country or sector.

The main empirical findings of the study are:

- The only other systematic evidence available on the international technological activities of firms, based on data on their R&D expenditures abroad, is entirely consistent with the evidence on patenting activities abroad.
- There has been an increase in the proportion of activities undertaken by firms from overseas locations, with Europe becoming a relatively more important location for firms from most countries.
- The largest increases in internationalisation have occurred as a result of mergers and acquisitions and not by means of organic growth.
- UK large firms are the most internationalised in terms of their technological activities and the pattern they have followed is in general different from that followed by firms from other European countries: for UK firms the United States is the most important foreign location (and has become increasingly so); for other European firms Europe as a whole (and Germany in particular) is relatively more important.
- Japanese firms remain the least internationalised (with overseas activities accounting for around 1%), and Japan remains unimportant as a location for firms from other countries.
- The most internationalised firms are not in the 'high-tech' product groups that are normally labelled as having a 'world mandate'. On the contrary they are in product groups where adaptation for serving local markets is important.

These results confirm the prediction in our earlier paper (Patel and Pavitt, 1991) that there would be increases in the amount of technological activities located outside the home base. The fact that Europe has become a relatively more important location is also consistent with some of the evidence from surveys reported in Section 1: namely, that an important set of factors in the decision to locate technology outside the home country are labelled 'political'. In this case the most important influence, especially for the non-EEC European countries, has been the setting up of the European internal market. Our results also suggest that the process of internationalisation of technology is influenced more by the need to adapt products for market conditions in the host country than by supply-side reasons concerned with exploiting the science and technology base of the host country.

The evidence presented here is also consistent with the early analyses of Vernon (1966) and Porter (1989), and the more recent literature on the 'national systems of innovation' (Lundvall (1992) and Nelson (1992)), in that, far from being irrelevant, what happens in home countries is still very important in the creation of global technological advantage for firms. Thus for policy analysis it becomes important to understand the nature of the country-specific factors that have an influence in creating national technological advantage, including the competitive climate, the financial system, and education, training and basic research institutions.

Finally, it also becomes important to improve our understanding of the reasons why firms involved in producing for a world market may keep most of their technology production close to the home base. Two key features related to the launching of major innovations may help explain the advantages of geographic concentration: the involvement of inputs of knowledge and information that are essentially 'person-embodied', and a high degree of uncertainty surrounding outputs. Both of these are best handled through intense and frequent personal communications and rapid decision making, i.e. through geographic concentration. Thus it may be most efficient for firms to concentrate the core of their technological activities in the home base with international 'listening posts' and small foreign laboratories for adaptive R&D.

Bibliography

Archibugi, D. and Michie, J. 1995. The globalisation of technology: a new taxonomy, *Cambridge Journal of Economics*, vol. 19, no. 1

Cantwell, J. 1992. The internationalisation of technological activity and its implications for competitiveness, in Granstrand, O., Hakanson, L. and Sjolander, S. (eds), *Technology Management and International Business: Internationalisation of R&D and Technology*, Chichester, Wiley

Chesnais, F. 1989. Technical co-operation agreements between firms, *STI Review*, Paris, OECD

Chesnais, F. 1992. National systems of innovation, foreign direct investment and the operations of multinational enterprises, in Lundvall, B. Å. (ed.) (1992)

Dorrenbacher, C. and Wortmann, M. 1991. The internationalisation of corporate research and development, *Intereconomics*, vol. 26, no. 3

Dunning, J. 1992. Multinational enterprises and the globalisation of innovatory capacity, in Granstrand, O., Hakanson, L. and Sjolander, S. (eds), *Technology Management and International Business: Internationalisation of R&D and Technology*, Chichester, Wiley

Etemad, H. and Seguin Dulude, L. 1987. Patenting patterns in 25 multinational enterprises, *Technovation*, vol. 7, no. 1

Freeman, C. and Hagedoorn, J. 1992. 'Convergence and Divergence in the Internationalisation of Technology', Paper presented at the MERIT Conference 'Convergence and Divergence in Economic Growth and Technical Change', December

Granstrand, O. and Sjolander, S. 1992. Internationalisation and diversification of multitechnology corporations, in Granstrand, O., Hakanson, L. and Sjolander, S. (eds), *Technology Management and International Business: Internationalisation of R&D and Technology*, Chichester, Wiley

Hakanson, L. 1992. Locational determinants of foreign R&D in Swedish multinationals, in Granstrand, O., Hakanson, L. and Sjolander, S. (eds), *Technology Management and International Business: internationalisation of R&D and Technology,* Chichester, Wiley

Howells, J. 1990. The internationalisation of R&D and the development of global research networks, *Regional Studies*, vol. 24, no. 6

Lundvall, B. Å. (ed.) 1992. *National System of Innovation,* London, Pinter

Nelson, R. (ed.) 1992. *National Innovation Systems: a Comparative Analysis,* New York, Oxford University Press

Patel, P. and Pavitt, K. 1991. Large firms in the production of the world's technology: an important case of 'non-globalisation', *Journal of International Business Studies*, vol. 22, no. 1

Pearce, R. and Singh, S. 1992. *Globalising Research and Development*, London, Macmillan

Porter, M. 1990. *The Competitive Advantage of Nations*, London, Macmillan

Vernon, R. 1966. International investment and international trade in the product cycle, *Quarterly Journal of Economics*, vol. 80, no. 2

Vernon, R. 1979. The product-cycle hypothesis in a new international environment, *Oxford Bulletin of Economics and Statistics*, vol. 41, no. 4

Wortmann, M. 1990. Multinationals and the internationalisation of R&D: new developments in German companies, *Research Policy*, vol. 19, no. 2

[7]

Cambridge Journal of Economics 1995, **19**, 155–174

The globalisation of technology: what remains of the product cycle model?

John Cantwell*

This paper re-examines two hypotheses associated with earlier versions of the product cycle model. The first hypothesis, that innovations are almost always located in the home country of the parent company, is rejected on the basis of evidence drawn from 100 years of US Patent Office data. The second hypothesis, that the international dispersion of activity is led by technology leaders, is historically valid. However, over the last 20 years, technology leaders have been ahead instead in the globalisation of technology—that is, in developing internal international networks to exploit the locationally differentiated potential of foreign centres of excellence.

1. Introduction

This paper re-examines two hypotheses associated with earlier versions of the product cycle model (Vernon, 1966), using new historical evidence on the international dispersion of corporate technological activity by large American and European industrial firms, based on patents granted to these companies at the US Patent Office since the turn of the century. The first hypothesis states that innovations are almost always located in the home country of the parent company, and usually close to the site of the corporate technological headquarters. The second hypothesis is that international investment is led by technology leaders, as a means by which they increase their share of world markets and world production. If the first hypothesis does not necessarily hold, then the second can be interpreted more specifically to state that the internationalisation of technological development is led by firms with the strongest records in innovation.

Contrary to the first hypothesis, a new literature on the recent internationalisation of industrial research (for example, Ronstadt, 1977; Lall, 1979; Pearce and Singh, 1991; Granstrand, Håkanson and Sjölander, 1992) has suggested that innovations may be geographically dispersed within multinational corporations (MNCs). US-owned MNCs have recently witnessed an increase in the degree of internationalisation of their research, from low levels in the 1960s. It is perhaps no accident that empirical support for the product cycle hypothesis and subsequent criticism of it was largely derived from data on US MNCs. The new trend towards the internationalisation of research and development

Manuscript received 31 May 1994; final version received 9 November 1994.

*University of Reading. I am grateful to the referees for helpful comments on an earlier draft, and to the Economic and Social Research Council (ESRC) for their support under research grant R000232250. I wish to thank also Pilar Barrera who worked on related aspects of the project, Jane Myers and Jim Hirabayashi of the US Patent and Trademark Office for their invaluable assistance, and Cathy Jones and her many fellow students for their tremendous efforts during the data collection.

0309-166X/95/010155+20 $08.00/0

(R&D) in US MNCs became evident by the 1970s (see, for example, Mansfield, Teece and Romeo, 1979).

Three kinds of theoretical justification were provided for the first product cycle hypothesis. First, there are economies of scale in the R&D function, and if they are strong enough R&D will be concentrated in a single centre. Second, there are locational economies of integration and agglomeration in innovation. In new product development a close interaction is normally required between research and production facilities and users, while if several firms are engaged in the development of related new products (or processes) geographical proximity will encourage a greater volume of beneficial spillovers between them. Third, the original product cycle model viewed innovation as a demand-led process (see Schmookler, 1966, for another contemporary account), and thus for US MNCs innovation would be located in the more conducive environment of their home country, stimulated by the particular characteristics of the demand of high income consumers and skill-intensive downstream production facilities.

However, more recent evidence has demonstrated that the internationalisation of technological activity in large MNCs is not simply the outcome of a new trend (Cantwell and Hodson, 1991; Patel and Pavitt, 1991). As early as the 1960s many European MNCs (especially those originating from the UK, The Netherlands, Belgium and Switzerland) had already been quite highly internationalised in their technological activity. In Europe, the largest French and German industrial MNCs come closest to the US pattern, recently increasing the extent of internationalisation of their R&D from comparatively low levels in the 1960s. By contrast, large Japanese firms owned few research facilities outside Japan in the 1960s. Although their international R&D has been expanding, research in Japan itself has been growing at least as quickly, so the degree of internationalisation of research in Japanese-owned MNCs has remained low.

Here, it is argued that the first product cycle hypothesis should be rejected, not so much as a result of recent trends, as on the basis of historical evidence. In particular, the leading US electrical equipment companies and European chemical firms enjoyed a significant international dispersal of their technological activity in the interwar period. The source of evidence for this view is the US patenting of the largest US and European industrial companies over time, which identifies the location of technological activity at the corporate level, and thus provides a measure of the geographical dispersion of corporate invention. It is this same source of evidence that has been used often in other similar international comparisons that relate to the recent period, and which has helped to create the re-evaluation of trends in the international R&D of MNCs just described. What is new about the data in this study is the extension of records of corporate patenting back in time: here they are considered back to 1920.

The second early product cycle hypothesis states that international investment (and here, the international dispersion of technological activity) is led by technology leaders. The theoretical justification for this hypothesis is that the most technologically competent companies enjoy lower operating costs than their competitors and provide higher product quality, which generates higher profits and rising international market shares. In their turn, different degrees of technological competence across companies are a consequence of the firm-specific and path-dependent characteristics of technological change (Cantwell, 1991B, 1994; Nelson, 1991, 1992; Dosi, Teece and Winter, 1992; Pavitt, 1992; Teece, Pisano and Shuen, 1990). The greater capability of the most competent or technologically leading firms (or their greater capability to release able management resources and team expertise to plan for and organise growth, as argued by

Penrose, 1959) enables them better to expand their activity in new fields or environments, and higher profits provide them with the financial wherewithal to offset the costs of doing so. An ancillary argument is especially important in the case of early internationalisation and is featured in the contemporary explanation of the product cycle hypothesis; that is, a technologically leading group of firms from a particular location may—past some point—find that the continued increase in its penetration of foreign markets is challenged by protectionist barriers, and by the gradual learning of local competitors in those markets. This raises the incentive to produce locally, and insofar as indigenous companies develop their own lines of related technological development there is also a greater incentive to internationalise innovation (in addition to the need to adapt products to the distinctive features of host country demand).

This hypothesis appears to be consistent with the historical evidence, except perhaps during phases of technological hegemony when innovative development is concentrated in the home country. However, the explanatory power of this hypothesis has been eroded by recent trends; the composition of firms involved in the internationalisation of technological activity has now been broadened, extending to cover a much wider range of companies. Therefore, an alternative version of the second hypothesis is proposed to fit the current situation. It is suggested that technology leaders are now ahead instead in the globalisation of technology—that is, in the development of international intra-firm networks to exploit the locationally differentiated potential of foreign centres of excellence. These networks are internal to the firm in order to build upon or extend its core technological competence through an internally coordinated learning process, but they are complementary to external inter-firm networks whose role is the exchange of knowledge and occasionally cooperation in learning through technology-based joint ventures (Cantwell, 1991B, 1994; Cantwell and Barrera, 1995).

The data employed are described further in Section 2, while Sections 3 and 4 relate to the re-assessment of the first and second product cycle hypotheses respectively. Section 3 examines the historical importance of the international dispersion of technological development in leading American and European firms. Particular attention is paid to firms in the chemical and electrical equipment industries broadly defined, as these are the industries which are most reliant on science-based technologies. Historically, US firms were strongest in the electrical equipment industry, while European (and especially German) firms were stronger in chemicals (Chandler, 1990; Cantwell and Barrera, 1993). Among leading companies such as these, technological activity is not always much more widely geographically dispersed today than it was in the interwar or early post-war periods. However, Section 4 considers what has changed recently for these firms, and terms this a new trend towards the globalisation of technology—that is, the emergence of internationally integrated structures for technological development. It is suggested that globalisation in this sense is a modern characteristic of corporate technology leaders. Some conclusions are set out in Section 5.

2. The data

Patenting is a measure of invention, and so corporate patenting is more a measure of wider technological activity (changes in production methods) in firms, and not just R&D as such. For larger firms like those covered here, R&D is the most important source of new knowledge and skills, and so for them the internationalisation of technological activity revolves around the internationalisation of research. However, production engineering is often an important complementary source of new inventions that are incorporated into technology. It should also be noted that the location of basic R&D that feeds into the development of productive applications at some other site may not itself be picked up separately in the patent statistics. For these reasons the title of the paper refers to the internationalisation of technological activity (the development of new methods of production), and not to the internationalisation of R&D as such.

Two types of information have been collected manually from the *US Index of Patents* and the *US Patent Office Gazette.* First, all patents were recorded that were assigned to a selection of large US-owned and European-owned firms between 1890 and 1968. From 1969 onwards equivalent information has been computerised by the US Patent Office. The firms selected for the historical patent search were identified in one of three ways. The first group consisted of those firms which have accounted for the highest levels of US patenting after 1969; the second group comprised other US, German or British firms which were historically among the largest 200 industrial corporations in each of these countries (derived from lists in Chandler, 1990); and the third group was made up of other companies which featured prominently in the US patent records of earlier years (a method that proved most significant for a number of French firms that had not been identified from other sources).

In each case, patents were counted as belonging to a common corporate group where they were assigned to affiliates of a parent company. Affiliate names were normally taken from individual company histories. In all, the US patenting of 857 companies or affiliates was traced historically; together these comprise 284 corporate groups. Owing to historical changes in ownership, 17 of the affiliates were allocated to more than one corporate group over the period as a whole. No significance has been attached to the location of the particular affiliate to which each patent is assigned, since this may be different from the location of the inventive activity that gave rise to the patent. However, the location of the parent company is an important dimension in the analysis, as this is treated as the home country or the country of origin of the corporate group. By consolidating patents attributable to international corporate groups, it is then feasible to examine the geographical distribution of technological activity within groups, and the possible formation of internal intra-group international networks (as opposed to external inter-firm networks, which are not the subject of this paper). Each corporate group is also allocated to an industry on the basis of its primary field of production; occasionally, firms have moved between industries historically, sometimes associated with changes in ownership, and this has been allowed for.

The company to which a patent has been assigned (if any), and the name and location of residence of the inventor responsible for the underlying invention, are both recorded separately in the US Patent Office data, including the earliest data. Where patents have been assigned to firms, the inventor is normally an employee of the company or is directly associated with it in some other way, but occasionally independent individual inventors do choose to assign their patents to firms (Schmookler, 1966). Assignments by independent individuals were more common in the nineteenth century but, at least from the inter-war years onwards, the typical assignor was a prominent member of a corporate research laboratory, or some other similar in-house company facility. Although it is normally difficult to trace these named individuals in secondary sources on the firms concerned (as they are not usually also senior managers), the location of assignors can be checked against business history sources on the international location of activity in particular firms. Such checks on a selection of large firms have confirmed that whenever a location has been responsible for significant numbers of patents being assigned to a company, that firm did indeed have some in-house facility in the location in question at the relevant time. Companies checked in this fashion include various US firms active abroad and European companies in the US (Stocking and Watkins, 1946; Beaton, 1957; Wilkins, 1974, 1989; Chandler, 1990), Courtaulds and British Celanese (Coleman, 1969), Du Pont and ICI (Hounshell and Smith, 1988), and General Electric and GEC (Reich, 1986; Jones and Marriot, 1971).

Second, using once again the *US Index of Patents* and the *US Patent Office Gazette*, for every patent granted in years between 1890 and 1962 the country of residence of the inventor has been recorded. From 1963 this information has already been computerised by the US Patent Office. Where patents are assigned to companies, these data on the country of origin of invention indicate the location of the R&D facilities (or other sources of technological improvement) that gave rise to each patent. As this information on the location of invention relates to individually numbered patents, it can be combined with a sectoral classification of the technological activity with which the patent is associated. This employs the system of patent classes used by the Patent Office; fortunately, as these classes change the Office reclassifies all earlier patents accordingly, so the classification is historically consistent. Although patents may be assigned to several fields, the primary classification was used in all cases. Various broad categories of technological activity were derived by allocating classes or sub-classes to common groups of activity.

Two distinctions between different aspects of these classifications of the data are worth re-emphasising. First, the sectoral classification of patents, in terms of the type of technological activity with which each patent is associated, is distinguished from the main industrial output or markets of the companies to which patents may be assigned, both of which have been recorded separately. Most large companies have engaged in at least some development in most of the general spheres of technological activity (for instance, chemical firms develop many mechanical technologies, including chemical machinery and equipment), irrespective of the industry in which they operate.

Second, the country of location of the invention, which for large firms typically represents or is allied to the location of corporate R&D, is distinguished from the location of the firm to which a patent is assigned, and from the location of the parent company which owns this firm. While no significance is attached to the location of the assignee, the location of research and the location of ultimate ownership (the parent company) are critical. The extent to which these locations differ over the total patenting of each corporate group is the measure of the degree of internationalisation of that group's technological activity. These distinctions are crucial to understanding that the measurement of the degree of internationalisation of technological activity in what follows is not to be mistaken for a measure of the international spread of patents taken out by firms (the patents counted were all those—and only those—granted by the US Patent Office), nor should it be mistaken for a measure of the international dispersion of the legal departments or agents responsible for making patent applications on behalf of the group (the actual assignee is ignored once it has been linked to a parent company of a large group in the dataset). Instead, what is measured is the internationalisation of the underlying technological activity that gave rise to the knowledge which subsequently led to a patent being granted to the group.

Table 1. *Shares of US patenting of the largest nationally owned industrial firms due to research located abroad (%)*

	1920–1939	1940–1968	1969–1990
US	6·81	3·57	6·82
Europe	12·03	26·65	27·13
UK	27·71	41·95	43·17
Germany	4·03	8·68	13·72
Italy	29·03	24·76	14·24
France	3·35	8·19	9·55
Netherlands	15·57	29·51	52·97
Belgium	95·00	53·90	60·60
Switzerland	5·67	28·33	43·76
Sweden	31·04	13·18	25·51
Total	7·91	8·08	14·52

Source: US patent data compiled at the University of Reading, with assistance from the US Patent and Trademark Office, US Department of Commerce.

3. The historical role of the internationalisation of technological activity

As just described, the degree of internationalisation of corporate technological activity is measured by the share of patenting that is attributable to research (or other sources of invention) located outside the home country of the corporate group or groups in question. The broad changes in the internationalisation of technological activity between 1920 and 1990 measured in this way are set out in Table 1, organised by national groups of firms according to the location of ultimate ownership. The share of foreign research in the total corporate technological activity of the largest US and European industrial firms considered together averaged about 8% in 1920–1939 and 1940–1968, before rising significantly to roughly 14·5% in 1969–1990. However, this recent average trend increase in the internationalisation of activity as measured by corporate patenting reflects primarily the rising share of US patenting accounted for by European firms (which are more internationalised) *vis-à-vis* their US counterparts.

There appear to be three categories of national groups of firms. In the first, German- and French-owned firms come closest to the standard view, in the sense that the degree of internationalisation of their technological activity was very low historically, but has been on a slowly rising trend, and has increased significantly recently. The second category comprises the historically more multinational British, Swiss and Dutch companies, whose technological activity became substantially internationalised after the Second World War, achieving an early increase in international scope that is reflected in the picture for large European firms as a whole. The third category is very different from the others. The technological activity of the largest US and Swedish firms was as highly internationalised historically as it is today; and, of course, for some individual companies (such as the American General Electric) the extent of internationalisation was much higher in the interwar period than it has been recently. The largest American and Swedish firms retreated from their international research operations after the Second World War, and they have only recently regained the position they held before that time.

Table 2. *Shares of US patenting of the largest nationally owned industrial firms due to research located abroad (%)*

	US	Europe	UK	Germany	France	Switzerland	Sweden	Total
1920–1924	3·32	8·45	7·61	2·89	3·02	4·00	39·77	4·03
1925–1929	6·61	8·19	14·26	5·42	2·67	3·73	29·88	6·87
1930–1934	8·17	10·83	25·88	4·79	4·12	6·38	27·08	8·80
1935–1939	6·94	15·11	34·63	2·83	3·33	6·04	30·23	8·86
1940–1959	3·79	27·10	40·84	9·46	8·90	22·24	11·98	7·67
1960–1964	2·85	24·22	39·52	8·26	7·46	27·28	13·24	7·62
1965–1968	3·69	28·02	46·44	8·21	7·76	39·56	15·22	9·42
1969–1972	5·22	28·21	42·66	12·65	7·71	45·86	18·14	12·21
1973–1977	5·98	25·42	40·09	11·03	6·44	44·67	21·15	13·18
1978–1982	6·65	24·60	39·17	12·14	6·66	43·68	27·25	13·82
1983–1986	8·51	27·13	45·32	14·83	9·20	41·02	29·84	15·92
1987–1990	8·95	30·86	50·55	17·76	18·26	42·74	31·49	18·56

Source: As for Table 1.

A more detailed periodisation of the records of each of these three groups of large firms can be gleaned from Table 2. For French and German companies this shows that the upward trend in the internationalisation of technological activity is not as gradual as might have been supposed from Table 1, but in fact is based on some discrete jumps. As for other European firms, there was an increase in internationalisation immediately after the Second World War, following the nationalist retreat of the late 1930s, and the expropriations of German-owned firms abroad in the 1940s. There was then a further rise in the early 1970s in German firms, and in the late 1980s for German companies again, but more especially for French firms (Cantwell and Kotecha, 1994). British and Swiss firms increased to much higher rates of internationalisation of technological activity after the war than did French or German companies. In the British case the origins of the process can be traced to a trend increase throughout the interwar period, which was consolidated after the war, rising close to a peak by the late 1960s, before recovering again in the 1980s. The degree of internationalisation of Swiss firms reached a peak in the early 1970s, which has not quite been matched since.

For the largest US firms the internationalisation of technological activity peaked in the early 1930s (at a level they have only very recently equalled), while Swedish firms began in the early 1920s from a high point that they have not yet since recaptured. The internationalisation of research in both the US and Swedish groups fell sharply after the war, but picked up again from the late 1960s onwards. It is this latter trend that has been quite widely commented upon, perhaps not surprisingly, especially by American and Swedish scholars (such as Ronstadt, 1977; Mansfield, Teece and Romeo, 1979; or Granstrand, Håkanson and Sjölander, 1992). However, this contrasts with the more general European experience. The internationalisation of technological activity in large European firms considered as a whole increased after the Second World War (as did all the constituent European national groups except the Swedish), and again from the mid-1980s (again this applies to all groups except the Swiss).

Besides their national groups, firms may also be allocated to industrial groups. The broad industrial classification adopted here groups industries in accordance with the prevailing type of technological activity in the sector in question. Thus, firms in the

Table 3. *Shares of US patenting of the largest nationally owned industrial firms due to research located abroad, grouped by industry (%)*

	US	Europe	UK	Germany	France	Switzerland	Sweden	Total
				1920–1939				
Chemical	2·75	12·42	41·48	4·88	4·64	5·02	4·44	6·88
Electrical	10·13	3·21	1·98	2·58	0·63	8·20	8·89	9·42
Mechanical	5·15	20·18	29·27	3·56	2·93	1·94	36·45	9·17
Transport	1·61	4·95	8·02	2·76	4·89	n.a.	0·00	2·01
Total	6·81	12·03	27·71	4·03	3·35	5·67	31·04	7·91
				1940–1968				
Chemical	2·24	39·91	66·42	12·51	15·65	31·65	12·40	11·74
Electrical	6·03	14·15	8·81	6·01	5·16	14·31	8·40	7·26
Mechanical	3·27	23·54	37·94	4·15	12·72	17·12	15·70	7·85
Transport	1·41	4·26	5·52	2·36	2·22	n.a.	5·50	1·77
Total	3·57	26·65	41·95	8·68	8·19	28·33	13·18	8·08
				1969–1990				
Chemical	5·65	32·65	55·33	18·08	9·03	46·84	14·52	17·91
Electrical	9·09	27·32	27·11	11·01	8·43	34·22	29·32	14·65
Mechanical	6·51	23·94	50·31	7·67	16·83	33·23	27·75	13·58
Transport	4·95	10·16	10·55	8·47	4·98	n.a.	12·36	6·53
Total	6·82	27·13	43·17	13·72	9·55	43·76	25·51	14·52

Note: n.a.=not applicable.
Source: As for Table 1.

chemical, pharmaceutical and coal and petroleum products industries all rely mainly on chemical and related technologies; firms in the electrical equipment and office equipment or computer industries base themselves principally on electrical technologies; and firms in the motor vehicle, aircraft and rubber and plastic products industries are concerned with the major transport technologies (engines and tyres). For ease of exposition these are each referred to collectively as the chemical, electrical equipment and transport industries. Firms in all other industries rely mainly on more traditional mechanical technologies, so the mechanical group is a much more heterogeneous mixture.

As shown in Table 3, the US electrical equipment firms were much more internationalised in their research than were European firms in the equivalent industry in the interwar period, despite the much higher overall rate of internationalisation of technological activity in the European group. In Britain, Germany and France, large chemical firms were more internationalised historically than those in the electrical equipment sector, although in Switzerland the electrical companies featured more strongly, and in Sweden it was mechanically-based firms that were responsible for the very high overall internationalisation of research in the largest nationally owned firms at that time. The mechanical group were also highly internationalised in the UK (notably the textile companies British Celanese and Courtaulds), and British and French transport firms (particularly the tyre companies Dunlop and Michelin) appear more prominently than the other members of that group.

A similar discrepancy between the relative extent of internationalisation of research in chemical and electrical equipment firms in the US and Europe still existed in recent years, but the difference is less marked than it was. Among European companies the stronger internationalisation of chemical than electrical firms also holds now for Switzerland, but not for Sweden. For Swedish firms the mechanical group does not dominate international research as it once did, but in the UK and France the mechanical group is relatively more prominent than it used to be. What this reflects in each case is a broadening of the range of firms engaged in the internationalisation of technological activity across a wider spectrum; for example, while the mechanical engineering firm Alfa-Laval accounted for the bulk of the foreign research of Swedish companies historically, its share has steadily declined as other firms have begun to engage in technological development abroad (Zander, 1994).

For US and European firms as a whole a more detailed periodisation for the major industrial groups can be found in Tables 4 and 5. Table 4 shows that US-owned electrical equipment firms in the 1930s had a higher degree of internationalisation of technological activity than they have had at any time since, including the 1980s. About 12% of their research was located abroad in the 1930s, compared to roughly 11% in the 1980s. Two American-owned companies that contributed especially heavily to European-located research and production in the interwar years were General Electric and RCA. The degree of internationalisation of technological activity in General Electric was 18·3% in 1920–1939 but only 2·4% in 1968–1990, while in RCA the equivalent proportions were 20·4% and 5·4%.

In contrast, the largest European firms sustained a sizeable increase in the extent of the internationalisation of their research between the early 1930s and the 1950s. This increase was led by companies in the science-based sectors, and especially those in chemicals. The share of foreign research in the leading European chemical firms rose from 8·2% in 1930–1934 to 42·6% in 1940–1959, while for electrical equipment companies it increased from 5·1% to 11·0% (see Table 5). For the largest European chemical firms the immediate post-war rate of internationalisation proved to be a peak that has not since been surpassed (for them the foreign share stood at 34·8% in 1987–1990), but in the post-war period the major European electrical equipment companies steadily increased their international research, to a point where the foreign share is now similar to that in chemicals.

The European story partly reflects the British experience, which is described in Table 6. For UK-owned companies the increase in the internationalisation of technological activity between the early 1930s and the 1950s (from 25·9% to 40·8%) was also largely associated with a very strong internationalisation of research in the chemical industry (from 23·3% to 66·7%). The prominent contribution of the chemical companies is consistent with the view that, historically, internationalisation was linked to technological competence. European firms were technologically strongest in the chemical fields, in which areas they internationalised their research early; while American firms were relatively stronger in the development of electrical technologies, and in this field they led the early internationalisation of technological activity.

Thus, at least for technology leaders, the internationalisation of technological activity is not a new phenomenon. Of course, the first product cycle hypothesis might be rescued by a restatement to the effect that as a general rule the home country centre has been and remains the single most important site for the technological development of MNCs. Given that, as reported in Tables 1 to 6, the foreign share of technological activity has rarely been greater than two-thirds and is usually much less, MNCs are not 'stateless corporations' (Patel and Pavitt, 1991). The national origins of MNCs have been and

Table 4. *Shares of US patenting of the largest US-owned industrial firms due to research located abroad, grouped by industry (%)*

	1920–24	1925–29	1930–34	1935–39	1940–59	1960–64	1965–68	1969–72	1973–77	1978–82	1983–86	1987–90
Chemical	0·96	1·90	4·03	2·36	1·87	2·12	3·08	4·37	5·21	5·33	6·36	7·71
Electrical	4·14	8·50	12·27	11·99	7·01	3·81	5·23	7·29	7·94	9·00	11·60	10·90
Mechanical	3·31	5·72	5·83	5·06	2·87	4·00	3·58	4·61	5·39	6·71	9·36	9·34
Transport	0·12	1·37	1·94	1·73	0·90	1·69	2·36	3·45	4·21	4·51	5·81	7·59
Total	3·32	6·61	8·17	6·94	3·79	2·85	3·69	5·22	5·98	6·65	8·51	8·95

Source: As for Table 1.

Table 5. *Shares of US patenting of the largest European-owned industrial firms due to research located abroad, grouped by industry (%)*

	1920–24	1925–29	1930–34	1935–39	1940–59	1960–64	1965–68	1969–72	1973–77	1978–82	1983–86	1987–90
Chemical	7·07	7·52	8·15	18·22	42·55	37·17	38·46	35·40	31·25	30·51	32·30	34·77
Electrical	4·12	2·81	5·14	1·73	11·01	13·87	19·96	26·90	24·51	21·44	28·17	34·10
Mechanical	10·84	12·91	23·00	24·74	25·73	18·73	23·31	25·45	22·25	23·98	23·87	24·59
Transport	3·75	9·15	4·69	3·04	4·77	3·34	4·45	7·12	7·40	7·69	12·76	16·19
Total	8·45	8·19	10·83	15·11	27·10	24·22	28·02	28·21	25·42	24·60	27·13	30·86

Source: As for Table 1.

Table 6. *Shares of US patenting of the largest UK-owned industrial firms due to research located abroad, grouped by industry (%)*

	1920–24	1925–29	1930–34	1935–39	1940–59	1960–64	1965–68	1969–72	1973–77	1978–82	1983–86	1987–90
Chemical	22·22	19·44	23·27	52·12	66·66	64·19	67·90	57·79	49·61	50·03	58·82	62·50
Electrical	7·14	4·67	0·55	0·67	5·61	6·94	17·37	27·09	33·56	24·85	19·53	26·53
Mechanical	6·55	16·34	36·51	32·13	35·57	38·53	44·70	50·06	48·10	45·47	51·14	59·88
Transport	5·45	14·46	8·13	5·68	5·75	4·37	6·03	7·75	7·65	7·77	13·29	20·31
Total	7·61	14·26	25·88	34·63	40·84	39·52	46·44	42·66	40·09	39·17	45·32	50·55

Source: As for Table 1.

continue to be critical in determining the geographical and sectoral composition of their technological activity, based on their path-dependent evolution from a nationally differentiated expertise (Kogut, 1987, 1990). While globalisation has been defined here as the international integration of MNC networks, some authors have used the term to mean the loss of national identity by companies, and when defined in their unhelpful way, globalisation has not taken place.

It is also clear that from the perspective of the MNC as a whole, or from that of its home country, the internationalisation of research is generally less than the internationalisation of production (Patel and Pavitt, 1991). One possible way of thinking about this is that most MNCs constitute an internal locational hierarchy of activity (Hymer, 1975). To simplify matters for the sake of exposition, suppose that the production of each firm is divided into the technologically sophisticated and the simple or assembly-type. Then, technologically sophisticated production tends to become geographically concentrated in certain locations, and assembly-type production tends to agglomerate in others (Cantwell, 1987). The home country operations of the MNC stand at the pinnacle of its hierarchy, as a base for technologically sophisticated production, as hinted at by both Vernon and Hymer. Since the siting of research facilities is normally linked to the local support of technologically sophisticated production while little or no research accompanies assembly-type production, it follows that production in total is more widely geographically dispersed than research. So, while the home country is the single most important site for innovation, it does not necessarily follow that production in the home centre is more research-intensive than in the other advanced centres in which the MNC establishes technologically sophisticated production.

This argument suggests that, of the original theoretical underpinnings of the first product cycle hypothesis, locational economies of integration and agglomeration play a greater role than do economies of scale in the R&D function. Previous criticisms of the first product cycle hypothesis had shown that, while there is a minimum efficient scale for R&D facilities, the effect of firm size on the degree of internationalisation of R&D is ambiguous (Mansfield, Teece and Romeo, 1979). By contrast, a good deal of evidence has been gathered in support of the importance of economies of agglomeration or local clustering in the location of production (Krugman, 1991; Porter, 1990; Dahmén, 1970; Sölvell, Zander and Porter, 1991), and especially in the geographical location of innovation (Jaffe, 1989; Jaffe, Trajtenberg and Henderson, 1993; Feldman, 1993; Cantwell, 1991A) and in the location of the technologically sophisticated production of MNCs (Cantwell, 1987). The lesson seems to be that economies of locational agglomeration are important, but that for MNCs they may occur in various centres and not exclusively in the home country, although the home base is the most significant single such centre.

4. The new globalisation of technology by corporate leaders

It has been shown that technological activity in some US MNCs was more widely dispersed internationally in the interwar period than it is today, while many European MNCs were already geared up to technology creation abroad by the 1950s. The companies that achieved the greatest internationalisation of activity historically were generally technology leaders. This is what is suggested by the second product cycle hypothesis as formulated above, although there is a qualification to an acceptance of the hypothesis. This is, that where technology leaders are in a very strong or hegemonic

Table 7. *The cross-firm coefficient of variation (expressed as a percentage) of the share of patenting due to research located abroad*

	US	UK	Germany	France	Switzerland	Sweden
Chemical industry group						
1920–1939	217·5	135·1	177·3	89·0	88·6	190·9
1969–1990	121·1	84·6	47·4	90·4	25·1	21·1
Electrical equipment industry group						
1920–1939	144·7	232·5	165·6	223·6	38·9	141·4
1969–1990	148·5	31·7	95·9	74·7	86·1	73·5
Mechanical industry group						
1920–1939	289·2	181·3	193·6	189·3	141·4	219·7
1969–1990	106·0	63·3	119·8	118·3	36·1	115·3
Transport industry group						
1920–1939	202·4	219·4	117·4	264·6	n.a.	n.a.
1969–1990	101·0	157·1	126·2	207·1	n.a.	99·1

Note: n.a.=not applicable.
Source: As for Table 1.

position through rapid innovation at home, the home centre may exercise such a strong attraction for further research that the internationalisation of technological activity is weak—a situation that seems to have applied to US companies in the early post-war years, and to Japanese firms today. Yet despite the historical significance of the geographical dispersion of innovation by MNCs, it is only relatively recently that the literature on MNCs has devoted much attention to the international creation of technology, as opposed to international technology transfer (Cantwell, 1994).

The more recent broadening of the internationalisation of technological development to a wider range of firms is partly responsible for finally drawing attention to this issue. The reduction in the industrial focus of internationalisation mentioned previously offers some indirect evidence consistent with the broadening of internationalisation across firms, but Table 7 provides direct evidence of this process within industrial groups. This shows that the extent of variation across firms in the degree of internationalisation of their technological activity has been generally on a downward trend. A wider range of companies has now engaged in foreign technological development, in what was once mainly the province of a smaller number of leading firms. However, there are occasional exceptions to this trend. In the US electrical equipment group, ITT has remained highly international, and Sperry (prior to the formation of Unisys) also became strongly committed to foreign research, leaving others in their wake. In the Swiss electrical equipment group it was Brown Boveri (prior to the formation of ABB) that pulled ahead of the pack in its foreign operations, and in the German transport group the motor vehicle component producer Robert Bosch played this role, in each case sustaining the extent of cross-firm variation in the degree of internationalisation. Elsewhere, lower transport and communication costs contributed to a general expansion across large firms in the internationalisation of technological activity.

The other recent change which has often been discussed in other contexts is that formerly local market oriented affiliates have been increasingly integrated into international networks within their respective MNCs, such networks coming to resemble 'heterarchies' more than hierarchies (Hedlund, 1986; Doz, 1986; Porter, 1986; Bartlett and Ghoshal, 1989; Dunning, 1992). In technological activity, too, the location-specific capabilities of internationally dispersed MNC affiliates may have become more closely integrated than in the past, linked to a strategy for technology creation in the MNC as a whole, and not only with separate reference to each of the geographical parts of the company's business. This can be termed a new globalisation of technological innovation.

The theoretical rationale for the recent international integration of productive activity is that the economic benefits attributable to a more refined locational division of labour within the MNC have often come to outweigh the costs of being less nationally responsive in each market, costs associated with adverse political repercussions and the continued national differentiation of demand (Doz, 1986). In an integrated MNC network each affiliate specialises in accordance with the specific characteristics of local production conditions, technological capabilities and user requirements. The network benefits from economies of scale through the local concentration of particular lines of activity (increasing returns from local research in a specialised field as opposed to research in general), economies of locational agglomeration through an interchange with others operating in the same vicinity in technologically allied fields as suggested earlier, and economies of scope through the international intra-firm coordination of related but geographically separated activities. The experience acquired in a specialised activity in one location creates technological spillovers that can be passed on to other parts of the MNC network elsewhere. It has been shown that since the 1970s, in industries in which such net advantages to multinational integration were available, multinationality has been a source of competitive success and faster growth (Cantwell and Sanna-Randaccio, 1993).

The transformation of the MNC, from a mainly multi-domestic structure of separate affiliates each serving their local markets, to an integrated internal network structure, has relevance to other aspects of the product cycle model apart from the hypotheses mentioned earlier. The wider product cycle model extended beyond the two hypotheses discussed here, to various other hypotheses and assumptions, such as that foreign direct investment can be treated as essentially local market-oriented, and that firms can be thought of as akin to single product producers. It can be argued that this wider model broke down largely because of globalisation in the sense just described, entailing the international integration of MNC networks. A global scanning for (new sources of) innovation and greater international linkages between production facilities imply an interactive flow of products and technological knowledge between countries (Vernon, 1979). This contrasts strongly with the original product cycle perspective of innovation and technological knowledge essentially flowing outwards from a single major centre, namely the home country.

The new globalisation of the responsibilities of affiliates to the MNC as a whole can be illustrated with reference to the shift that has occurred in the pattern of technological specialisation of foreign-owned research facilities. To examine this issue the patent data were used to construct a measure of the distribution of technological specialisation across various types of activity, for different groups of firms within their industry. It is also possible to distinguish between the pattern of specialisation in domestic technological activity and in research abroad, with reference to a corporate group or groups. The index of technological specialisation across different fields of activity that has been calculated is often termed an index of revealed technological advantage (RTA). The RTA value of a selected group of firms in a particular sector of technological activity is given by its share of US patents in that sector granted to companies in the same industry, relative to that

Table 8. *The RTA values in selected sectors of technological activity, of US-owned firms in the electrical equipment industry, and German-owned firms in the chemicals industry*

	1920–1968		1969–1990	
	At home	Abroad	At home	Abroad
(a) US-owned firms in the electrical equipment industry				
Lighting and wiring	0·96	1·30	0·94	0·77
General industrial equipment	1·07	0·67	0·97	1·96
(b) German-owned firms in the chemicals industry				
Bleaching and dyeing processes	2·43	1·43	1·91	0·71
Organic chemicals (dyestuffs)	1·46	1·47	1·33	0·97
Pharmaceuticals	1·22	3·16	1·22	2·04

Source: As for Table 1.

group's overall share of all US patents assigned to firms in the industry in question. Denoting as P_{ij} the number of US patents granted in the field of technological activity i (defined with reference to the patent class system, as described earlier) to the selected group of firms j in a particular industry, then the RTA index is defined as follows:

$$RTA_{ij} = (P_{ij} / \Sigma_j P_{ij}) / (\Sigma_i P_{ij} / \Sigma_{ij} P_{ij})$$

The index varies around unity, such that values greater than one suggest that a group of firms is comparatively specialised in the activity in question relative to other firms in the same industry, while values less than one are indicative of a lack of specialisation by the standards of the industry (see Cantwell, 1993, for a further discussion). For the purposes of historical comparison, attention is focused on the two groups of corporate technology leaders most prominent historically, each originating from major centres of innovation in their respective industries—that is, US-owned firms in the electrical equipment industry, and German-owned companies in the chemical industry. For each group the RTA index is separately calculated for activity located in its home research and its foreign research facilities, and for each of the two broad periods 1920–1968 and 1969–1990. In order to avoid the problems associated with low numbers of patents, the analysis is restricted to sectors of technological activity in which all large firms in the industry in question were granted 900 US patents or more in 1920–1968. This involved 19 sectors in the electrical equipment industry, and 20 in the case of chemicals. The values of the RTA index calculated for a few selected sectors are shown in Table 8.

The European-located research of US-owned electrical equipment firms was not historically (1920–1968) geared to local European strengths, but instead represented the local development of fields related to the core technologies that had been pioneered at home (in telecommunications and general electrical systems, including lighting). For these purposes, the pattern of technological specialisation of countries and broader regions such as Europe relative to one another was observed using an analogous RTA measure for countries rather than national groups of firms, grouping all patents (and not just those assigned to the largest firms) by the location of invention—indeed, it was in this country-specific (as opposed to firm-specific) form that the RTA index was originally proposed (Soete, 1987; Cantwell, 1989). In more recent years (1969–1990), while the domestic activity of US-owned firms in this sector has continued to concentrate on general electrical equipment, the focus of their foreign research has shifted to specialised machinery and general industrial equipment, both of which are areas of European advantage. Thus, the technological activity of US-owned foreign affiliates (which is still mainly conducted in Europe) has shifted towards an attempt to exploit the technological potential of the location in which it is carried out.

Similarly, the largest German chemical firms were strongly specialised historically (between 1920 and 1968) in the same fields in developing technology both at home and abroad—namely, in bleaching and dyeing processes, in organic chemicals including dyestuffs, and to a lesser extent in pharmaceuticals. Thus, these large German companies quite directly exploited their major strengths abroad, further developing these new products and techniques for local industries and markets in other countries. In foreign research today (since 1969) these firms are no longer specialised in their corporate strengths of dyes and dyeing, but they retain a stronger focus on pharmaceuticals in which they are not quite so strong at home. Rather like the American electrical companies, the German chemical firms have shifted their international research strategies away from the pure exploitation of their own strengths adapted to the needs of each particular local market, and towards an attempt to tap into foreign centres of expertise, in their case mainly in the development of pharmaceuticals in the US, Britain and Switzerland.

Other recent evidence also suggests that this type of internationally integrated or globalised strategy for innovation characterises corporate technology leaders today. The extent to which the affiliates of MNCs specialise within their industry across national boundaries in accordance with the comparative advantage of local expertise seems to depend upon the pattern of locational hierarchy that exists between alternative centres. In the European chemical industry Germany is the dominant centre, the UK is a second-order centre, and Italy is of the third order. In this sector, German MNCs (the leaders) are technologically specialised in the other European centres in line with host country strengths, and the same is true of British chemical MNCs in Italy (Cantwell and Sanna-Randaccio, 1992).

However, when operating in Germany, British chemical companies follow a pattern of technological specialisation that accords with their own comparative advantages in the industry and those of their home centre, the UK. They do not appear to be especially prone to try and tap into the areas in which Germany expertise is relatively greatest, but rather treat Germany as a general reservoir of skills that can be used principally to extend those lines of operation on which they are already focused in their home base. Technology leaders originating from higher-order centres tend to establish a more extensive locationally specialised network of technological activity in support of an international innovation strategy than has been developed as yet by firms that originate from lower-order centres.

It is true that this result depends upon an analysis of cross-border technological specialisation conducted at a fairly broad level of aggregation—across 20 or 30 sectors of technological activity, as described earlier. It may well be that when companies from a lower-order centre locate research in a higher-order one to engage in the same broad lines of activity in which they are specialised at home, there is still some geographical specialisation at the more detailed level of particular products or processes, and at the more finely disaggregated patent class level (Zander, 1994). In this event, the distinction between corporate technology leaders and other firms would be more a matter of the degree of locational specialisation that they have managed to achieve in technological development, rather than the existence of such a strategy of specialisation across centres.

The globalisation of technological innovation in MNCs, in the sense here of an international integration of geographically dispersed and locally specialised activities, tends to reinforce and not to dismantle nationally distinctive patterns of development or national systems of innovation (Nelson, 1993). Contrary to what is sometimes alleged, globalisation and national specialisation are complementary parts of a common process, and not conflicting trends (see Archibugi and Michie, 1993, for a further discussion). The incentive to organise affiliate specialisation is the desire to tap into the locally specific and differentiated stream of innovation in each centre, but by specialising in accordance with these local strengths the latter are reinforced. The creation of tacit capability is localised and embedded in social organisations (Nelson and Winter, 1982), and this organisational distinctiveness has a location-specific as well as a firm-specific dimension. The particular path of innovation followed in each country or region has historical origins (Rosenberg, 1976, 1982). In the period of globalisation since the late 1960s the general tendency has been for MNCs to become more technologically diversified as they establish newly-integrated technological systems, while countries or locations have become more specialised in their technological activity (Cantwell, 1993).

5. Conclusions

One aspect of the product cycle model is now rather discredited, namely the idea that demand-led innovation (together with economies of scale in R&D) in the home country dictates the geographical restriction of corporate research and the most technologically sophisticated production to the site of the parent company. For one thing, the demand-led view of innovation that was prevalent in the 1960s is now more widely acknowledged to be one-sided and potentially misleading (Mowery and Rosenberg, 1979). Innovations generally rely on a firm-specific learning process that interacts with both the growth of demand and the creation of new scientific and technological knowledge. In a region or country that enjoys technological leadership, high incomes and demand are as much a consequence of that leadership (high technological capability and thus high productivity) as they are a cause of it. For another thing, the peculiarities of foreign production conditions and demand have required leading MNCs historically to develop innovations abroad, related to those that had been pioneered at home.

For this latter reason, another aspect of the model can be extended to provide a further historical application that remains relevant. That is, the product cycle view that outward industrial investment is most widely associated with technology leaders, generally in conjunction with their holding a strong export position; so, too, the earliest internationalisation of technological development was largely due to such leaders. Historically, some highly competent US electrical equipment firms were considerably more international in their technology creation then than they are now. In Europe large chemical firms were relatively technologically stronger, and it was they that led the historical internationalisation of corporate research facilities.

In more recent times technology leaders have altered the nature of international technology creation by pioneering the international integration of MNC facilities into regional or global networks. Globalisation in this sense involves the establishment of new international structures for technology creation. In the past, foreign technological activity exploited domestic strengths abroad, it was located in response to local demand conditions, it assisted in the growth of other high income areas, and its role ranged from the adaptation of products to suit local tastes through to the establishment of new local industries. At that time the capacity to develop internationally dispersed innovations derived from a position of technological strength in the firm's home country base, and led to similar lines of technological development being established abroad. By contrast, today, for companies of the leading centres, foreign technological activity now increasingly aims to tap into local fields of expertise, and to provide a further source of new technology that can be utilised internationally in the other operations of the MNC. In this respect, innovation in the leading MNCs is now more genuinely international or, in the terminology used here, it has become 'globalised'.

There are two similarities between the theoretical rationale provided for the product cycle model (which applied best to the US and to US-owned MNCs in the early post-war period) and that suggested for the current globalisation of technological activity in MNCs. Both explanations rely on the role of the economies of locational agglomeration, and on the leadership exercised by the most technologically competent firms. The essential difference is that in the product cycle model just one pre-eminent centre for innovation was recognised, whereas in the globalisation story there are multiple locations for innovation, and even lower-order or less developed centres can still be sources of innovation. Hence, the theoretical concepts also used in the product cycle approach apply differently in the account of globalisation; locational agglomeration occurs in the clusters of distinctive innovations that occur in many centres and not only in one unique centre, while the greater capability of the most competent firms manifests itself not just in the wider geographical dispersion of their investments (a more important consideration historically), but in the broader degree of cross-border specialisation that they are able to manage.

The product cycle view of the MNC as a locational hierarchy also remains relevant, although this too needs extending. It is still true that the home country is generally the single most important site for corporate technological development. More interestingly, the form of locational hierarchy in the leading firms of the most advanced centres is now much more complex than it used to be, and more complex than is the equivalent hierarchy of other MNCs. The affiliates of the leading companies in other major centres may be thought of as constituting an interactive network. Cross-investments between the major centres in the most technologically dynamic industries (Dunning, 1988; Cantwell, 1989; Cantwell and Sanna-Randaccio, 1992) have probably helped to reinforce the existing pattern of geographical specialisation, and the importance of these centres as locations for innovation. Having been the first to establish an international spread of technological activity, MNCs from the leading centres in a given industry now exploit locational diversity in paths of innovation to a greater extent than do other firms.

Bibliography

Archibugi, D. and Michie, J. 1993. 'The Globalisation of Technology: Myths and Realities', University of Cambridge Research Papers in Management Studies, No. 18, May

Bartlett, C. A. and Ghoshal, S. 1989. *Managing Across Borders: The Transnational Solution,* Boston, MA, Harvard Business School Press

Beaton, K. 1957. *Enterprise in Oil: A History of Shell in the United States,* New York, Appleton-Century-Crofts

Cantwell, J. A. 1987. The reorganisation of European industries after integration, *Journal of Common Market Studies,* vol. 26, 127–151

Cantwell, J. A. 1989. *Technological Innovation and Multinational Corporations*, Oxford, Basil Blackwell

Cantwell, J. A. 1991A. The international agglomeration of R&D, in Casson, M. C. (ed.), *Global Research Strategy and International Competitiveness*, Oxford, Basil Blackwell

Cantwell, J. A. 1991B. The theory of technological competence and its application to international production, in McFetridge, D. G. (ed.), *Foreign Investment, Technology and Economic Growth*, Calgary, University of Calgary Press

Cantwell, J. A. 1993. Corporate technological specialisation in international industries, in Casson, M. C. and Creedy, J. (eds), *Industrial Concentration and Economic Inequality*, Aldershot, Edward Elgar

Cantwell, J. A. 1994. Introduction, in Cantwell, J. A. (ed.), *Transnational Corporations and Innovatory Activities*, London, Routledge

Cantwell, J. A. and Barrera, M. P. 1993. 'The Rise of Corporate R&D and the Technological Performance of the Largest European Firms from the Interwar Years Onwards', University of Reading Discussion Papers in Economics, no. 271, September

Cantwell, J. A. and Barrera, M. P. 1995. Inter-company agreements for technological development: lessons from international cartels, *International Studies in Management and Organisation*, forthcoming

Cantwell, J. A. and Hodson, C. 1991. Global R&D and UK competitiveness, in Casson, M. C. (ed.), *Global Research Strategy and International Competitiveness*, Oxford, Basil Blackwell

Cantwell, J. A. and Kotecha, U. 1994. L'internationalisation des activités technologiques: le cas français en pespective, in Sachwald, F. (ed.), *Les Défis de la Mondialisation: Innovation et Concurrence*, Paris, Masson

Cantwell, J. A. and Sanna-Randaccio, F. 1992. Intra-industry direct investment in the European Community: oligopolistic rivalry and technological competition, in Cantwell, J. A. (ed.), *Multinational Investment in Modern Europe*, Aldershot, Edward Elgar

Cantwell, J. A. and Sanna-Randaccio, F. 1993. Multinationality and firm growth, *Weltwirtschaftliches Archiv*, vol. 129, 275–299

Chandler, A. D. 1990. *Scale and Scope: The Dynamics of Industrial Capitalism*, Cambridge, MA, Harvard University Press

Coleman, D. C. 1969. *Courtaulds: An Economic and Social History (Volume II: Rayon)*, Oxford, Oxford University Press

Dahmén, E. 1970. *Entrepreneurial Activity and the Development of Swedish Industry, 1919–1939*, Homewood, American Economic Association Translation Series

Dosi, G., Teece, D. J. and Winter, S. G. 1992. Towards a theory of corporate coherence: preliminary remarks, in Dosi, G., Giannetti, R. and Toninelli, P. A. (eds), *Technology and Enterprise in a Historical Perspective*, Oxford, Oxford University Press

Doz, Y. 1986. *Strategic Management in Multinational Companies*, Oxford, Pergamon

Dunning, J. H. 1988. *Multinationals, Technology and Competitiveness*, London, Unwin Hyman

Dunning, J. H. 1992. *Multinational Enterprises and the Global Economy*, Wokingham, Addison-Wesley

Feldman, M. P. 1993. An examination of the geography of innovation, *Industrial and Corporate Change*, vol. 2, 451–470

Granstrand, O., Håkanson, L. and Sjölander, S. (eds) 1992. *Technology Management and International Business: Internationalisation of R&D and Technology*, Chichester, John Wiley

Hedlund, G. 1986. The hypermodern MNC: a heterarchy?, *Human Resource Management*, vol. 25, 9–25

Hounshell, D. A. and Smith, J. K. 1988. *Science and Corporate Strategy: Du Pont R&D, 1902–1980*, Cambridge, Cambridge University Press

Hymer, S. 1975. The multinational corporation and the law of uneven development, in Radice, H. (ed.), *International Firms and Modern Imperialism*, Harmondsworth, Penguin

Jaffe, A. B. 1989. Real effects of academic research, *American Economic Review*, vol. 79, 957–970

Jaffe, A. B., Trajtenberg, M. and Henderson, R. 1993. Geographical localization of knowledge spillovers as evidenced by patent citations, *Quarterly Journal of Economics*, vol. 108, 577–598

Jones, R. and Marriot, O. 1971. *Anatomy of a Merger: A History of GEC, AEI and English Electric*, London, Cape Ltd

Kogut, B. 1987. Country patterns in international competition: appropriability and oligopolistic agreement, in Hood, N. and Vahlne, J. E. (eds), *Strategies in Global Competition*, London, Croom Helm

Kogut, B. 1990. The permeability of borders and the speed of learning among countries, in Dunning, J. H., Kogut, B. and Blomström, M. (eds), *Globalisation of Firms and the Competitiveness of Nations*, Lund, Lund University Press
Krugman, P. R. 1991. *Geography and Trade*, Cambridge, MA, MIT Press
Lall, S. 1979. The international allocation of research activity by US multinationals, *Oxford Bulletin of Economics and Statistics*, vol. 41, 313–331
Mansfield, E., Teece, D. J. and Romeo, A. 1979. Overseas research and development by US based firms, *Economica*, vol. 46, 187–196
Mowery, D. C. and Rosenberg, N. 1979. The influence of market demand upon innovation: a critical review of some recent empirical studies, *Research Policy*, vol. 8, 103–153
Nelson, R. R. 1991. Why do firms differ and how does it matter?, *Strategic Management Journal*, vol. 12, 61–74
Nelson, R. R. 1992. The role of firms in technical advance, in Dosi, G., Giannetti, R. and Toninelli, P. A. (eds), *Technology and Enterprise in a Historical Perspective*, Oxford, Oxford University Press
Nelson, R. R. (ed.) 1993. *National Innovation Systems*, Oxford, Oxford University Press
Nelson, R. R. and Winter, S. G. 1982. *An Evolutionary Theory of Economic Change*, Cambridge, MA, Harvard University Press
Patel, P. and Pavitt, K. L. R. 1991. Large firms in the production of the world's technology: an important case of 'non-globalisation', *Journal of International Business Studies*, vol. 22, 1–21
Pavitt, K. L. R. 1992. Some foundations for a theory of the large innovating firm, in Dosi, G., Giannetti, R. and Toninelli, P. A. (eds), *Technology and Enterprise in a Historical Perspective*, Oxford, Oxford University Press
Pearce, R. D. and Singh, S. 1991 *Globalising Research and Development*, London, Macmillan
Penrose, E. T. 1959. *The Theory of the Growth of the Firm*, Oxford, Basil Blackwell
Porter, M. E. 1986. Competition in global industries: a conceptual framework, in Porter, M. E. (ed.), *Competition in Global Industries*, Boston, Harvard Business School Press
Porter, M. E. 1990. *The Competitive Advantage of Nations*, New York, The Free Press
Reich, L. S. 1986. *The Making of American Industrial Research: Science and Business at GE and Bell, 1876–1926*, Cambridge, Cambridge University Press
Ronstadt, R. C. 1977. *Research and Development Abroad by US Multinationals*, New York, Praeger
Rosenberg, N. 1976. *Perspectives on Technology*, Cambridge, Cambridge University Press
Rosenberg, N. 1982. *Inside the Black Box: Technology and Economics*, Cambridge, Cambridge University Press
Schmookler, J. 1966. *Invention and Economic Growth*, Cambridge, MA, Harvard University Press
Soete, L. L. G. 1987. The impact of technological innovation on international trade patterns: the evidence reconsidered, *Research Policy*, vol. 16, 101–130
Stocking, G. W. and Watkins, M. W. 1946. *Cartels in Action*, New York, Twentieth Century Fund
Sölvell, Ö., Zander, I. and Porter, M. E. 1991. *Advantage Sweden*, Stockholm, Norstedts
Teece, D. J., Pisano, G. and Shuen, A. 1990. 'Firm Capabilities, Resources and the Concept of Strategy', University of California at Berkeley Consortium on Competitiveness and Cooperation Working Papers, No. 90-8
Vernon, R. 1966. International investment and international trade in the product cycle, *Quarterly Journal of Economics*, vol. 80, 190–207
Vernon, R. 1979. The product cycle hypothesis in a new international environment, *Oxford Bulletin of Economics and Statistics*, vol. 41, 255–267
Wilkins, M. 1974. *The Maturing of Multinational Enterprise*, Cambridge, MA, Harvard University Press
Wilkins, M. 1989. *The History of Foreign Investment in the United States to 1914*, Cambridge, MA, Harvard University Press
Zander, I. 1994. *The Tortoise Evolution of the Multinational Corporation*, Stockholm, Institute of International Business, Stockholm School of Economics

[8]

mir vol. 37, 1997/1, pp. 5–25

mir
Management
International Review

Marina Papanastassiou/Robert Pearce

Technology Sourcing and the Strategic Roles of Manufacturing Subsidiaries in the U.K.: Local Competences and Global Competitiveness

Abstract

- As global competitiveness intensifies MNEs need to get new products into all major markets quickly and in ways that respond distinctively to the needs of these markets.
- In allowing subsidiaries to implement this responsiveness an additional advantage is that these enhanced roles permit them to extend their use of local technological expertise and widen the knowledge scope of the group.

Key Results

- A widening of their sources of technology inputs underwrites creative roles in subsidiaries, but this usually remains within the established technological trajectory of the group.

Authors

Marina Papanastassiou, Research Fellow Department of Economics, University of Reading, Reading, U.K.
Robert Pearce, Reader in International Business, Department of Economics, University of Reading, Reading, U.K.

Manuscript received December 1994, revised November 1995, revised June 1996.

Marina Papanastassiou/Robert Pearce

Introduction

The new competitive conditions of the global economy demand that MNEs adopt a genuinely decentralised perspective on not only the use of their technological and managerial competences (the way they are applied), but also on the way they are generated. This development involves a thorough redefinition of the roles of overseas subsidiaries in the process of modernising the competitive organisation of the MNE.

Whilst the demand situation facing subsidiaries has become more intense and distinctive, their supply situation has also developed increasingly original conditions which allow them the opportunity to acquire competences within their MNE group. Rising incomes in many host countries have allowed for the rise of more discriminating, and less submissive, consumer tastes, manifested through an increasing demand for national or regional products. The strengthening of indigenous industry in many host countries, alongside the emergence of new MNEs (from Japan and similar countries) made the industrial environment vastly more competitive. This increased diversification of tastes and emergence of new forces of competition was reinforced by the generation of managerial and other skills, and in particular of technological capacities and specialised competences, in an increasing range of countries, which therefore became available to MNEs' subsidiaries in these countries. Thus creative subsidiaries are increasingly allocated extensive responsibilities to develop their own original capacities in order to supply effectively their own original and distinctive markets (which are now likely to extend beyond their host country). [1]

Central to the emergence of such creative subsidiaries is the development of their own technological capabilities within the portfolio of technologies available to the group. In achieving this the subsidiary will build up a certain degree of technological capacity within its own operations (i. e. a R & D laboratory or an engineering unit) but may also establish collaborative arrangements with other elements in the host-country science and technology base. In this way the technological profile of a creative subsidiary will reflect the distinctive specialised technological capacities of the country in which it is located. This, in turn, represents a realistic view of the need for a MNE to take advantage of all the technological opportunities available to it throughout its global operations, in order to generate and retain competitive advantages.

These perspectives on contemporary MNE operations are investigated in this paper through an analysis of replies to a questionnaire survey sent to manufacturing subsidiaries of foreign companies operating in the U.K. [2] In total the questionnaire was sent to 812 such subsidiaries with satisfactory replies being received from 190 of them. [3] In the next section evidence is provided on the relative prevalence of four different types of subsidiary role. The third section evaluates the importance in these subsidiaries' operations of seven different sources of technology, and relates their relative position to the roles played by the subsidiaries.

Roles of Subsidiaries

Respondents to the questionnaire were asked to evaluate the importance of four possible roles in their operations as being either 'our only role', 'our predominant role', 'a secondary role', or 'not a part of our role'. Overall 39 indicated that they focused on a unique role, whilst 216 evaluations of particular roles as being predominant ones were provided. The latter number reflects the reluctance of some respondents to separate two roles in terms of the lead (predominant) position in the subsidiary's overall operations. The subsidiaries of European MNEs emerged as rather more inclined to possess distinct lead roles, and/or to be able to delineate clearly the hierarchy of roles in their operations. Thus whilst the proportion of European subsidiaries that discerned a particular role as being their only one was very close to the average for all respondents, an above average proportion reported certain roles as absent from their operations and a below average proportion considered roles to be their predominant one.[4]

The first role investigated was defined as 'to produce for the U.K. market products that are already established in our MNE group's product range'. This type of import-substituting role (supplying to the host-country market goods that are already established elsewhere in the MNE's operations, so that they could otherwise have been supplied through trade) is traditionally perceived as one that emerged in eras that were characterised by high levels of protectionism, which artificially fragmented markets. With the lowering of protectionism through GATT rounds and the emergency of free-trade areas, and the general intensification of international competition, this role is often considered to be in decline,[5] to be replaced by subsidiaries that are more oriented to specialised positions in the supply of the wider markets that are now accessible. Within the context of the more contemporary approaches to global competition, however, local-market-focused subsidiaries could still persist where a strong motivation to adapt established products to meet distinctive local-market needs and characteristics exists.

Of the 185 subsidiaries that evaluated their operations in terms of this first role 15 (8.1%) considered it to be their only one, 69 (37.3%) rated it a predominant role, 50 (27.0%) felt it only took a secondary position and 51 (27.6%) did not include it. Once these replies are summarised in terms of average responses (ARs) Table 1 shows that this is still, very marginally, the most pervasive of the four roles. It is also marginally the relatively most prevalent role amongst European MNEs' subsidiaries in the U.K.[6], and takes the equal leading role for the

Table 1. Relative Importance of Roles Played by MNE Subsidiaries in the U.K., by Industry and Home Country

	Importance of roles* (average response**)			
	A	B	C	D
By industry				
Food	2.56	2.33	1.11	2.67
Automobiles	2.24	2.33	1.65	2.35
Aerospace	2.33	2.17	1.33	2.17
Electronics and electrical appliances	2.33	2.53	1.41	1.96
Mechanical engineering	2.08	2.27	1.33	2.36
Instruments	2.33	2.10	1.33	2.36
Industrial and agricultural chemicals	2.14	2.03	1.48	2.17
Pharmaceuticals and consumer chemicals	2.27	2.55	1.45	1.91
Metal manufacture and products	2.09	1.90	1.22	2.20
Other manufacturing	2.44	1.73	1.20	1.80
Total	2.26	2.24	1.38	2.15
By home country				
U.S.A.	2.28	2.24	1.52	2.27
Japan	2.31	2.41	1.32	2.03
Europe	2.13	2.07	1.29	2.09
Total***	2.26	2.24	1.38	2.15

Role of subsidiaries:

A – to produce for the U.K. market products that are already established in our MNE group's product range.

B – to play a role in the MNE group's European supply network by specialising in the production and export of part of the established product range.

C – to play a role in the MNE group's European supply network by producing and exporting component parts for assembly elsewhere.

D – to develop, produce and market for the U.K. and/or European (or wider) markets, new products additional to the MNE group's existing range.

* Respondents were asked to evaluate each role as (i) our only role, (ii) our predominant role (iii), a secondary role, (iv) not a part of our role.

** The average response is calculated by allocating a value of 4 to 'our only role', 3 to 'our predominant role', 2 to 'a secondary role' and 1 to 'not a part of our role'.

*** Includes subsidiaries of MNEs from Australia and Canada.

U.S. Food is the industry in which this role is most prevalent in subsidiaries in the U.K. It is least prominent in two industries, mechanical engineering and metal manufacture and products, where the frequent need to customise items for individual consumers mitigates against routine supply of established standardised (or even slightly adapted) products.

The second type of subsidiary position investigated was 'to play a role in the MNE group's European supply network by specialising in the production and export of part of the established product range'. This can be perceived as a contemporary evolution of the first role in terms of one of its two key characteristics, i. e.

its target market. Thus this second type of role reflects the ease of trade within the EU, in particular, so that MNEs can allow individual subsidiaries to focus on the cost-effective production of a limited part of the relevant product range and supply this to markets throughout the region. In terms of the other defining characteristic, however, there is no substantive evolution, in that the subsidiary still produces goods that are already in position as part of its MNE's established product range. Thus the new role may enhance the effectiveness of the subsidiary's contribution to group operations by increasing the efficiency of its use of standard production inputs (by focusing on a narrower range of products and achieving economies of scale that were not available when it was limited essentially to its host-country market), but it still does not provide scope for the use of the creative managerial, technological and other talents that might be accessible to it.

Again 185 subsidiaries evaluated their operations in terms of this role, with 6 (3.2%) rating it their only one, 86 (46.5%) their predominant one, 40 (21.6%) a secondary one and 53 (28.6%) believing it to be absent. As Table 1 shows, when these results were converted into an average response this role virtually matched the first as the most prevalent in the current operations of MNE subsidiaries in the U.K. This type of role appears to have emerged most decisively in the operations of Japanese subsidiaries.

Operations of this second type are found most extensively in two industries, pharmaceuticals and electronics, where it might have been expected to encounter some constraints. In the case of pharmaceuticals the frequently observed importance of host-country regulations, and also distinctive elements of consumer taste, might have been expected to fragment the market in such a manner that the need to adapt products to meet particular national requirements would severely compromise the ability of one supply location to meet the wider European demand for a product. In fact the U.K. subsidiaries of European pharmaceutical MNEs (an AR of 2.80) and also those from the U.S. (AR of 2.40) defy this proposition. A key element in this may be that the reasons for product adaptation may be easy to both discern and communicate in the pharmaceutical industry. Thus the specific requirements of government regulations should be set out in a clearly documented form which can be easily communicated, as perhaps can the relevant aspects of consumer tastes (e. g. for drugs in tablet, capsule or liquid form). In this way a subsidiary producing a particular product can differentiate it for a range of separate national markets without needing the close association with consumers in each of them that is often considered necessary in other industries.

In electronics it might have been anticipated that unique European standards (e. g. TV transmission systems and electricity supply characteristics) might have meant that U.S. and Japanese companies would need to develop unique new versions of their products for the European market, with these taking on distinctive characteristics that go beyond the mere adaptation of those that originated in the home country. In fact the Japanese subsidiaries to a very notable degree (AR of

2.77 for this role), and also the U.S. to a quite significant extent (AR of 2.33), were able to use their U.K. operations to supply an established part of the MNEs' product range to the wider European market. By contrast European electronics MNEs' subsidiaries in the U.K., were very strongly oriented to the first, local-market, role (AR of 2.71), with little commitment to the second (AR of 1.86). This may mean that European electronics companies are mainly interested in niche markets of limited concern to U.S. and Japanese enterprises, and their U.K. subsidiaries seek to adapt such products to particular local requirements.

As a complement to the second role the next was 'to play a role in the MNE group's European supply network by producing and exporting component parts for assembly elsewhere'. Once again the assumption is that the emergence of such roles in a European network of MNE subsidiaries would seek to optimise the more static dimensions of efficiency by achieving economies of scale or by producing particular components in locations that are especially favourable in terms of input availabilities or costs. This role turned out to be by far the least prevalent of the four, with 127 (70.2%) of the 181 subsidiaries who evaluated it considering it was not a part of their operations, only 11 (6.1%) rating it a predominant role and 2 (1.1%) focusing on it entirely. This may mean either that MNEs generally do not make extensive use of such decentralised component-supply networks in their European operations, or that the U.K. is not often considered a particularly relevant site for that type of activity. The role does seem to be rather more developed in U.S. subsidiaries in the U.K. than in those of Japanese or European MNEs. It is thus U.S. subsidiaries that contribute particularly strongly to those industries that have a relatively strong orientation to this role, with ARs of 1.88 in automobiles, 1.75 in chemicals and 1.60 in pharmaceuticals.

The final role analysed was defined as being 'to develop, produce and market for the U.K. and/or European (or wider) markets, new products additional to the MNE group's existing range'. The key evolutionary aspect of this role is therefore the widening of the creative scope of the subsidiary by allowing it to move beyond the range of established products and derive its own distinctive contribution to the group's capabilities. As in the second and third roles the subsidiary's position is again predicated on specialization, but here this is secured not so much from a static advantage in the use of an existing technology (embodied in the established goods produced) but more from a dynamic ability to move the group's technology itself forward in the direction of new applications in distinctive additional products. When they escape from situations which mainly involve the effective implementation of allocated roles in existing supply networks and instead are allowed to win higher-value-added positions based around their own capabilities, subsidiaries immediately enhance the value of their own human capital by enabling management, engineering, scientific and other personnel to more fully realise their potential. The alleviation of the sense of frustration that is likely to beset such personnel when their scope is limited to the dependent implementation of externally-determined group-level production programmes is another likely benefit of their accession to a more autonomous and creative situation. The possibility of a country being host to subsidiaries playing this role is then less likely to be decisively influenced by the availability and cost of production inputs, with the scope and quality of creative assets (e. g. managerial and marketing personnel, the wider scientific community) emerging as more relevant determinants. With its emphasis on product development the definition of this fourth role is agnostic as to the market focus of the operations. However, it is very likely that a subsidiary that played the role successfully would be expected to create products that could secure a share of wider markets (here at least the European) than that of their host country.

Of the 184 responding subsidiaries that evaluated the relevance of this role in their operations 16 (8.7%) said it was their only one, 50 (27.2%) their predominant role, 63 (34.2%) that it took a secondary position and 55 (29.9%) that it was absent. When these figures are converted into an average response (Table 1) it is clear that these more contemporary creative type of operations have now taken their place among the key roles played by MNE subsidiaries in the U.K. Indeed only for Japanese subsidiaries does the presence of this role not yet match that of the first two. Thus the Japanese MNEs' U.K. operations encompass very strongly one evolutinary development, a notable orientation to supply the wider European market (role two), but allocate much less prominance to the other in the form of creative responsibility at the subsidiary level (role four). Both of these differences may reflect the relative newness of the Japanese facilities.[7] Thus the majority of these subsidiaries have been set up at a time when the need for a European perspective on supply programmes is fully understood in many industries. On the other hand the viability and benefits of the incorporation of creative scope within subsidiaries may only be understood after the success of the more routine roles has been secured. The extension of operations to the fourth role may still be an evolutionary one, only perceived as viable when the less ambitious roles have allowed subsidiaries to fully settle in an environment and comprehend its needs and potentials. On this basis it may well be the case that this role may soon match the others in Japanese subsidiaries, as indeed it may move into clearer predominance in U.S. or European operations.

Food industry subsidiaries are the most strongly committed to this fourth role, with an especially extensive involvement of U.S. subsidiaries (an AR of 2.80). This parallels their strong evolution of the first, U.K.-market focused, role (AR of 2.80) so it may be that U.S. subsidiaries initially needed to adapt their established food products quite extensively to meet U.K. tastes and conditions and in doing so created types of expertise within the subsidiaries that encouraged the complementary implementation of more extensive product development. The strong product development role in mechanical engineering, which reflects the pervasive need to supply customised capital goods to meet individual firms' specifications, is most notable in U.S. subsidiaries (AR of 2.71) but also quite prevalent in Japanese and European (AR of 2.22 in both cases). The result for instruments mainly represents the behaviour of U.S. subsidiaries (AR of 2.43) and probably reflects the need of these firms to apply their existing technology to European conditions more competitively by developing products that embody it in ways that meet distinctive requirements of particular scientists, health services, firms, etc. The strength of the product development role in automobiles reflects the behaviour of U.S. (AR of 2.50) and Japanese (AR of 2.43) subsidiaries in the U.K., rather than those from European MNEs (AR of 1.50). Thus in the latter case the U.K. mainly provides a supply base for cars and components that have already been designed to fully incorporate European needs, but the U.S. and Japanese subsidiaries in the U.K. have a much more extensive responsibility to contribute to the derivation of specifically European parts of their MNE's product range.

Pharmaceuticals and electronics are the two industries in which the product development role is least extensive in subsidiaries in the U.K. Taken alongside its prevalence in food and mechanical engineering it is clear that participation in a technologically-dynamic industry is neither necessary nor sufficient to generate product development activity in subsidiaries. Implementation of such operations is clearly more likely to reflect a wide range of subsidiary and host-country endowments and motivations, rather than predominantly respond to the degree of technological opportunity endemic in the industry. Indeed in industries such as pharmaceuticals and electronics, where leading contemporary technological paradigms (biotechnology; micro-electronics and information technology) are very influential, the desire of leading MNEs to maximise the benefits from participation in such scientific areas may lead to greater centralised control over the firm's own technological trajectory with less room for relatively autonomous dispersed creative initiatives. In pharmaceuticals it therefore seems likely that most product development is centralised (or at least carried out in labs with no direct association with a particular production unit) and can be relatively easily and effectively communicated to specialised production subsidiaries. These in turn can, as noted earlier, supply the product to a range of markets using their own adaptation capabilities of differentiate it in ways needed to respond to any specific requirements of individual national markets in their target region. In electronics too it may be that the factors that might provoke the need for U.S. or Japanese companies to develop specific European product variants can also take a clearly documented or codified form (regulations or technical specifications), rather than being the type of diffused and evolving taste-related characteristics that need subsidiary-level marketing expertise to detect, and in-house technological capacity to respond to adequately and sensitively. If the relevant specifications or characteristics are defined relatively precisely then development of a European product can again be carried out away from the specialised production facility, either in the home country or in a dedicated R & D unit serving all European operations.

Sources of Technology in the Subsidiaries

Traditionally it has been accepted that the role played by overseas subsidiaries in MNEs has been determined by factors that are predominantly related to market orientation and the availability and cost of physical inputs into the production process. Once such influences have determined the role that is to be allocated to a subsidiary within the MNE's overall operations, its basic technological needs will follow and can then be supplied from within the group's established knowledge capacity. In these circumstances the upper limit to the individual technological capacities required within a subsidiary would be the ability to assimilate the relevant knowledge effectively, and to discern the need for, and to secure the effective implementation of, any product or process adaptation necessary to optimise the standard technology's applicability to its particular market and operating conditions. Now the ability of individual subsidiaries to develop their own creative competences, often reflecting the wider scientific and knowledge (educational) background of their host country, may to some extent be reversing the direction of causation.

The emergence in subsidiaries of creative competences (managerial, technological, marketing) that provide a scope that extends beyond that needed in the more traditional roles underwrites their pursuit of, and accession to, the higher-value-added product development activities that increase both their own autonomy and the wider competitiveness of their group. Ultimately the skills and technology available to the subsidiary will determine its role, rather than an externally allocated role determining its needs with regards to such assets. Thus subsidiaries go through a creative transition [8] in which distinctive attributes increasingly differentiate their position in the group, whose overall competitive scope is thereby extended. The ability of a subsidiary to achieve an effective creative transition both depends upon, and enhances the return to, a host country's knowledge (scientific and educational) background.

Against this background we proceed by analysing the relative importance of seven possible sources of technology that could be used in MNEs' overseas subsidiaries, using information derived from the survey of such operations in the U.K. In Table 2 the replies are summarised with average responses in order to show the variation in the use of the sources according to industry and the home country of the MNE. Table 3 extends the analysis by looking at the manner in which use of particular technology sources varies according to the role of the subsidiary. Thus for each source of technology an AR is calculated for subsidiaries that reported a particular role as either their only one or their predominant one.

The list of potential technology sources begins with two that reflect the fact that, even when they are able to create strong elements of unique creative competence within their own operations, subsidiaries are likely to remain to a sub-

Table 2. Relative Importance of Sources of Technology in MNE Subsidiaries in the U.K., by Industry and Home Country.

	Importance of sources of technology* (average response**)						
	A	B	C	D	E	F	G
By industry							
Food	2.89	2.22	2.50	1.78	1.56	1.67	2.00
Automobiles	2.72	2.18	1.83	2.65	1.71	1.39	2.00
Aerospace	2.60	2.50	1.50	1.83	1.17	1.33	2.17
Electronics & electrical appliances	2.79	2.35	1.98	2.33	1.43	1.32	2.13
Mechanical engineering	2.85	2.46	1.85	2.00	1.42	1.62	2.19
Instruments	2.82	2.00	2.27	2.18	1.45	1.73	2.50
Industrial & agricultural chemicals	3.24	2.21	2.31	2.31	1.62	1.52	1.79
Pharmaceuticals & consumer chemicals	3.00	2.40	2.30	2.40	1.70	1.60	1.70
Metal manufacture & products	2.91	2.09	1.91	1.82	1.27	1.36	2.09
Other manufacturing	3.13	2.47	1.71	2.14	1.29	1.50	1.64
Total	2.86	2.30	2.02	2.22	1.48	1.48	2.02
By home country							
U.S.A.	2.78	2.25	2.15	2.18	1.57	1.62	2.06
Japan	2.88	2.35	1.78	2.23	1.31	1.27	2.02
Europe	2.91	2.30	2.12	2.25	1.57	1.55	1.88
Total***	2.86	2.30	2.02	2.22	1.48	1.48	2.02

Sources of technology:
A – existing technology embodied in established products we produce.
B – technology of our MNE group from which we introduce new products for the U.K./European market, that differ from other variants introduced in other markets.
C – R & D carried out by our own laboratory.
D – R & D carried out for us by another R & D laboratory of our MNE group.
E – R & D carried out in collaboration with another firm.
F – R & D carried out for us by local scientific institutions (e.g. Universities; independent labs; industry labs).
G – development and adaptation carried out less formally by members of our engineering unit and production personnel.
* Respondents were asked to grade each source of technology for their operation as (i) our only source, (ii) a major source, (iii) a secondary source, (iv) not a source.
** The average response is calculated by allocating a value of 4 to 'our only source', 3 to 'a major source', 2 to 'a secondary source', 1 to 'not a source'.
*** Includes subsidiaries of MNEs from Australia and Canada.

stantial degree within a pervasive technological scope that defines the MNE group of which they are a part. Thus respondents were firstly asked to grade as either 'our only source of technology', or 'a major source', 'a secondary source', 'not a source', 'existing technology embodied in established products we produce'. This

Table 3. Relative Importance of Sources of Technology in MNE Subsidiaries in the U.K., by Role of Subsidiary.

	Importance of sources of technology* (average response**)						
	A	B	C	D	E	F	G
Roles of subsidiaries							
To produce for the U.K. market products that are already established in our MNE group's product range.	2.88	2.27	1.85	2.29	1.41	1.39	2.01
To play a role in the MNE group's European supply network by specialising in the production and export of part of the established product range.	2.92	2.36	2.08	2.27	1.52	1.41	2.11
To play a role in the MNE group's European supply network by producing and exporting component parts for assembly elsewhere.	2.83	2.62	2.17	2.08	1.67	1.58	2.17
To develop, produce and market for the U.K. and/or European (or wider) markets, new products additional to the MNE group's existing range.	2.86	2.33	2.52	2.05	1.55	1.62	2.13

Sources of technology:

A – existing technology embodied in established products we produce.
B – technology of our MNE group from which we introduce new products for the U.K./European market, that differ from other variants introduced in other markets.
C – R & D carried out by our own laboratory.
D – R & D carried out for us by another R & D laboratory of our MNE group.
E – R & D carried out in collaboration with another firm.
F – R & D carried out for us by local scientific institutions (e. g. Universities; independent labs; industry labs).
G – development and adaptation carried out less formally by members of our engineering unit and production personnel.
* Respondents were asked to grade each source of technology for their operation as (i) our only source, (ii) a major source, (iii) a secondary source, (iv) not a source.
** The average response is calculated, for respondents that said that a particular role was either their only one or their predominant one, by allocating a value of 4 to 'our only source', 3 to 'a major source', 2 to 'a secondary source', 1 to 'not a source'.

then relates to the mature mainstream technology of the group, as already successfull embodied in the products whose market success has achieved levels that involve decentralised production in overseas subsidiaries. This emerges as being clearly perceived as the most significant operative source of their technology by the responding subsidiaries. Indeed its pervasiveness is such that Table 2 shows it to be the leading source in subsidiaries in every industry covered and from each of the home countries.

Such established embodied technology is shown in Table 3 to be the major source used in the three subsidiary roles that are either defined around existing products (the first two discussed in the previous section) or are likely to be locked into current supply networks (the third). Though this would have been a clearly predicted outcome, the equally strong prevalence of this source of technology in the operations of the fourth (product development) type of subsidiary is much more surprising. Of some influence on this result is the fact that a number of subsidiaries that rated the fourth role as a predominant one also provided the same rating to one of the others, and included such embodied technology as a major source because of this. This factor, however, is very unlikely to explain more than a small part of the position of this technology in the product development subsidiaries. An explanation that is inherent to the mode of operation of these subsidiaries can be derived from the perception of the evolutionary theory of technological change and the importance of a tacit element of knowledge.[9] Thus the skills and the technological knowledge that underpin the distinctive competences that earn product development subsidiaries their new position are likely to have emerged from, and will therefore have been conditioned by, what they have been doing in their previous dominant role. Though the broad motivation and capability of key personnel in subsidiaries that seek to transcend previous dependent roles clearly reflect their own talent and education, its detailed applicability to the nature of evolving operations will still encompass a strong element that derives from training in the MNE and from the subsidiaries earlier background. Similarly the degree of technological autonomy achieved in such product development activity is likely to reflect a strong evolutionary addition of local knowledge and expertise to the group's mainstream technology, rather than a revolutionary break from it. Established products previously (and perhaps still) produced by the subsidiaries are likely to be a crucial element through which they perceive the group technology which will remain a component in their more idiosyncratic evolution.

The second source of technology available to subsidiaries that reflects a MNE's current predominant technological trajectory was defined as 'technology of our MNE group from which we introduce new products for the U.K./European market, that differ from other variants introduced in other markets'. The key difference from the previous type of technology is that this is accessed by the subsidiary in a disembodied form, from which it can itself create the full detail of a product competitive in its own markets. This mode of behaviour suggests first that, at the group level, the technology underpinning a major new innovation emerges. Then, however, rather allowing the innovation process to be completed centrally, in the manner of the original product-cycle model,[10] and subsequently gradually emerge into foreign markets and overseas production, it is likely that the pressures of contemporary competition would dictate that the new good be immediately produced in major markets in forms that fully respond to local needs there.[11] Subsidiaries in such markets therefore pick up the technology that defines the outlines of the new product and use their own distinctive local competences (marketing and technological, etc.) to complete its competitive development. This type of scenario would clearly locate the use of the decentralised creative and scientific competences in the newer type of subsidiary within a technological programme that remains under strong central influence and supervision.

Though of quite widespread relevance (only 23% of respondents felt it played no role in their technology inputs) this source was clearly less significant than the first. Table 3 shows that the expectation that the emergent product development subsidiaries would be the most decisive users of this type of technology is somewhat confounded. It seems that in their behaviour so far these types of subsidiaries tend to build their distinctive competences as an evolution of their past activity (based around established group technology, as just discussed), rather more than as a systematic involvement with the development and application of new group-level technology. In fact this type of technology appears to be most applicable in the component supply operations of the third type of subsidiary. This provides the first indication from our results that this type of subsidiary may not be a purely dependent operation supplying well-defined inputs (and therefore almost exclusively using established embodied technology) to cost-effective supply networks for a MNE's mature generations of products. Instead at least part of their motivation does seem to involve direct access to group-level technology and through this the creation of new input goods, either to help upgrade existing products in which they are incorporated or as a collaboration with those product development subsidiaries which themselves are working with new group technology in the manner hypothesised.

The reporting of this source of technology by subsidiaries of the first two types, which are mandated to manufacture products for already strongly established markets, may occur when they do sufficient adaptive work on the existing product to feel an involvement in the creative process. This perception of their position with regard to the technology of the products they supply may reflect a frustration in such subsidiaries at their dependent position, and indicate an incipient attempt to move towards the more fully-fledged product-development role. Also, in some industries, there may exist a phase in the application of a particular piece of knowledge where it is not easy to distinguish between the adaptive evolution of existing products and the creation of distinctly new ones.

The evidence so far, therefore, clearly indicates that overseas subsidiaries in MNEs still operate within the confines of well-defined group-level technology. However, it is a strong thesis of this paper that to sustain true global competitiveness it is desirable for MNEs to allow their overseas subsidiaries to mediate in the effective application of this group technology in their various market and production environments, either by adapting products which already embody established group knowledge or by developing additional products from its more recently created technology. The remaining five sources of technology covered in the survey therefore relate predominantly to ways in which subsidiaries can obtain the ability to add their own creative efforts to the core technology secured from within their group.

The first of this second group of sources was 'R & D carried out by our own laboratory'. One of the most visible manifestations of the widening of the geographical dimensions of technological activity in MNEs in recent years has been the growth in the number of their overseas R & D units, whilst the range of roles played by such facilities endorses the newer perspectives on global competitiveness pursued through subsidiaries.[12] Of the 179 respondents which evaluated this source 73 (40.8%) said it did not play any role in their technology, 36 (20.1%) rated it a secondary source, 64 (35.8%) believed it to be a major source of their technology and 6 (3.4%) the only one. This suggests that over half of the subsidiaries in the survey possessed their own R & D operation as a means of implementing group technology, and of extending their distinctive capabilities and capacities.

As would have been expected it is the product development type of subsidiaries that find the strongest position for a R & D laboratory within their operations (Table 3). The second place taken by the specialist component supply subsidiaries in terms of finding a need for their own R & D units reinforces the emerging perception that performance of this role is somewhat less technologically dependent than might have been expected, and takes on its own distinct element of creativity. The less strong position for their own R & D in the two types of subsidiaries that supply established products reflects the fact that their needs should be limited to product adaptation and assistance in implementing an existing production process, so that where they have a lab its contribution is likely to be quite constrained.[13]

The use of their own R & D is much less well developed in the Japanese subsidiaries in the U.K. than in those with U.S. or European parents (Table 2). This parallels the relatively limited emergence of the product development role in these usually recently established operations, and, as already speculated, greater familiarity with the local environment may enable them to move towards the incorporation of more technological scope and complementary increase in subsidiary creativity and autonomy.[14] The industry in which respondents reported the strongest use of their own R & D was food. In line with earlier observations it may be that the U.S. subsidiaries that are prevalent in the sector set up labs early to assist in the adaptation of established products for the U.K. market, and the dynamic generated by their success then underpined a move towards the complementary incorporation of the product development role, which in turn increased the importance of the labs within their overall operations. The importance of such in-house R & D in instruments and industrial and agricultural chemicals again parallels the relatively strong presence of the product development role in the subsidiaries. Use of their own R & D is also very important in pharmaceutical subsidiaries despite the relative weakness of the product development role. We speculated earlier, however, that the strong export role in this industry might involve the supply of adapted variants of a centrally-created product to meet particular government regulations and health service requirements in several different markets. Implementing these adaptations in such a technologically sophisticated industry may then need the commitment of a dedicated R & D unit to support the production facility.

The survey also investigated the relevance as a source of subsidiary technology of 'R & D carried out for us by another R & D laboratory of our MNE group'. Of 181 respondents that evaluated this 48 (26.5%) did not use it, for 55 (30.4%) it served as a secondary source of technology, 69 (38.1%) considered it a major source and 9 (5.0%) said it was their only one. The derived AR of 2.22 thus places this source somewhat ahead of subsidiaries' own R & D (2.02). R & D from other labs may have secured this position either by substituting for in-house R & D, by enabling a subsidiary to secure certain knowledge inputs for which it would otherwise have needed to create its own facility, or by complementing it by providing extra specialised expertise that can fill out a research programme. We have already noted that only 26.5% of respondents did not make any use of R & D supplied by another group lab, whilst 40.8% did not use R & D from a lab of their own. This would suggest that until the need for R & D support becomes extensive or persistent subsidiaries seek to secure it elsewhere and avoid the expense of an in-house unit. It can also be observed that amongst respondents that did use their own R & D results, only 34.0% considered this only a secondary source of technology, whilst 41.4% of those that received R & D from another lab felt its use was limited to the secondary level. This would provide modest backing for the view that once a subsidiary has its own lab this is then more likely to supply key technological inputs, with externally acquired R & D becoming more of a supplementary source.

A logical extension of this view is that a subsidiary's own laboratory is likely to develop R & D programmes that support its wider objectives, and may complement this by commissioning specialised support from other group facilities when relevant. Thus the R & D from other labs may substitute for a subsidiary's own in-house operations up to a certain level of technological need or ambition, after which a lab of its own may move towards the core of the subsidiary's technical activity, with the acquisition of R & D from elsewhere now serving as a complementary source. Overall these perspectives suggest that, even as subsidiaries increasingly pursue certain degrees of creative autonomy, the scope for technological interdependence throughout a MNE group remains an important asset. Creative autonomy should not mean technological autarchy.

It emerges in Table 3 that R & D carried out in another laboratory of the MNE group takes an important position, and clear precedence over a subsidiary's own R & D, where the main roles are the two involving supply of established products. This is likely to be a case where subsidiaries may need assistance in assimilating and/or adapting existing technology, but often do not feel that the problems are sufficiently deep or extensive to merit the creation of their own lab. In addition they may be able to secure this support from another group lab which already has experience of the technology at issue, either having been involved in its creation or having assisted in its implementation in another context. In product development subsidiaries, and to a lesser extent in those producing components, the use of R & D from other labs takes second place to in-house work. Thus once a subsidiary's level of ambition moves from supply of a standardised good to product creation its own R & D laboratory takes a position at the centre of its technology, with continued access to inputs from elsewhere in the MNE group's R & D network now serving as a useful auxilliary source.

Use of R & D inputs from another group lab is of notable importance in automobiles, an industry where the substantial presence of product development subsidiaries does not result in a strong commitment to in-house R & D. It seems likely that in this case the product development responsibility in U.K. subsidiaries (which also complements a strong position for the export of established products) is often allocated by Europe-level planning in MNEs (rather than resulting from the creative initiative of the subsidiaries), and is implemented through close and systematic collaboration with key centralised labs which play leading roles in supporting their groups' Europe-wide operations. A preference for strong centralised labs supporting a Europe-wide network of supply facilities is also likely to explain the more extensive use of R & D inputs from elsewhere in their MNE group by electronics subsidiaries in the U.K. The significant position of this source both in industrial and agricultural chemicals and in pharmaceuticals suggests that a relative ease of knowledge communication in these industries enables in-house facilities to collaborate with other group labs to ensure the most effective assimilation and adaptation of their MNE's staple technologies by U.K. subsidiaries.

As a source of technology for MNE subsidiaries in the U.K. 'R & D carried out in collaboration with another firm' emerged in a very minor position. Thus 106 (58.9%) of 180 respondents said it played no role, and 63 (85.1%) of those that did allocate it some position in their technology supply considered it to be only a secondary one. Though inter-firm technology collaboration has grown in importance in recent years [15] it is likely that when such alliances involve joint R & D work this serves a predominantly strategic function, [16] mainly focusing on basic and precompetitive research, with the programmes articulated through central-laboratory planners and not involving labs in subsidiaries (or directly supporting their production operations).

The use of local scientific institutions (e. g. universities; independent or industry-funded labs) to carry out R & D for the subsidiary also takes a relatively limited position in their technology supply. Of 181 respondents that evaluated this source of technology 102 (56.4%) considered that it had no role, and a further 73 (40.3%) that it was restricted to a secondary position. Use of this source of

R & D does appear in Table 3 to be somewhat more prevalent in the product development and component supply subsidiaries than in those that are more oriented to the manufacture of established products. Thus, as would be anticipated, R & D of local universities etc. seems more likely to be called into play to supplement efforts that are already established within subsidiaries, than as a source of direct solutions to their technology problems. This tends to be confirmed by the fact that 4 of the 5 industries that have an above average use of this source of R & D are the 4 that provide distinctively the strongest role for R & D carried out by their own labs (Table 2). In fact earlier analysis has shown [17] that the MNE labs that are most likely to establish links with local scientific institutions tend to be standalone facilities (not committed to support an associated production unit), carrying out basic/precompetitive research as part of group-supported programmes.

The final source of technology that the subsidiaries were asked to evaluate was 'development and adaptation carried out less formally by members of our engineering unit and production personnel.' With 1 (0.6%) of 178 respondents considering this its only source, 41 (23.0%) rating it a major source, 97 (54.5%) a secondary source and only 39 (21.9%) believing it played no role, it emerges in Table 2 with an identical AR to the other in-house source of technology, i. e. 'R & D carried out by our own laboratory'. Table 3 shows that, compared with the more formal in-house R & D, the creative use of engineering and production personnel was somewhat more prevalent in the subsidiaries supplying established goods to U.K. markets, of equal relevance in those exporting established goods or components, and relatively of less importance in those doing product development. Talented engineers and production personnel are likely to embody tacit knowledge derived predominantly from their training and day-to-day experience within the MNE group. This type of knowledge is therefore strongly firm specific, and of considerable value when the subsidiary is applying established group technology. Even when the technology to be used is new to the subsidiary it will be within a tradition such personnel understand, so they can therefore usually make a significant contribution (sometimes in association with a laboratory) to its effective assimilation. When a subsidiary's motivation is more decisively oriented to product development and the desire to make a distinctive technological contribution that extends the scope of the group, these extra dimensions (because they to some degree transcend the tradition that the engineers and production personnel are familiar with) often need to be pursued through a more extensive formalised R & D effort. Thus, whereas this role of engineers adds to group technological cohesion and coherence, it may also be the first step towards the emergence of new subsidiary-level knowledge scope.

Two industries in which this source of technology input is of above average importance, and which also report an above average orientation to product development work but a relatively limited use of in-house R & D, are mechanical engineering and metal manufacture and products. In both these cases it seems likely that the predominant form of product development will be the derivation of customised goods for individual clients. Where this occurs the goods will usually be created from a staple established technology, rather than needing a research-based extension of the industry's scope, so that it can be implemented through engineering, prodution, design or other shop-floor personnel who have a sustained involvement with the company's knowledge background. With the exception of instruments, the industries that have extensive in-house R & D (food, chemicals, pharmaceuticals) find relatively limited need to depend on creative inputs from engineering and production personnel.

Conclusion

An important recent development in the literature of the MNE has been an understanding of the range of roles that can be played by subsidiaries in the context of their group's wider strategies. [18] Within this strand of the literature a very relevant emphasis is now often placed on the ability of subsidiaries to differentiate their technological scope and competences in ways that earn them distinctive positions within their MNE group. [19] This type of subsidiary positioning both augments the range of knowledge sources available to the group and also facilitates an increasingly effective application of this technological scope to the needs of global competitiveness. It is therefore perceived that the level of response to the needs of the more dynamic global competitive environment can valuably be increasingly decentralised, with creative subsidiaries extending the range of response to market signals and enhancing the scope of new technological inputs in the group. The survey evidence discussed in this paper indicates that product development is now a clearly established role in MNE subsidiaries in the U.K.

It is also suggested by the survey results that the distinctive competences which provide creative product development subsidiaries with their enhanced positions emerge from their background and experience within the group, rather than constituting more radical perspectives. These subsidiaries add an increased range to decentralised competences which build on, and increase the effective implementation of, the MNE's established technological trajectory rather than comprising a significant challenge to it. Nevertheless the acquisition of an in-house R & D unit does seem to be a characteristic of many such product development subsidiaries, so that their emergency in MNEs should widen the range and scope of the science available to them worldwide and thereby perhaps affect the longer-term evolution of the group's technology. As the experience and aims of the in-house subsidiary R & D labs deepen it may be that they will mediate more in a closer collaboration with other elements in the local scientific community (e. g. universities, other firms' labs), but to date this seems quite limited. Rather more developed so far seems to be the practice of securing R & D support from elsewhere in the MNE group, either to supplement their own lab's work in product development subsidiaries, or to help with the implementation of established products in less technologically ambitious subsidiaries. Overall then the picture is one of substantial evolution in the technological behaviour of MNEs, with new global challenges bringing forth more globalised perspectives on the creation and use of their knowledge base.

Endnotes

1 See also Bartlett (1986), Bartlett and Ghoshal (1986), Ghoshal and Nohria (1989), Hedlund (1986), for discussion of the emergence of heterogeneous subsidiary competences in the contemporary MNEs' strategic evolution.

2 The questionnaire was sent, in late 1993/early 1994, to all the relevant subsidiaries that could be identified from the National Register Publishing Company's *International Directory of Corporate Affiliations*. It was addressed, in the first instance, to the Managing Director, though there is evidence that in some cases the responsibility for reply was delegated to senior subordinates.

3 By industry the highest response rates were achieved in electronics and electrical appliances (30.5%), pharmaceuticals (27.9%) and instruments (25.5%) and the lowest in food (19.1%) and mechanical engineering (19.4%). By home country of subsidiary Japanese respondents provided clearly the best response rate (33.7%) with the U.S. (21.0%) and Europe (19.5%) both slightly below average. Since it also appears that the response rate was generally rather higher amongst larger and longer established subsidiaries the higher response rate in Japanese subsidiaries may suggest a proportionately greater extension into relatively smaller and newer subsidiaries. If this is so it may reinforce the view in the paper that Japanese subsidiaries are less far (compared to U.S. and European) down the path of implementation of in-house technology and adoption of product development roles.

4 Overall 5.8% of European subsidiaries' replies evaluating particular roles said they were the only one, compared with 5.3% of all replies. However only 23.7% of European subsidiaries' replies related to predominant roles compared with 29.4% of all replies. By contrast 45.1% of European replies were 'not a part of our role' compared with 34.0% of U.S. subsidiaries replies and 40.4% of Japanese.

5 See Pearce (1992) and Eden (1991) for discussion of the changes in the competitive environment that underwrite this role for subsidiaries.

6 This is compatible with other evidence from the survey which shows that the subsidiaries of European MNEs were more likely than U.S. or Japanese to see the U.K. market as their only one, and were also clearly the least export oriented overall.

7 Only 17.6% of Japanese subsidiaries had been set up before 1980. This compares with 76.5% of U.S. and 71.1% of European.

8 See Papanastassiou and Pearce (1994 a).

9 See Nelson and Winter (1982), Cantwell (1991), Dosi (1988).

10 See Vernon (1966).

11 See Papanastassiou and Pearce (1994 b) for discussion of globalised approach to innovation in MNEs.

12 See Pearce (1989), Pearce and Singh (1992).

13 Of the subsidiaries playing (either exclusively or predominantly) the role of supplying established products to the U.K. market 48.1% did not consider their own R & D to be a source of technology, and 19.0% only rated it a secondary source. For those exporting established goods

to European and other markets the comparable figures were 35.6% and 24.1%, and for the component supplying subsidiaries 33.3% and 16.7%. Only 22.2% of the product development subsidiaries did not use their own R & D as a source of technology and only 11.1% felt it was limited to a secondary source.

14 In fact JETRO data (see Papanastassiou and Pearce 1994 c) indicates a notable increase in Japanese MNEs' R & D units in the U.K. in the last few years.

15 See Dunning (1993), Chesnais (1988) and Hagedoorn (1993) for discussion of the emergence of strategic technology alliances.

16 See Hagedoorn and Schakenraad (1991) for evidence on the position of R & D-driven alliances amongst all strategic technology alliances.

17 Pearce and Singh (1992).

18 For a valuable delineation of the position of analysis of subsidiary roles in the wider evolution of MNE literature see Birkinshaw (1994).

19 Such creative (product mandated) subsidiaries are discussed in Etemad and Séguin Dulude (1986), and are also covered in the work of White and Poynter (1984), Roth and Morrison (1992) and Pearce (1989, 1992).

References

Bartlett, C.A., Building and Managing the Transnational: The New Organisational Challenge, in Porter, M.E. (Ed), *Competition in Global Industries,* Harvard: Harvard Business School Press 1986, pp. 367–401.

Bartlett, C.A./Ghoshal, S., Tap Your Subsidiaries for Global Research, *Harvard Business Review,* 64, 6, 1986, pp. 87–94.

Birkinshaw, J.M., Towards Heterarchy – A Review of the Literature on Multinational Strategy and Structure, *Advances in Comparative Management,* 9, 1994, pp. 111–114.

Cantwell, J.A., The Theory of Technological Competence and its Application to International Production, in McFetridge, D. (ed.), *Foreign Investment, Technology and Economic Growth,* Calgary: The University of Calgary Press 1991, pp. 33–70.

Chesnais, F., Technical Cooperation Agreements Between Firms, *STI Review,* 1988, pp. 51–120.

Dosi, G., The Nature of the Innovative Process, in Dosi, G. et al. (eds.), *Technical Change and Economic Theory,* London: Printer Publishers 1988, pp. 221–238.

Dunning, J.H., *The Globalisation of Business,* London: Routledge 1993.

Eden, L., Multinational Responses to Trade and Technology Changes: Implications for Canada, in McFetridge, D. (ed.), *Foreign Investment, Technology and Economic Growth,* Calgary: The University of Calgary Press 1991, pp. 133–172.

Etemad, H./Séguin Dulude, L. (ed.) *Managing the Multinational Subsidiary,* London: Croom Helm 1986.

Ghoshal, S./Nohria, N., Internal Differentiation within Multinational Corporations, *Strategic Management Journal,* 10, 1989, pp. 323–337.

Hagedoorn, J., Understanding the Rationale of Strategic Technology Partnering: Interorganizational Modes of Cooperation and Sectoral Differences, *Strategic Management Journal,* 14, 1993, pp. 371–385.

Hagedoorn, J./Schakenraad, J., *The Role of Interfirm Cooperation Agreements in the Globalisation of Economy and Technology,* The MONITOR-FAST Programme Prospective Dossier No. 2. Globalisation of Economy and Technology, 8, Commission of the European Communities 1991.

Hedlund, G., The Hypermodern MNC-A Heterarchy?, *Human Resource Management,* 25, 1, 1986, pp. 9–35.

Nelson, R./Winter, S., *An Evolutionary Theory of Economic Change,* The Belknap Press of Cambridge, MA: Harvard University Press 1982.

Papanastassiou, M./Pearce, R.D., Host-country Determinants of the Market Strategies of U.S. Companies' Overseas Subsidiaries, *Journal of the Economics of Business,* 1, 2, 1994 a, pp. 199–217.

Papanastassiou, M./Pearce, R.D., The Creation and Application of Technology by MNEs' Subsidiaries in Europe, and Their Role in a Global-innovation Strategy, University of Reading, Department of Economics, Discussion Papers in International Investment and Business Studies, No. 189, 1994 b.

Papanastassiou, M./Pearce, R.D., The Internationalisation of Research and Development by Japanese Enterprises, *R & D Management*, 24, 2, 1994 c, pp. 155–165.

Pearce, R.D., *The Internationalisation of Research and Development by Multinational Enterprises*, London: Macmillan 1989.

Pearce, R.D., World Product Mandates and MNE Specialisation, *Scandinavian International Business Review*, 1, 2, 1992, pp. 38–58.

Pearce, R.D./Singh, S., *The Globalisation of Research and Development*, London: Macmillan 1992.

Roth, K./Morrison, A.J., Implementing Global Strategy: Characteristics of Global Subsidiary Mandates. *Journal of International Business Studies*, 23, 1992, pp. 715–736.

Vernon, R., International Investment and International Trade in the Product Cycle, *Quarterly Journal of Economics*, 80, 1996, pp. 190–207.

White, R.E./Poynter, T.A., Strategies for Foreign-owned Subsidiaries in Canada, *Business Quarterly*, 49, 1984, pp. 59–69.

Part II
Technology-Based Inter-Company Alliances

[9]

Multinational enterprises and the international diffusion of technology

François Chesnais

Directorate for Science, Technology and Industry, OECD, Paris, and Département des Sciences Economiques, Université Paris X, 92001 Nanterre

Introduction

The last chapter in this section discusses some contemporary aspects of the relationships between science and technology and international investment (generally designated as foreign direct investment (FDI)), with special emphasis on the world-wide operations of multinational enterprises (MNEs). It is concerned with recent or ongoing changes in the overall pattern of FDI; in the international sourcing, creation and transfer of technology by MNEs; and finally in inter-firm technical cooperation agreements also involving such firms.

Considerable research and policy-oriented discussion have been directed towards the international dissemination or *transfer* of technology by MNEs (for a review and the appropriate references, see, *inter alia*, Caves, 1982). Rather *less* research and discussion have been devoted to the international *sourcing* of technology, and more broadly of scientific and technological knowledge, by MNEs, although this deficiency has now begun to be recognised by some (cf. Dunning and Cantwell, 1986; Bertin, 1986). This chapter will show that the international sourcing of technology includes, of course, the organisation of corporate R & D on an international level with laboratories in several countries, but covers also a wide variety of processes and mechanisms through which MNEs can organise the centralisation and appropriation of technology and technical knowledge. Today such mechanisms include a variety of inter-firm technical cooperation agreements which large corporations belonging to the MNE category are setting up with other enterprises, notably small, knowledge-intensive firms in 'high technology' industries, as well as with universities.

The chapter relates the new developments in the technological strategies of MNEs, both the ones they develop individually and those they organise through collective action, to the changes which have occurred internationally since the mid-1970s, notably: (i) the emergence and rapid generalisation of 'international' or 'world' oligopoly as the dominant form of supply structures in all R & D intensive and scale-intensive industries; (ii) the world economic situation which has dominated since 1975; and (iii) last but not least, the important changes in science and technology with which the whole of this book is concerned.

Two theoretical traditions and a step towards a synthesis

The focus of the analysis is essentially on international investment and the MNEs. The setting of the analysis, however, is the wider process of internationalisation. This notion designates the wide set of economic mechanisms and relationships whereby previously fairly separate national economies become increasingly interrelated and interdependent with one another in all areas of economic activity. These mechanisms and relationships include the export and import of goods and services; outward and inward flows of direct investment and financial capital; outward and inward flows of embodied and disembodied technology; international movement of skilled personnel and transborder information flows; and, of course, the internationalised monetary and financial system.

The reference to the notion of internationalisation warrants a few indications concerning the theoretical underpinnings of this chapter. The conceptual approach which underlies the analysis is at the junction point between the most significant Anglo-Saxon work on foreign direct investment and the MNE, and the dominant French approaches towards the analysis of 'internationalization' (de Bernis, 1977), 'accumulation at world level' (Amin, 1970; Palloix, 1975), or 'world capitalism' (Michalet, 1976).

The approaches used by American and British scholars are essentially based on an extension of the decision to invest abroad either of the theory of the *firm* (Coase, 1937, as extended by Hymer, 1970; Horst, 1974; Buckley and Casson, 1976, with 'internalisation' and the interplay between 'markets' and 'hierarchies' (Williamson, 1975) as the key concepts, or else of the theory of *industrial organisation* and market structure theory (*inter alia*, Bain, 1956; Scherer, 1970; Caves, 1971; Knickerbocker, 1973; Vernon, 1974; and, more recently, Newfarmer, 1983, 1985). Here the extension of domestic oligopoly beyond national boundaries, prior to the emergence of global or international oligopoly, represents the main unifying trend in the analysis. Within the Anglo-Saxon tradition, Dunning's 'eclectic theory' 1981) represents an attempt to combine the two trends and impose a recognition of the *specificity* of the MNE as opposed to the traditional view of FDI, while seeking at the same time to bridge the gap between the theory of international investment and that of international location and trade.

French approaches are all different blends of what may be broadly defined as a 'neo-marxist' tradition. In contrast to the Anglo-Saxon approaches, all the French approaches essentially involve an extension at an international level of the theory of capitalist development (cf. Dobb, 1963; Sweezy, 1946), with the accumulation of capital and the related processes of (industrial) concentration and (financial) centralisation of capital as the key concepts. In all neo-marxist approaches government, e.g. the state, is also given a prominent central role as an institution shaping all aspects of modern capitalist society (cf. Mandel, 1975).

The blending of the French and the Anglo-Saxon approaches consists basically in the recognition that while MNEs are obviously active agents in the process of internationalisation and even architects of some aspects of the process, and *must consequently be analysed in their own right*, they are, nonetheless, *responding to an overall set of factors* over which they have in fact *little or no control*, and which *all stem from the basic mechanisms* driving the historical process of capitalist development. One of these mechanisms is the development (in a contradictory, antagonistic and unequal manner) of the forces of production, among which science and technology play an increasingly quite central role (see, *inter alia*, Rosenberg's essay on 'Marx as a student of technology', 1982, and Harvey, 1982, Chapter 4).

Once they are set against the background of the process of concentration and centralisation of capital and a situation where forces of production (notably those directly shaped by science and technology) have overrun national boundaries, a number of observations and concepts derived from the economics of technical change, the theory of the firm and the economics of industrial organisation and market structure acquire their full intelligibility and become necessary components of a global analysis. Set against such a background, empirical observations relating, for instance, to the close relationships between R & D intensity, industrial concentration and the scale of FDI in R & D- and scale-intensive industries take on a new dimension and can be interpreted as expressing a real *objective* constraint on firms in such industries to adopt the *world market* as the *only possible market* on which they can deploy and fully reap 'ownership advantages'.

The merging of the Anglo-Saxon approaches with the understanding derived from the extension at the world level of the process of capitalist development must *not* be interpreted, however, as implying a one-way process. For instance, while it is hard to establish a sound foundation for the analysis of global or international oligopoly without an understanding of the way in which the concentration and centralisation of capital will begin to take place at the international level with its tendency to englobe the advanced capitalist countries as a whole, through a process of 'mutual raiding' (Erdilek, 1985), international cross investment (Mucchielli and Thuillier, 1982), acquisitions and mergers, *conversely* the analysis of *firm behaviour in conditions of international oligopoly* (each firm backed by its home country government in one way or another), represents the *only way* of making *progress towards a better understanding* of the working of 'monopoly capital' and 'inter-imperialist rivalry'.

Similarly, once the concept of *internalisation* is set against the background of the twofold process of industrial concentration and financial centralisation of capital (as defined and distinguished in the marxist approach), and MNEs are defined in relation to their *group* structures and their *combined* industrial and financial attributes, it becomes possible to appreciate fully what Dunning has to say about the *organisation* of 'market failure' by large firms:

Where, for example, enterprises choose to replace, or not to use, the mechanism of the market, but instead allocate resources by their own control procedures, *not only do they gain* but, depending on the reason for internalisation, *others* (notably their customers and suppliers prior to vertical integration, and their competitors prior to horizontal integration) *may lose*. Internalisation is, thus, a powerful motive for takeovers or mergers, and a valuable tool in the strategy of oligopolists. [Dunning, 1981, p. 28]

In the same order of ideas, Dosi (1984) points out the importance of not considering internalisation as a passive response to 'market failure':

It would be misleading to consider internalization simply as an effect of and a reaction to some kind of 'market imperfection'. It is probably more accurate to consider it as one of the inner trends (and one of the 'rules of the game') in oligopolistic rivalry towards the transformation of the untraded features of technical change into proprietor assets which, as such, also represent entry barriers and differential advantages *vis-à-vis* other competitors.

The 'distinctive nature' of the multinational enterprise

A reference has been made to the financial dimensions of MNEs. This is a somewhat underlooked dimension of their scope and power, which has however been the object of a fair amount of attention and research in France and more recently in the United Kingdom (different French approaches include Morin, 1974, Pastré, 1980, Chesnais, 1979, and Grou, 1983; in the United Kingdom see Scott, 1979, 1986). These features are complementary to and indeed often an outcome of conglomerate organisation and are expressed *inter alia* in the 'portofolio' approach to assets, including industrial assets and the scale and significance of operations made by industrial MNEs as lenders in short-term international money markets (eurocurrencies) (Cohen, 1980). In this approach MNE must be viewed as a specific form of 'finance capital' (Hilferding, 1910; Lenin, 1915), e.g. a form which is predominantly engaged in industrial activity but in ways which are increasingly shaped by purely financial strategies as developed by financial analysts, financial market operators and bankers (see Minsky, 1982). This characteristic will be enhanced by the transformation of parent companies into 'holding' corporations, but is present today in all large multidivisional and multinational firms. It is a central factor in the emergence of the 'hollow corporation' as coined by *Business Week* which is now at the centre of debate in the United States (see Cohen and Zysman, 1987).

If due recognition is given to this dimension, and if the particular capacity of the MNEs to use world-based information technologies and systems to their own best advantage is also recognised (Antonelli, 1984), then Dunning's account of the 'distinctive nature' of the MNE (Dunning, 1981, p. 27) is a very acceptable one, namely that MNEs enjoy three combined and cumulative sets of advantages:

(i) those which large firms may have over others 'producing in the same location' and which 'stem from size, monopoly power and better resource capability and usage';

(ii) those pertaining to the advantage a branch plant or subsidiary can derive from belonging to a group, thus benefiting from many of the endowments of the parent company, 'for example, access to cheaper inputs, knowledge of markets, centralised accounting procedures, administrative experience, R & D, etc., at zero low marginal cost; or the *de novo* form will normally have to bear the full cost. The greater the non-production overheads of the enterprise, the more pronounced this advantage is likely to be'; and

(iii) the particular type of advantage 'which arises specifically from the multinationality of a company, and is an extension of the other two. The larger the number and the greater the differences between economic environments in which an enterprise operates, the better placed it is to take advantage of different factor endowments and market situations'.

If these three aspects are used in combination, the MNE will be viewed as a firm possessing a very wide range of opportunities simply not offered to smaller domestic or regional firms, for *entry* into as well as *exit* from given activities and markets; or again for 'internalising externalities', in our case, for instance, the results of government or university-financed and executed R & D. These firms also have special opportunities for reaping 'appropriable rents' or quasi-rents (Klein *et al.*, 1978), and for exerting forms of monopolistic or monopsonistic 'market power' (falling short of outright monopoly or monopsony) which can allow them to impose quasi-integration-type contractual arrangements of an essentially *predatory* nature on smaller supplier firms, including specialised 'knowledge-intensive' small firms.

But such firms are also oligopolists, in their home markets of course, and also today in international markets, where oligopoly prevails in all 'science-based' and in many 'scale-intensive' industries. The market power of each firm is bounded by that of the other rivals which make up the oligopoly. 'Mutual recognition' and oligopolistic reaction and interaction, which are the hallmarks of oligopoly, will take place in all spheres of activity, including technology, and may lead to situations where in the area of R & D and the development manufacturing and marketing of high-technology products a considerable amount of cooperation between oligopolists occurs. Once international oligopoly sets in, as is the case today, the 'endogenous generation of market structure and technological performance' (Nelson and Winter, 1982) will likewise start to become an international process. Before turning to these hypotheses in the later part of the chapter, we first discuss other important, albeit more traditional, aspects of the relationships between science and technology, FDI and MNEs.

Patterns of international investment in the 1980s

Major changes in the overall international pattern of foreign direct investment began to occur in the early 1970s. The fact that they started by *coinciding* with the deep changes in the world economic situation, before being later *accelerated* by the developments of the 1980s—in particular the collapse of profitable direct investment activity in most developing countries from 1981–2 onwards (the only exception being the Asian NICs) and the progressive extension of the new paradigms in science and technology and also manufacturing processes (notably Japanese—see Sciberras, 1980, for colour TV, and Jones, 1987, for automobiles)—helps to *explain the delay there has been in recognising the new international patterns* of FDI and in isolating them prior to adequate analysis.

The new patterns are the following: (i) a decline, later followed by a severe reduction, in the flow of FDI towards 'market economy' developing countries, in contrast with the People's Republic of China which has received the largest fraction of DC-orientated investment during the 1980s (UNCTNC, 1987); and (ii) more important still, the *recentring or reconcentration of the flow of FDI within the OECD area*, along with two major related changes, namely (a) the progressive shift in the role of the US economy from a 'home' to a 'host' country for FDI as European and later Japanese MNEs have developed their investment in the US market, and (b) the emergence since the late 1970s of Japan as a major 'home' country for MNEs and a large source of FDI directed towards the United States but also towards Europe.

The second, if not the first, of these two major developments was fairly foreseeable. It was announced in a Hymer and Rowthorn (1970) article, in which the authors predicted the emergence of global oligopoly as a result of the growth of European and Japanese FDI. The central finding of the Hymer–Caves strand of analysis (see also Knickerbocker, 1973) is that once a major firm in a concentrated industry has started to invest and manufacture abroad, its oligopolistic rivals are obliged to follow suit. While this may *start* as a rather unilateral process of oligopolistic rivalry between large firms based in the same economy (as it did in the 1950s and 1960s in the case of US-based foreign direct investment), *as soon as accumulation and the concentration and centralisation of capital have developed (or developed again) in other countries*, regions or poles of the internationalised capitalist world economy, the situation will *necessarily* increasingly be one of mutual cross-investment by MNEs, into one another's home markets and domestic technology bases.

This is the process that occurred increasingly in the 1970s and 1980s as a result of large-scale FDI by OECD-based MNEs into the US economy. Since the earlier 1980s, the United States has for the first time this century become a net importer of capital again. Part of this capital has had purely financial destinations, but a significant fraction has taken the form of investment into US manufacturing industry either through mergers or 'green-field' investment. At the same time as inward foreign direct investment has grown, outward US investment has somewhat declined. Consequently, if intra-Europe FDI is deducted from the European total, Europe and the United States now host approximately the same level of foreign investment (Fouguin, 1986). As in the case of trade flows, where Japan's imports are much lower than its exports, notably in high-technology industries, Japan's place in multipolar international investment is significantly different from that of Europe and the United States. Compared with Europe and the United States the level of foreign direct investment into Japan still remains extremely low. By contrast, Japanese *outward* foreign investment has been growing rapidly and will quite certainly be further accelerated by the changes in foreign exchange rates. Up to now a large part of Japanese investment in Europe and in the United States has been through joint ventures and inter-firm cooperation agreements of various sorts, one dimension of which is, as in the case of several US-based joint ventures, technical cooperation and the joint development of new technologies, and another the desire to overcome political and social sensitivities towards the entry of Japanese firms within the markets and domestic supply structures of US and European industry. When these sensitivities disappear, however, the preference of Japanese firms for majority-owned affiliates tends to reassert itself, as in the case of the United Kingdom (Dunning, 1985).

Trade, foreign investment and international flows of technology: the paucity of sound knowledge

In many instances, foreign direct investment has clearly recognisable trade-substituting effects: delocated production within a country replaces the exports previously made to that country from the MNE's home base. In other instances, foreign direct investment will create trade, along with new forms of dependencies and interdependencies. This is the case in particular when firms investing in a foreign economy continue to source capital goods and intermediary products from their home economy (as is often the case for Japanese firms at the moment) or when MNEs use affiliates to ship products (generally intermediary goods) to the parent company or to other affiliates.

An indication of the magnitude of the trade-substituting impacts which a unified theory would have to give an account of is given by UK data which show that in the case of the fifty largest corporations in 1981–2 the relation between overseas production of sales and export both as a percentage of sales was 3 to 1. Recent US Department of Commerce data show likewise that in 1983 the overall ratio of sales by the foreign affiliates of US MNEs to exports from the United States was a little over 2 to 1. This represents a drop in comparison to earlier periods and is due to the fall-back in US foreign direct investment since the end of the 1970s. However, in R & D-intensive industries where internationalisation has developed for many decades through direct investment rather than through exports, such as pharmaceuticals and chemicals, the ratio remains much higher: nearly 5 to 1 in pharmaceuticals and 3 to 1 in chemicals.

Even if the trade-substituting effects of foreign direct investment are quite substantial in some industries, in retrospect it is quite clear that the spectacular growth of exports and of internationalisation through trade would never have taken place on the scale it did in the absence of foreign direct investment. The trade-barrier-jumping capacity of foreign direct investment probably represents the strongest available deterrent to trade barriers, both tariff and non-tariff barriers. Foreign direct investment when coupled with trade also has a capacity to lower industrial entry barriers which trade alone does not possess. In the transitory period preceding the emergence of situations of world or international oligopoly in 'global market' industries and sectors, FDI in *combination with trade* succeeded in *lowering entry barriers*, reducing domestic concentration levels and weakening positions of domestic or regional oligopoly. In Europe notably, FDI increased United States competition significantly and had a positive effect on the competitiveness of European firms (Dunning, 1982). The need to *combine* the discipline stemming from low trade barriers with that resulting from *foreign direct investment* must be stressed. When foreign direct investment takes place behind high trade barriers, foreign affiliates, as shown by many industrial and country case studies and highlighted in Unger's chapter, tend to adapt themselves to reigning conditions of production and competitiveness and become a part of domestic oligopolistic supply structures. This, for instance, occurred in Spain in the 1960s and 1970s and of course in large Latin American countries like Brazil and Mexico (Connor, 1977, and Unger's chapter).

Another important theoretical issue in this area is the *relationship between intra-industry investment and intra-industry trade*. Discussing the case of industries in which 'there are a number of countries each home to a group of highly innovative firms' (e.g. in our terminology industries of the R & D-intensive, international oligopoly category), Dunning and Cantwell observe that:

> Over the last 25 years, this is the kind of industry that has been characterised by the rise of the cross-hauling of investments between those countries harbouring the strongest firms. Such countries become hosts to the greatest levels of international production as well as being homes to MNEs of their own. This phenomenon is known as intra-industry trade and production; and, we might add, intra-industry technology trade and diffusion as well. [Dunning and Cantwell, 1986]

This, of course, remains an extremely general statement and amounts to a *plea for research* into those now extremely frequent situations where *intra-industry trade and intra-industry FDI occur simultaneously*, along with related two-way international flows of technology. In 1978, Dunning had already indicated that 'any theory of intra-industry trade must now take account of intra-industry investment. *We are only at the borders of research in this area*' (Dunning, in Giersch, 1979, p. 70, author's emphasis). Nearly ten years later we are still about at the same point: trade analysis *continues* to be carried out *separately* from that of investment. This was again the case for recent research at OECD during the structural adjustment study (OECD, 1987b), with resulting limitations on the usefulness of the findings. The latter confirm the continuation and strengthening of trends noted in earlier studies, but they would require an integration of the international investment variable to become really *meaningful*. The findings are the following (OECD, 1987b)

(i) 'OECD trade had increasingly involved products characterised by significant economies of scale in production, extensive product differentiation or close links to the science-base. Thus, from 1962 to 1985, the share of scale or production-intensive and science-based goods in total OECD manufacturing trade (exports plus imports) increased from 53 to 65 per cent, with growth rates of the latter category being particularly high in the most recent period.'

(ii) The growth of trade has mainly entailed specialisation within industries. Using an index of intra-industry trade, based on the ratio of gross trade to net trade, the study finds that 'with the important exception of Japan, this index increased significantly for all the countries listed, the average for the group as a whole *rising by more than two-thirds* over the period 1959–1985'.

(iii) This pattern of specialisation was accompanied by major shifts in the geographical structure of trade. 'Thus, throughout the 1950s and 1960s, trade occurred primarily between industrialised countries, sharing broadly similar factor endowments and patterns of demand. Trade grew particularly rapidly between closely located countries, mainly the United States and Canada on the one hand and the European OECD countries on the other'.

(iv) Perhaps the most interesting finding, however, is the one concerning recent trends in the NICs which by 1985 accounted for fully 8 per cent of OECD manufactured imports. Regarding these countries, the study finds that 'the extension of interdependence to new trading partners did not weaken the major tendencies affecting the structure of manufactured trade. Thus, by the end of the period, nearly 40 per cent of OECD imports from the NICs were not in labour or resource intensive products, but in products characterised by significant scale economies or extensive product differentiation. And intra-industry trade—which had been a marginal factor in OECD trade with developing countries in the late 1960s—accounted for fully 31 per cent of the NIC's manufacturing trade with the OECD area in 1985: confirming that they too were increasingly drawn into the dynamic functions of trade'.

Such situations simply *cannot* be explained independently of FDI and the effects it has on the structure of trade. These effects are not limited to trade substitution (*inter alia* Vernon, 1979), but extend to other less well recognised phenomena, in particular:

(i) the role FDI plays in gradually homogenising the organisation of production (pattern of industrial output and levels of productivity) in those countries in which the overall effect of such investment, through the combined influence of MNE strategies and host government policy, is (as in the Asian NICs) to integrate previous DCs into the 'North';
(ii) the role of international sub-contracting (Germidis, 1981) and other long-term commercial arrangements bearing on the supply of manufactured inputs to industrial production; and, of course,
(iii) 'intra-firm' trade, e.g. trade between affiliated firms within multinational corporate structures (Helleiner, in Giersch, 1979).

It is this type of structural impact, along the particular forms of integration of domestic economies into the world market it entails, which some authors have viewed as leading potentially to what Michalet (1976) has called 'le capitalisme mondial', i.e. a world wide set of economic relationships possessing systemic features, which might supersede the present pattern of national economies and international market relationships. In other recent work (Chesnais, 1988), we have argued that the end of Bretton Woods and the subsequent international monetary system, the onset of deep economic and trade instability (whether one decides or not to use the term economic crisis), along with the full restoration of the 'cumulative oppressive power' of *rentier* money capital (Keynes, 1936, pp. 375–76), have offset the emergence of any such system. Very strong liquidity preferences by money capital have been accompanied by increasingly 'footloose' strategies by MNEs and a premium on the part of FDI to industries and countries where *exit* barriers (Porter, 1985) are low. Contemporary microelectronics technology has given additional scope for such strategies by MNEs and wrecked previous industrialisation policies in almost all developing countries and most Latin American NICs (see Kaplinsky, 1987 for automobiles; Mytelka, 1988 for garments, as well as the chapter below by Unger). But this does not modify the essential fact that the structure of international trade *cannot be analysed independently of FDI and the worldwide strategies of MNEs*.

The internationalised sourcing of technology: an under-researched dimension of MNE operations

The organisation by MNEs of the *creation* and/or the *acquisition and appropriation* of technology on an *international* level, across national boundaries, is by no means a new phenomenom. It dates back to the period when attention was almost exclusively concentrated on the international *transfer* (i.e. dissemination) of technology by and in particular *within* the MNE. If little is known about world-wide sourcing operations by MNEs for technology, this is mainly because it has been the object of little research (see also Bertin, 1986): MNEs were thought of as simply (or mainly) transferring technology *outwards* and not also transferring it *inwards*. Discussing overseas R & D spending by MNEs, Caves (1982), for instance, simply mentions in passing that the 'basic research of MNEs is much more footloose than is applied research, and that some of it goes abroad to seek out particular scientific specialists'.

In a number of industries (cf. the studies produced at OECD in the late 1970s, Michalet and Delapierre, 1978; Chesnais, in OECD, 1979; Burstall, Dunning and Lake, 1981) there is evidence that MNEs, for instance, the European MNEs in pharmaceuticals and food processing and the US MNEs in the electrical, electronics and computer industries, were early to understand that foreign direct investment and internationalised group structures could form the basis not only for internalised transfer of technology conducive to the most advantageous exploitation of 'firm-specific advantages' across national frontiers, but also for the *sourcing and centralisation of scientific and technical knowledge and resources on an international scale*. The findings of a recent US survey (Fusfeld, 1986, pp. 132–3) regarding the motivations of US MNEs for setting up foreign laboratories have provided further information regarding the strategies of MNEs in this area. The most frequently quoted objective has been the desire to 'have a window on foreign science', first and foremost in Europe. Other objectives include access to special skills not easily available in the home country, developing new sources of technical concepts or simply establishing on an international stage of operations of corporations concerned with science and technology.

The telecommunications and computer and data processing industries have probably been those where the world-wide organisation of corporate R & D and sourcing of scientific and technical resources has experienced the greatest development. As early as the mid-1970s several large firms had a kind of international technical system with a foot in several national systems, but with identifiable autonomous features of their own, ensuring the international flow of technology within international group structures. IBM has, of course, often been studied in this respect (see Michalet and Delapierre, 1978). By the mid-1970s IBM had set up a world-based set of R & D activities organised independently of its manufacturing and marketing affiliates, in which the tasks assigned to laboratories did not necessarily match those of the subsidiary's production units to which given laboratories belonged formally. Few laboratories within the IBM group were engaged competitively or simultaneously in research, and then only in the exploratory stage before any heavy development expenses were involved. Their work was organised so as to maximise complementarities. At the end of the 1970s, IBM had three laboratories performing fundamental research, two in the United States and one in Switzerland. Development tasks were distributed on a world basis among all the other laboratories, i.e. fourteen major laboratories in the United States and eight in other countries: by order of establishment, The Netherlands, Germany, United Kingdom, Japan, Sweden, Austria, Canada and France.

To ensure that the R & D undertaken in affiliate laboratories geographically far removed from each other was consistent, IBM had set up a vast telecommunications network for regularly pooling all the group's technological resources, where laboratory computers were interconnected on a world-wide basis and a single data bank set up at the corporate R & D headquarters. In this respect again IBM has been exemplary. In particular since the emergence of world-wide telecommunication networks, its experience has been studied and partially followed by an ever more important number of firms. This is confirmed by the findings of the survey undertaken by Antonelli (1984) covering forty US and European MNEs, where replies show that among the seven main reasons given by firms for adopting the new international telecommunication equipment in their operations and undertaking the organisational changes required in corporate structures were the new possibilities offered for the 'international implementation of R & D capacities generated by increased interaction among affiliates and headquarters and greater division of labour, according to the technical requirements and scientific endowments of countries and affiliates'.

The centralisation of external scientific and technological knowledge is not limited to 'science-based' or R & D-intensive industries, but has been equally important in industries such as food processing, where innovation relies heavily on inter-industry transfers of technology. In our study of the technological activity of MNEs in this industry, we laid particular stress on the *horizontal coordination and management of inter-industrial technology transfers* by firms such as Unilever, Nestlé or CPC International and the role of engineering departments (Chesnais in OECD, 1979, Chapter XI).

The need to organise the sourcing of technological resources on an international basis and, in particular, to establish 'a window' on the world's most advanced science base, namely that of the United States, explains, of course, why large European firms, rapidly followed later by Japanese ones, have followed the road taken by the US MNEs in the preceding period. In the 1970s and 1980s non-US MNEs have increasingly sought to create laboratories in the United States as part of their investment in the US economy, in some instances as an important, if not a major component of the overall objective of this investment. This is particularly true in the chemical and chemicals-related industries, in particular in *pharmaceuticals*, where the need to gain direct access to the US science and technology base in the field of biotechnology has been a motive for much of the US investment made by non-US MNEs, which has generally taken the form of take-overs of existing firms, along with the acquisition of their R & D facilities which the new parent firms have often expanded. The process of 'international cross-investment in R&D', in the form of laboratories and other forms of scientific investments, including research contracts with universities, has been documented up to 1982 by Burstall in the case of the pharmaceutical industry (Burstall and Dunning, 1985). Since 1982, further European and Japanese investment has occurred in the United States, Japanese investment has grown in Europe, as has US and European investment in R & D facilities in Japan. Most firms still normally conduct the most sensitive and demanding types of work in centres located in their country of origin. Only still relatively few of the world's top thirty pharmaceutical MNEs have laboratories of the highest capacity situated elsewhere, but almost all now have some kind of 'window' open on the scientific capacity in their main competitors' home economies. This is all the more true since the emergence of biotechnology in the United States at the end of the 1970s and the expansion of inter-firm technological co-operation agreements, *inter alia* with the small genetic engineering firms, as a new mechanism for the acquisition of technology external to the large established MNEs (Chesnais, 1986; US Congress Office of Technological Assessment, 1984).

Increased pressures for an external technology sourcing by firms and the growth of inter-firm technical cooperation

Discussing the international flow of technology and technological balances of payments, the authors of the OECD Science and Technology Indicators No. 2 note that

> problems of interpretation are raised not only by the mixed contents of the TBPs, but also by the elusive character of certain international flows of technological knowledge, i.e. those for which there is no visible form of payment (among others: cross-licensing, transfer of knowledge to a subsidiary, international co-operation of a non-commercial type). [OECD, 1986a, p. 54]

Today, on account of the rapid growth of inter-firm agreements, probably a very large part of outwards and inwards transfers of technology by firms now take this non-visible form.

In discussing the contemporary growth of inter-firm agreements it is important, of course, to keep a sense of *historical perspective*. International cross-licensing between large firms, which remains a fairly basic form of technical cooperation agreement in concentrated R & D intensive industries, was already a significant feature of the chemical and heavy electrical equipment industries in the 1930s (Newfarmer, 1978, 1985). Little-publicised technical cooperation has been a long-standing feature of the relationships between firms like Siemens and Philips, ICI and DuPont, or the three Swiss pharmaceutical majors in Basle.

Today, in addition to the deeply troubled world economic situation, there appears nonetheless to be *two series of major driving forces* explaining why present trends probably represent a qualitative development which will not be easily reversed. The first series of factors relate to the *point reached in the overall process of internationalisation*, along with the implications this has for firms, in the form of a continually growing requirement to wage what the business management literature calls 'global competition'. In all R & D-intensive industries, as well as in industries where scale economies are decisive, competition (e.g. oligopolistic rivalry) now takes place: (i) between a relatively small number of large firms (ranging from three in large commercial aircraft, and two or at best three in large civil aircraft engines, to ten or so in automobiles or in the main segments of the electronics sector); (ii) in a world arena which includes the respective home and host markets of rival MNEs as well as third markets, within and outside OECD; and (iii) through a wide array of means by which firms can gain access to technology and markets (notably 'reserved' government procurement markets). Such rivalry implies 'mutual recognition' and a variety of combinations between *competition* and *cooperation*.

The second set of factors concern contemporary developments in science and technology which help to explain why the access to a wide science and technology base that was an *advantage* in earlier phases is now a *necessity*. The overall trend is one where: (i) basic scientific knowledge is playing an increasingly crucial role in opening up new possibilities of major technological advance, or, to put it in another way, where the knowledge base of technology of firms is increasingly founded in basic science; (ii) many recent breakthroughs have occurred as a result of cross-fertilisation between scientific disciplines; (iii) technology has acquired stronger systemic features. These features are the hallmark not only of spectacular developments in space technology, telecommunications or military systems, but also of more mundane, albeit revolutionary, technologies in CAD/CAM, new materials, etc.

Ongoing paradigmatic changes in science and technology are accentuating all three aspects. Major innovations are based even more strongly on scientific knowledge; synergies and cross-fertilisation, both between scientific disciplines and between scientific and technological advances, play an ever more important role, notably through the advances continually occurring in computing technologies. The massive entry of computing into instrumentation has further strengthened the role played by the latter. The extension of the systemic features of technologies to a larger number of areas is a necessary and inevitable outcome and expression of these developments. Alongside these processes and as an expression of their pressure on firms, there has been a general tendency towards increases in R & D costs and outlays. This has been particularly noticeable in computers, electronics and components and pharmaceuticals, but would also be identifiable in areas such as new materials if detailed company data on costs were available.

In combination these developments in science and technology have created what Fusfeld (1986, p. 143) calls a 'capability squeeze' on firms, marked by: 'the increase in the number of technical fields relevant to corporate growth' and 'totally new requirements for significant technical advances'. These pressures can be met *partly* by increasing *in-house R & D* within corporate structures, both nationally and internationally, or by the establishment in cooperation with other firms of joint-venture corporations solely dedicated to R & D. In other cases, they will require the *external acquisition of knowledge*, know-how and skills located in *other* organisations, whether universities (when the knowledge is still close to basic research) or firms. These are the processes lying behind the observations made by Dunning and Cantwell (1986) that:

'there has been a historical shift away from technology being viewed by the firm as a specific and single purpose input towards the development of integrated technological systems which require coordinated governance for their economic deployment. The rise of such technological interdependence means that it is no longer appropriate to think simply in terms of a sequence which runs from technology creation to transfer and diffusion. The successful creation and application of new technology has become much more dependent upon the earlier dissemination of related technologies within the firm, and of the parallel technologies developed elsewhere by other firms.'

A wide range of interfirm agreements

The external acquisition of technology can be organised either through 'arm's-length' operations, notably straightforward licence agreements, through mergers and hierarchies (Williamson, 1975) (e.g. the outright acquisition of the firms possessing the desired technology or accumulated scientific and technical knowledge), or on the basis of a *wide range of inter-firm agreements* falling 'between markets and hierarchies', (Mariti and Smiley, 1983). The range of the main types of agreements established by firms with a view to producing, acquiring and/or commercially exploiting technology in common are set out in Table 23.1.

A. University-based cooperative research projects are collective R & D undertakings established and financed by firms in universities with or without public support. The distinctive characteristics of this type of agreement are location of the R & D in academic structures and extensive support and direction by firms (as opposed to governments) even if some public support is offered.
B. Government–industry cooperative national or international research projects are collective R & D undertakings with strong government backing, located in firms but also in universities and public research institutes. Here government initiative and financial support (national as in the case of Alvey or by the EC as in that of Esprit), and a much greater variety in the location and execution of R & D, are distinctive features.
C. Research corporations are *private joint-venture companies* financed on

Figure 23.1 Inter-firm R & D, technology and manufacturing co-operation agreements and the R & D, to marketing spectrum

Pre-competitive stage Research and development cooperation			Competitive stage Technological cooperation			Manufacturing and/or marketing cooperation		
A	B	C	D	E	F	G	H	I
University based cooperative research financed by associated firms (with or without public support)	Government industry cooperative R & D projects with universities and public research institute involvement	Research and development corporations on a private joint-venture basis	Corporate venture capital in small high-tech. firm (by one or by several firms, otherwise competitors)	Non-equity cooperative research and development agreements between two firms in selected areas	Technical agreements between firms concerning completed technology *inter alia*: * technology-sharing agreements; * second-sourcing agreements; * complex two-way licensing; * cross-licensing in separate product markets	Industrial joint venture firms and comprehensive R & D, manufacturing and marketing consortia	Customer–supplier agreements, notably partnerships	One-way licensing and/or marketing agreements (including OEM sales agreements)
Many partners	Several partners		Few or very few partners			Few or very few partners		

Source: Chesnais in DSTI, 1986, *Technical Co-operation Agreements Between Firms: Some Initial Data and Analyses*, on the basis of typologies by Hacklisch (1986) and Ricotta and Mariotti (1986).

a shareholder basis by a number of firms, generally MNEs. Programmes focus on generic technology relating directly to the competitive interests of the joint-venture partners. The research results are proprietary and form a kind of pool of patents and know-how for partners.

D. Agreements involving the use of corporate venture capital (CVC) are used by large corporations, notably MNEs, to make small but strategic investments in new innovative processes, still belonging to small firms. Used in combination with R & D contracts, such investments prepared acquisition without interfering immediately with the activities of the small, innovative firms; these remain autonomous in their R & D and management decisions (Ricotta and Mariotti, 1986).

E. Non-equity cooperative research agreements are flexible forms of cooperation of limited duration, without shareholder participation, established between a very small number of firms (usually only two) to deal with specific technical problems. If the R & D is successful it can either be exploited commercially by the participating firms each in their own business, or lead to the establishment of a separate joint venture.

F. Technological agreements bearing on existing 'proven technologies' can take a variety of forms, depending on the specific characteristics of industries and technologies. They include: (i) technology-sharing agreements bearing on complementarity technologies; (ii) second-sourcing agreements between large MNE producers, otherwise competitors (as in semiconductors), involving access to proprietary technology; (iii) two-way exchanges of licences in directly complementary areas with clauses providing for the continuous exchange of improvements and developments; and (iv) two-way exchanges of licences or of scientific and technical knowledge produced by firms as a 'by-product' of their main lines of scientific and commercial specialisation.

G. Comprehensive R & D, manufacturing and marketing consortia are joint ventures with a number of partners, formed with the aim of creating, testing, producing and commercialising a product all the way from the R & D to the final market.

H. Customer–supplier partnerships, notably between MNEs, which represent 'the formalisation of a link that reflects a significant reliance of the partners on one another. It may or may not involve an equity interest or an exclusive relationship . . . It responds to the intensifying *systemic* dimension of components and provides a mechanism for leveraging critical technical and financial resources for the partners (Hacklisch, 1986).

I. Licence agreements and technology transfers forming part of a long term relationship between two firms, as in OEM agreements.

In search of useful interpretative analytical frameworks

The information base concerning inter-firm technical cooperation agreements includes a variety of industry case studies, some very detailed (Mowery, 1986; Hacklisch, 1986; Payne, 1986), others more limited in scope; a motley collection of statistical data bases; one or two Ph.D. theses containing some empirical testing (Hladik, 1984); articles written from the standpoint of the theory of the firm (Mariti and Smiley, 1983), and a rapidly growing but fairly heterogeneous literature by business management economists (Harrigan, 1985; Ricotta and Mariotti, 1986; Doz *et al.*, 1986). At present, there is really only one conclusion to be drawn from this material—that inter-firm agreements occur within a very wide variety of concrete situations, on the basis of a wide range of corporate strategies and with very differing consequences for the future of firms and the maintenance and/or strengthening of domestic technology bases.

Progress towards the development of appropriate conceptual frameworks for the interpretation of the existing data and the elaboration of new and hopefully more significant and/or easily usable studies than many of those made at present can take place along several alternative paths, on the basis of the different approaches discussed in earlier sections. The main theoretical question (which to our knowledge has not yet been really tackled) bears on the explanation of why MNEs, after giving priority to internalised modes of creation and dissemination of technology, have moved towards a much greater recourse to cooperation with other firms, e.g. why, *after* moving from the *market* to *hierarchies* (Williamson, 1975), they are now moving from *hierarchies* to the interfirm *organisation mode* (Imai and Itami, 1984).

Whatever the approach chosen, one essential condition must be satisfied, which is an adequate *understanding and account of technology and innovation*, along with the *particular constraints*, they place on firms to enter into agreement and the *particular advantages* which can be created by the *successful centralisation and appropriation of technology through such agreements*. While a purely scientific and technological interpretation developed independently of any theoretical basis *can only yield descriptive results at the best*, conversely no proper analysis of agreements can sidestep the economics of innovation and technical change.

Not many studies have yet succeeded in combining the reference to a well-defined economic approach with a sufficient detailed analysis of the technological constraints and opportunities pressing towards the establishment of agreements. Teece (1986) offers one example written from the standpoint of the theory of corporate strategy. Teece's central hypothesis is that:

> in almost all cases, the successful commercialisation of an innovation requires that the know-how in question be utilised in conjunction with the services of other assets. Services such as marketing, competitive manufacturing, and after-sales support are almost always needed.

In some cases, firms (notably MNEs) will have internalised the specialised complementary assets through previous investments and mergers. In others, firms (notably small and medium enterprises or purely domestic or regional firms) will lack one or several of the necessary complementary assets and will be forced to find adequate partnerships and/or establish cooperation agreements. Other factors shaping corporate strategy are technological and rest on contemporary innovation theory as developed *inter alia* by Abernathy and Utterback (1978), Nelson and Winter (1982) and Dosi (1986). They concern in particular: (i) the appropriability regime, e.g. the degree to which an innovation can be protected (ranging from 'tight' regimes where technology is extremely difficult to imitate, to very 'weak' regimes where it is almost impossible to protect); and (ii) the degree to which a dominant design has been developed and imposed by one or several firms in an industry or, on the contrary, where there is a state of technological flux (a pre-paradigmatic stage with respect to dominant design).

In the Teece approach, as in all similar approaches influenced by the work of Porter (1984), the key strategic issue is the struggle 'to avoid handing over the lion's share of the profits from its innovation to imitators and/or owners of specialised and co-specialised complementary assets'. Which are the precise key assets required will obviously *vary* from industry to industry or even from product to product, while the degree of external dependence of firms and the risk that they fare badly in the distribution of value added and the flow of rent from innovation will obviously depend on their size, diversification and degree of multinational expansion.

Another analytical framework is the one developed by Bertin (1986) for the interpretation of data from the AREPTIT-SPRU questionnaire to MNEs concerning the *inward* as well as the outward transfer of technology through agreements involving patents (see Bertin and Wyatt, 1986, for a full account of this study). According to Bertin, MNEs must attempt to avoid two pitfalls: 'one is the transfer of advanced or sensitive technology to partners who might turn into dangerous competitors in the future; the second is the acquisition of such technology from a partner who may gain some type of control of a significant share of the firm's activity through its use'. In the face of rising R & D costs and the need to acquire technologies outside the firm, MNEs may be expected to develop strategies along two contrasting models:

In the *first model*, the firm holds a competitive position which is strongly centered in a main field of activity. It allocates the major share of its R & D expenses to this main field. It restricts itself to few transfers of technique, mostly with internal partners—subsidiaries or associates, whether foreign or domestic.

In the *second type*, closer to what we know as the industrial conglomerate model, the firm holds no such definite and strong competitive position or it has several distinct ones. Accordingly its R & D is not as specialised and it turns to external as well as internal partners for frequent technical transfers to complement its own research activity or to valorise its own technical output.

MNE strategies will also be shaped by certain features of the technology concerned, its availability (along with possible limitations set by sellers), the cost of acquisition and the expected returns. Bertin tests his model and comes up with results which show *extremely strong industry-specific features*, partially outriding the initial assumptions:

(i) External transfer partners play a significant role in industries of the world-oligopoly type, such as automobiles and electronics. What is rather new is that such partners are also significant in the chemical industry. The increasing cost of research may exercise a strong influence on the firm's strategy, as suggested in various interviews, compelling the firm to increase technology exchange with competitors.

(ii) This is especially true of the electronics industry where more than two-thirds of firms have the same partners on sales and purchases of technology; the importance of cross-licensing is one possible explanation for this result, assuming a small number of suppliers of new technology.

(iii) Except for world oligopoly sectors, internal partners are given a strong preference in what could be called the core activities of MNEs . . . External partners are preferred in activities where they offer new potentialities for growth in the near future (markets) or in the distant one (new research fields). Uncertainty and heavy costs of entry are grave concerns in new fields of activity such as new materials, communications and advanced computing. These concerns induce firms, even the largest ones, to look for extensive external technical contacts and joint-ventures in research.

A taxonomy with the role of government, supply structures and key features of technology as parameters

As can be anticipated from the earlier sections of this chapter, our own attempt at establishing an interpretative analytical framework of taxonomy does not take the individual firm at its starting point of a theory of corporate strategy, but seeks to interpret inter-firm technical cooperation agreements in the context of the main 'exogenous' factors shaping the individual firm's environment: the progress of internationalisation, the tendency towards concentration and centralisation, the overall fall in the rate of return to capital (OECD, *Economic Outlook*, 1986c), the role of government, and the pace and direction of technological change.[1]

Ever since beginning to work on inter-firm technical cooperation agreements, we have been struck by *the extremely strong industry and/or technology-specific differences* in the type and range of agreements met within different industrial sectors (see OECD, 1986b). These must be related, of course, to the particular features of technology in different

Table 23.2 Main parameters shaping the types of agreements most frequently used by firms in different categories of industry

Main parameters / *Types of agreement*	Role of government in structuring the industry and its markets	International supply structure	Degree of novelty and sophistication of technology	Location of R & D and of 'technological capital' accumulation	Size of investment and rate of capital depreciation	Degree and nature of the systemic attributes of final product	Representative industries
Industrial consortia and international joint ventures aimed at world markets	Important or very important; industries considered as strategic	Highly concentrated (tending towards monopoly or duopoly)	New or very new and/or very sophisticated technologies	Large firms with privileged relations with government large public R & D labs	Very large investments with very long depreciation periods	Strong systemic with high technological content	Space launchers and satellites; long distance civil aircraft; jet engines; weapons and military systems; telecom. equipment
Complex international agreements pertaining to engineering, technology, financing and manufacture for domestic or regional market projects	Important in DCs, limited in advanced countries (except for export credits and foreign policy support during commercial negotiations)	World oligopoly with or without a dominant firm with trends towards cartel-type situations or fairly organised oligopolistic rivalry	Fairly well-known technologies requiring large accumulated know-how	Large firms belonging to the world oligopoly; engineering firms and heavy equipment manufacturers	Large investments and long or very long depreciation	Systemic with medium or low technological content	Large-scale bulk non-ferrous metal electrical power stations (including nuclear); large-scale public infrastructures
R & D joint-venture firms leading to patent pools open to corporate partners	Large buyer and/or purveyor of R & D funds	World oligopoly without a dominant firm and with very acute rivalry	Constantly and rapidly evolving technology	Large firms belonging to the world oligopoly	Large investments in particular in R & D requiring cooperation in generic technology	Not very systemic but sometimes significant technological complementarities	Electronic components
Technology sharing agreements; complex two-way licensing, cross-licensing in separate product markets; second sourcing agreements	As above	As above	As above	As above	As above	As above	Electronic components
	Weaker role	Medium to strong oligopolistic rivalry	Technology changing rapidly after a period of slow development	Ditto	Large investments still within the scope of large firms	Tech. complementarities and/or specialisation in sub-markets	Pharmaceuticals; specialised chemicals (high-grade plastics, composites)
R & D contracts between large and small firms, with or without the use of corporate venture capital	As above	As above	Rapidly evolving technology on the basis of radical paradigm change	Small knowledge-intensive high-tech firms with specialised skills; univ. or public laboratories	Investments in innovation not very costly but with very high risk	As above	Biotechnology; specialised electronics and software; certain new materials

sectors, and also to the nature of supply structures, the role of government in support, the very existence of certain industries, the subsidisation of R & D, and the establishment of non-tariff protection around 'reserved' public procurement markets (Chesnais 1987).

We have also observed that inter-firm technological cooperation across national frontiers invariably *combines* several dimensions or, again, *meets several objectives*, notably access to technology in combination with guaranteed access to certain semi-protected or threatened markets. In many industries, the exchange of technology is also an effective way of consolidating 'mutual recognition' and 'mutual forbearance' between oligopolists and of laying the basis for a strengthening of entry barriers, possibly paving the way for quasi-cartel-like behaviour. In instances where the main objective is access to technology, the type of agreement chosen will depend on the nature of the firm or institution where the technology is located and the steps to be taken to appropriate it. As a result, Table 23.2 has been built using six parameters:

(i) The first concerns the role played by governments in shaping the structures of the industry (through R & D subsidies, large-scale procurement, open involvement in the organisation of the industry, and the division of tasks between firms).

(ii) The second parameter relates to the international supply structure (e.g. situations of near monopoly on duopoly: concentrated world oligopoly with a dominant firm (e.g. computers); concentrated world oligopoly with several firms of near equal strength; loose world oligopoly in combination or not with concentrated oligopoly in domestic or regional markets, etc.).

(iii) The *speed* of technical change, and the degree to which it has a *novel* or *radical* character and is therefore likely to have appeared outside established firms requiring them to resort to external acquisition.

(iv) The country location of the largest R & D outlays and the institutions (firms, government agencies and their laboratories, universities) within which the accumulation of technology has been proceeding. In combination, (iii) and (iv) *influence* the easiness or, on the contrary, the obstacles to *external appropriation*.

(v) The level of R & D and capital investment thresholds and the pressure they put on firms to cooperate.

(vi) The extent to which R & D- and scale-intensive products have *systemic* features, authorising a division of tasks and hence cooperation between firms in their conception and manufacture: this is an important dimension of some type of cooperation well illustrated by Mowery's study (1986) of the aircraft and engine industries.

Using the taxonomy: two examples

Industrial consortia and international joint ventures aimed at world markets (for instance, the General-Electric–SNECMA (France) joint ventures for the development production of the CFM 56 jet engine, see Baranson, 1976, and Mowery, 1986) are the form in which inter-firm cooperation (with which governments are also generally involved directly or indirectly) invariably takes place in industries such as *space products* (launchers and satellites), commercial *aircraft* of the large inter-continental categories, *jet engines*, and increasingly the *defence* industries, where international inter-firm agreements backed or initiated by governments have become more and more frequent.

This form of cooperation involves a carefully negotiated division of industrial responsibilities and workloads, financial risk, assets and profits. Agreements are almost invariably publicised in detail since they often involve the use of public funds. Membership is defined in terms of *percentages* of assets and liabilities and a precise definition of industrial roles.

The use of industrial cooperation as the preferred, if not exclusive form of cooperation can be directly related to

(i) the 'strategic' and 'political' character of the industries concerned;

(ii) the way in which the pace and nature of world demand (including the size of the market and the particularities of market access), coupled with the extremely high costs and complexity of scientific and technical requirements for production, have created a situation where only a very limited number of producers are active, and where the threat of *outright* or *near-monopoly situations* is a real one and where alliance is the only choice open to almost all countries and to practically all firms; but also

(iii) where the strong *systemic*, but at the same time *high-technology features* of the end product allow a division of tasks, either among equals or near-equals (as in the case of the main participants in Airbus industries or Ariane), with all the complex problems of rivalry and coordination which this raises, or between firms in an unequal situation with one firm acting as prime contractor with sole final responsibility (as in the Boeing, TWR or Hughes Aircraft-led US consortia).

The semiconductor industry is one of those where the *greatest number and widest range of forms of agreements are used* by the large firms. The situation reflects, in part, the pace of technological progress in the industry (cf. Dosi, 1984, for a complete interpretation of the early 1980s). The close network of agreements which the major firms have established among themselves must, however, be related to the specific *supply structure of the industry*. The internationalisation and diversification strategies followed by all large electronics firms now place major manufacturers in a situation of oligopolistic rivalry generally involving competition on a combination of identical and/or complementary product markets. Electronics and in particular semiconductors are industries *par excellence* 'characterised by strong oligopolistic and technological competition' where 'MNEs . . . need a direct presence in all those countries which hold leading positions in the development of the industry and of associated technology' (Dunning and Cantley, 1986). As a result one finds:

- at the *stage of precompetitive R & D* a number of *national* alliances between oligopolists belonging to the same country: e.g. the Japanese example of the large industry—MITI projects, e.g. VLSI, the '5th generation computer project' (Sigurdson, 1986) which the United States and Europe have recently tried to imitate (e.g. the three US cooperative arrangements in microelectronics, the Semiconductor Research Corporation (SRC), the Microelectronics Center of North Carolina (MCNC), and the Stanford University Center for Integrated Systems (CIS), which belong to category A of Table 23.1; the UK Alvey, the Esprit project (examples of category B), or again the new private R & D joint-venture corporations (type C), such as the European Computer Research Center (ICL, Bull and Siemens) and, of course, the US Microelectronics and Computer Technology Corporation (MCC);
- at the *competitive stage* a considerable focus *on F-type agreements*, between oligopolists *of different nationality* bearing on already developed technologies. In semiconductors alliances very rarely take the form of consortia of joint ventures, but rather of *bilateral* agreements between oligopolistic rivals. Faced with continually increasing R & D costs, MNEs see at present considerable advantages in sealing mutual recognition through agreements involving a two-way exchange of technology, thus reducing some R & D outlays and at the same time raising the stakes for non-participants in the agreements. In combination and on account of their number, their bilateral agreements create a web-like network of agreements, at various points of which one finds (in a sort of nodal position) the firms which have established the largest number of linkages.

The most carefully documented research on this industry (Hacklisch, 1986) shows that technology-sharing agreements, second-sourcing with technological exchange and joint and/or complementary development accords are the forms of agreement most frequently encountered in semiconductors. Technology-sharing agreements generally involve a two-way exchange of comparable, but *complementary* technical expertise. Major examples are the US–Japan agreements in which US design capabilities have been shared in exchange for Japanese competence in CMOS fabrication technology. A major finding by Hacklisch is that 'The respective strengths of US companies in design and software microprocessor technology and of Japanese firms in CMOS are a prime characteristic in a number of US–Japan agreements in the area of MUCs/MPUs'. Examples given include the Intel–Oki, Intel–Fujitsu, Motorola–Hitachi, Zilog–NEC and Zilog–Toshiba agreements.

The pooling of patents (which will occur if corporations like MCC are successful) and reciprocal licensing confront potential competitors, notably smaller firms or firms contemplating entry into the industry, with formidable competitive stakes and entry barriers. The situation in semiconductors today presents both *similarities* and important *differences* with earlier examples of the 1930s (in heavy electrical equipment and chemicals). Technology-sharing and exchange between the largest firms in the industry are quite certainly increasing the advance of these firms over those which cannot hope to be parties to the most important types of partnerships. They represent one of the factors lying behind the continual increase in concentration at world level. At the same time, however, and this, of course, is the novel aspect, rivalry continues to be extremely acute within the small group of leading world firms, notably between the main US and Japanese manufacturers. This particular combination of cooperation and competition is one of the foundations for the fear sometimes expressed by some observers in the United States (cf. Reich and Mankin, 1986) that the technology agreements set up with US firms are, in fact, providing Japanese firms with an accelerated access to the technology of their US competitors.

Concluding remarks

These must necessarily be extremely brief. The first concerns our belief that there is need for a multi-form theoretical and analytical attack in all the areas discussed from the third section onwards, where important lacunae in our knowledge, due to gaps in theory which urgently require bridging, have been identified.

The second observation concerns the overall message stemming from our analysis, which is not extremely optimistic for developing countries (see also the chapter by Unger), nor indeed for the smaller or less developed OECD countries (Walsh, 1987). Much more research would be required to understand the medium and long-term effects of a situation where the *largest* and *most advanced firms*, technologically speaking, are exchanging *between themselves*, vital, *complementary* technologies. On the basis, however, of available data in the light of the understanding built up by earlier work on *barriers to entry* (Bain, 1956) and *technology gaps* (OECD, 1970), it can safely be stated that such cooperation creates *formidable new entry barriers* at the heart of the industry and with respect to its 'core technology' base (US National Research Council, 1983); thus creating new conditions of interfirm and intercountry dependencies, in the form of a whole new web of dependent technological links *vis-à-vis* the industry leaders. This is felt even by advanced small and medium-size OECD countries. It is certain to affect developing countries very strongly in the future. As Dunning and Cantwell (1986) have put it:

The change in the international environment has further compelled a move towards *integrated technological systems*. The advantages which stem from *such centralised co-ordination* has helped to strengthen the position of many MNEs *vis-à-vis* uninational firms, who are constrained to follow seriously only those activities in which their own countries have an existing or potential locational advantage. [author's emphasis]

The identification of such systems and the development of appropriate corporate strategies, (see the work of LAREA, GEST (1985) on 'Technology cluster Strategies') has already begun to create yawning new technological and industrial gaps between firms and countries, even within the most advanced OECD countries. The capacity on part of firms and countries to recognise (i) that core technologies may lie at the heart of technology *systems* (see Gille, 1978 and also with application to the Japanese situation Imai; 1984); (ii) that centralised coordination must start from a sound and firm domination of technological advance in these core core technologies; and (iii) that successful competition requires appropriate diversified large company industrial group structure, (of which the Japanese 'keiretsu' represents today the most successful example, see Freeman, 1987), *may represent* in coming years one of the most powerful instruments of *inequal development* at work in the world economy.

Note

1. *All* these factors of course are *endogenous* to the overall movement of the capitalist economy. The term 'exogenous' is used simply to designate the incapacity of any given firm, however large it may be and however much it contributes to their acceleration, to control the factors shaping the overall international economic, technological and political context in which it operates. The environment becomes increasingly 'exogenous' (e.g. constraining) as the size of firms become smaller, but *even the largest* MNE can never do more than *adapt*, with major advantages over small firms, *to the changes and challenges of the environment*.

References

Abernathy, H.L. (1978), *The Productivity Dilemna: Roadblock to Innovation in the Automobile Industry*, Baltimore, Johns Hopkins University Press.

Abernathy, H.L. and Utterback, J.M. (1978), 'Patterns of industrial innovation', *Technology Review*, no. 80, June–July.

Adam, G. (1973), 'Multinational Corporations and Worldwide Sourcing', in H. Radice (ed.) (1975), *International Firms and Modern Imperialism*, London, Penguin.

Amin, S. (1970), *L'accumulation à l'échelle mondiale*, Paris, Maspero.

Antonelli, C. (1984), *Cambiamento tecnologico e impresa multinazionali: il ruolo deli reti telematiche nelle strategie globale*, Milano, Franco Angeli.

—— (1987), 'L'impresa rete', *CESPE Papers*, CESPE Fondazione, Rome, mimeo.

Bain, J. (1956), *Barriers to New Competition*, Cambridge, Mass., Harvard University Press.

Baranson, J. (1976), *International Transfers of Industrial Technology by US Firms*, Lexington, Mass., Lexington Books.

Beaud, M. (1987), *Le système national mondial hiérarchisé*, Paris, La Découverte.

Bertin, G.Y. (1986), 'Multinational enterprises: transfer partners and transfer policies', in A.E. Safarian and G.Y. Bertin (eds), *Multinationals, Governments and International Technology Transfer*, London, Croom Helm.

Bertin, G.Y. and Wyatt, S. (1986), *Multinationales et propriété industrielle*, Institut de Recherche sur les Multinationales (IRM), Paris, PUF.

Buckley, P.J. and Casson, M. (1976), *The future of the Multinational Enterprise*, London, Macmillan.

Burstall, M., Dunning, J.H. and Lake A. (1981), *Multinational Enterprises Governments and Technology: The Pharmaceutical Industry*, OECD, Paris.

Burstall, M.L. and Dunning, J.H. (1985), 'International investment in innovation', in N. Wells, (ed.), *Pharmaceuticals Among the Sunrise Industries*, Croom Helm, London.

Caves, R.E. (1971), 'International corporations: the industrial economics of foreign investment', *Economica*, vol. 38, p. 149.

—— (1974), 'International trade, international investment, and imperfect markets', special paper in International Economics No. 10 (November) Dept. of Economics, Princeton University.

—— (1982), *Multinational Enterprise and Economic Analysis*, Cambridge, Cambridge University Press.

Centre de Prospective et d'Evaluation (CPE) (1985), *Rapport sur l'état de la technique*, numéro spécial de *Sciences et Techniques*.

Chesnais, F. (1979), 'Capital financier et groupes financiers: recherche sur l'origine des concepts et leur utilisation actuelle en France' dans *Internationalisation des Banques et des Groupes Financiers*, Séminaire CEREM, Nanterre, novembre 1979, Paris, Éditions du CNRS (1981).

Chesnais, F. et Michon-Savarit, C. (1980), 'Some observations on alternative approaches to the analysis of international competitiveness and the role of technology factor', Conférence sur les indicateurs de la science et de la technologie, OCDE, Paris, miméo.

—— (1984), 'Quelques remarques sur le contexte mondial de la dette des PED et la nature du capital prêté', *Tiers-Monde*, Tome XXV, no 99, juillet–septembre 1984 (numéro spécial sur La Dette du Tiers-Monde).

—— (1986a), 'Technological cumulativeness, the appropriation of technology and technological progressiveness in concentrated market structures', paper presented to the Conference on Technology Diffusion, Venice, March 1986.

—— (1986b), 'Science, technology and competitiveness', *DSTI Review*, no. 1, Autumn.

—— (1987), 'Les accords de coopération technologique et les choix des entreprises européennés', communication au colloque Europrospective, CPE, FAST, Commissariat général au Plan, La Villette, Paris, avril 1987.

—— (1988), 'Internationalisation, changement technique radical et compétitivité des systèmes productifs nationaux', dans J. Noisi (ed.), *Technologie et compétitivité internationale*, Actes du colloque international, Oligopoles, innovations technologiques et concurrence internationale, CREDIT, Montréal, octobre 1987, mimeo.

Coase, R.H. (1937), 'The nature of the firm', *Economica*, no. 4, November, CEPII (1987).

Cohen, S. and Zysman, J. (1987), *Manufacturing Matters: the Myth of the Post Industrial Economy*, New York, Basic Books.

Connor, J.M. and Mueller, W.F. (1977), *Market Power and Profitability of Multinational Corporations*, Report to the Senate Subcommittee on Multinational Corporations, Washington, DC, Government Printing Office.

Cotta, A. (1978), *La France et l'impératif mondial*, Paris, Press Universitaires de France.

Destanne de Bernis, G. (1977), *Relations économiques internationales*, 4e édn., Paris, Dalloz.

Dobb, M. (1963), *The Economic Development of Capitalism*, London, Routledge.

Dosi, G. (1982a), 'Technology industrial structures and international economic performance', paper prepared for the OECD Study on Technology and competitiveness, Paris, mimeo.

—— (1982b), 'Technical paradigms and technological trajectories: suggested interpretation of the determinants and directions of technical change', *Research Policy*, vol. 11, pp. 147–62.

—— (1984), *Technical Change and Industrial Transformation*, London, Macmillan.

Doz, Y., Hamel, G. and Prahalad, C.K. (1986), 'Strategic partnership: success or surrender: the challenge of competitive collaboration', mimeo, INSEAD (Fontainebleau), London Business School, University of Michigan, mimeo.

Dunning, J.H. (1981), *International Production and the Multinational Enterprise*, London, George Allen & Unwin.

—— (1985), *Japanese Participation in British Industry*, London, Croom Helm.

Dunning, J.H. and Cantwell, J.A., (1986), 'The changing role of multinational enterprises in the international creation, transfer and diffusion of technology', paper presented to the Conference on Innovation Diffusion, Venice, March 1980.

Erdilek, A. (ed.) (1985), *Multinationals as Mutual Invaders: Intra-Industry Direct Foreign Investment*, London, Croom Helm.

Fouquin, M. (ed.), (1986), *Industrie mondiale: La compétitivité à tout prix*, Centre d'Etude et Prospectives Internationales, Paris, Economica.

Freeman, C. (1982), *The Economics of Industrial Innovation*, 2nd edn., London, Frances Pinter.

—— (1987), *Technology Policy and Economic Performance: Lessons from Japan*, London, Pinter Publishers.

Fusfeld, H.T. (1986), *The Technical Enterprise, Present and Future Patterns*, Cambridge, Mass., Ballinger.

Germidis, D. (1981), *International Subcontracting: A New form of Investment*, Paris, OECD Development Centre.

Giersch, H. (ed.) (1979), *Proceedings of the 1978 Kiel Symposium on the Economics of Intra-Industry Trade*, Tübingen, J.C.B. Mohr.

—— (1982), *Proceedings of the 1981 Kiel Symposium on emerging Technology: Consequences for Economic Growth, Structural Change and Employment in Advanced Open Economies*, Tübingen, J.C.B. Mohr.

Gille, B. (1978), *Histoire des techniques*, Paris, La Pléïade.

Grou, P. (1983), *La Structure financière du capitalisme multinational*, Paris, Fondation des Sciences Politiques.

Haklisch, C.S. (1986), 'Technical alliances in the semiconductor industry', Center for Science and Technology Policy, New York University, mimeo.

Harrigan, K.R. (1985), *Strategies for Joint Ventures*, Lexington, Mass. Lexington Books.

Harvey, D. (1982), *The Limits to Capital*, Oxford, Blackwell.

Hilferding R. (1970 ed.), *Le capital financier*, Paris Éditions de Minuit.

Hladik, K.J. (1984), *International Joint Ventures*, Lexington, Mass., Lexington Books.

Horst, T. (1974), 'The theory of the firm', in J.H. Dunning (ed.), *Economic Analysis and the Multinational Enterprise*, London, George Allen & Unwin.

Hymer, S.H. (1960), *The International Operations of National Firms: A Study of Direct Foreign Investment*, Cambridge, Mass., MIT (published by MIT Press, 1976).

Hymer, S.H. and Rowthorn, R. (1970), 'Multinational corporations and international oligopoly: the non-American challenge', in C.P. Kindleberger (ed.), *The International Corporation*, Cambridge, Mass., MIT.

Imai, K. and Itami, B. (1984), 'Mutual infiltration of organization and market—Japan's firm and market in comparison with the US', *International Journal of Industrial Organization*, vol. 2, no. 4.

Imai, K. (1984), *Japan's Industrial Policy for High Technology Industries*, Conference on Japanese Industrial Policy in comparative Perspective, New York.

Jones, D. (1987), 'Structural adjustment in the automobile industry', *STI Review*, no. 3, Winter.

Kaplinsky, R. (1987), *'Technological Revolution' and the International Division of Labour in Manufacturing: A Place for the Third World?*, EADI Conference on New Technologies and the Third World, Institute of Development Studies, University of Sussex, Brighton.

Kemp, J. (1967), *Theories of Imperialism*, London, Routledge.

Keynes, J.M. (1936), *The General Theory of Employment, Interest and Money*, London, Macmillan.

Klein, B. (1979), 'The slowdown in productivity advances: a dynamic explanation', in C.T. Hill and J.M. Utterback (eds), *Technological Innovation for a Dynamic Economy*, New York, Pergamon Press.

Klein, B., Crawford, R.G. and Alchian, A.A. (1978), 'Vertical integration, appropriable rents and the competitive contracting process', *Journal of Law and Economics*, October.

Knickerbocker, F.T. (1973), 'Oligopolistic reaction and multinational enterprise', Graduate School of Business Administration, Harvard University.

LAREA/GEST (1985), *Grappes technologiques et stratégies industrielles*, Étude CPE no. 57, Centre de Prospective et d'Évaluation, Paris.

Lenin. V.I. (1970 ed.), 'Imperialism: the highest stage of capitalism', in *Selected Works*, Moscow, State Editions (First edition, Zurich, 1915).

Lundvall, B.A. (1986), *Product Innovation and User-producer Inter Action*, Industrial Development Research Series, no. 31, Aalborg University Press.

Mandel, E. (1975), *Late Capitalism*, London, New Left Books.

Mariti, P. and Smiley, R.H. (1983), 'Co-operative agreements and the organisation of industry', *Journal of Industrial Economics*, vol. XXXI, no. 4, June.

Michalet, C.A. (1976), *Le capitalisme mondial*, Paris, nouvelle édn. entièrement

refondue, 1985, Paris, PUF.
Michalet C.A. and Delapierre, M. (1978), 'The impact of multinational enterprises on national scientific and technological capacities in the computer industry', mimeo.
Minskey, H.P. (1982), *Inflation, Recession and Economic Policy*, Brighton, Wheatsheaf Books.
Momigliano, F. (1981), 'Technological innovation, international trade and direct foreign investment: old and new problems for economic theory and empirical research', paper prepared for the OECD Study on Technology and Competitiveness, Paris, mimeo.
Momigliano, F. and Dosi, G. (1983), *Tecnologia e organizzazione industriale internazionale*, Bologna, Il Mulino.
Morin, F. (1974), *La structure financière du capitalisme français*, Paris, Presses Universitaires de France.
Mowery, D.C. (1986), 'Multinational joint ventures in product development and manufacture: the case of commercial aircraft, department of Social Sciences, Carngie Mellon University, Pittsburg, mimeo.
Muccielli, J.L. and Thuillier, J.P. (1982), *Multinationales européenes et investissements croisés*, Paris, Presses Universitaires de France.
Murray, R. (1971), 'The internationalisation of capital and the nation state', in H. Radice (ed.) (1925), *International Firms and Modern Imperialism*, London, Penquin.
Mytelka, L.K. (1988), 'New modes of competition in the textile and clothing industries: some consequences for the Third World', dans J. Niosi (ed.), *Technologie et compétitivité internationale*, Actes du colloque international, Oligopoles, innovations technologiques et concurrence internationale, CREDIT, Montréal, octobre 1987, mimeo.
Nelson, R. (1980), 'Competition, innovation, productivity growth and public policy', in H. Giersch (ed.), *Towards an Explanation of Economic Growth*, Symposium 1980, Tübingen, J.C. Mohr, 1981.
Nelson, R. and Winter, S.G. (1982), *An Evolutionary Theory of Economic Change*, Cambridge, Mass., The Belknap Press of Harvard University Press.
Newfarmer, R.S. (1983), 'Multinationals and the marketplace magic', in C.P. Kindleberger (ed.), *The Multinational Corporation in the 1980s*, Cambridge, Mass., The MIT Press.
—— (1985), *Profits, Progress and Powerty: Case Studies of International Industries in Latin America*, Indiana, University of Notre Dame Press.
OECD (1968), *Gaps in Technology, Analytical Report*, Paris, OECD.
—— (1979), 'Multinational enterprises and national scientific and technological capacities in the food processing industries', mimeo.
—— (1986a), *Science and Technology Indicators No. 2: R & D, Invention and Competitiveness, Paris, OECD.*
—— (1986b), 'Technical co-operation agreements between firms: some initial data and analysis', mimeo.
—— (1986c), *Economic Outlook*, no. 39, May, Paris, OECD.
—— (1987a), 'The contribution of science and technology to economic growth', mimeo.
—— (1987b), *Structural Adjustment and Economic Peformance*, Paris, OECD.
Ohmae, K. (1985), *Triad Power: The Coming Shape of Global Competition*, New York, The Free Press.

Oman, C. (1984), *New Forms of International Investment in Developing Countries*, Paris, OECD Development Centre.

Palloix, C. (1975), *L'internationalisation du capital*, Paris, Maspéro.

Papon, P. (1983), *Pour une prospective de la science: recherche et technologie, les enjeux de l'avenir*, Paris, Seghers.

Pastré, O. (1980), *La strategie internationale des groupes financiers américains*, Paris, Le Seuil.

Payne, B. (1986), *Co-operation and Technological in the International Machine Tool Industry: An Explaratory Analysis of Commercial and Technical Agreements*, London, Technical Change Center, mimeo.

Pavitt, K. (1984), 'Sectoral patterns of technical change: towards a taxonomy and a theory', *Research Policy*.

Porter, M.E. (1985), *Competitive Advantage*, New York, The Free Press.

Reich, R.B. and Mankin, E.D. (1986), 'Joint ventures with Japan give away our future', *Harvard Business Review*, March/April.

Ricotta, E. and Mariotti, S. (1986), *Diversification Agreements Among Firms and Innovative Behaviour*, Paper presented at the Conference on Innovation Diffusion, Venice, March 1986.

Rosenberg, N. (1976), *Perspectives on Technology*, Cambridge, Cambridge University Press.

—— (1982), *Inside the Black Box: Technology and Economics*, Cambridge, Cambridge University Press.

Sciberras, E. (1980), 'Technical innovation and international competitiveness in the television industry', Brighton, Science Policy Research Unit, mimeo.

Scherer, F.M. (1970), *Industrial Market Structure and Economic Performance*, (2nd ed., 1980). Chicago, Rand. McNally College Publishing Company.

Schumpeter, J.A. (1943), *Capitalism, Socialism and Democracy*, New York, Harper, (2nd ed. 1947).

Scott, J. (1979), *Corporations, Classes and Capitalism*, London, Hutchinson.

—— (1986), *Capitalist Property and Financial Power*, Brighton, Wheatsheaf Books.

Sigurdson, J. (1986), *Industry and State Partnership in Japan: The Very Large Scale Integrated Circuits (VLSI) Project*, Discussion Paper no. 168, Lund, Research Policy Institute.

Sylos-Labini, P. (1962), *Oligopoly and Technical Progress*, Cambridge, Mass., Harvard University Press (rev. ed., 1969).

Teece, D.J. (1986), 'Capturing value from technological innovation: integration, strategic partnering and licensing decisions', paper presented at the Conference on Innovation Diffusion, Venice, March 1986.

Telesio, P. (1979), *Technology Licencing and Multinational Enterprises*, New York, Praeger.

UNCTNC (United Nations Centre on Transnational Corporations) (1987), *Recent Developments Related to Transnational Corporations and International Economic Relations*, United Nations Economic and Social Council, New York, January, mimeo.

US National Research Council (1983), *International Competition in Advanced Technologies, Decision for America*, National Academy, Washington, DC.

US Congress, Office of Technology Assessment (1984), *Commercial Biotechnology: An International Analysis*.

US National Research Board (1985), *Science Indicators for 1985*, Washington, DC.

Vernon, R. (1966), 'International investment and international trade in the product cycle', *Quarterly Journal of Economics*, May.

—— (1974), 'The location of economic activity' in J.H. Dunning (ed.), *Economic Analysis and the Multinational Enterprise*, London, George Allen & Unwin.

—— (1979), 'The product cycle in a new international environment', *Oxford Bulletin of Economics and Statistics*, November.

Walsh, V. (1987), 'Technology, competitiveness and the special problems of small countries', *STI Review*, no. 2, September.

Williamson, O.E. (1975), *Markets and Hierarchies: Analysis and Antitrust Implications*, New York, Free Press.

[10]

International and domestic collaboration in research and development

Chapter 8 described the changes that have occurred during the past three decades in the international competitive and technological environment of U.S. manufacturing firms. These changes, which include the development of Japanese technological competition, have led to considerable experimentation and innovation in the U.S. R&D system. One of the most widespread types of innovation is the effort by firms to develop external sources of research and development expertise. These efforts have resulted in considerable expansion of collaboration in R&D that involves U.S. and foreign firms, as well as U.S. universities. Collaborative ventures obviously represent a considerable shift away from the in-house research and development that dominated industrial research during the first seventy-five years of this century.

The causes and consequences of these new organizational structures are still uncertain, since in most cases collaboration is less than ten years old. Nevertheless, there appear to be important contrasts in structure and motives between the international and domestic research and development ventures into which U.S. firms have entered. One must distinguish among at least three broad categories of research collaboration: collaborative ventures between U.S. and foreign firms; research collaboration among U.S. firms; and domestic university–industry research collaboration. International collaborative ventures focus mainly on development, production, and marketing, rather than precommercial research. Thus far, domestic collaborations among U.S. firms have been concerned with research that is less applied in character and less closely linked to a specific commercial product. In general, however, these domestic collaborative ventures do not focus primarily on basic research despite the intentions and founding aspirations of several of them. Indeed, their role and activities are likely to change over time. Finally, university–industry research collaboration appears to incorporate scientific and engineering research of a more fundamental character. These different forms of collaboration are not substitutes but complements, and their effectiveness depends on a strong in-house research capability.

Public funding from the federal and state governments has supported much of the recent growth in university–industry research collaboration.

With the significant exceptions of Sematech and the National Center for Manufacturing Sciences (see Chapter 6), however, public funding has not been important in recently established U.S. research consortia involving only private firms, nor have U.S. public funds supported the international collaborative activities of U.S. firms. This state of affairs contrasts with that in Western Europe, where public funds have been very important in such collaborative research programs as Alvey in Great Britain or Eureka, Esprit, and Brite in the Common Market countries. Public funds have supported the international Airbus Industrie consortium that involves Western European firms. Research collaboration among industrial firms in Japan also has benefited from public funding.

Advantages of research collaboration

Many of the widely cited advantages of collaborative research are grounded in the neoclassical analysis of R&D investment discussed in Chapter 1 and elsewhere. As Bozeman, Link, and Zardkoohi (1986) note, if research collaboration among firms can lower the costs to any single firm of R&D and if the results of such research are made available to the participants, cooperative research can reduce the severity of the market failure stemming from the limited appropriability of research results. Since the profits from basic research are particularly difficult to appropriate, this argument suggests that cooperation among firms in basic research should be especially effective.

Access to research results at lower cost has been an important motive for the establishment of numerous cooperative research programs and university–industry cooperative research programs in recent years. These programs allow participants to monitor developments in specific technologies or scientific disciplines at lower cost than the development of an independent in-house research capability. The importance of such a monitoring capability has increased in recent years with growth in the costs of research and development. As we noted in Chapter 8, technological convergence, which has made unfamiliar technologies in many industries central to the competitive future of firms in these industries, also has increased the importance of this monitoring function.

Another argument in favor of cooperative R&D concerns its efficiency. Duplication among the research activities of competitor firms (Hirshleifer, 1971) can be reduced through collaboration in research. Moreover, if the research process is characterized by economies of scale, collaboration can lower the costs per unit of R&D. By spreading the costs of R&D among firms, cooperative R&D also lowers the costs to any single firm of the failure of a research program, reducing another disincentive to invest in R&D.

Disadvantages of collaborative research

The benefits of collaborative research rest on a number of assumptions. Critical evaluation of these assumptions suggests that cooperative research ventures are not a panacea and will not be equally useful in all industries. Research collaboration may aid the rapid commercialization or development of many technologies, but collaboration alone is insufficient for commercial success or the realization of an economic payoff.

The first limitation of cooperative research rests on the criticism of the neoclassical framework we set out in earlier chapters. The fruits of research do not consist solely of information that can be utilized for commercial purposes at minimal cost. The output of cooperative research must be absorbed by the participant firms and transformed into commercially relevant knowledge. This transformation typically requires considerable intrafirm expertise (Mowery, 1983; Cohen and Levinthal, 1987). Cooperative research thus is not a substitute for, but a complement to, in-house research. In order to exploit externally performed research, whether this research is performed in a multifirm consortium, a federal laboratory, or a university, participant firms must have some in-house expertise. Some duplication of the in-house research investments of firms thus is inevitable even when these firms participate in collaborative research.

Moreover, to the extent that basic research requires a larger investment in this in-house "absorption mechanism," as the Balfour Committee referred to it (Chapter 5), participants in cooperative research may react to the high costs of exploiting collaborative basic research by shifting the collaborative research agenda away from basic research. These pressures appear to have led such well-established cooperative research institutions as the Electric Power Research Institute to reduce their commitment to basic research, as the discussion in Chapter 1 noted.

The structure of cooperative research programs also may reduce the productivity of the research performed within them. Nelson (1961) has argued that diversity in research projects provides an important hedge against the possibility that any single research project will be fruitless. Yet if all or most firms in an industry participate in a cooperative research scheme, the number of independent lines of inquiry in specific problem areas may decline. Heavy reliance on cooperative research may reduce the diversity of research projects in a sector or technology (although this outcome is not a necessary result of cooperative research), reducing the productivity of the larger research effort. Scott's analysis (1987) of preliminary evidence suggests that the National Cooperative Research Act (NCRA) may have had this effect.

The existence of the research economies of scale that are widely cited as a source of efficiency gains through research collaboration also is not well supported by empirical evidence. The average costs of research experiments at large, specialized research facilities, such as wind tunnels or advanced instrumentation (discussed later in this chapter), will decline with more intensive use. There exists no strong evidence, however, suggesting that all industrial R&D exhibits significant economies of scale (Fisher and Temin, 1974), nor are there indicators that enable one to identify the industries or technologies characterized by R&D scale economies.

Cooperative research that focuses exclusively on basic research offers relatively few opportunities for collusion among competitors in the marketplace. Collaborative research does, however, create some potential for anticompetitive behavior in research and in other areas of business activity.[1] There is little evidence that cooperative research programs have resulted in such collusion in research, development, or other areas of business behavior (e.g., pricing or output decisions), nor need collusion among U.S. firms result in significant market power within an open economy. Nevertheless, such collusion could restrict the diffusion of a technology in the early stages of its development or prevent the development of a diversified portfolio of approaches within a technology. The potential for such abuses is greater in domestic and international ventures that involve applied research, product development, or manufacturing, rather than basic research.

International collaborative ventures

As we noted in Chapter 8, during the past two decades the relative economic and technological strengths of the United States have declined.[2] These strengths are now more evenly distributed among U.S. and foreign firms. Combined with changes in product and process technologies and in the industrial and trade policies of the United States and foreign governments, this new international economic environment has contributed to growth in the number and importance of collaborative ventures between U.S. and foreign firms in product development, manufacture, and marketing, activities that are generally somewhat further "downstream" than those typically included in U.S. domestic research collaborations.

Many of these interfirm alliances involve significant exchanges of technology, and several assessments have expressed concern about the long-term implications of international collaboration for the competitiveness of

[1] Collaboration among U.S. automobile firms during the 1960s in applied research on anti-pollution technologies was the target of an antitrust suit by the U.S. Department of Justice that was based in part on these concerns (Office of Technology Assessment, 1981; Brodley, 1982; Alic, 1986).

[2] This section draws on Mowery (1988c).

U.S. industry.[3] Other accounts, as we saw in Chapter 8, assert that collaborative ventures will produce a "hollow corporation," not engaged in producing goods, employing a small number of highly paid financial and marketing executives and few if any production workers ("Special Report," *Business Week*, March 3, 1986). Still others view international collaboration as an essential part of global competitive strategy (Perlmutter and Heenan, 1986; Porter and Fuller, 1986).

International collaborative ventures are not completely novel undertakings for U.S. firms, but those into which U.S. firms have entered during the past decade differ in some significant ways from the joint ventures of the 1950s and 1960s. International collaboration is in part a response to the same factors that have given rise to increased domestic collaboration in research among U.S. firms. Why and how have these factors contributed to the growing resort to international collaborative ventures? How do they influence the types of collaboration between U.S. and foreign firms that have developed in recent years?

Collaboration between U.S. and foreign firms assumes many forms. Many collaborative ventures fit a narrow definition of a joint venture, including separate incorporation as an entity in which equity holdings are divided among the partners. Others, such as partnerships between "risk-sharing" subcontractors and prime contractors, or the purchase by one firm of an equity share in another, do not. In some industries, joint ventures and other forms of collaboration are extensions of subcontracting relationships that cover product development and manufacture; others focus on marketing of products manufactured largely by one partner. The following discussion uses the terms "collaboration" and "joint venture" interchangeably and therefore occasionally denotes as joint ventures organizational structures that do not fit the legal definition of this entity.[4]

An international collaborative venture may be defined as ***interfirm collaboration in product development, manufacture, or marketing that spans national boundaries, is not based on arm's-length market transactions, and includes substantial contributions by partners of capital, technology, or other assets.*** This definition excludes other forms of international economic activity, such as export, direct foreign investment (which implies

[3] Reich and Mankin (1986) argue that "the U.S. strategy [in joint ventures with Japan] seems shortsighted. In exchange for a few lower skilled, lower paying jobs and easy access to our competitors' high-quality, low-cost products, we are apparently willing to sacrifice our competitiveness in a host of industries – autos, machine tools, consumer electronics, and semiconductors today, and others in the future" (pp. 78–9).

[4] The formal structure of collaborative ventures, however, often has little to do with either their management or their success, as Gullander (1976) has noted: "There are indications that the difference between a contractual and an equity relationship is highly exaggerated; sophisticated 'cooperators' seem to downplay the importance of ownership control as compared to management control or control through other means" (p. 86).

complete intrafirm control of production and product development activities), and the sale of technology through licensing.

Joint ventures have long been common in extractive industries such as mining and petroleum production (see Stuckey 1983) and account for a significant share of the foreign investment of U.S. manufacturing firms since World War II. Several features of recent collaborative ventures, however, differentiate them from prior cases. The number of collaborations, both those involving only U.S. firms and those between U.S. and foreign enterprises, appears to have grown. Joint ventures now appear in a much wider range of industries.[5]

International alliances involving U.S. firms now are appearing in industries in which direct foreign investment or multinational corporate organization was the exception rather than the rule. Commercial aircraft, for example, is an industry in which direct foreign investment was minimal (Hirsch, 1976). Yet as we made clear in Chapter 7, this industry now is among the leaders in international collaboration (Mowery, 1987). Other U.S. industries in which these ventures have assumed considerable importance in recent years include both "mature" (e.g., steel and automobiles) and "young" (biotechnology) industries.

Many recent collaborative ventures include activities, such as research, product development, and production for world markets, that were absent from the ventures of the pre-1975 era, which focused primarily on production and marketing for a specific foreign market. Collaborations that include R&D and product development may result in significant technology transfer. In some cases (e.g., commercial aircraft), most of the technology transfer within international alliances consists of exports of technology from the United States. In other industries, however, such as steel or autos, these ventures are associated with significant U.S. imports of technology.

International collaborative ventures are based on an exchange relationship, but the commodities exchanged often differ from those exchanged

[5] Harrigan (1984) found that domestic joint ventures involving U.S. firms had grown during the previous decade. In the 1960s, joint ventures were concentrated in chemicals; primary metals; paper; and stone, clay, and glass industries, but now extend beyond these sectors (see Harrigan 1985). Hladik (1985) found significant growth from 1975 to 1982 in the number of international joint ventures involving U.S. firms, a growth trend that almost certainly continued through 1985. Hladik's conclusions disagree with those of Ghemawat, Porter, and Rawlinson (1986), who compiled a time series of "international coalition announcements" (joint ventures, license agreements, supply agreements, and "other long-term interfirm accords" [p. 346]) covering 1970–82 that displays no upward trend. The differences are likely to be more apparent than real; the Ghemawat et al. data base includes a number of interfirm mechanisms of collaboration that are excluded by Hladik. The possibility thus exists that a shift in the mix of the different forms considered by Ghemawat et al. occurred during 1970–82, as joint ventures and other collaborative agreements increased in importance relative to such alternatives as licensing.

through licensing or export. Some collaborations allow U.S. and foreign firms to pool their technological assets in a single product or product line without merging all of their activities into a single corporate entity. Other collaborative ventures combine one firm's technological capabilities with the marketing or distribution assets of another for a single product. Why are collaborative ventures, rather than alternative mechanisms such as licensing, employed for this exchange?

Alternative channels for the exploitation of technological and other firm-specific assets

Alternatives to direct foreign investment for the exploitation of firm-specific assets include licensing and collaborative ventures. These different channels for the realization of the value of an asset may be substitutes or complements, and firms frequently utilize multiple channels. The advantages and disadvantages of each of these methods depend on the characteristics of the asset in question. These factors also affect the activities undertaken within collaborative ventures, an issue that receives attention later in this chapter.

In many instances, collaborative ventures provide an alternative to licensing for the exchange of corporate assets that are based on a firm's technological capabilities.[6] Markets for the licensing of advanced technologies that exploit firm-specific knowledge often are very thin, however, with few buyers or sellers. The dearth of alternative outlets for sellers or alternative sources for buyers means that opportunistic behavior may hamper the operation of markets for technology licenses.[7] Other impediments to licensing technological assets include the tacit nature of much of the knowledge necessary to exploit the technology, the need to reveal a great deal about a technological advance in order to convince a prospective licensee of the value of the license (Arrow, 1962; Teece, 1986), and the problems of regulating licensor and licensee behavior in a dynamic and uncertain world.

Licensing is likely to be preferred to either direct foreign investment or collaborative ventures in technologies that are not complex, have strong,

[6] Caves (1982) notes that "as indicators of these [firm-specific] assets, economists have seized on the outlays for advertising and research and development (R&D) undertaken by firms classified to an industry. That the share of foreign subsidiary assets in the total assets of U.S. corporations increases significantly with the importance of advertising and R&D outlays in the industry has been confirmed in many studies" (p. 9).

[7] In Williamson's terminology, a "small numbers condition" characterizes such markets: "The transactional dilemma that is posed is this: it is in the interest of each party to seek terms most favorable to him, which encourages opportunistic representations and haggling. The interests of the *system*, by contrast, are promoted if the parties can be joined in such a way as to avoid both the bargaining costs and the indirect costs (mainly maladaptation costs) which are generated in the process" (1975, p. 26, emphasis in the original).

well-enforced patents, are relatively mature, and do not rely on "user-active" innovation (von Hippel, 1976), with its requirements for strong links between marketing and product development. Industries in which these features are not characteristic of the technology and means other than licensing are important in realizing value from technology include steel, automobiles, commercial aircraft, and telecommunications equipment. Licensing is an important alternative to international collaboration, however, in pharmaceuticals. Licensing and collaboration often complement one another in segments of the microelectronics and robotics industries and in biotechnology.

Direct foreign investment supports the exploitation of firm-specific technological capabilities but involves high risks and costs. The cost penalties of establishing multiple production facilities argue against direct foreign investment in industries in which production technologies exhibit a large minimum efficient scale or strong plant-specific learning and cost-reduction effects; commercial aircraft (primarily airframes) and steel are examples of such industries. The high fixed costs of establishing an offshore production, distribution, and marketing network may preclude direct foreign investment. Licensing, direct foreign investment, and export all force the innovating firm to bear all of the costs and risks of research and development, which have grown considerably in many industries, as we argued in Chapter 8.

Uncertainties about economic and political conditions in many foreign markets reduce the attractiveness of direct foreign investment. Political barriers have impeded the establishment of wholly owned production facilities in high-technology industries in a number of nations; both the Japanese and the U.S. governments, for example, have opposed direct foreign investment in their domestic semiconductor industries. International joint ventures have substituted for, and in some cases have complemented, Japanese and U.S. firms' direct foreign investment in this and other industries. A number of collaborative ventures in the integrated circuit, automobile, and pharmaceuticals industries have led to the establishment of a wholly owned offshore production or marketing facility.[8]

What advantages does international collaboration have over licensing or direct foreign investment? Many of the contractual limitations and transactions costs of licensing for the exploitation of technological capabilities can be avoided within a collaborative venture. The problem of determining the value of partners' contributions can be reduced through

[8] This is consistent with Porter and Fuller's observation that collaborative ventures centered on marketing "may be particularly unstable, however, because they frequently are formed because of the access motive on one or both sides. For example, one partner needs market access while the other needs access to product. As the foreign partner's market knowledge increases, there is less and less need for a local partner" (1986, p. 334).

collaboration. Partner firms make financial commitments to a collaborative venture that back their claims for the value of the assets they contribute; such financial commitments can substitute for the complete revelation of the value and characteristics of the asset that could be necessary to complete a licensing agreement.[9]

The noncodified, "inseparable" character of firm-specific assets that may preclude their exploitation through licensing need not prevent the pooling of such assets by several firms within a joint venture, or the effective sale of such assets by one firm to another within a joint venture. Technology transfer may also be controlled or regulated more effectively within collaborative ventures than in licensing transactions. Joint ventures enable partner firms to "unbundle" their portfolios of technological assets and transfer components of this portfolio, individual components of which may be worthless in isolation, to a partner. The transfer of technology through a joint venture from a technologically advanced firm to a less advanced enterprise may therefore enable the technologically "senior" firm to reap some financial returns to portions of its portfolio of technological capabilities. In many cases, these returns cannot be attained through licensing because of the difficulties of unbundling the firm's technological portfolio through a license. Moreover, monitoring the behavior of the recipient of technology within a joint venture reduces the risk that the transferor will not benefit from any improvements in transferred technologies made by the recipient. Examples of collaborative ventures that illustrate this argument may be found in steel (National Steel and Nippon Kokan), automobiles (Toyota and General Motors), and commercial aircraft (General Electric and SNECMA).

Collaborative ventures offer an alternative to the complete merger of firms as a means of pooling assets. Such ventures fall short of a complete merger of the partner firms, covering only a limited range of functions or products. Partner firms may well be competitors in other product areas.

[9] Brodley (1982) summarizes the advantages of joint ventures, defined as separate corporate entities in which all partners hold equity shares, over mergers or market transactions as follows: "By providing for shared profits and managerial control, joint ventures tend to protect the participants from opportunism and information imbalance. The problem of valuing the respective contributions of the participants is mitigated, because they can await an actual market judgment. The temptation to exploit a favored bargaining position by threatening to withhold infusions of capital or other contributions is reduced by the need for continuous cooperation if the joint venture is to be effective. Moreover, a firm supplying capital to the joint venture can closely monitor the use of its contributed capital and thereby reduce its risk of loss. Common ownership also provides a means of spreading the costs of producing valuable information that could otherwise be protected from appropriation only by difficult-to-enforce contractual undertakings. Finally, joint ventures can effect economies of scale in research not achievable through single-firm action. Because of these advantages, joint ventures are especially likely to provide an optimal enterprise form in undertakings involving high risks, technological innovations, or high information costs" (pp. 1528–9).

Collaboration may provide established firms a faster and less costly means than internal development to gain access to new technologies that are not easily licensed. This "technology access" motive for collaboration between established and young firms has been particularly important in industries based on new technologies, such as biotechnology and robotics.

By comparison with direct foreign investment, licensing, or export, collaborative ventures also reduce the financial and political risks of innovation and foreign marketing. The products of a collaborative venture between a U.S. and a foreign firm may well encounter fewer political impediments to market access in the domestic market of the foreign firm than would direct exports from the U.S. firm.

Nonetheless, the potential difficulties of collaborative ventures should not be minimized. Management of these undertakings has proved to be extremely difficult. The amount of technology transfer occurring within collaborations that focus on new product development, for example, may cause conflict in ventures that involve firms with different technological capabilities. The incentives of the partners in such ventures often are opposed, as the senior firm wishes to minimize, and the junior firm to maximize, the amount of technology transfer.

This problem can be illustrated by the case of International Aero Engines, a consortium developing the V2500 jet engine that includes Pratt and Whitney, Rolls Royce, Fiat, Motoren-Turbinen Union, Ishikawajima-Harima Heavy Industries, Kawasaki Heavy Industries, and Mitsubishi Heavy Industries. Pratt and Whitney and Rolls Royce, the "senior partners" within the consortium, attempted to minimize the transfer of engine technology within the consortium by designing the engine in modular form and assigning the development of different modules to different participants. Serious problems in the integration of the engine components, however, have led to delays in the delivery of the V2500 engine and to a loss of orders ("U.S., Europeans Clash . . . ," "Pratt and Whitney Expands Role . . ."; Carley, 1988).

Joint product development ventures between erstwhile competitors that cover only one product line among the many produced by the partner firms may also be unstable. Technological change can cause independently manufactured products to become competitors with those jointly developed within the partnership. Such encroachment contributed to the 1977 demise of the collaboration of Pratt and Whitney and Rolls Royce that was intended to develop a high-bypass, high-thrust engine (the JT10D), and contributed to the collapse of a joint venture between Fokker and McDonnell Douglas in 1982. Collaborative ventures that focus on marketing by one partner of the products of another may not endure, as Porter and Fuller (1986) note, since one partner's knowledge of market conditions and demand may lose its value to the other partner.

Why have international collaborative ventures, which represent a hybrid of interfirm and intrafirm modes for the exchange or sharing of technological and other assets, assumed such importance recently, and why do these ventures now incorporate a wider range of activities? The basic answer is simple: Changes in the technological and policy environment (some of which were discussed in Chapter 8) mean that the potential contributions of foreign firms to collaborative ventures now are much more attractive to U.S. firms.

Causes of increased reliance on joint ventures

Changes in the technical capabilities of foreign firms and in the nature of product demand have increased U.S. firms' demand for foreign partners in collaborative ventures. The enhanced technological capabilities of many foreign firms make them more attractive partners in joint ventures with U.S. firms. Foreign firms now are better able to absorb and exploit advanced technologies from U.S. firms in industries in which there remains a substantial technology gap between U.S. and foreign firms. In other industries, foreign firms either are more advanced or are the technological equals of U.S. firms and therefore can contribute managerial or technological expertise to joint ventures with U.S. firms. U.S. firms in the automotive and steel industries and some U.S. firms in the microelectronics industry now collaborate with foreign firms to gain access to superior foreign technologies.

The growth of nontariff trade barriers that was discussed in Chapter 8 has also increased the incentives for U.S. firms to seek foreign partners. Tariff barriers tend to favor direct foreign investment or joint production ventures as means for market penetration, since they affect only the relative prices of domestically produced and imported goods. Nontariff barriers, however, especially procurement policies or technology transfer requirements, favor the use of collaborative ventures that incorporate product research, development, and marketing, as well as manufacture. Purchase decisions of foreign governments play a major role in the export markets for such goods as commercial aircraft and telecommunications equipment. These decisions can be influenced by offsets, the production (or development and production) of components for the purchased product by domestic firms in the purchase nation.[10] Foreign governments also

[10] Recent efforts by a number of European governments to sell off publicly owned enterprises, such as the British government's divestiture of British Airways, British Aerospace, and British Telecom, as well as discussion of plans by the Dutch, West German, and Swedish governments to sell all or part of their state-owned airlines, may prefigure a significant change in the character of major foreign markets in some industries. This "trend," however, is a very modest ripple at present and will not dissolve the informal ties and sources of financial pressure and support that enable governments to exert considerable influence on the purchase decisions of "private" corporations.

frequently provide development funding and risk capital to domestic firms as part of industrial development policies. Combined with high and rapidly increasing product development costs in many U.S. industries, such as microelectronics, commercial aircraft, telecommunications equipment, and robotics, the availability of capital from public sources for foreign firms has enhanced their attractiveness as partners in product development ventures.

Just as foreign government trade and industrial policies have created incentives for U.S. firms to collaborate with foreign firms in export markets, nontariff restrictions on foreign access to the U.S. market have led to increased collaboration between U.S. firms and foreign firms wishing to sell in the United States. In several protected U.S. industries, a foreign production presence has been achieved through a joint venture. Examples include the Toyota–General Motors and Nippon Kokan–National Steel ventures. In the wake of the United States–Japan Semiconductor Agreement and Fujitsu's failure to acquire Fairchild Semiconductor, joint ventures may become a more important means for Japanese semiconductor producers to establish a U.S. production base.[11]

Paradoxically, the pursuit by the United States and other industrialized nations of nationalistic policies of support for domestic industries has encouraged the development of consortia spanning national boundaries. Although the postwar growth of multinational firms and direct foreign investment raised the prospect in some assessments of "global firms" to whom national boundaries would mean little or nothing, much of the current wave of international joint venture activity reflects the opposite phenomenon. National governments are able to influence not only production but, increasingly, the product development and technology transfer decisions of firms through the use of trade and other policies.

Many of the technological factors we discussed in Chapter 8 have also influenced recent decisions of U.S. firms to collaborate with both domestic and foreign firms. Firms in industries in which product cycles are becoming shorter, a group including microelectronics, telecommunications equipment, and robotics, may seek rapid market penetration at low cost through collaboration with a foreign firm with an established marketing network. In addition, advanced communications and computer-assisted design and production technologies facilitate the operation of international collaborative ventures.

Factors influencing the structure of international collaborative ventures

One of the most important motives for collaboration is access to markets, whether in the United States or in a foreign nation. The asset provided in

[11] See Borrus (1988) for a similar argument.

exchange for market access is technology. The form in which the technology is provided, which is determined by both the motives of the participants and the characteristics of the product, plays a central role in structuring the collaboration. Thus, where the technology is provided in essentially "embodied" form (i.e., in a finished product), collaboration either is unimportant or focuses largely on marketing, as in the pharmaceutical industry (outside of biotechnology).

In more mature high-technology industries, such as telecommunications equipment, microelectronics, and commercial aircraft, the high costs of new product development, demanding requirements for systems integration, and the nature of political barriers to market access all mean that many collaborative ventures have focused on product development. A similar focus characterizes collaborative ventures in robotics between user and supplier firms that exchange proprietary data between the partners. Increasingly complex requirements for technology integration in robotics have also led to a number of domestic and international collaborative ventures among suppliers of robotics and factory-automation equipment that deal primarily with product development. In commercial aircraft, telecommunications equipment, and segments of the microelectronics industry, the desire of U.S. firms for risk-sharing partners and access to foreign capital or technology provides an additional motive to focus collaboration on research and product development.

In other high-technology industries with a large number of new firms, such as biotechnology, much domestic and international collaboration focuses on the marketing and distribution by established firms of the technologies developed by new entrants. This type of collaboration aids entry into new domestic or foreign markets and need not always incorporate joint product development. The major international collaborations in the automotive and steel industries center on the exchange of foreign process technology, managerial expertise, and production systems for access to U.S. markets. These joint ventures accordingly deal with production.

Since technology transfer is at the center of many joint ventures, the management by partner firms of both technological development and technology transfer is critical to the success or failure of international collaboration. The interests of technological leaders, reluctant to allow the transfer of key technological capabilities, differ from those of technological followers, whose participation may hinge on the amount of technology transfer. The success of a project may be undercut by the attempts of "junior" partners, motivated by their desire to maximize technology transfer and learning, to participate in all aspects of the project, rather than specializing in a particular area or activity.

The importance of technology transfer as a source of cohesion within ventures between a dominant and a subordinate firm nevertheless means that these ventures often are more durable and successful than those among technological equals (Killing, 1983). Evidence from the commercial aircraft industry, in which international collaboration has been widespread for more than a decade, strongly supports this hypothesis. Ventures of technological equals, such as those between Rolls Royce and Pratt and Whitney in the JT10D jet engine project, Fokker and McDonnell Douglas in the MDF100 commercial aircraft project, and Saab and Fairchild in the SF340 commuter aircraft project, frequently have failed to bring a product to market or have been unable to achieve commercial success with a product after its introduction. A major venture of technological equals in telecommunication equipment, AT&T-Philips, also has been rumored to be near dissolution ("AT&T: The Making of a Comeback," p. 62). Ventures between technologically dominant and subordinate firms, however, such as the CFM International venture between General Electric and SNECMA of France and the collaborative ventures involving Boeing and the Japan Commercial Transport Development Corporation appear to be more durable.

The organizational structure of international collaborative ventures, especially those involving joint research and product development, raises additional challenges. There is no optimal management structure for a collaborative venture; the appropriate design will depend, among other things, on the character and magnitude of the contributions of the participants. In collaborations of technological equals, an autonomous management structure charged with the responsibility for a wide range of design, marketing, production, and product support is preferable. This type of management structure often is costly, since it duplicates some or all of the management structure of the member firms. Nevertheless, the experience of recent collaborative ventures clearly indicates the importance of strong links between the product development and design team and the organization charged with responsibility for marketing and product support. The organization managing the collaborative venture, be it the single dominant firm or an independent hybrid of the parent firms, must retain control of a number of downstream activities. On the other hand, in collaborations involving a senior and a junior firm, financial and organizational structure appears to be less important, so long as the technologically more advanced firm retains overall control of technology and management decisions.

Finally, the case of the Anglo–French Concorde partnership, where total project costs rose from $450 million in 1962 to $4 billion by 1978, illustrates the need for building cost controls into the structure of a collaborative venture. This issue is important because of the weaker incentives for participants to minimize shared costs. In the Airbus and other joint ventures with less disastrous financial consequences than Concorde, fixed-price contracts between a central management organization and the partner firms have preserved incentives for partners to minimize costs. Profit sharing, rather than cost sharing, is crucial.

Public policy implications of international collaborative ventures

Although the full impact of international collaboration will not be apparent for some time, a comparative analysis of international collaboration in U.S. manufacturing (Mowery, 1988a) does not support the critical view of international joint ventures presented in such works as Reich and Mankin (1986); nor is a specific federal policy governing such collaborative ventures advisable. Technology transfer within these ventures is more modest in scope and less uniformly "outbound" than some assessments assume. Just as U.S. industries vary in their trade balances in goods, the net inflows or outflows of technology through international collaborations vary across industries. Requiring balance in technology transfer on an industry-by-industry basis makes no more sense than a requirement for such balance in goods trade.

Restrictions or controls on international collaborative ventures involving U.S. firms do not appear to be an effective means to improve U.S. international competitiveness and in fact might impair competitiveness. The complexity of international collaborative ventures, the fact that the pattern and impact of these ventures vary considerably across industries, and the historical evidence that restrictions on technology transfer are either ineffective or perverse in their impacts (Harris, 1986; COSEPUP Panel on the Impact of National Controls on International Technology Transfer, 1987) all argue against controls on collaborations involving nondefense technologies. Controls on the transfer of technologies and technical data are currently imposed on collaborative ventures in technologies with potential military applications for reasons of national security, and the case for additional controls on civilian technologies is not compelling.

Efforts to restrict international collaboration also overlook the fact that in a number of industries, including steel, automobiles, and portions of microelectronics, international collaboration can improve the international competitiveness of the U.S. participants. In other industries, such as robotics, the competitiveness of U.S. systems engineering and software firms and the ability of large U.S. firms to offer a "full line" of factory automation hardware and software depend on access to foreign hardware through joint ventures and licensing.

A policy concerned specifically with international collaborative ventures in civilian technologies thus is ill-advised. Nevertheless, the incentives for U.S. firms to enter into international collaborative ventures and the payoffs from these ventures for U.S. firms and the U.S. economy are heavily influenced by U.S. antitrust, trade, and research policies. Should these policies be revised in order to reduce the incentives for collaboration between U.S. and foreign firms?

There is little evidence to support the argument that U.S. antitrust policy is a central factor in the decisions of American firms to collaborate with foreign enterprises.[12] International collaborative ventures generally are not substitutes for the collaboration among U.S. firms that might develop in the absence of antitrust restrictions. In the semiconductor industry, for example, the firms that are active in developing domestic R&D collaborations and consortia are among the most active in international collaborative ventures. If U.S. firms were forced to choose between international and domestic collaborative ventures because of antitrust restrictions (see Langlois et al., 1988, for a discussion of the semiconductor case), the opposite relationship would hold, in which firms active in international collaboration did not pursue domestic collaboration.

The primary motive for many collaborative ventures – access to foreign markets – is not affected by restrictions on collaboration between U.S. firms. In other cases, such as the collaborative ventures in the automobile and steel industries, U.S. firms collaborate with foreign firms in order to gain access to technological and other assets that are not available from other U.S. firms. Antitrust policy also has little effect on precommercial research collaboration among U.S. firms in the wake of the National Cooperative Research Act of 1984, and antitrust rarely applies to vertical collaborations between firms and their suppliers.

Antitrust policy has played an indirect role, however, in international collaborative ventures in several industries. The 1983 Modified Final Judgment in the case of *U.S. v. AT&T* opened the domestic U.S. market to foreign telecommunications equipment suppliers and unleashed AT&T to compete in markets outside telecommunications and in foreign markets for telecommunications equipment. As a result, AT&T has entered into international collaborative ventures to penetrate foreign markets for equipment and has developed alliances with domestic (Sun Microsystems) and foreign (Olivetti) producers of computers to aid its entry into businesses outside telecommunications. Joint ventures between the Bell operating companies and foreign equipment producers also might develop as a means of aiding foreign access to the U.S. telephone equipment

[12] Nelson (1984) argues that "although I am less easy about joint design and production ventures than I am about generic research cooperation, it does not seem right that such an international venture would receive totally different treatment from that received by two or more U.S. firms in a design and production venture for which the market is clearly international. . . . It seems odd that we would discriminate against a national partnership if each partner judged this venture more promising economically than an international consortium" (p. 84).

market if current judicial restrictions on the activities of the U.S. operating companies were lifted.

Antitrust policy has affected international collaboration in other industries as well. The limited duration of the NUMMI joint venture between General Motors and Toyota is a direct result of the settlement reached in the antitrust suit against NUMMI filed by the Chrysler Corporation. There is some evidence in the commercial aircraft engine industry that the U.S. Department of Justice has allowed international collaboration between competing firms that would not have been allowed between the major U.S. firms, based on an inappropriate distinction between the domestic and international markets for these products (see Mowery, 1987). The evidence is not conclusive in this case, however, and the example appears to be an isolated one.

Restrictions on U.S. imports also have resulted in collaboration between U.S. and foreign firms that was prohibited among U.S. firms for antitrust reasons. The conjunction of antitrust and trade restrictions influenced the joint venture between National Steel and Nippon Kokan. That collaboration was established after the Justice Department quashed the sale of a number of National Steel plants to the U.S. Steel Corporation, based on the department's judgment that the domestic U.S. market was protected from imports.[13] The consequences of this collaboration, however, do not appear to be harmful to the long-term competitiveness of the U.S. firms involved. Indeed, in the absence of such a collaborative venture, National Intergroup (parent firm of National Steel) might have withdrawn entirely from the steel industry. The tendency of trade restrictions to create such incentives for international collaboration is another reason to avoid protectionist policies, rather than an argument for the relaxation of antitrust restrictions in protected industries.

Research and product development subsidies are another potential inducement to engage in international collaboration. But with the exception of spectacular cases of subsidized product development in the commercial aircraft industry (Airbus Industrie, which by some estimates received billions of dollars in public subsidies between 1968 and 1982; see Krugman, 1984), public financial support for recent Japanese and European technology development programs is modest, both absolutely or by comparison with U.S. programs. Moreover, these programs typically focus on precommercial research, rather than the development of specific prod-

[13] Hufbauer and Rosen (1986) have noted this tendency, arguing that "a pattern has arisen in which transnational associations raise fewer antitrust concerns than combinations involving two domestic producers. . . . The numerous equity purchases plus one joint venture in the auto industry, as well as several joint ventures in steel, exemplify this phenomenon. While this is the logical consequence of import restraints, it was probably an unintended effect" (p. 62).

ucts. Where the results of these publicly supported research programs are published in the open scientific literature, such programs can in fact benefit nonparticipant U.S. firms, which may gain access to many of the results of the basic research without having to bear the full costs of such research. This possibility, however, requires a greater investment by U.S. firms in monitoring foreign scientific and engineering research and, possibly, a greater willingness to utilize scientific and technological advances developed outside of the firm, as Japanese firms have done with considerable success.[14]

Two of the three major areas of trade policy that affect collaboration – public subsidies and government procurement – are now covered by multilateral agreements ("Codes") under the General Agreement on Tariffs and Trade that were negotiated during the Tokyo Round of trade talks (see Chapter 10 for additional discussion). In addition, the commercial aircraft industry is covered by the Agreement on Trade in Civil Aircraft, which specifies acceptable practices for procurement and development funding. To the extent that international collaboration poses a long-term threat to U.S. international competitiveness and is motivated by the trade-distorting practices of foreign governments, both the Codes and the Agreement on Trade in Civil Aircraft must be strengthened.

The Uruguay Round of trade talks that began in 1986 will consider the treatment of foreign investment, including the "right of establishment" by firms of production and other operations in foreign markets. Since government restrictions on foreign investment provide a significant motive for international collaborative ventures, the development within these negotiations of multilateral rules for the treatment of foreign investment could alter the incentives for international collaboration in some industries.

The development and enforcement of stronger forms of intellectual property protection, another subject of the Uruguay Round negotiations, may also facilitate the licensing of process and product technologies to foreign firms, increasing the attractiveness to U.S. firms of an alternative channel for the exploitation of technical capabilities in such industries as biotechnology. In view, however, of the limited importance of patent, copyright, and trade secret protection for the establishment or preservation of a technological advantage (Levin et al., 1987), stronger intellectual property protection probably will have a modest impact on international collaboration in most U.S. manufacturing industries.

The growing role of bilateral negotiations in U.S. trade policy, especially negotiations over access to foreign (frequently Japanese) markets, may lead to additional collaborative ventures between U.S. and foreign

[14] See Chapter 8; also see Mansfield (1988) and Rosenberg and Steinmueller (1988).

firms. Recent bilateral negotiations between the U.S. and Japanese governments, according to some observers, have produced understandings that stipulate, among other things, that U.S. firms shall attain a specific market share in the foreign market (see Bhagwati, 1987; Prestowitz, 1987; World Bank, 1987). Achievement of this outcome in many markets, especially the domestic Japanese market, requires either direct foreign investment by U.S. firms, which is slow, risky, and costly, or the formation of long-term collaborations between U.S. and foreign firms. The bilateral talks in semiconductors that led to the United States–Japan Semiconductor Agreement of 1986 have been followed by negotiations over construction services and may lead to further talks covering other industries. If these talks and the associated government monitoring of trade outcomes expand, so too will international collaborative ventures.[15] Trade policy influences international collaboration between U.S. and foreign firms, but the results of such collaboration increasingly may affect trade policy as well. As the national origins of the components in a central office telephone switch, jet engine, or robot become more blurred and complex, national trade policies that are predicated on the protection or promotion of goods with a high domestic content will be unworkable, as we note in Chapter 10.

Which U.S. industries are likely to rely on international collaboration as a permanent component of their global competitive strategies? Industries in which collaboration in product development is motivated by a desire to reduce risk, to gain access to technology, or to deal with the effects of a significant government role in procurement or in the resolution and regulation of international trade disputes, will continue to rely on international collaboration. These industries include telecommunications equipment, commercial aircraft, and (potentially) semiconductors. The reluctance of Japanese firms to assume ownership of additional production capacity and the continued restrictions on foreign access to the U.S. market mean that international collaboration in the steel industry is likely to continue. International collaborative ventures will remain significant in the U.S. auto parts industry as a result of the substantial obstacles that U.S. firms face in gaining access to Japanese management and production systems and to Japanese markets.

In other U.S. industries, international collaboration may be more transitory. The wave of collaborations between major Japanese and U.S. automobile firms that expanded U.S. production of Japanese auto designs, for example, may have peaked. The major Japanese producers now have

[15] According to one account, the March 1988 agreement between the U.S. and Japanese governments easing access by U.S. firms to Japanese public construction projects has resulted in an extensive set of joint ventures between U.S. and large Japanese construction firms (Rubinfien, 1988).

established production facilities in the United States, bypassing the barriers to their access to this market. International collaboration that is motivated by market access concerns in industries in which governments are not major customers may also decline in importance in the long run. Industries in which these motives figure prominently include biotechnology, segments of the robotics industry, and pharmaceuticals.

Domestic research collaboration

Along with international collaboration in product development and manufacture, domestic collaboration in research has expanded dramatically within the U.S. economy during the past decade. Research collaboration between industry and universities has also grown rapidly, although industry's financial contribution to university research still is dwarfed by that of the federal government. Private research collaborations also have sprung up – examples include the Microelectronics and Computer Technology Corporation, the Software Productivity Consortium, and numerous smaller domestic joint ventures. Although the proximate causes of these collaborative undertakings may resemble some of those responsible for international collaboration, domestic research collaborations appear to differ considerably from international ventures in focus and structure. In this section, we discuss university–industry research collaboration and compare research collaboration among firms and collaboration between industry and universities.

University–industry cooperation

During the past decade, financial support from industry has established a number of research facilities on university campuses to conduct research with potential commercial value. Stanford University has the Center for Integrated Systems, with financial support from twenty corporations, which is devoted to developing large-scale integrated microelectronic circuits; MIT has the Whitehead Institute, with a huge private endowment, which is devoted to biomedical research; MIT also has a ten-year research contract with the Exxon Research and Engineering Company to support research in the field of combustion. Hoechst AG of West Germany has given the Massachusetts General Hospital, a teaching arm of the Harvard Medical School, $50 million for the support of basic research in molecular biology.

Important initiatives are coming from the federal government as well as private industry. The National Science Foundation has embarked on a program to establish a number of engineering research centers on university campuses. The centers focus on engineering research that is in many cases interdisciplinary, often linking engineering research with such traditional scientific disciplines as biology and physics, and emphasizing advanced computer applications. The financial structure of these centers, which resembles that of the NSF university–industry research centers, also is relatively novel, since it combines "seed-money" support from the federal government (as well, in many cases, as state and local governments) with major contributions from private corporations that are affiliated with the centers. A number of these centers already exist. Columbia University has a center for telecommunications; the University of Maryland and Harvard University, centers for the application of advanced computers to the design of communications systems; MIT, for the improvement of manufacturing processes in the biotechnology industry; and Purdue University, for automated manufacturing systems.

The development of these novel institutions has raised concerns within universities that the university's commitment to basic research will be eroded by an increasing focus on short-term, less fundamental research problems because priorities will be determined more and more by commercial considerations. Closely connected is the concern that scientists will lose their intellectual autonomy, which is to say their freedom to determine their own research agenda, as private industry comes to play a more prominent role in financial support. Protection of this autonomy has long been a serious concern of science policymakers; in fact, peer reviews and other protective arrangements were introduced when the federal government became a massive patron of university research after the Second World War, in response to very similar concerns (we discuss these later).

In considering the advantages and disadvantages of research collaboration between U.S. universities and industry, several points must be kept in mind:

1. The phenomenon of the university–industry research collaboration is not new, as we have noted in Chapters 4 and 6. Research collaboration between universities and enterprises in both the agricultural and manufacturing sectors is more than one hundred years old in the United States. Indeed, industrial funds have accounted for a declining share of university research expenditures through much of the postwar period. In 1953, industry financed 11 percent of university research, a share that declined to 5.5 percent in 1960 and 2.7 percent in 1978. By 1985–6, estimates suggest that industrial funds accounted for 5–7 percent of university research.[16] Industrial

[16] See National Commission on Research (1980), pp. 8–9; cf. Government-University-Industry Research Roundtable (1986), 15–16.

funds are unlikely to exceed 8–10 percent of total U.S. university research budgets in the foreseeable future.

2. In several respects, the recent growth in university–industry research cooperation represents a renewal of linkages of long standing. As we noted in Chapter 4, American state universities have a long history of performing highly applied engineering research and extension services for industrial and state government sponsors. The significant role of state governments as sponsors of recent cooperative research initiatives is also not new, as Chapter 4's discussion of the significant state government role in the pre–World War II era noted. To a great extent, the recent development of closer research ties between universities and industry represents the restoration of a linkage that has been weakened in recent years, rather than a fundamental departure.
3. There is a vast array of forms of research collaboration between universities and industry, making generalizations virtually impossible. Moreover, as we note later, the relationship between university research and commercial technology varies considerably across industries. No single model or description of the constraints, advantages, and disadvantages of such collaboration is likely to be accurate for all university–industry collaborations.
4. Many of the issues raised by research collaboration between industry and universities are not new, but resemble those raised in earlier periods by the expansion of external research support from other sources. For example, concerns over intellectual property rights, excessive growth of salaries for faculty specializing in particular disciplines, and other issues raised in recent discussions of university–industry research relationships were important topics of debate in the context of research contracts with nonprofit foundations and the federal government.[17]

A fundamental motive for closer ties between university and industrial research is the fact that U.S. universities account for a growing share of total U.S. basic research. The relative importance of university and industrial performers of basic research has shifted dramatically during the postwar period. In 1953, industry accounted for 58 percent of the com-

[17] The Association of American Universities (AAU) noted in a 1985 report that a major policy statement on faculty conflicts of interest between externally sponsored research and other responsibilities of university faculty was issued by the American Association of University Professors and the American Council on Education in 1964 as a result of concerns over research sponsored by the federal government (AAU, 1985, p. 10).

bined basic research budget (from all sources) of the universities and industry; in 1978, universities accounted for 76 percent of the combined basic research budget of universities and industry (National Commission on Research, 1980, pp. 8–9).

Moreover, in recent years such flagships of industrial basic research as the Du Pont Central Research Laboratories, the RCA Sarnoff Laboratories, General Electric's central R&D facility, and even Bell Telephone Laboratories (in the wake of the 1984 divestiture of the operating companies) have either been closed or have reduced their commitment to basic research. Increasingly, universities occupy the central position in the performance of basic research within this economy. The importance of sustained interaction among the various stages of the invention or innovation processes means that new organizational linkages between universities and industry are needed, in order to increase the economic rewards from the growing basic research investment within universities.

From private industry's point of view, there is a growing awareness that an increasing number of fields of research at the universities hold out significant promise of generating findings that may be of great commercial significance. Close proximity to the university scientific community offers firms better access to knowledge that may have great value in the marketplace. The ability to tap into such new knowledge sources, and to convert that knowledge into new and improved products before one's competitors or potential competitors, is an immense attraction. Even in the absence of any proprietary control over the new research findings, a lead time over the competition of several months can often be converted into a decisive competitive advantage.

There are a number of fields in which university research is working at intellectual frontiers that have great potential commercial payoffs. Recent discoveries in high-temperature superconductivity are one example, although the commercialization prospects of these discoveries are perhaps the most distant of the fields that are currently the object of university research. A partial list of other fields would include artificial intelligence, robotics and remote sensing, computers, microelectronics, lasers, composite and other new materials, and biotechnology.[18] Past and prospective research in each of these fields provides a basis for strong industrial interest in strengthening its connections with university science.

The connection between university research and commercial technology appears to be particularly close in the area of biotechnology, which

[18] According to the *Economist*, the commercial importance of university biotechnology research has led the U.S. pharmaceuticals industry to increase its financial support of biotechnology research in universities by 90 percent per year in recent years: "American industry now pays for 40% of the $13 billion-a-year biomedical research bill" ("Mismanaging Drug Research," Nov. 21, 1987, p. 67).

influences the character of many university–industry research relationships in this field. This close relationship is due in part to the nature of biotechnology. Recombinant DNA and genetic engineering techniques in many ways represent radical scientific breakthroughs that are being transferred to industry and reduced to practice. In Gomory's terminology, biotechnology is a "ladder" technology, that is, a case in which "the new idea is dominant and the product forms itself around the new idea or new technology. Those who understand that idea or technology are often scientists, and they therefore play leading roles in its introduction" (Gomory, 1988, p. 11).[19] A great many other industrial technologies, by contrast, develop through a process of cyclic, incremental improvement (a process in which many Japanese firms excel, as we pointed out in Chapter 8), in which the application of basic science is more gradual and less direct.

A central influence on the establishment of these new facilities is the growing necessity of a multidisciplinary approach to important problems in both science and technology. The growing importance of multidisciplinary research approaches is another reflection of the phenomenon of technological convergence we discussed in Chapter 8. Multidisciplinary research has developed historically from the emergence of problems at the research frontier of a particular discipline (e.g., cell biology) that required a better understanding of the role of certain processes that were the specialty of scientists in a different discipline (e.g., chemistry). The outcome – biochemistry – was a natural outgrowth of the changing requirements of an expanding body of research knowledge. Similarly, geophysics emerged as an independent subdiscipline of geology when it became possible to apply the methodologies that had been developed in physics to the understanding of the structure and the dynamics of the earth (as well as other planets). The introduction of new techniques of instrumentation has led, time and again, to a beneficial crossing of traditional boundaries between physics and chemistry.[20]

[19] Another example of a ladder technology cited by Gomory is the transistor. In contrast to biotechnology, of course, the transistor was first developed within industry. The different origin of these two major scientific discoveries may reflect the shifting role of industry and universities as basic research performers noted earlier.

[20] "The advent of new instrumentation of unprecedented power has required chemists either to develop instrumentation in the traditional fashion of physicists or else to enter into close collaboration with physicists in order to address the microscopic properties of complex molecules, materials, and interfaces. For their part, condensed-matter physicists, propelled by instrumentation, theory, and technology toward materials of increasing complexity, find themselves often confronting questions requiring increased chemical insight and techniques for their answers. Finally, new complex materials displaying remarkable new chemical and physical properties have nucleated a coalition among synthetic chemists, physical experimentalists, and theoretical physicists from deep within their respective fields, far from the traditional interface. Each of these currents has already created new fields and produced new discoveries in established fields. As the principals continue

The need to exploit the knowledge and the methodologies of more than one discipline has become apparent not only at the level of basic science but in applied sciences and engineering as well. In recent years, medical science has benefited not only from such "nearby" disciplines as biology, genetics, and chemistry, but has drawn on research in nuclear physics (magnetic resonance), electronics, and materials science and engineering. Important recent advances in pharmaceuticals have been based on such fields as biochemistry, molecular and cell biology, immunology, neurobiology, and scientific instrumentation. The new pattern of innovation is, by its very nature, multidisciplinary. Success requires close cooperation among an increasing number of specialists.[21]

The development of multidisciplinary university research institutes provides important support for these new cross-disciplinary research problems and opportunities. Moreover, these institutes can train young scholars and prospective industrial researchers in new techniques and research perspectives. Interdisciplinary research institutes are hardly a novel feature in the postwar landscape of American higher education. Major research initiatives in area studies, in computer science, and in materials science, for example, were launched with federal research funding during the 1950s and 1960s. Many of the research institutes spawned by these initiatives, especially in computer and materials sciences, have emerged as freestanding departments in new "disciplines." The novel aspect of recent interdisciplinary research institutes, as in university–industry research collaboration generally, is the source of the research funds, rather than the presence of significant extramural research funding.

The changes in the competitive and technological environment we discussed in Chapter 8 have also increased the payoff to university–industry research collaboration. Faced with spiraling R&D costs, greater demands to monitor a broader array of scientific and engineering fields, and increased competitive pressure from other U.S. and foreign firms to get products to market rapidly, U.S. firms have had to develop research relationships with an array of external institutions that could complement

to collaborate and to learn one another's language, there is reason to expect much more" (Physics Survey Committee, 1986, p. 54).

[21] "This new environment is an intensely interdisciplinary one in which the intuition and skill of the chemist are combined with the knowledge of biochemists, pharmacologists, and computer scientists as well as other researchers in universities, industry, and the government. . . . One of the clearest trends is that the biological sciences will play a pivotal role in drug discovery and development. . . . Developments in biochemistry have encouraged greater exploitation of enzymes as specific targets for drugs. The recent success in synthesizing angiotensin-converting enzyme inhibitors as antihypertensives is one of the first examples of drug design based upon structural and functional information on enzymes" (*Outlook for Science and Technology*, pp. 591–2).

enhance the payoff from their in-house R&D activities.[22] Such rela-ships provide lower-cost windows on emerging technologies, allow ıs to detect emerging commercial opportunities more rapidly, and spread the risks of failure among a larger number of research performers and research budgets. In addition, rotation of research personnel from the firm to external research institutions provides an important means to transfer research know-how and findings from external sources to the firm and can improve the attractiveness of the firm's laboratories to researchers.

Teaming with a university, rather than with a consortium of other firms or a publicly funded national research laboratory, allows the firm additional advantages beyond those already noted. By virtue of their unique mission as educational as well as research institutions, U.S. universities are critically important sources of scientific and engineering personnel. Participant firms can employ these collaborative ventures as "filters" for hiring research personnel, observing the performance of potential researchers before making employment commitments. Moreover, the demonstrated importance of people as a key vehicle for the transfer of scientific and technological knowledge means that the hiring by firms of the graduates of these programs facilitates the transfer of knowledge and technology from university to industry even more effectively than does the rotation of industry personnel through university research facilities.

The increasingly interdisciplinary character of technological and research challenges makes this training and hiring benefit particularly important. Increasingly, firms in the semiconductor, biotechnology, or robotics industries need individuals with interdisciplinary research experience and training. Such interdisciplinary research and educational activities are notoriously difficult to establish within universities with internal funds alone, although the departmental structure of U.S. universities makes this task easier than in many Western European universities.[23] Industrial

[22] A recent OECD study quotes a Xerox Corporation research executive's description of the firm's investment in the Center for Integrated Systems at Stanford University: "Xerox's contribution to CIS is very small compared to what we are investing internally in the same kind of research. For little additional investment we enlarge our perspective by participating in a broad program of basic research. We envision opportunities for joint interaction with the university and with other companies, as well as the ability to recruit students. On a per-dollar basis it should be a good investment" (quoted in OECD, 1984, p. 47).

[23] "Among the factors cited to explain West Germany's slow entry into commercial biotechnology is an educational system that prevents the kind of interdisciplinary cooperation that is viewed by most experts as essential to the development of this field. In particular, the traditional separation of technical faculties from their arts and sciences counterparts means that process technicians, usually located in the technical schools, rarely come into contact with colleagues holding university appointments in biochemistry or microbiology" (Office of Technology Assessment, 1984, p. 424).

research funds, like federal government research funds in the 1960s and 1970s, provide important resources for universities to establish such interdisciplinary research and education programs. Finally, by providing research and instrumentation funds for university faculty who otherwise might depart for industrial employment, industry research funds sustain the supply of scientific and engineering personnel in a way that private firm consortiums cannot.

Omitted from this discussion of the industrial benefits of university–industry collaboration are technological innovations developed in university laboratories and "handed off" to industrial collaborators. Although the wide variety of university–industry collaborative programs makes generalization hazardous, the weight of anecdotal evidence (see the Office of Technology Assessment, 1984, 1985; Government-University-Industry Research Roundtable, 1986; and other reports) suggests that collaborative ventures that focus on applied development work or on technological "deliverables" in many cases are less successful. The benefits of collaboration discussed earlier are longer-term benefits that rely on the establishment of a relationship lasting a number of years.[24] Although many firms are unable to adopt such a lengthy time horizon in their internal R&D planning and strategy, the lower costs of external research collaboration may support a longer-term perspective in at least this portion of the firm's research portfolio. Within collaborative relationships of this type, intellectual property issues often are of secondary importance, in contrast to the predictions of the neoclassical analytic paradigm.[25]

As we noted earlier, one exception to this pattern appears to be biotechnology. Blumenthal et al. (1986) report that small biotechnology firms do appear to rely heavily on university research as a basis for patent ap-

[24] Hercules and Enyart (1983) report from their survey of chemicals companies and academic researchers that the following four areas of collaboration had very high potential payoffs and currently lacked sufficient activity: (1) lectureships by academic scientists at industrial sites; (2) student interns at industrial sites; (3) continuing education programs at industrial sites; and (4) corporate support for employees to obtain advanced degrees (p. 7). None of these involve significant transfers from universities to industry of intellectual property or other deliverables.

[25] Consistent with this assessment, Gray and Gidley (1986) found in their analysis of evaluations of the first six NSF-funded university–industry research centers that "both faculty and industry respondents rated general expansion of knowledge as the most important goal and the more short-term goals of patent and product development as the least important" (p. 26). Describing firms' expectations at the inception of these cooperative research undertakings, Gray and Gidley found that "the benefits seen [by industry respondents] as most likely to accrue to companies were improved research projects in the company (mean, 2.60; 1 = scarcely likely, 4 = almost certain) and better personnel recruitment (mean, 2.54). Patentable products (1.62) and commercialized products (1.75) were seen as benefits which were 'somewhat' to 'scarcely likely' to accrue through Center participation" (p. 29).

plications and trade secrets. This is consistent with Gomory's description of biotechnology as a "ladder" technology. The authors concluded that

> small firms seem to constitute the greatest gamble for universities. On the one hand, the financial benefits of these relationships could be very large, since these UIRRs [university–industry research relationships] seem to produce many more patent applications per dollar invested than do relationships with large firms. . . . On the other hand, it is far from certain that applications for patents held by universities will ever produce profitable licenses. Moreover, the amount of research support provided by these relationships is small compared with UIRR's with larger firms, and the potential threats to university values seem greater. Compared with large companies in our study, small firms are more likely to support faculty with significant equity in their companies, are more likely to report the occurrence of trade secrets, and tend to fund projects of very short duration. (p. 246)

Just as the spiraling costs of R&D in industry during an era of financial stringency have increased industrial interest in closer research relationships with universities, the combination of soaring equipment and facilities costs and diminishing federal support has aroused academic interest in research collaboration with industry.[26] Federal outlays on R&D plant and equipment have been declining for many years. Such outlays reached their peak in the middle of the 1960s and then declined drastically, rising again in real terms in the late 1970s (although never approaching the earlier peak levels) and then declining again in the 1980s (see Table 9.1). Indeed, for some forms of research capital, federal support has totally collapsed. Although there was considerable support for research facilities of a bricks-and-mortar kind during the 1960s, such programs essentially ended about twenty years ago.[27] Nevertheless, private industry financial

[26] Although the soaring costs of research instrumentation pose some threat to the health of academic scientific research, many of the most widely used scientific instruments originated within the university scientific research community where they were developed by working scientists or engineers in response to some specific experimental or observational requirement. Once developed, these instruments were taken up by commercial manufacturers and made available for use in other scientific disciplines and in private industry. The computer is, of course, a classic example, and is distinguished mainly by being the most widely diffused of all scientific instruments. Many of the developers of this technology were academics attempting to develop a technology for reducing the extremely time-consuming computational requirements of their research, typically involving solutions to systems of differential equations. See Ritchie (1986).

[27] Continued deterioration in the quality of such facilities could erode the substantial benefits that flow from the close integration of graduate training and high-level research that has distinguished the U.S. system. The well-equipped European research laboratories

Table 9.1. *Federal outlays for R&D plant, fiscal years 1960–83 (dollars in millions)*

Year	Current dollars	Constant 1972 dollars[a]
1960	443.8	637.9
1961	539.1	766.2
1962	555.2	777.9
1963	673.6	928.3
1964	948.1	1,292.0
1965	1,077.4	1,436.9
1966	1,047.8	1,361.5
1967	792.7	997.9
1968	723.8	879.4
1969	657.0	762.5
1970	578.9	635.9
1971	612.7	640.8
1972	564.4	564.4
1973	638.4	611.2
1974	704.6	628.8
1975	829.7	673.1
1976	800.6	607.1
1977	800.2	568.5
1978	1,107.8	736.9
1979	1,202.8	735.8
1980	1,481.7	833.3
1981	1,606.7	822.8
1982	1,376.1	639.7
1983	1,299.9	581.9
1984	1,639.2	707.8
1985	1,688.9	706.4
1986	1,481.2	603.6
1987 (est.)	1,769.9	702.3
1988 (est.)	1,888.4	723.5

[a]GNP fiscal year implicit price deflators used to convert current dollars to constant 1972 dollars.

Source: National Science Foundation, *Federal Funds for Research and Development: Fiscal Years 1986, 1987 and 1988*, and earlier years.

support of university research remains modest. Industry provided about 19 percent of the funds for all basic research (Table 9.2) in 1984, but it funds a much smaller percentage of the basic research performed at the universities – around 6 percent of the total.

A comparison of two forms of domestic research collaboration in microelectronics

In this chapter we have discussed and compared two broad categories of domestic cooperative research activity – consortia of private firms and research cooperation between industry and universities. In this section, we briefly compare the two types of cooperative research in an industry in which research collaboration has grown rapidly in recent years, namely, microelectronics.

University–industry collaborative research programs in microelectronics include the Semiconductor Research Corporation (SRC), which funds research at a number of universities, and the Stanford University Center for Integrated Systems (CIS), which is based in a single university. Roughly sixty firms participated in the SRC and twenty were members of the CIS as of late 1986, according to the Government-University-Industry Research Roundtable (1986). Member firms in both organizations are granted access to all research findings and results through licenses, briefings, and so forth, and often are encouraged to send technical personnel to university research facilities for extended periods of time. Rather than developing applications for immediate transfer to industry, these programs perform research, train technical personnel for industry, and expose firm employees to advances in the scientific and technological frontiers.

Unlike research collaboration between universities and small firms in biotechnology, the SRC does not appear to focus on tangible "deliverables" from university research. In many ways, collaborative research organizations like the SRC operate as "catch-up" mechanisms for U.S. private industry, enabling more rapid and effective utilization of advanced scientific and technological findings by participant firms. Just as many Japanese cooperative research projects operated to bring Japanese firms up to a technological or scientific frontier that was historically defined by foreign firms and laboratories, university–industry cooperative research projects within the United States can bring U.S. firms closer (through

such as CERN, or the national research laboratories such as the Max Planck Institutes in West Germany, do not train beginning research personnel. This has contributed to a relative shortage of qualified, high-level European researchers. The loss of university access to instrumentation of the highest quality in America or the concentration of such instrumentation in a very small number of locations would be a serious threat to the system for training research scientists that has been built up in the United States in the postwar years.

Table 9.2. *Sources of funds for basic research by sector: 1953, 1960, 1965–84 (dollars in millions)*

Current dollars					
Year	Total	Federal government	Industry	Universities and colleges	Other nonprofit institutions
1953	441	251	153	10	27
1960[a]	1,197	715	342	72	68
1965[b]	2,555	1,809	461	164	121
1966	2,814	1,978	510	197	129
1967	3,056	2,201	492	223	140
1968	3,296	2,336	535	276	149
1969	3,441	2,441	540	298	162
1970	3,549	2,489	528	350	182
1971	3,672	2,529	547	400	196
1972	3,829	2,633	563	415	218
1973	3,946	2,709	605	408	224
1974	4,239	2,912	651	432	244
1975	4,608	3,139	705	478	286
1976	4,977	3,436	769	475	297
1977	5,537	3,823	850	527	337
1978	6,392	4,445	964	605	378
1979	7,257	5,044	1,091	711	411
1980	8,089	5,559	1,265	805	460
1981	9,217	6,236	1,585	909	487
1982	9,886	6,588	1,805	983	510
1983 (est.)	10,610	6,970	2,025	1,075	540
1984 (est.)	11,850	7,775	2,270	1,220	585

Constant 1972 dollars[c]					
Year	Total	Federal government	Industry	Universities and colleges	Other nonprofit institutions
1953	742	421	259	17	45
1960[a]	1,729	1,030	497	103	99
1965[b]	3,416	2,415	620	219	162
1966	3,660	2,571	665	256	168
1967	3,853	2,774	622	281	176
1968	4,001	2,837	649	335	180
1969	3,985	2,829	623	346	187

Table 9.2. *Cont.*

Constant 1972 dollars[c]					
Year	Total	Federal government	Industry	Universities and colleges	Other nonprofit institutions
1970	3,895	2,733	578	384	200
1971	3,836	2,644	570	418	204
1972	3,829	2,633	563	415	218
1973	3,766	2,589	573	391	213
1974	3,757	2,589	567	386	215
1975	3,720	2,540	562	388	230
1976	3,770	2,604	581	360	225
1977	3,939	2,718	607	374	240
1978	4,250	2,956	641	402	251
1979	4,438	3,085	667	435	251
1980	4,548	3,128	709	453	258
1981	4,727	3,199	813	466	249
1982	4,749	3,160	872	471	246
1983 (est.)	4,886	3,205	938	494	249
1984 (est.)	5,229	3,427	1,006	537	259

[a] Data for 1954–59 can be found in *National Patterns of R&D Resources, 1953–1977* (NSF 77-310).
[b] Data for 1961–64 can be found in *National Patterns of Science and Technology Resources, 1981* (NSF 81-311).
[c] Based on GNP implicit price deflator.
Source: National Science Foundation.

such mechanisms as personnel transfers and the hiring of program graduates) to a scientific frontier defined by the basic research within university laboratories. This point is an important one because it reduced many of the severe difficulties of defining a research agenda and minimizing "free rider" problems that have afflicted some recent collaborative research ventures in both the United States and Japan that do not enlist the participation of universities. The point also underscores one of our central arguments about interfirm research collaboration: It may well prove most effective in supporting the diffusion and utilization, rather than primarily the generation, of new scientific and technological knowledge. The contrast between the SRC and the Microelectronics and Computer Technology Corporation (MCC) also supports this assessment.

The structure and goals of two of the leading private research consortia,

MCC and Sematech, contrast with those of university-based collaborative research ventures. MCC was established in 1983 as an independent research organization in which firms could purchase shares and participate in programs of research in integrated-circuit (IC) packaging, software engineering, very large scale integration, and advanced computer architecture. Member firms are not allowed access to the results or findings of MCC programs in which they are not participants.

MCC thus has a somewhat anomalous structure and goals: Established to support precommercial, cooperative research, the corporation now performs proprietary research for its members. MCC has encountered some difficulties in defining its research agenda, transferring technology to member firms, and enlisting the full cooperation of member firms. Although it was designed to operate with staff drawn largely from member firms, some firms sent their less able technical employees to staff MCC. This policy was abandoned, and MCC now is staffed largely by its own employees (see Sanger, 1984). One of the primary reasons for the original staffing plan was to ensure rapid and effective transfer and utilization of technology by MCC member firms. Reliance by MCC on its own research staff has made technology transfer more difficult, and this issue has created problems for the consortium. In contrast to university-based collaborative research operations, MCC cannot offer its members the attraction of recruiting graduates from the laboratory and transferring technology and know-how through this channel.

Serious differences have also developed among members in research strategies and goals (for a less pessimistic assessment of MCC, see Peck, 1986). Some member firms have expressed displeasure with the lengthy time frame of MCC's research activities, pushing instead for applied research projects that would yield applications more quickly. Turnover in MCC's membership has been high, the price of MCC shares has been reduced, and the leadership of the Corporation has changed hands since its foundation (Lineback, 1987a, 1987b). As we noted in Chapter 1, MCC appears to be shifting its research agenda somewhat from basic to near-term research.

Many of the difficulties encountered by MCC may prove equally acute for Sematech (see Chapter 6 for a brief overview of the structure and origins of this consortium), which is also designed to operate as a free-standing research consortium with no direct university ties and its own research facilities. The problems of conflicting research priorities and time horizons that have hampered MCC may well afflict Sematech as well.[28]

[28] According to one account, however, a much larger fraction of Sematech's research staff is drawn from member firms than is true of MCC. According to Corcoran (1989), roughly one-half of Sematech researchers are on secondment from member firms for two-year assignments, substantially higher than the 15 percent of MCC research staff drawn from member firms. Personnel rotation, if it can be sustained, will facilitate the transfer of findings and results from the consortium's research facilities to the member firms.

The proprietary nature of much of the process technology employed within the semiconductor industry, which is the putative focus of Sematech research, further complicates the consortium's role within this industry. In addition, the establishment of applied research priorities within the venture poses considerable risks, in view of the rapid evolution of process and product technologies and the resulting uncertainties within the industry over future technological opportunities (Waldman, 1987). Member companies may well want different things. Some members of Sematech already have .5 micron facilities, at least in their laboratories, while other members do not. Small firms tend to have different priorities than large ones. There is thus a real danger, inherent in a consortium whose members have different needs and capabilities, that the eventual output of the R&D activity will not be of real commercial value to any large segment of participants.

Although its prospects and ultimate structure are uncertain at present, the research agenda of Sematech does not resemble the precommercial, basic research often advocated by supporters of collaborative research. Moreover, the Department of Defense is a major financial supporter of Sematech, and this inevitably raises concern that its research thrust will be heavily influenced by the requirements of military, rather than civilian, markets.

Conclusion

Changes in the economic and technological environment of U.S. firms, growing national concern over U.S. competitiveness, and increased interest within state governments in the use of innovation for economic development have all led to a considerable increase in international and domestic R&D collaboration. Despite their differences in structure and focus, both domestic and international R&D collaboration have grown rapidly during the past decade. R&D collaboration is relatively new within the U.S. R&D system, and its ultimate consequences are difficult to predict at this time.

International collaboration reflects the rising costs and risks of unassisted product development, the growth of significant technological strengths within foreign firms, and the increasingly prominent role of nontariff trade barriers and other government support for the development of advanced technologies. International collaboration has grown dramatically in a diverse array of U.S. manufacturing industries, although its long-term prospects appear brighter in some industries than in others. The consequences for U.S. competitiveness of such collaboration thus far appear to be fairly benign – in most cases, international collaboration is a response to, rather than a cause of, declining U.S. competitiveness. In other industries, collaboration has assisted U.S. firms in strengthening their technological and production skills.

The growing number and importance of international collaborative ventures do not reflect the discouragement of domestic collaboration by the U.S. antitrust statutes. International collaboration ventures create some complex policy implications, however, in both the trade and technology policy spheres. Along with numerous other factors, international collaborative ventures will accelerate international technology transfer in the global economy of the future. Nevertheless, the technology transfer operating within international collaborative ventures involving U.S. firms will remain a two-way flow for the foreseeable future. Trade policy clearly influences the development of international collaborative ventures. Increasingly, however, trade policy formulation and implementation will be affected by the operation and existence of these ventures – notably, the internationalization of sources of R&D and components. We discuss these policy issues at greater length in Chapter 10.

Much of the domestic research collaboration that has grown in recent years involves the renewal of linkages among state governments, publicly supported universities, and industry that had been languishing for a number of years. The erosion of these links during the postwar period was in part a result of the massive expansion in federal funding of research within both universities and industry. Closer research collaboration between universities and U.S. industry could hasten the translation of basic research findings into commercial products and processes: As we argued in this chapter, interfirm collaboration may assist the absorption and more rapid utilization by firms of the research advances of academic scientists and engineers. This absorption nevertheless requires investment by firms in their internal research operations and personnel as well as in the external research consortium.

The huge size, decentralized structure, and research-intensive character of the U.S. higher education system are unique among industrial nations, and these factors increase the potential payoff from closer research collaboration between universities and industry. Nevertheless, university–industry collaboration, when most successful, does not appear to be a short-term proposition, nor is it (nor should it be) a relationship in which clearly defined "deliverables" figure prominently. Indeed, as Blumenthal et al. (1986) noted, the biotechnology research collaborations that appear to focus most intensively on such deliverables pose the greatest challenges to academic ideals, challenges that in many ways resemble those created two decades earlier by classified defense-related research on university campuses.

Thus far, university–industry collaboration appears to have been slightly more successful than private interfirm consortia in supporting long-term research of a generic, precommercial variety. This conclusion reflects the tendency of such collaborative ventures to shift the research agenda toward applied, near-term research, as well as the absence of the important personnel flows between sponsor firm and consortium that typify university–industry research collaboration. The mixed results of U.S. interfirm collaboration in research, however, stem from factors that are also likely to reduce the attractiveness of such research collaboration within Japanese industries in which individual firms are already at the technological frontier.

As we pointed out in Chapter 8, leading Japanese firms have become increasingly reluctant to furnish their most expert research personnel to such ventures and are far less willing to pool their knowledge with competitors that may be able to quickly employ such knowledge against its provider. The heyday of Japanese research collaboration may already have passed: Such ventures function most effectively as instruments for the diffusion of precommercial scientific and technological knowledge among participants, and therefore are most successful where all participants are on a roughly equal footing and well behind the technological frontier. In other words, interfirm research collaboration may be most effective in an environment of technological "catch-up," rather than one of leadership. U.S. firms typically differ substantially in the level of their in-house expertise, and some members of an industry are likely to be far closer to the frontier than others. Both factors complicate the operation of private interfirm research consortia in the United States, just as they increasingly hamper such consortia within Japan. Both factors may favor the development of university–industry research consortia within the United States.

[11]

Leading companies and networks of strategic alliances in information technologies *

John Hagedoorn and Jos Schakenraad

MERIT, Faculty of Economics, University of Limburg, 6200 MD Maastricht, Netherlands

Final version received May 1991

Strategic alliances, joint ventures and a wide array of cooperative R&D pacts have received considerable attention in recent years. In this paper an attempt is made to understand not only some recent trends in inter-firm cooperation in information technologies but also to reveal the role played by a large group of companies. The research is based on a large databank with information on thousands of alliances and their participating companies. Applying a number of statistical techniques it is possible to identify the major players within information technology and its sub-fields. The analysis enables us to recognize the major international networks of inter-firm alliances, the changes over time and different positions taken by world leading companies in several industrial and technological settings.

1. Introduction

In recent years strategic technology partnerships and inter-firm alliances in general have raised an increasing interest, both in the popular business press as well as in more academically inclined journals. In this paper we will attempt to contribute to the understanding of this phenomenon with analysis of some historical trends, industrial patterns and networks of inter-firm agreements in a relatively large field of technology, i.e. information technologies. Our present analysis covers over 1,700 strategic technology alliances in information technology with its sub-fields computers, industrial automation, microelectronics, software and telecommunications [1]. We have chosen information technology for this particular paper not only because it is a major field of technology in general. but also because a very large share of strategic technology alliances are related to information technology. In our data nearly 42 percent of all worldwide strategic technology partnerships are found in information technology which make this by far the largest field of the alliances and also the sector where companies seems to have the longest experience with this phenomenon [7].

Our analysis will be limited to those agreements made by firms in which the transfer of technology or the creation of new technology through R&D or other innovative efforts are central to the agreement, thereby omitting a wide range of exclusively marketing. production or sales agreements. In essence we restrict our analysis to strategic technology partnerships such as joint ventures for which joint R&D or technology sharing is a major objective. research corporations, joint R&D pacts, and minority holdings coupled with research contracts [5]. We only analyze alliances made by companies on a "private"

* This paper is one of a series of papers in a research project on "Inter-company Cooperation and Technological Developments" at MERIT. This research focuses on the empirical analysis of changes in industry structures and global trends in different modes of inter-firm agreements in a large number of fields of technology. It also addresses theoretical questions in this field of research as well as methodological issues concerning applied network and multivariate analysis of strategies and industry structures. Empirical analysis is based upon the CATI database which contains information on several thousands of worldwide cooperative agreements and the companies involved.
We are grateful to two anonymous referees for helpful comments on an earlier draft.

[1] Our research is partly based on our CATI databank which, at present, contains information on nearly 10,000 worldwide cooperative agreements. of which over 4,000 strategic technology alliances, in a large number of technologies and several thousands of participating companies (see also Appendix I).

Research Policy 21 (1992) 163–190
North-Holland

basis, so we do not include agreements made in the context of (inter)national shared-cost programmes such as ESPRIT and EUREKA. This enables us to achieve some understanding of patterns of "pure" inter-firm technology sharing, i.e. strategic partnering for which the incentive is found within firms and not (partly) induced by government programmes or international institutions. A further restriction is that we only study "strategic" alliances, defined as those inter-firm agreements that can reasonably be assumed to effect the long-term product market positioning of at least one partner. The effect of this particular constriction is that all agreements which we expect to have mainly cost-economizing consequences are excluded from the following analyses. [2]

The emphasis in this article will be on empirical analysis and the testing of some hypotheses related to historical patterns of strategic technology partnering, the evolution of inter-firm networks and the role played by leading companies in the sub-fields of information technology mentioned above. In short, the core of this contribution is built on the identification of basic trends in strategic technology alliances in information technologies, the main "actors" and their networks, as well as market structural aspects of these alliances. In order to reconstruct such networks of strategic alliances we will apply a multidimensional scaling technique and cluster analyses. We will also present a structural analysis of network density and test the stability in groups of "leading" partnering firms. As we are interested in the possible changes that took place over time we will look at both the first and the second half of the 1980s. Apart from this historical specification we apply a "sectoral" differentiation into sub-fields of information technology in order to achieve a less aggregated understanding of networks and their main "actors". Although our graphical presentation is somewhat space-consuming it enables us, contrary to most forms of statistical abstraction, to identify concrete networks and the major companies involved.

2. General trends in strategic technology alliances during the 1980s

A considerable number of studies suggests that the past decades of industrial development are, amongst other things, characterized by a substantial growth in the actual number of inter-firm strategic alliances, see for instance [2,3,6–10,12]. In the literature this increase is frequently explained by the present rapid changes in technological development, the necessity of pre-empting strikes, sharing of costs, monitoring of relevant technological developments and many other motives that play a role in particular cases. Given the relevance of these developments and their impact on modern capitalist development and some of our own previous research experience and findings by others we expect strategic technology partnerships to be with us for a long period. However, we also expect the increase of strategic partnering that began during the 1970s to gradually wear off [7]. In other words, we envision not so much a constant or exponential growth rate of new strategic technology alliances but a gradual flattening of the growth rate during the later years of the 1980s. As companies have increasingly been involved in strategic alliances we can assume them to experience that these strategic alliances do not provide a panacea for all their problems; alliances have to be carefully managed and firms run all sorts of risks with their partners in these alliances. As the phenomenon of strategic technology partnering has become a familiar aspect of strategic behaviour of a growing number of companies they will become somewhat more conscious of the implicit danger in alliances which will be translated in a gradual flattening and even a decline in the growth pattern of these agreements. This pattern should be very apparent in information technologies because this field of technology is found to have a relatively long tradition and extensive experience with strategic technology alliances [6.7].

As shown in fig. 1 the pattern of all newly established strategic technology alliances found in our databank demonstrates that the first years of the 1980s are characterized by a somewhat constant increase of new agreements. followed by a sharp rise during the mid-1980s, which is continued by a somewhat slower rate of increase during the final years of the 1980s. The increase

[2] See Hagedoorn and Schakenraad [7] for details about the analysis of both cost-economizing agreements and strategic partnerships and a description of the distribution of the most strategic modes of cooperation.

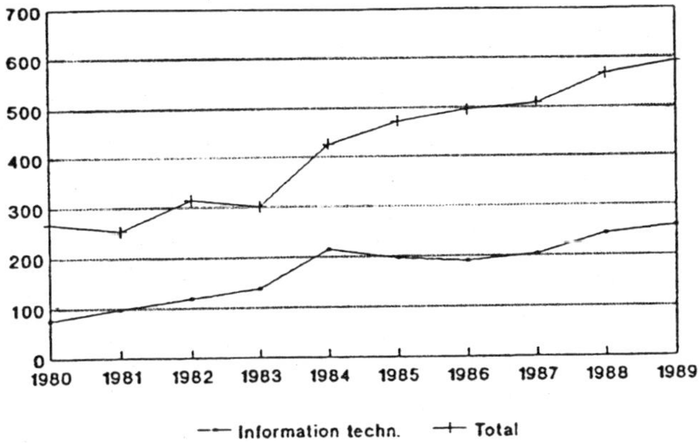

Fig. 1. Growth of numbers of newly established strategic technology alliances in general and in information technologies, 1980–89; *source*: MERIT–CATI.

of strategic technology partnerships in information technologies shows a somewhat different pattern with a clear growth of new agreements in the first half of the 1980s, a stable number of new alliances in the following years and another increase during the final years of the decade again. In other words, this pattern, in particular the one in information technology, does not appear to provide any verification of our hypothesis of a flattening of the growth rate in strategic technology partnering.

However, if we look at the pattern in different sub-fields of information technology (fig. 2), we can take the analysis one step further and distinguish some differences in patterns of growth of new alliances. In telecommunications there is a gradual increase in the number of new technology alliances, although this pattern has clear annual ups and downs. In computers we see a relatively constant number of new alliances with two periods of accelerated growth, one during the mid-1980s and one at the end of the period. In microelectronics there is a clear and rapid growth in new alliances until 1984 followed by a period of an almost constant number of new agreements. Strategic technology alliances in industrial au-

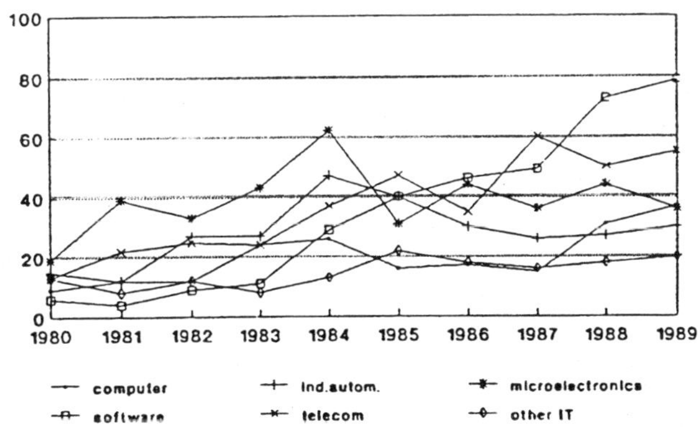

Fig. 2. Growth of newly established strategic technology alliances in information technologies, 1980–89; *source*: MERIT–CATI.

tomation increased during the first half of the 1980s after which the number of new agreements gradually decreased again. A clear pattern of growth with a steep rise of new strategic technology alliances is visible for the field of software where partnering appears to have "boomed" since 1983.

So, with the exeption of software this general trend indicates that strategic alliances in information technology have come into a more or less stable situation. An explanation for the growth in technology partnerships in software has to be found in the crucial role that software plays in linking different sub-fields of information technologies and the complementarity between hardware manufacturers and software producers. Recent developments in information technologies stress the possibilities of building complex networks of once relatively separated sub-fields such as telecom and computers and industrial automation and computers. As these systems become potentially more interlinked, software with its intermediary function becomes vital to further development. Specialized software developing firms have particular capabilities that turn them into attractive but also compulsory partners in strategic technology alliances. Hence, strategic technology alliances between software developers and a wide variety of companies from different fields of information technology blossom in a period when the general increase of strategic technology alliances in this broad field of technology has become relatively stable.

3. The general structure of networks of strategic technology partnering in information technologies

So far contributions from economics and strategic management suggest a number of interpretations regarding the structure of inter-firm networks. In very general terms it can be claimed that these networks are applied by large companies in their strategies for further internationalization, diversification and complementing their existing technological competence (see Hagedoorn [5] for a survey of the literature). From a theoretical position there are at least two confronting lines of thought, one supporting the notion that market leaders take nodal positions in inter-firm networks, the other maintains that strategic partnering is an activity in which in particular "second division" companies are engaged. Following Casson [1] strategic partnering could be seen in the light of oligopolistic rivalry in which companies meet in a number of markets, in some of which collaboration is a viable option. In particular leading companies, i.e. market leaders, would be able to set up networks of strategic partnering in which they act as nodal companies. Others, for instance Porter, take a different point of view when describing these strategic alliances in terms of weakness and not strength of participating firms. So, from this perspective strategic alliances "... appear to be most common among second-tier competitors..." [14, p. 67]. We, however, take a somewhat intermediary position. We expect leading companies, which are often of a diversified nature, to create flexible networks that enable them to capitalize on economies scope generated by changing networks for a number of their businesses. Market leaders have, almost by definition, the discretion to choose their partners from a relatively large group of companies and, if they are diversified, they can also spread their alliances over a number of their activities and not just their core business. This flexibility implies that we do not expect a strict correlation between the relative market position of companies in distinct businesses and their position in networks of strategic technology alliances. However, we also do not expect strategic technology partnering to be a game of "second-tier competitors", if this term implies competitors that are at the loosing end of (inter)national competition. It is obvious that there are examples of combinations of such companies, but in general there seems little indication to expect strategic technology partnering to be dominated by "losers". A more subtle interpretation would suggest that leading companies forge alliances with different partners in order to create spill overs from a variety of joint projects and R&D activities. If market leaders do not necessarily concentrate their strategic partnerships in their core activity, their overall presence in alliances for all their interests is still expected to stand out among the general population of companies engaging in strategic technology partnering.

In this section as well as in the next the concrete analysis of networks of strategic partnering can provide some first understanding of the position taken by the most prominently cooperating companies in a number of industrial and technological settings. The analysis of the general structure of networks of strategic technology partnering in this section is of a fairly descriptive nature. However, the analysis does enable us to do more than just give some measurement of increased cooperation or other rather abstract indicators of inter-firm technology partnering as it actually facilitates the identification of both major "actors" in several sub-fields of information technology and some possible clusters of companies, as well as the changes that have taken place during the 1980s.

In the previous section we have seen that strategic technology partnering has increased during the 1980s. Despite a relative slow down in the growth of new partnerships in recent years about 60% of all agreements made during the 1980s have been established during its second half. If some companies play a central role in strategic alliances and consequently strategic technology partnerships are unevenly distributed towards "nodal" companies, we can expect the linkages of these firms to have escalated during the second half of the 1980s, which should show up in the analysis of intensified inter-firm partnerships in the following paragraphs.

In order to reconstruct networks of strategic alliances we have applied both multidimensional scaling and cluster analyses. A non-metric multidimensional scaling (MDS) technique will be applied to give an overview of the network structure (see Appendix II). We already introduced an MDS technique for the analysis of inter-firm alliances in previous work, see Hagedoorn and Schakenraad [6,7], and also Kruskal and Wish [11], and Wilkinson [15]. Cluster analysis will be applied to detect groupings of collaborating firms (see Appendix II). A very common way of presenting the results of clustering is by means of a so-called dendrogram. We, however, will put a summary of the results of the cluster analysis in table 1 and discuss them together with the MDS solutions.

For information technologies in general the intensity and structure of technology cooperation in two periods is pictured in fig. 3a and 3b. Cluster analysis applied to the 45 most cooperating firms shown in fig. 3a results in three distinct clusters. The first cluster consists of American chip and computer manufacturers, the second contains some European firms, and the third one is a Japanese cluster. [3] On the left-hand side of dimension one in fig. 3a we notice a concentration of US companies such as RCA, NCR, Allied, Hewlett-Packard, Harris, DEC, Schlumberger (including Fairchild), United Technologies (Mostek), Honeywell, CDC, Advanced Micro Devices and General Electric. On the right-hand side, in the first quadrant, we see a concentration of a number of Japanese companies such as Mitsubishi, NTT, Hitachi, Oki Electric, Toshiba and Fujitsu. European companies also have a high proximity.

To improve our interpretation of MDS pictures, it is useful to draw lines between companies. [4] From fig. 3a we learn that during the first half of the 1980s there was a number of companies with rather strong interrelationships. For example: the networks around CDC, Sperry, Intel, Motorola, Siemens, NEC, Fujitsu and Philips. All these firms rank among the top ten of most cooperating companies in information technologies in that period (see also table 3).

If we take a look at the situation in information technologies during the second half of the 1980s (fig. 3b), we notice that there is still a concentration of intra-Japanese strategic partnering with an almost identical group of companies as during the previous period. From the bottom upwards to the centre we find US firms. On the right-hand side of dimension one we see a concentration of intra-European cooperation in particular by companies such as GEC, CGE, Daimler-Benz (especially through its acquisition of AEG), Bosch, IRI (represented by many of its units such as STET, CSELT, DEA, Consultel), Plessey, Nokia, Ericsson. This notion of rather strong geographical concentration is supported by the results of cluster analysis. In the centre of

[3] For an explanation of company codes in this paper we refer to Appendix III.

[4] We have drawn lines between every pair of companies of which the proximity exceeds some threshold value. Fat solid lines indicate very strong cooperation (7 cooperative agreements or more), normal lines stand for strong cooperation (5 or 6 agreements), while dashed lines represent moderate cooperation (3 or 4 agreements). The position of companies which are not connected to other companies through lines is by no means truly peripheral. Only their agreements are spread over a number of companies without having more than two agreements with one of the other companies.

both dimensions we find the companies that have most international strategic partnerships with a number of companies. Here we find a number of the world leading companies in information technologies such as IBM, Unisys, Siemens, AT&T, Thomson, Bull, Philips, STC, GE, Toshiba, Intel, HP and DEC. Compared to the previous period we notice that most of the largest US and European companies, led by Siemens, have moved towards the centre, which indicates their pivotal role in worldwide partnering.

During the second half of the 1980s a number of US companies have "disappeared". Some companies merged, such as Sperry and Burroughs,

Table 1
Results of cluster analysis for information technologies in 1980–84 and 1985–89

	1980–84	1985–89
Information technologies (figs. 3a. 3b)	Three clusters: –Europe, US, Japan	Basically the same pattern as in 1980–84
Computers (figs. 4a. 4b)	Five clusters: –TRW, AMP, etc. group including Sperry, Toshiba, Mitsubishi, IBM –Cluster around CDC Smaller groups with Japanese firms in the centre: –Fujitsu –NEC –Matsushita, Oki, Kanematsu	Seven clusters: –Honeywell, Bull, NEC, Hitachi –Fujitsu and STC –Olivetti cluster (with Stratus. etc.) –CDC group (with Elbit, Cray) –Philips, IRI, GEC, Nixdorf –AT&T, Unisys, Toshiba, Mitsubishi –Cogent, Sequent, Ncube, Float PS
Industrial automation (figs. 5a. 5b)	Several small groups: –GM, Westinghouse etc. –Dainichi Kiko etc. –Siemens, Fanuc etc. –Bosch and Shoun –Olivetti and Allen-Bradley etc. –IBM group –Cincinnati Milacron cluster –Yaskawa, CGE, Toshiba	Relative large groups: –Rockwell cluster (with DEC, Koreans. Olivetti, Fiat, Honeywell) –GM group (with Fanuc, Fujitsu. Tandem) –Shoun, Toyota, Nippon Autom.
Micro-electronics (figs. 6a. 6b)	Two main clusters: –Intel, Fujitsu, Matra-Harris. NEC. Siemens, AMD. IBM –Motorola, Thomson, Philips. Matsushita. Schlumberger, Hitachi.	Three main clusters: –a combination of the Intel/Motorola groups from the previous period plus Toshiba –a cluster with AT&T (and SUN. Samsung. Lucky. CTNE, Olivetti) –The HP. RCA, Kodak, TI. GE. IBM. GE. etc. consortium
Software (figs. 7a. 7b)	Some rather separate groups: –US consortium with Sperry, NCR. DEC, etc. –European firms (Philips. Siemens. STC. Bull) –Microsoft, IBM, Olivetti –Hitachi, Telex etc. –Digital Research, Exxon. Gould. NEC	Four main clusters: –a US aerospace group (with Boeing, Northrop, etc.) –the US groups with Unisys. DEC. NCR, etc. (same group as in previous period) –a Japanese cluster –US/Europe group with several subgroups: –Volmac, CAP-Gemini, Sema –Microsoft. IBM, Olivetti –AT&T, SUN –Philips, BSO, Nixdorf, BULL, STC
Telecom (figs. 8a. 8b)	Some largely national groups: –Japanese firms (plus IRI, Rolm. Mitel, IBM,) GTE) –German/Scandinavian cluster plus ITT –AT&T plus Philips and Olivetti –UK firms (Racal, Plessey, GEC. British Telecom) –DEC, XEROX, Intel	More international groups: –IRI, AT&T, Toshiba, Olivetti –Siemens. IBM, Philips, FIAT, Bosch. CGE. Daimler, ITT –Racal, Matra, Ericsson, GEC, Plessey –NT, DEC. BT, STC –Some Baby Bells –Japanese firms and Bellcore, GTE and C&W.

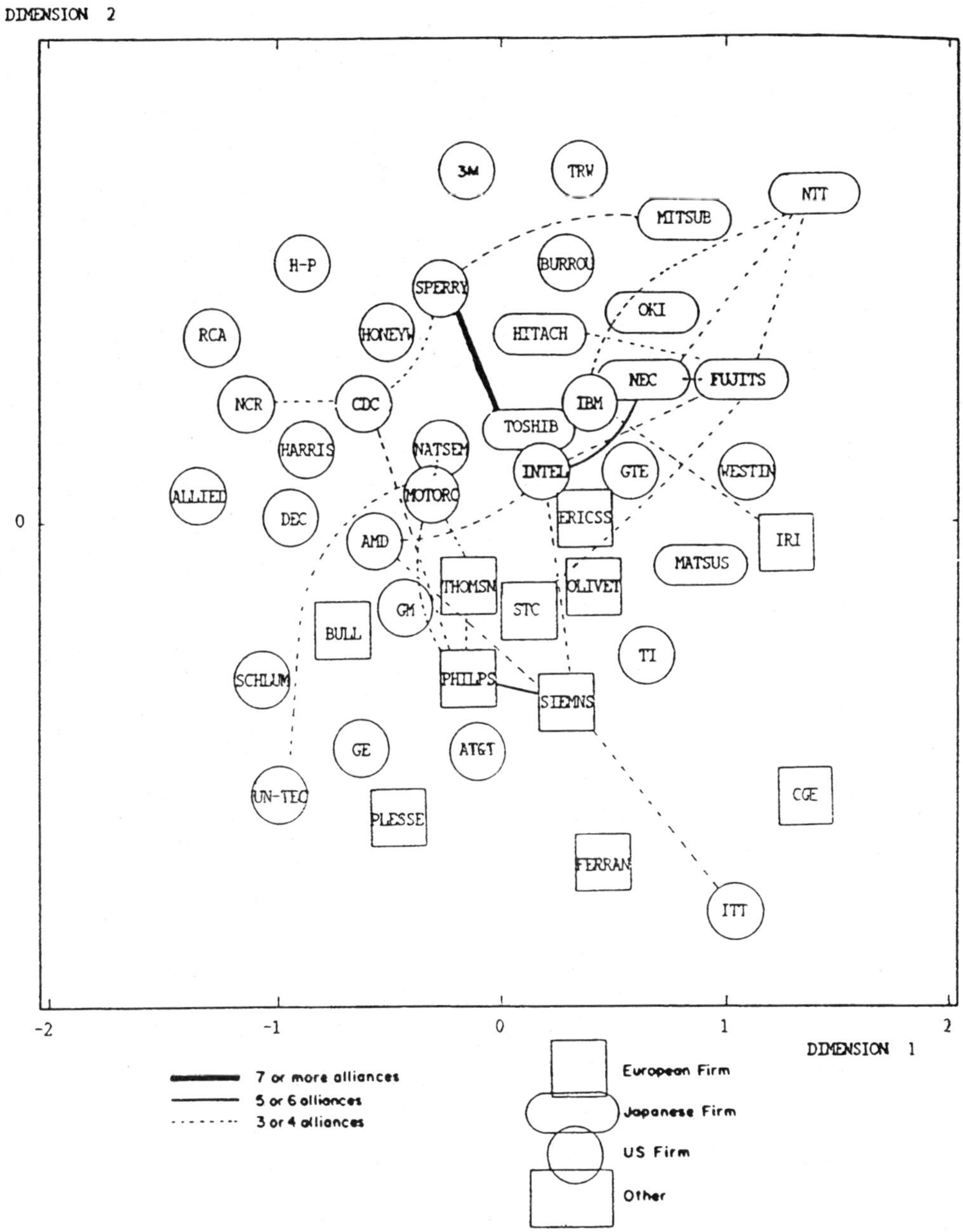

Fig. 3a. The structure of strategic partnering in information technologies. 1980–84.

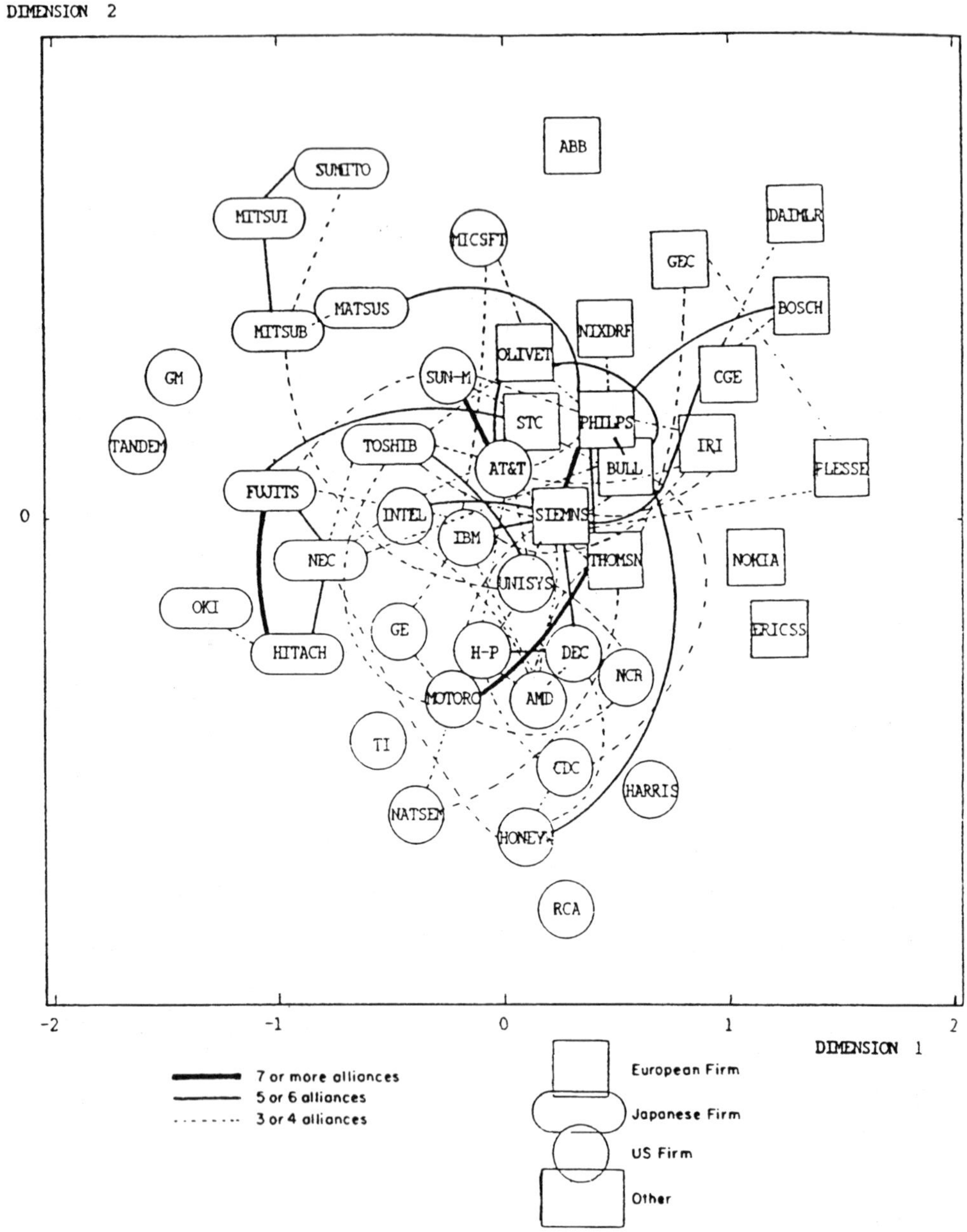

Fig. 3b. The structure of strategic partnering in information technologies, 1985–89.

others divested (parts of) their activities in information technologies, such as ITT's divestment of its telecom division to Alcatel (CGE), and United Technologies' selling of Mostek to Thomson.

From fig. 3b we learn that a number of very strong tie-ups are apparent, such as: Siemens with IBM, Intel, Philips, CGE, DEC; Philips with Siemens and Thomson; AT&T with Sun Microsystems; Fujitsu with STC, Hitachi and NEC, Mitsubishi and Mitsui, and, in particular before Honeywell divested its I.T. business, a very strong link between that company and Bull of France.

So for information technologies at large we see that firms have formed many linkages with a wide variety of partners, which nevertheless generate separate clusters dominated by either Japanese, European or US companies. It is also evident that many leading US and European information technology firms take a nodal position in these networks for which the linkages have increased during the past decade.

3.1. Networks of strategic alliances in computers

In fig. 4a and 4b we present the MDS solutions for computer technology in both periods. [5] Interpreting fig. 4a and looking at the results of the cluster analysis in table 1, we first discover a dense cluster of American firms at the foot of the right-hand side. During the early 1980s all of these companies were member of the "Optical Circuitry" research corporation, which started some research on super fast optical computers. Of all the participants in this research corporation only IBM and Sperry had substantial links with other firms and already held central positions. The strategic position of Control Data Corp. (CDC) is also evident; this firm had close ties with Philips (CD memory systems), Honeywell (disk systems), Sperry, STC and NCR (for computer peripherals), with Elbit from Israel, Hitachi and Hewlett-Packard (several links in the area of minicomputers). It also made strategic investments in promising start-ups such as Trilogy and Centronics Data Computer. Other key positions are held by two Japanese firms, NEC and Fujitsu. The latter had tied up with STC (in the field of mainframes), Amdahl (plug compatible mainframes), Computer Consoles and Signal-daughter Ampex (peripheral equipment and tapes). NEC is a dominant partner in arrangements with Siemens, BASF (and with the Siemens-BASF joint venture Comparex), National Advanced Systems (NAS), Honeywell, STC, and Olivetti (all on the field of mainframes). NEC also established technical links with 3M (on information storage), IBM, Burroughs and Honeywell (for mainframes).

Turning to the second half of the 1980s (fig. 4b) we notice that companies such as Fujitsu, NEC, Unisys (the Sperry–Burroughs merger), and CDC still take central positions during the second half of the 1980s. Others, such as Olivetti, Bull and Siemens, have improved their network positions slightly. Philips collaborates with Nixdorf, IRI and GEC on unconventional computer structures. The world's largest computer manufacturer and market leader IBM has moved somewhat to the periphery. Newcomers at the top of the network are Apple, Cray, the cluster with Ncube, Cogent, Floating Point Systems and Sequent (brought together through their membership of a parallel computing consortium), and Sun Microsystems which rapidly created its own network.

3.2. Networks of strategic alliances in industrial automation

Strategic partnering in industrial automation during the early 1980s (fig. 5a) can hardly be described in terms of a dense network. Instead, few medium-sized clusters and many couples characterize the picture. The first and largest cluster is built around Fanuc and Siemens and its tentacles reach Philips, KUKA, IRI and even Apollo and Shoun. The second cluster which has no ties with the first one, comprises GM, Westinghouse, Schaudt, etc. Dainichi Kiko, Aritmos of Sweden, DAF, Hoogovens and Sykes from the third cluster. The picture contains some couples of rather intense collaborating firms such as Bosch and Shoun, Allen-Bradley and Olivetti, and the Swiss duo Georg Fischer–ADC. Many isolated firms have no links with any of the firms pictured, but they do belong to the group of most cooperat-

[5] Due to the smaller numbers of alliances between firms at the level of sub-fields we adapted our legend. For all of the following MDS plots a very fat solid line indicates that we registered 4 or more strategic alliances between a couple of firms. A fat line stands for 3 alliances; a normal line represent 2 alliances and a dashed line indicates one interfirm agreement.

ing firms; their less cooperating partners, however, are not incorporated in the MDS procedure.

Turning to the next period (fig. 5b), the first thing to mention is that several links between already existing, separate networks have been

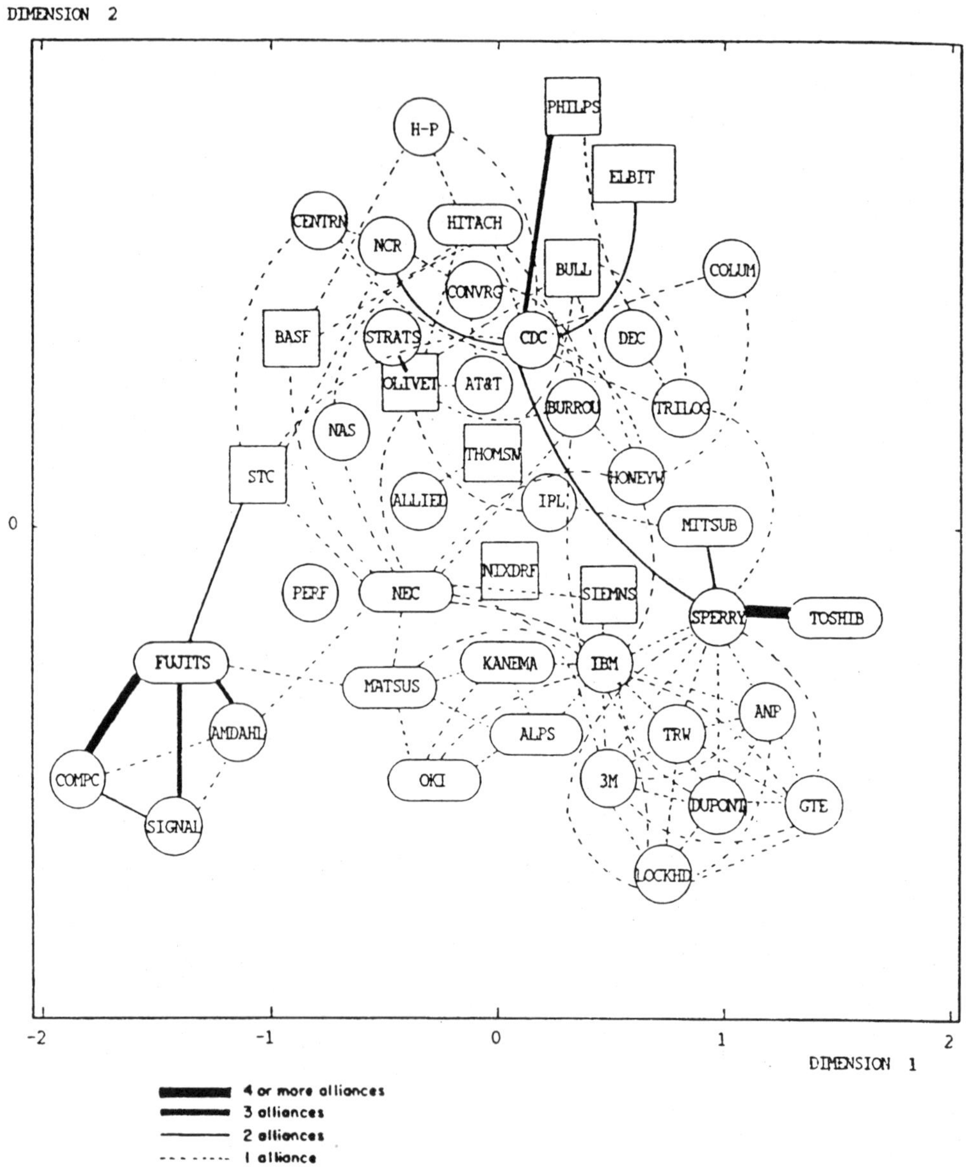

Fig. 4a. The structure of strategic partnering in computers, 1980–84.

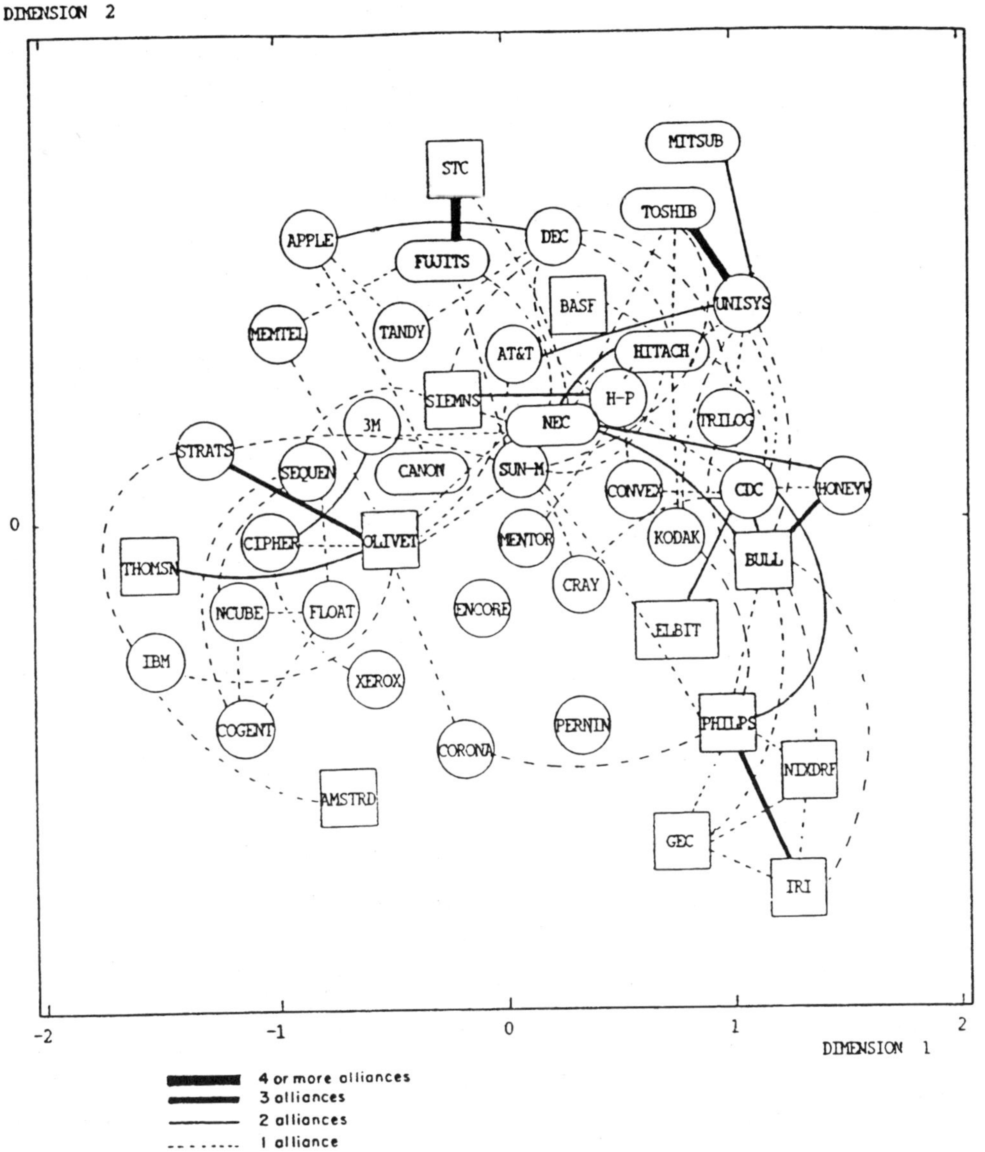

Fig. 4b. The structure of strategic partnering in computers, 1985–89.

forged, although some floating teams still exist. In particular Rockwell (which has taken over Allen-Bradly), GM and IBM have been active along these lines. The Korean companies Kia and Daewoo together with Rockwell formed a CNC manufacturing joint venture in South Korea. GM strengthened its network position by creating a joint venture with Fanuc, called GMF, which itself is also active in the field of strategic partnering. IBM follows a cooperative strategy in industrial automation, as well. It has links with IRI, ABB, Tandem, Cincinnati Milacron, etc. Important newcomers are DEC, Tandem, and the recently established R & D joint venture of KUKA, Chrysler, Litton, Du Pont and Cincinnati Milacron.

3.3. Networks of strategic alliances in microelectronics

Cluster analysis for microelectronics in the first half of the 1980s (see fig. 6a and table 1) basically reveals two large clusters. One cluster is headed by Intel and includes large integrated chip manufacturers such as Fujitsu, Matra-Harris, NEC, Siemens, AMD and IBM, the other one is being dominated by Motorola and has Thomson, Philips, Matsushita, Schlumberger (Fairchild) and Hitachi among its most important members. The emergence of two clusters, one around Motorola and another around Intel can be explained as a result of both cooperation and fierce competition in microelectronics during the first half of the 1980s, when second-sourcing arrangements for RAM chips and micro-processors were the most common way of partnering. Intel and Motorola produced highly competitive processors, Intel supplied its 8084, 8086 and 80286 series, Motorola its 68000 family. These products were subject to many second-sourcing agreements which ultimately led to the clustering described above. During the second half of the 1980s the Intel and Motorola clusters have converged somewhat, although still no important ties between both firms are registered, see fig. 6b. Toshiba got a strong foothold to the Intel–Motorola cluster by teaming up with Siemens and Motorola. A large number of research links between dominant firms were forged during this period. New networks were being formed with AT&T, Sun Microsystems, Samsung, Lucky Goldstar, CTNE and Olivetti in active roles. On the right-hand side of fig. 6b we see many members of the Semiconductor Research Corp., established in 1986 to support basic semiconductor research in the US; among the major members of this consortium are Harris, RCA, DEC, and CDC.

3.4. Networks of strategic alliances in software

Strategic partnering in software demonstrates some rather separate groups in the period from 1980 till 1985 (see fig. 7a). The first group is the MCC consortium from the USA with 13 of its members lying on the right-hand side of the picture. Philips. Siemens, STC and Bull are representatives of a European cluster. We also have to mention the positions of Microsoft and Digital Research, two influential software firms.

During the second half of the 1980s the network density has increased dramatically as one can learn from fig. 7b. The MCC consortium is still operating, it can be found in the first quadrant, but another large joint venture has been created; the Software Productivity Consortium which develops tools for writing complex computer software. Its members originate from the automotive, aeronautics and defense industry (Ford, Lockheed, Harris). Siemens joined indirectly through its acquisition of Bendix. Mitsubishi, Hitachi, NEC and Fujitsu are brought together by the TRON project which is aimed at developing a new computer operating system. Boeing, Harris and Siemens occupy key positions by having access to different major groups. The remaining part of the picture consists of a US–Europe cluster with several subgroups such as Volmac, CAP-Gemini, Sema; Microsoft, IBM and Olivetti; AT&T and SUN Microsystems; Philips, BSO, Nixdorf, Bull and STC.

3.5. Networks of strategic alliances in telecommunications

In the telecommunications network for 1980 up to 1985 some largely national groups can be discovered (see fig. 8a). On the left-hand side, we find Japan's NTT accompanied by some of its traditional suppliers Oki, Hitachi. Fujitsu and NEC. On the opposite side a German–Scandinavian cluster is located with MBB. ANT, AEG, Siemens, Bosch, Nokia, Ericsson, and the, at that time, European telecom unit of ITT. Standard Elektrik Lorenz. The UK quartet Plessey, Racal, GEC and British Telecom lies at the foot of the picture and can be expanded with STC. In contrast to these nationally oriented clusters are the clearly international positions of the US firms AT&T, GTE, IBM and Motorola.

Although we found that during the second half of the 1980s some shift has taken place from public digital exchanges to PABX, mobile telecom, and telematics as popular items for collaboration, this has not really changed the structure of strategic networks with national and regional

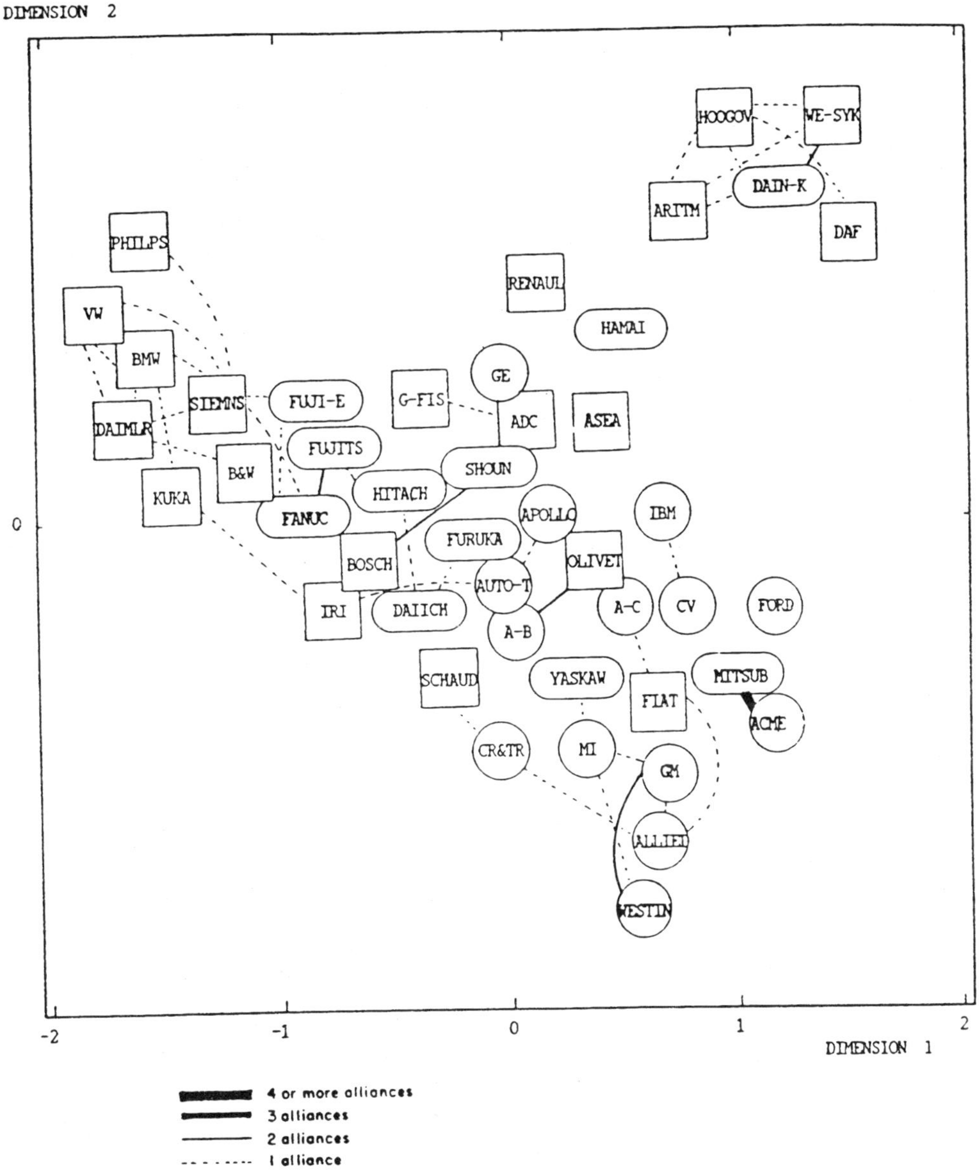

Fig. 5a. The structure of strategic partnering in industrial automation, 1980–84.

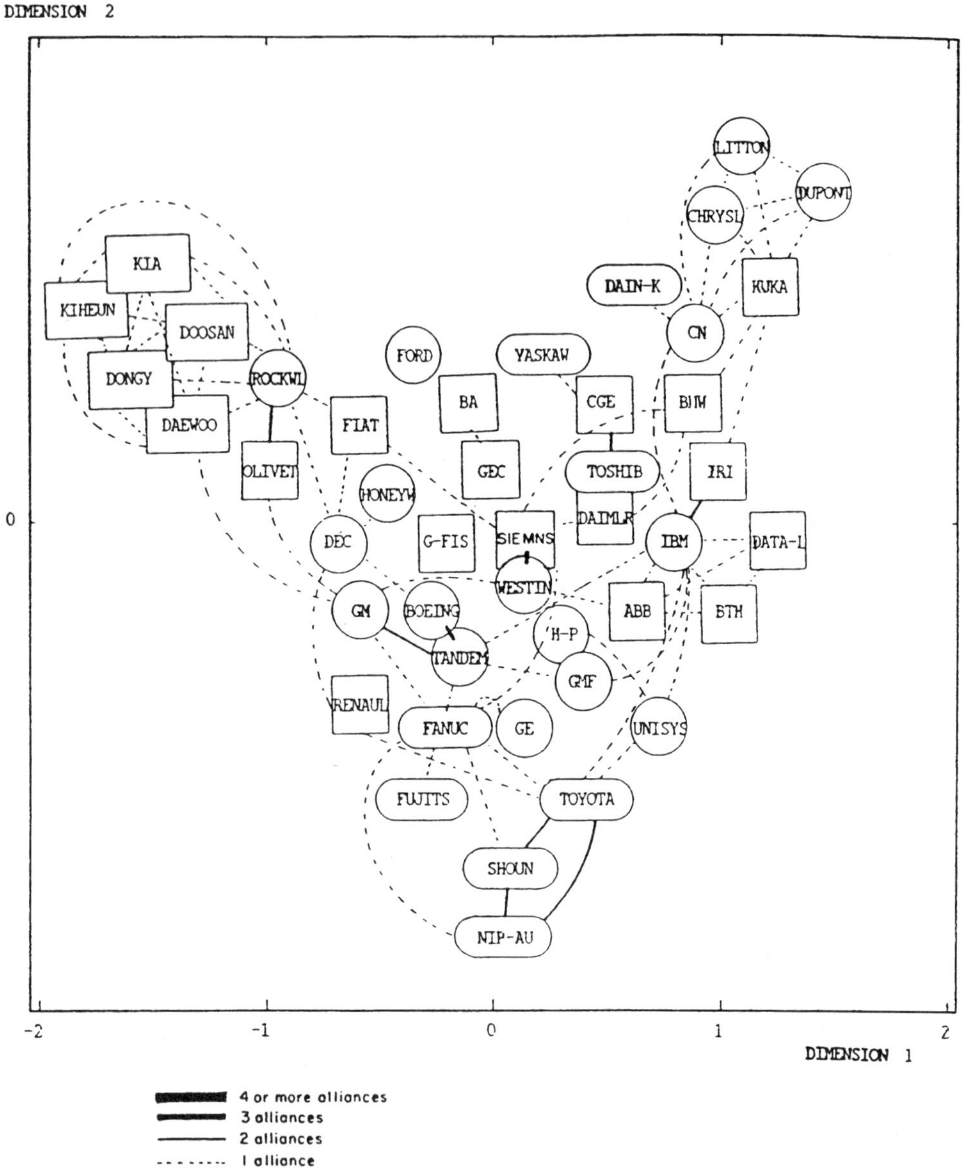

Fig. 5b. The structure of strategic partnering in industrial automation, 1985–89.

groups and a few truly international cooperating firms (see fig. 8b). One can quickly recognize Japanese (NEC, Fujitsu), European (Bosch, Daimler, CGE) and US clusters (the Baby Bells). AT&T, IBM, GTE and Motorola still keep truly international, "triadic" links, but some Japanese firms and Siemens, Cable & Wireless, and Bellcore are trying to succeed in this way as well. Finally, we have to mention the strategic positions of Daimler (after absorbing AEG and MBB), CGE (after taken over ITT's European telecom businesses), and Ericsson. The so-called Baby

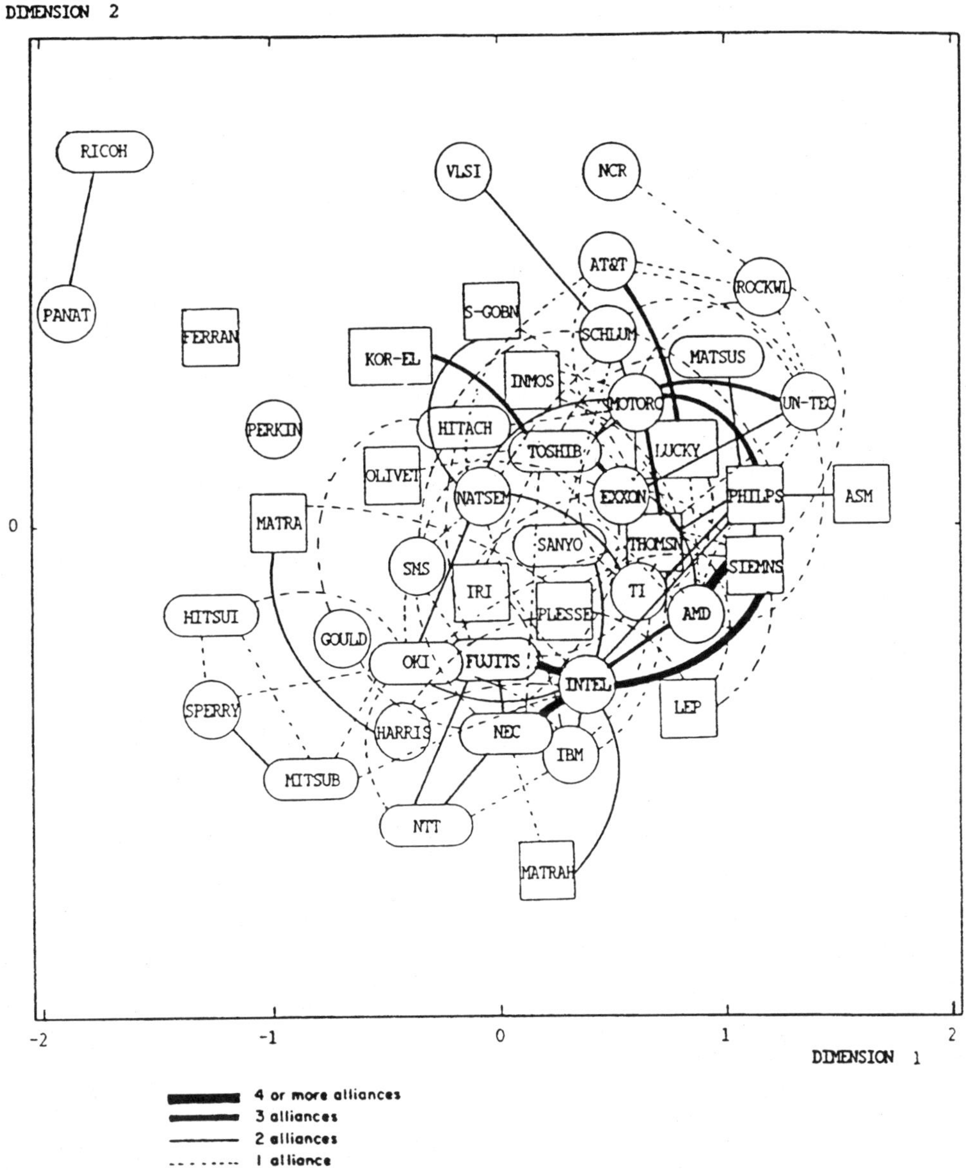

Fig. 6a. The structure of strategic partnering in microelectronics, 1980–84.

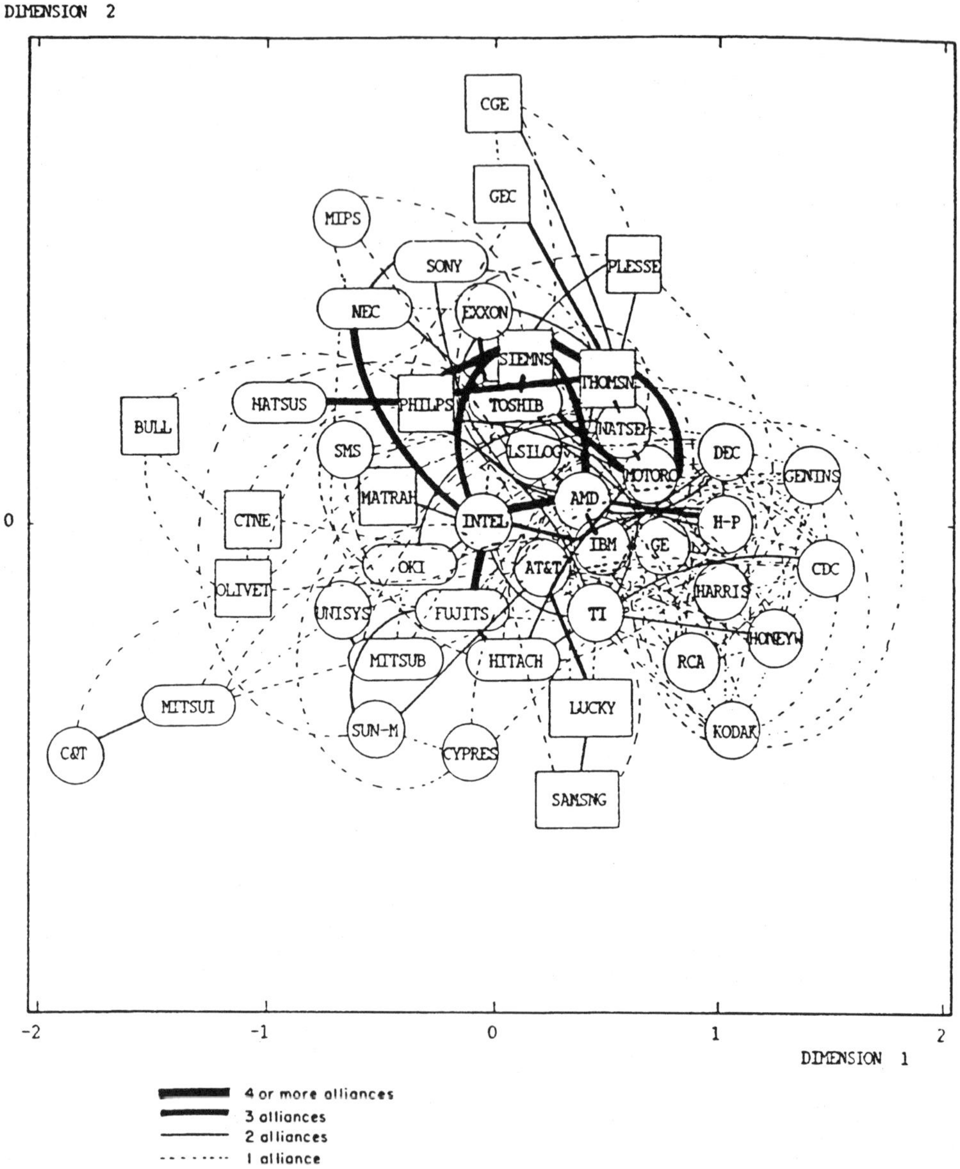

Fig. 6b. The structure of strategic partnering in microelectronics, 1985–89.

Bells (Nynex, Bell South, Pacific Telesis etc.) and their R&D joint venture Bell Communications Research (Bellcore) have gradually become more internationally oriented, which is reflected in their growing number of international alliances.

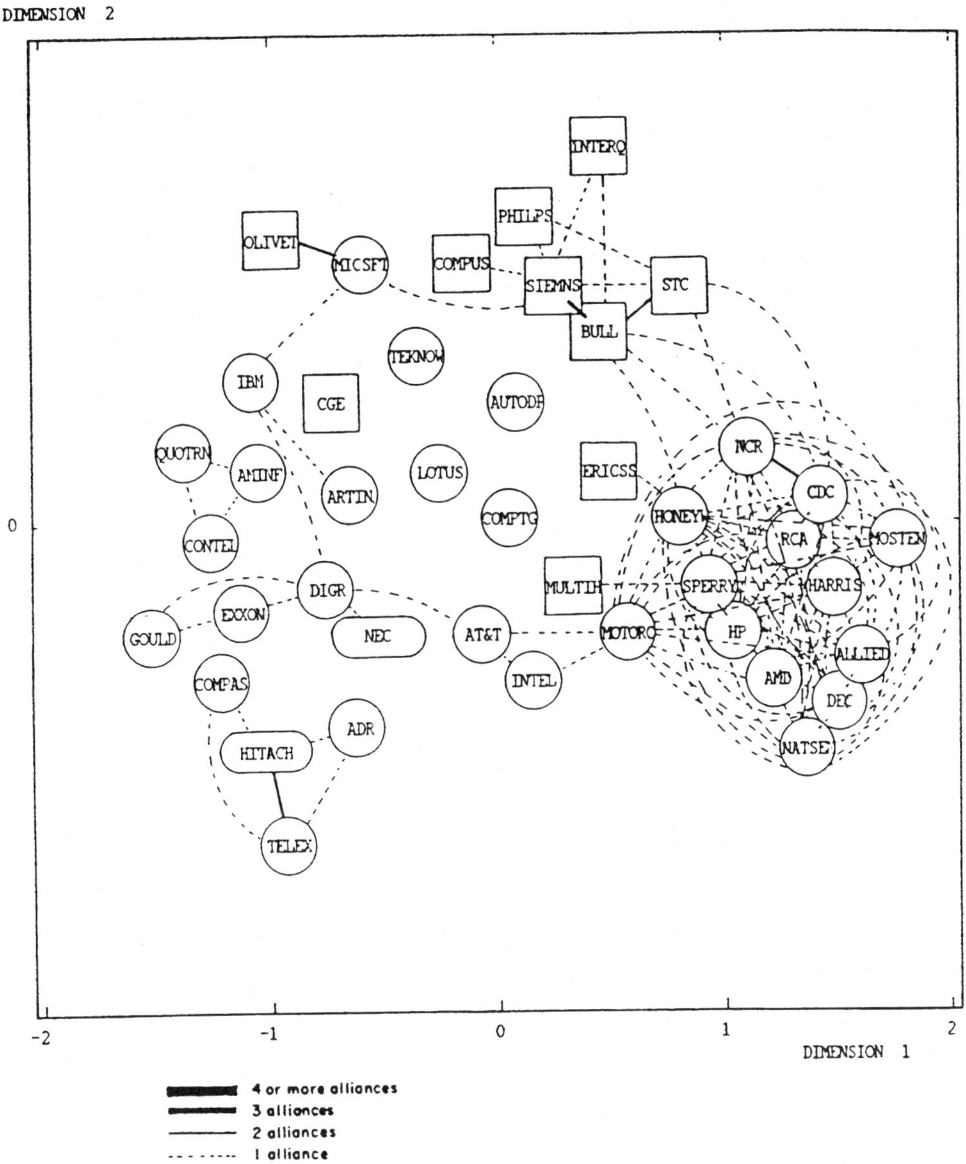

Fig. 7a. The structure of strategic partnering in software, 1980–84.

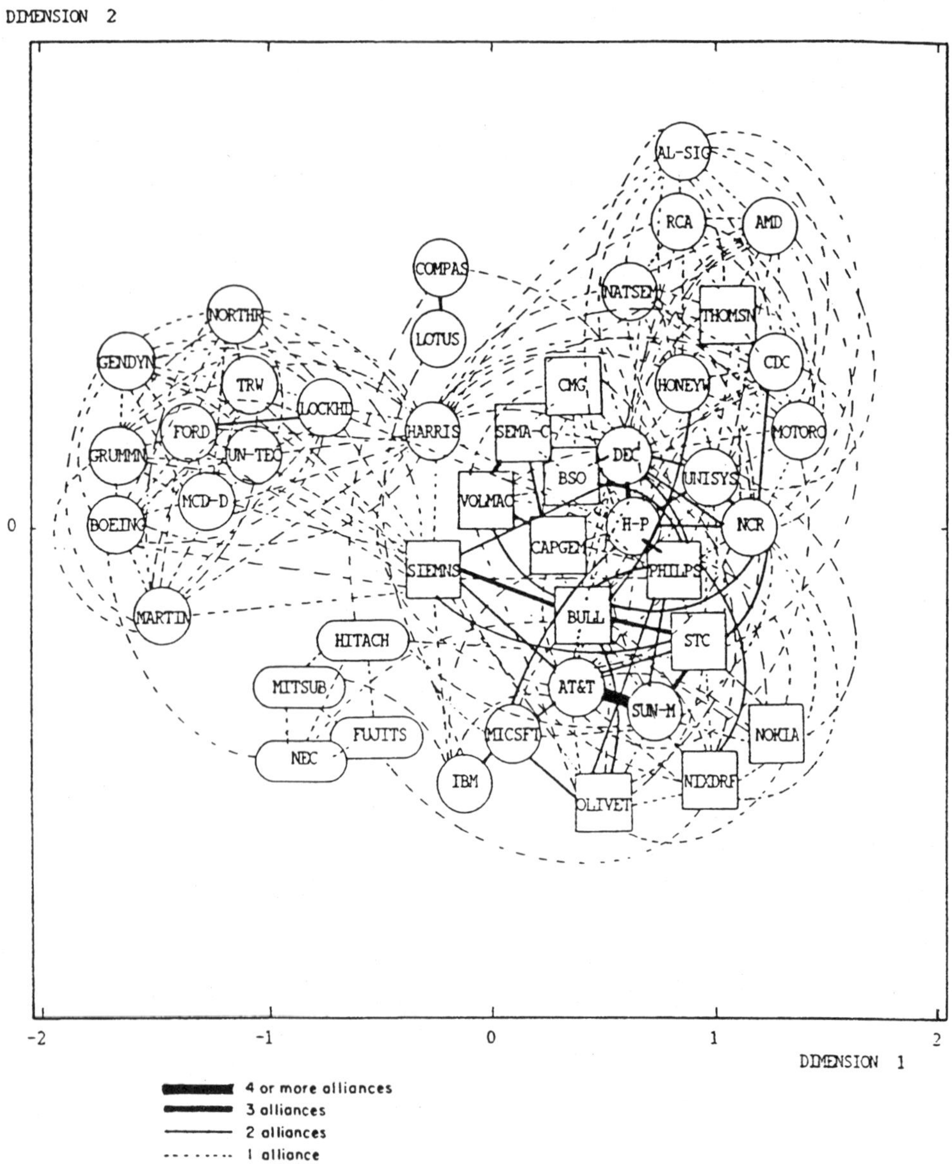

Fig. 7b. The structure of strategic partnering in software, 1980–89.

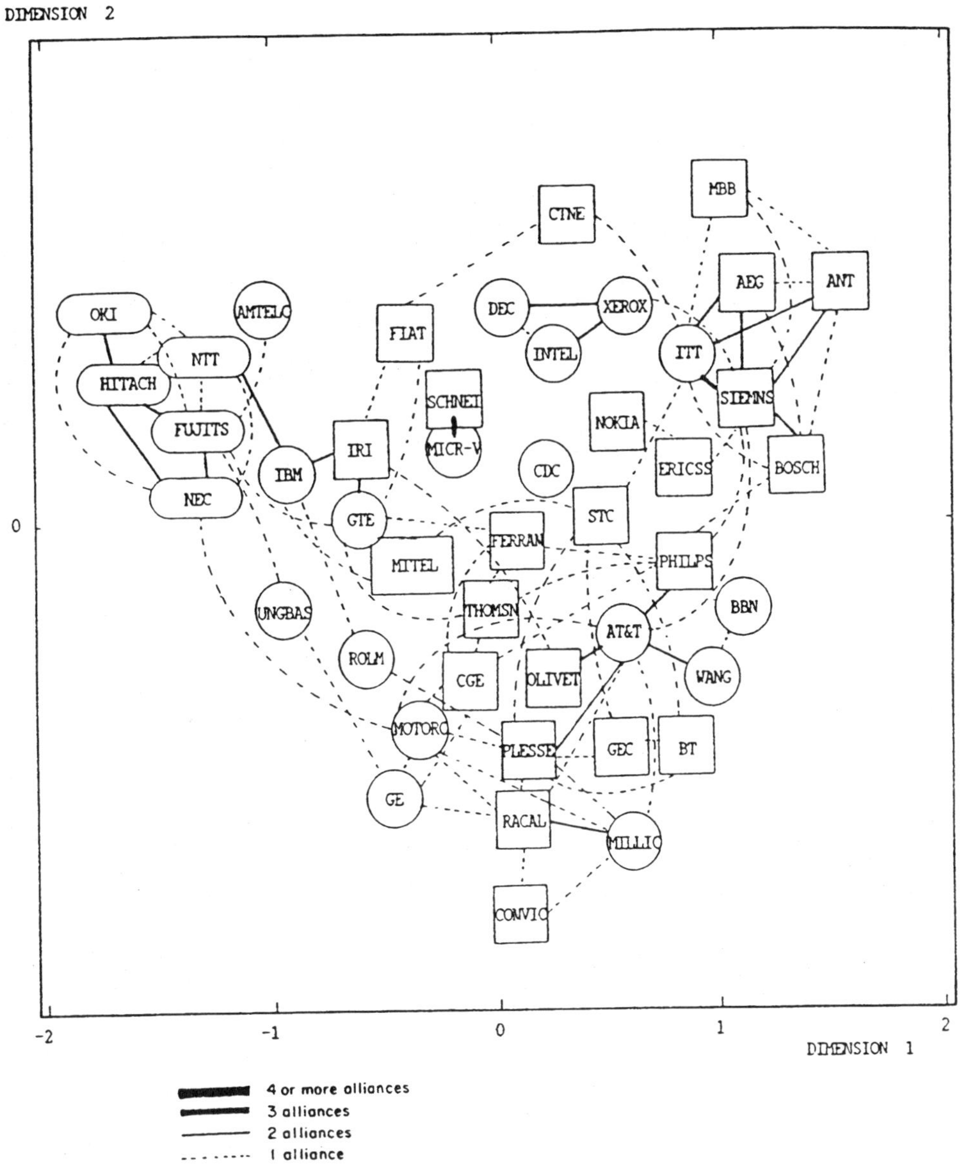

Fig. 8a. The structure of strategic partnering in telecommunications, 1980–84.

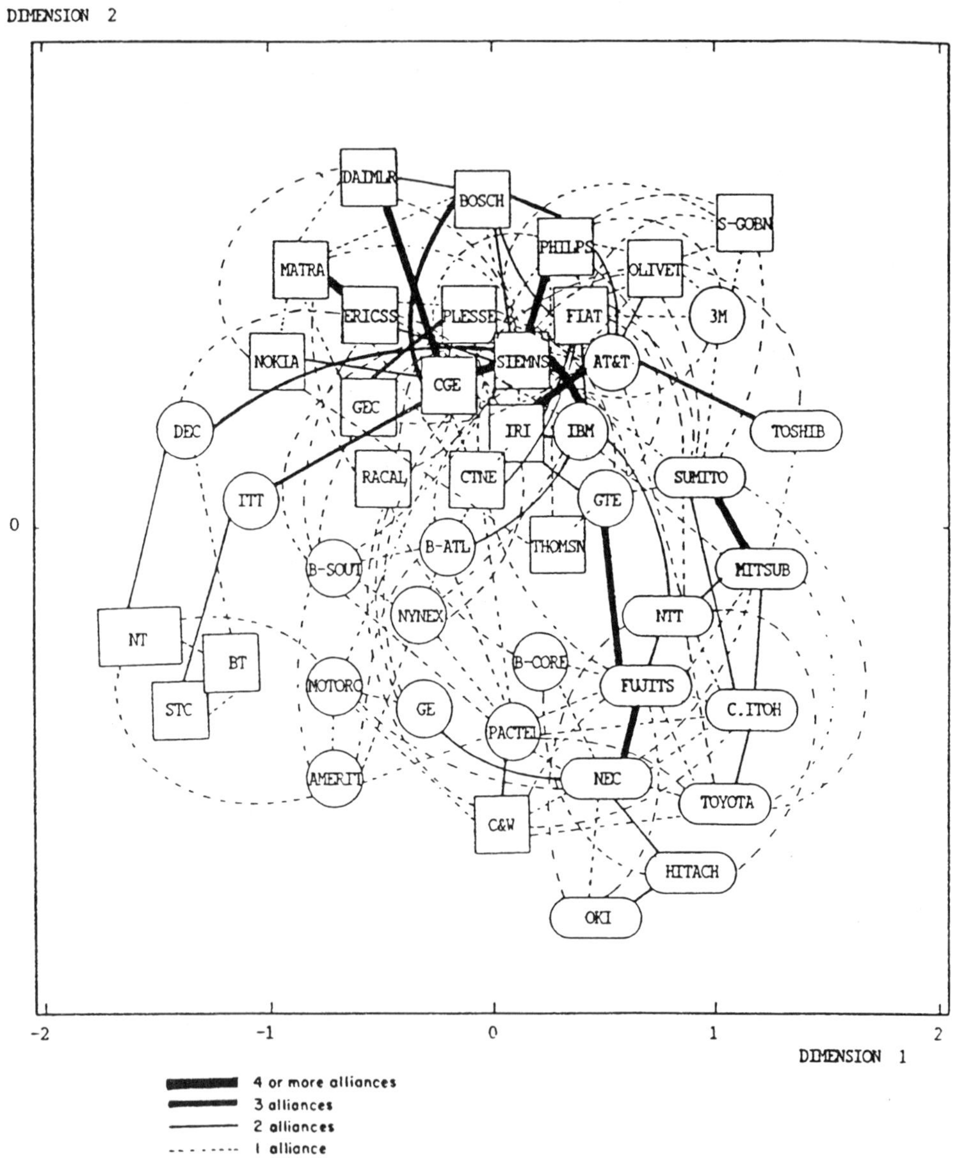

Fig. 8b. The structure of strategic partnering in telecommunications, 1980–89.

4. A semi-aggregated analysis of networks in information technologies

Expanding the analysis from the previous sections we can formulate the hypothesis that an intensification of strategic technology partnering during the past decade should show up in an increased network density for both information technology at large and its main sub-fields. To uncover some aspects of structural centrality of a network we computed a network density index. This density index is defined as the ratio of the actual number of links between companies (k) to the possible number of links $1/2n(n-1)$ where n denotes the number of points in the network. The network density in information technologies (see table 2) changed considerably from 23 percent in the first half of the 1980s to 40 percent in the second half. This means that in the years since 1985 40 percent of all theoretically possible links between the 45 most cooperating firms are also empirically observed. In other words, one can speak of a very intensive, dense network in information technologies. Of course, this high density is partly explained by the rather broad scope of the combined fields of information technologies. Therefore, it is interesting to focus on networks in the sub-fields of information techologies. It appears that these sub-fields all show increased density except for computer technology where network density declined to 8 percent. For microelectronics and software we see the highest density growth, resulting in a 23 percent and 28 percent density during the second half of the 1980s.

The next step in our analysis is to examine the stability in the group of most cooperating firms. In case strategic technology partnering would be part of some sort of static oligopolistic structure the group of most cooperating companies would resemble a relatively closed system with little or no entry and a more or less stable rank order if one would compare these cooperating companies at different time intervals. Based on our conceptualization of networks of strategic technology alliances as moderately flexible market institutions, we can expect inter-firm networks that are partly open to new entrants but still with some moderate domination by companies that make the core of these networks, otherwise the increase of network density would be difficult to explain. In other words, we expect a relatively stable group of intensively cooperating companies which, however, does not exclude entry or changes in the rank order.

This hypotheses is supported by the findings of rank correlations presented in table 2. If we look at the presence of companies amongst the group of leading cooperating companies during both periods of the 1980s we find that for information technologies in general 33 out of 45 firms, that is 73 percent remain on the list of most cooperative companies. For sub-fields we recorded the lowest percentage of "stayers" in industrial automation (42%) and the highest for telecommunications (64%). The correlation between the ranks of the "stayers" in both periods, however, failed to reach any level of significance in general information technologies. Although the proportion of "stayers" was very high (nearly three-quarter) there were considerable shifts in their rankings. In sub-fields of information technologies, where the number of potentially relevant firms can be expected to be smaller, we found significant rank correlations, with the exception of software. This means that in computers, industrial automation, microelectronics, and telecommunications some firms indeed do leave and others enter the group of most collaborative companies, but the rankings of the remaining companies did not change significantly.

In addition to this we still have to analyze strategic technology partnering in terms of a possible domination by market leaders. In table 3 one can find the lists of companies having most strategic links. If we keep in mind the most cooperating firms in information technologies in both periods and check how regularly they appear on lists for sub-fields, it becomes clear that seven firms are among the top of cooperating firms for at least three different fields of technology. These "leading" companies are Siemens (industrial automation, chips, software, telecom), IBM (industrial automation, chips, telecom), Philips (computers, chips, software, telecom), Fujitsu and NEC (computers, chips, telecom), Olivetti (computers, software, telecom), and finally, AT&T (chips, software, telecom).

It is obvious that all these companies are very well known, large and also market leaders in very general terms, although not always for the fields in which they stand out as major strategic tech-

Table 2
A comparison of some network aspects for 1980–84 and 1985–89 in information technologies

	Network density		Rank correlation 1980–84 and 1985–89		
	1980–84	1985–89	*N*	(% of 45)	*r*
Information technologies	23%	40%	33	73%	0.17
Computers	10%	8%	23	51%	0.52 *
Industrial automation	4%	8%	19	42%	0.56 *
Microelectronics	11%	23%	25	56%	0.73 *
Software	12%	·28%	22	49%	0.27
Telecommunications	9%	17%	29	64%	0.48 *

* significant at 0.05 level.

Table 3
A comparison of the top ten firms with the most strategic links in information technologies in 1980–84 and 1985–89 (numbers in brackets)

Information technologies				Computers			
1980–84		1985–89		1980–84		1985–89	
1. Motorola	(53)	Siemens	(134)	1. Sperry	(27)	Olivetti	(22)
2. Siemens	(51)	Philips	(127)	2. IBM	(19)	CDC	(19)
3. IBM	(48)	Olivetti	(110)	3. CDC	(18)	Unisys	(17)
4. Sperry	(47)	IBM	(108)	4. Olivetti	(17)	Bull	(14)
5. Fujitsu	(46)	HP	(96)	5. Fujitsu	(15)	Philips	(13)
6. Olivetti	(42)	DEC	(95)	6. NEC	(12)	Fujitsu	(12)
7. CDC	(41)	AT&T	(90)	7. Burroughs	(11)	NEC	(12)
8. Intel	(41)	Thomson	(83)	8. Toshiba	(10)	SUN-Micr.	(11)
9. Philips	(40)	Fujitsu	(78)	9. Du Pont	(10)	DEC	(10)
10. NEC	(39)	Motorola	(68)	10. 3M	(10)	Hitachi	(10)
Industrial automation				**Microelectronics**			
1980–84		1985–89		1980–84		1985–89	
1. GM	(8)	GM	(20)	1. INTEL	(34)	Thomson	(51)
2. Mitsubishi	(8)	IBM	(20)	2. Motorola	(23)	INTEL	(46)
3. Dainichi	(6)	ABB	(13)	3. Philips	(20)	AMD	(42)
4. Siemens	(6)	Dainichi	(11)	4. Thomson	(19)	Motorola	(40)
5. Westingh.	(6)	Tandem	(11)	5. Toshiba	(18)	Philips	(39)
6. ACME-C.	(5)	FANUC	(10)	6. Siemens	(17)	TI	(37)
7. Asea	(5)	Rockwell	(10)	7. Fujitsu	(16)	Siemens	(36)
8. Daimler	(5)	Siemens	(10)	8. NEC	(16)	IBM	(30)
9. FANUC	(5)	Westingh.	(10)	9. EXXON	(15)	Toshiba	(27)
10. IBM	(4)	C.Milacron	(9)	10. AMD	(14)	AT&T	(26)
Software				**Telecommunications**			
1980–84		1985–89		1980–84		1985–89	
1. CDC	(18)	HP	(47)	1. Siemens	(17)	Siemens	(45)
2. NCR	(16)	DEC	(45)	2. AT&T	(15)	CGE	(32)
3. Honeywell	(14)	Siemens	(36)	3. ITT	(14)	Sumitomo	(29)
4. Motorola	(14)	Bull	(34)	4. Fujitsu	(10)	Mitsubishi	(28)
5. HP	(13)	AT&T	(33)	5. IBM	(10)	Fujitsu	(27)
6. Sperry	(13)	Philips	(32)	6. Plessey	(10)	AT&T	(26)
7. Allied	(12)	SUN-Micr.	(31)	7. Hitachi	(9)	Philips	(26)
8. AMD	(12)	NCR	(29)	8. ANT	(8)	IBM	(24)
9. DEC	(12)	Volmac	(29)	9. NEC	(8)	NEC	(23)
10. Harris	(12)	Olivetti	(28)	10. Olivetti	(8)	Ericsson	(20)

Source: MERIT–CATI databank.

nology collaborators. For a more precise determination of the relation between strategic technology partnering and the market position of companies additional analysis is compulsory. Following the brief discussion at the beginning of section 2, we can expect that neither the leading firms in terms of market position will completely dominate the list of leading cooperators, nor do we anticipate that "second tier" competitors play a decisive role in strategic technology partnering.

At the present level of analysis we can only attempt to find some very indirect indicators to test such a hypothesis, an attempt which is severely hampered by the lack of systematic statistical data. It is obvious from table 3 that all the main partnering companies belong to the group of leading companies in each sub-field. In order to take the analysis one step further we have constructed three lists of indicative top-ten sales for computers, microelectronics and telecommunications. [6] For these three fields of information technologies one can see that about half of the leading companies in strategic partnering are also found in the lists of leading companies according to market share. In computers six out of ten of the most cooperating companies in the second half of the 1980s also belong to the top ten of suppliers, although it is clear that there is no statistical relation between both rankings. In computers IBM has disappeared from the list of most cooperating companies during the second half of the 1980s and therefore this market leader plays only a moderate role in partnering in its core-business, whereas its position in fields such as telecommunications, industrials automation and microelectronics is quite remarkable. In microelectronics 50 percent of the leading suppliers are also part of the list of most cooperating companies but we have to acknowledge that the leading firms, according to market share, are not represented or play only a moderate role in the list of major partnering companies. In telecommunications there is a coverage of six out of ten for the second half of the 1980s. In general the conclusion has to be that there is no clear correlation between both rankings, it is obvious that the leading companies are well represented among the most collaborating companies, without quantitatively dominating the general network of strategic technology alliances.

5. Some conclusions

Our analysis of strategic technology alliances demonstrates that this phenomenon has become an important issue in company behaviour during the 1980s, although its growth appears to have levelled off again during more recent years. The growth in the number of strategic alliances in information technologies parallels an increasing intensity and complexity of inter-firm technology cooperation in which particular companies appear to take a more active role than others. It has become clear that network-density of inter-firm collaboration has increased in nearly all fields of information technology and reached a relatively high level. Many of the leading cooperating companies in most sub-fields prevailed throughout the decade. This indicates a certain stability in the international pattern of strategic alliances, although the exact order of the leading cooperators might have changed considerably for some fields, and companies have also entered and left this group of leading cooperating firms. In other words, the network of strategic technology partnering that we investigated support the notion that the pattern that emerges is one of an only relatively stable character.

Our study also suggests that most of the leading, diversified companies play a prominent role in strategic partnering in information technologies. This does not suggest that market leaders are concentrating their alliances in their core activities or that they dominate strategic technology partnering across the board. A major conclusion from our research so far is that many market leaders do play a substantial role in strategic technology partnering, but not for all leading firms is their partnering necessarily close to their core activity. However, we also found little evidence that supports the notion that "second-tier competitors" control the scene of strategic alliances. In general one can state that, based on

[6] The ten leading manufacturers of computers, microelectronics and telecommunications, ranked in order of sales during the second half of the 1980s. Computers: IBM, DEC, Unisys, Fujitsu, NCR, NEC, CDC, HP, Siemens, Hitachi. Microelectronics: NEC, Toshiba, Hitachi, Motorola, TI, Intel, Fujitsu, Mitsubishi, Matsushita, Philips. Telecommunications: ATT, CGE, Siemens, LME, NT, NEC, Fujitsu, Telit, GTE, Plessey.

the analysis of the structure of inter-firm strategic technology partnering in distinct fields of information technology, it appears that the pattern of strategic alliances only partly mirrors the existing market structural hierarchy.

Appendix I – The cooperative agreements and technology indicators (CATI) information system

The CATI databank is a relational database which contains separate data files that can be linked to each other and provide (des)aggregate and combined information from several files. So far information on nearly 10,000 cooperative agreements involving some 3,500 different parent companies has been collected. In this appendix we will elaborate on sources of information and limitations of databank.

Systematic collection of inter-firm alliances started in 1987. If available, many sources from earlier years were consulted enabling us to take a retrospective view. In order to collect inter-firm alliances we consulted various sources, of which the most important are newspaper and journal articles, books dealing with the subject, and in particular specialized journals which report on business events. Company annual reports, the *Financial Times Industrial Companies Yearbooks* and Dun & Bradstreet's *Who Owns Whom* provide information about dissolved equity ventures and investments, as well as ventures that we did not register when surveying alliances.

This method of information gathering which we might call "literature-based alliance counting" has its drawbacks and limitations:

- In general we have only come to know those arrangements that are made public by the companies themselves.
- Newspaper and journals reports are likely to be incomplete, especially when they go back in history and/or regard firms from countries outside the scope of the journal. Furthermore, in earlier years some journals simply did not exist whereas existing periodicals might grasp the collaboration subject less thoroughly.
- A low profile of small firms without well-established names is likely to have their collaborative links excluded.
- Some journals emphasize fashionable items, such as superconductivity or HDTV, while interest for "outdated" topics such as solar and wind energy seems to fade away.
- The fact that we read mainly articles written in English probably causes some bias and distortion, too.
- Another problem is that information about the dissolution of agreements is not systematically published. This is in particular true for licensing and customer–supplier relationships. On the other hand, research contracts and joint product developments have often disclosed time schedules. Equity joint venture and dissolutions of investments are published rather systematically in specialized journals.
- One final problem is that the number of customer–supplier relations and licensing agreements is subject to a fierce underestimation due to the fact that these more casual agreements are little reported publicly, even in the professional literature.

All together, these handicaps in the first place lead to a skewed distribution in the modes of cooperation, followed by some geographic – i.e. Anglo-Saxon – bias. Next, we have to reckon with a possible underestimation of certain technological fields not belonging and finally, there is some overrepresentation of large firms.

Despite these shortcomings, which are largely unsolvable even in a situation of extensive and large-scale data collection, we think we have been able to produce a clear picture of the joint efforts of many companies. This enables us to perform empirical research which goes beyond case studies or general statements. Some of the weakness of the database can easily be evaded by focusing on the more reliable parts, such as strategic alliances (see Appendix III).

The databank contains information on each agreement and some information on companies participating in these agreements. The first entity is the inter-firm cooperative agreement. We define cooperative agreements as common interests between independent (industrial) partners which are not connected through (majority) ownership. In the CATI database only those inter-firm agreements are being collected, that contain some arrangements for transferring technology or joint research. Joint research pacts, second-sourcing and licensing agreements are clear-cut examples. We also collect information on joint ventures in which new technology is received from at least one of the partners, or joint ventures having some R&D programme. Mere production or marketing joint ventures are excluded. In other words, our analysis is primarily related to technology cooperation. We are discussing those forms of cooperation and agreements for which a combined innovative activity or an exchange of technology is at least part of the agreement. Consequently, partnerships are omitted that regulate no more than the sharing of production facilities, the setting of standards, collusive behaviour in price-setting and raising entry barriers – although all of these may be side effects of inter-firm cooperation as we define it.

We regard as a relevant input of information for each alliance: the number of companies involved; names of companies (or important subsidiaries); year of establishment, time-horizon, duration and year of dissolution; capital investments and involvement of banks and research institutes or universities; field(s) of technology; modes of cooperation; and some comment or available information about progress. Depending on the very form of cooperation we collect information on the operational context; the name of the agreement or project; equity sharing; the direction of capital or technology flows; the degree of participation in case of minority holdings; some information about motives underlying the alliance; the character of cooperation, such as basic research, applied research, or product development possibly associated with production and/or marketing arrangements. In some cases we also indicate who has benefitted most.

The second major entity is the individual subsidiary or parent company involved in one (registered) alliance at least. In the first place we assess the company's cooperative strategy by adding its alliances and computing its network centrality. Second, we ascertain its nationality, its possible (majority) owner in case this is an industrial firm, too. Changes in (majority) ownership in the 1980s were also registered. Next, we determine the main branch in which it is operating and classify its number of employees. In addition, for three separate subsets of firms time-series for employment, turnover, net income, R&D expenditures and numbers of assigned US patents have been stored. The first subset is based on the *Business Week* R&D scoreboard, the second on *Fortune's* International 500, and the third group was retrieved from the US Department of Commerce's patent tapes.

Appendix II – A short note on MDS and cluster analysis

The core of the MDS technique can be explained as follows. MDS is a data reduction procedure comparable to principal component analysis and other factor analytical methods. One of the main advantages of MDS is that usually, but not necessarily, MDS can fit an appropriate model in fewer dimensions than can factor analytical methods. This increases the possibility of easy interpretable two-dimensional pictures. MDS offers scaling of similarity data into points lying in an X-dimensional space. The purpose of this method is to provide coordinates for these points in such a way that distances between pairs of points fit as closely as possible to the observed similarities. In order to facilitate interpretation the solution is given in two dimensions, provided that the fit of the model is acceptable. A stress value indicates the goodness-of-fit of the configuration. For all MDS solutions presented in this paper the stress values, as given in table AII.1, range from acceptable to very good.

The total number of strategic partnerships between two companies is taken as a measure of similarity between those two companies. A large similarity, as found in a similarity matrix, indicates intensive cooperation. [7]

Unfortunately our MDS software can only analyze similarity matrices to a maximum of 45

Table AII.1
Stress values of MDS solutions

	1980–84	1985–89
Information technologies	0.12	0.13
computers	0.05	0.06
industrial automation	0.01	0.03
microelectr.	0.06	0.09
software	0.02	0.08
telecom	0.05	0.07

[7] These similarity matrices are not reproduced in this paper.

rows, which means that for each field of technology the analysis is restricted to a maximum of 45 companies with the largest numbers of strategic partnerships. MDS solutions are presented for two periods, the years from 1980 to 1984, and from 1985 to 1989. For the first period (1980–84) we have taken those alliances established in that particular period plus those alliances made before 1980 that were not already discontinued in 1980. For the second period, the years since 1985, we follow the same procedure; we have taken all alliances forged in that period and added those alliances from the earlier period which were not already discontinued in 1985. During each period alliances of subsidiaries and divisions are assigned to the parent company. Also within each period, the still existing alliances of companies taken over by others or partnerships made by merging companies will be assigned to the acquiring company or the new corporation.

The same similarity data were used for a cluster analysis. The objective of cluster analysis in this study is simply to group firms on the basis of numbers of alliances between them. The result is a series of clusters comprised of intensely cooperating companies. The particular technique employed here is known as hierarchical cluster analysis [4,15]. Clusters are formed by grouping firms into a small but highly cohesive clusters. These groups are subject to a subsequent clustering procedure until each firm is a member of a single cluster. The basic characteristic of hierarchical clustering is that once a firm is a member of a cluster, it remains part of that cluster. As the term suggests, each cluster is embedded within a larger cluster. In one extreme case the "network" can be seen as comprised of 45 clusters which all contain only one firm, while in another extreme case the network can be seen as one cluster having all 45 firms in it. It is obvious that we are interested in some intermediate level of clustering. Before applying cluster analysis a number of additional decisions were made. First, to what extent firms are alike is measured by Euclidean distance. Second, as a criterion for combining clusters we chose the "average linkage" method. This method combines clusters in such a way that the average distance between all firms in the resulting cluster is as small as possible. In other words: the average number of alliances between all firms should be as high as possible.

Appendix III – List of companies

Label	*Company's full name*	*Country*
3M	Minnesota Mining & Mfg. Co. (3M)	USA
A-B	Allen-Bradley Co.	USA
A-C	Allis-Chalmers Corp.	USA
ABB	ABB Asea Brown Boveri A.G.	SWI
ACME-C	Acme Cleveland	USA
ADC	Ateliers des Charmilles	SWI
ADR	Applied Data Research (ADR)	USA
AEG	AEG	FRG
AI-SIG	Allied-Signal Inc.	USA
ALLIED	Allied Corp.	USA
ALPS	Alps Electric Co.	JPN
AMD	Advanced Micro Devices Inc.	USA
AMDAHL	Amdahl Corp.	USA
AMERIT	Ameritech	USA
AMINF	American Information	USA
AMP	AMP Inc.	USA
AMSTRD	Amstrad Plc.	UK
AMTELC	American Telecom	USA
ANT	ANT Nachrichtentechnik	FRG
APOLLO	Apollo Computer	USA
APPLE	Apple Computer Inc.	USA
ARITM	Aritmos	SWE
ARTINT	Artificial Intelligence	USA
ASEA	Asea A.B.	SWE
ASM	ASM Int.	NET
AT&T	Am. Telephone & Telegraph Co. (AT&T)	USA
AUTO-T	Auto-Trol Technologies	USA
AUTODP	Automatic Data Processing	USA
B&W	Burkhard & Weber	FRG
B-ATL	Bell Atlantic	USA
B-CORE	Bellcore	USA
B-SOUT	Bell South Corp.	USA
BA	British Aerospace	UK
BASF	Basf A.G.	FRG
BBN	Bolt Beranek & Newman	USA
BMW	Bayerische Motor Werke A.G.	FRG
BOEING	Boeing Aerospace Co.	USA
BOSCH	Bosch	FRG
BSO	BSO	NET
BT	British Telecom	UK
BTH	BT Industries	SWE
BULL	Bull Groupe S.A.	FRA
BURROU	Burroughs	USA
C&T	Chips & Technologies	USA
C&W	Cable & Wireless Plc.	UK
C.ITOH	C.Itoh	JPN
CANON	Canon Inc.	JPN
CAPGEM	Cap Gemini Sogeti	FRA
CDC	Control Data Corp. (CDC)	USA
CENTRN	Centronics Data Computer	USA
CGE	Cie Générale d'Électricité (CGE)	FRA
CHRYSL	Chrysler Motor Corp.	USA
CIPHER	Cipher Data Products Inc.	USA
CM	Cincinnati Milacron Inc.	USA
CMG	Computer Management Group (CMG)	USA
COGENT	Cogent Research	USA
COLUM	Columbia Data Products	USA

COMPAS	Computer Associates Int. Inc.	USA
COMPC	Computer Consoles Inc. (CCI)	USA
COMPTG	Computer Task Group	USA
COMPUS	Computer Service Holland	NET
CONTEL	Continental Telecom	USA
CONVEX	Convex Computer	USA
CONVIC	Convick	SWE
CONVRG	Convergent Technologies Inc.	USA
CORONA	Corona Data Systems	USA
CR&TR	Cross & Trecker	USA
CRAY	Cray Research	USA
CTNE	Compañia Telefónica Naçional de E.	SPA
CV	Computer Vision Corp.	USA
CYPRES	Cypress Semiconductor	USA
DAEWOO	Daewoo	SK
DAF	DAF Trucks N.V.	NET
DAIICH	Dai-ichi	JPN
DAIMLR	Daimler-Benz A.G.	FRG
DAIN-K	Dainichi Kiko	JPN
DATA-L	Datalogen A.B. (Data Logic)	SWE
DEC	Digital Equipment Corp. (DEC)	USA
DIGR	Digital Research Inc.	USA
DONGY	Dong Yang Electric Discharge M.Co.	SK
DOOSAN	Doosan Mfg. Co. Ltd	SK
DUPONT	Du Pont de Nemours	USA
ELBIT	Elbit Computers	ISR
ENCORE	Encore Computer	USA
ERICSS	Ericsson A.B.	SWE
EXXON	Exxon Corp.	USA
FANUC	Fanuc Ltd.	JPN
FERRAN	Ferranti	UK
FIAT	Fiat SpA.	ITA
FLOAT	Floating Point Systems	USA
FORD	Ford Motor Co.	USA
FUJI-E	Fuji Electric	JPN
FUJITS	Fujitsu Ltd.	JPN
G-FIS	Georg Fischer	SWI
GE	General Electric Co.(GE)	USA
GEC	GEC	UK
GENDYN	General Dynamics Corp.	USA
GENINS	General Instrument	USA
GM	General Motors	USA
GMF	GMF Robotics	USA
GOULD	Gould Inc.	USA
GRUMMN	Grumman	USA
GTE	General Telephone & Electric (GTE)	USA
H-P	Hewlett-Packard Co.	USA
HAMAI	Hamai	JPN
HARRIS	Harris	USA
HITACH	Hitachi Ltd.	JPN
HONEYW	Honeywell Corp.	USA
HOOGOV	Hoogovens N.V.	NET
IBM	Int. Business Machines Corp.(IBM)	USA
INMOS	Inmos	UK
INTEL	Intel Corp.	USA
INTERQ	Intertechnique	FRA
IPL	IPL Systems	USA
IRI	IRI	ITA
ITT	Int. Tel.& Telegraph Corp.(ITT)	USA
KANEMA	Kanematsu Semiconductor	JPN
KIA	Kia Machine Tool	SK
KIHEUN	Kiheung Machinery Works	SK
KODAK	Eastman Kodak Co.	USA
KOR-EL	Korea Electronics	SK
KUKA	Kuka	FRG
LEP	LEP	FRA
LITTON	Litton Industries Inc.	USA
LOCKHD	Lockheed	USA
LOTUS	Lotus Development Corp.	USA
LSILOG	LSI Logic	USA
LUCKY	Lucky Group Ltd.	SK
MARTIN	Martin-Marietta Corp.	USA
MATRA	MATRA S.A.	FRA
MATRAH	Matra-Harris Semiconducteurs	FRA
MATSUS	Matsushita Elect. Industrial Co.Ltd.	JPN
MBB	Messerschmitt-Bölkow-Blohm (MBB)	FRG
MCD-D	McDonnell Douglas Corp.	USA
MEMTEL	Memorex Telex	USA
MENTOR	Mentor Graphics	USA
MI	Machine Intelligence	USA
MICR-V	Micro-V	USA
MICSFT	Microsoft Corp.	USA
MILLIC	Millicom	USA
MIPS	Mips Computer Systems	USA
MITEL	Mitel	CAN
MITSUB	Mitsubishi Corp.	JPN
MITSUI	Mitsui & Co.	JPN
MOSTEK	Mostek	USA
MOTORO	Motorola Inc.	USA
MULTIH	Multihouse	NET
NAS	National Advanced Systems (NAS)	USA
NATSEM	National Semiconductor Corp.	USA
NCR	National Cash Register Corp. (NCR)	USA
NCUBE	Ncube	USA
NEC	Nippon Electric Corp.(NEC)	JPN
NIP-AU	Nippon Automation	JPN
NIXDRF	Nixdorf	FRG
NOKIA	Nokia Oy.	FIN
NORTHR	Northrop	USA
NT	Northern Telecom	Can
NTT	Nippon Telegraph & Telephone (NTT)	JPN
NYNEX	Nynex Enterprises	USA
OKI	Oki Electric Industry Co.	JPN
OLIVET	Olivetti	ITA
PANAT	Panatec R&D Corp.	USA
PERF	Perfect Data Corp.	USA
PERKIN	Perkin-Elmer	USA
PHILIPS	Philips, Gloeilampenfabriek N.V.	NET
PLESSE	Plessey Co.	UK
QUOTRN	Quotron Systems	USA
RACAL	Racal	UK
RCA	RCA (Radio Co. of America)	USA
RENAUL	Renault	FRA
RICOH	Ricoh	JPN
ROCKWL	Rockwell Int. Corp.	USA
ROLM	Rolm	USA
S-GOBN	Saint-Gobain S.A.	FRA
SAMSNG	Samsung Co.Ltd.	SK

SANYO	Sanyo Electric Co.Ltd.	JPN
SCHAUD	Schaudt Maschinenbau GmbH.	FRG
SCHLUM	Schlumberger N.V.	USA
SCHNEI	Schneider S.A.	FRA
SEMA-C	Sema-CAP	UK
SEQUEN	Sequent Computer Systems Inc.	USA
SHOUN	Shoun Machine Tool	JPN
SIEMNS	Siemens A.G.	FRG
SIGNAL	Signal Companies Inc.	USA
SMS	Standard Microsystems Corp.	USA
SONY	Sony Corp.	JPN
SPERRY	Sperry	USA
STC	Standard Telephone Co.	UK
STRATS	Stratus Computer	USA
SUMITO	Sumitomo Group	USA
SUN-M	Sun Microsystems	USA
TANDEM	Tandem Corp.	USA
TANDY	Tandy Corp.	USA
TEKNOW	Teknowledge	USA
TELEX	Telex	USA
THOMSN	Thomson S.A.	FRA
TI	Texas Instruments Inc.	USA
TOSHIB	Toshiba Corp.	JPN
TOYOTA	Toyota Motors	JPN
TRILOG	Trilogy Systems	USA
TRW	Thompson Ramo Woolridge Inc.	USA
UN-TEC	United Technologies Corp.(UTV)	USA
UNG-BA	Ungermann-Bass	USA
UNISYS	Unisys	USA
UNISYS	Unisys Corp.	USA
VLSI	VLSI Technology	USA
VOLMAC	Volmac	NET
VW	Volkswagen A.G.	FRG
WANG	Wang Laboratories Inc.	USA
WE-SYK	WE Sykes	UK
WESTIN	Westinghouse	USA
XEROX	Xerox	USA
YASKAW	Yaskawa Electric	JPN

References

[1] M. Casson, *The Firm and the Market* (Blackwell, Oxford, 1987).

[2] F. Chesnais, Technical Co-operation Agreements between Firms, *STI Review* 4 (1988).

[3] F.J. Contractor and P. Lorange, Why should Firms cooperate? The Strategy and Economics Basis for Cooperative Ventures, in: F.J. Contractor and P. Lorange (eds.), *Cooperative Strategies in International Business* (Lexington Books, Lexington, 1988).

[4] B.S. Everitt, *Cluster Analysis* (Heineman, London, 1980).

[5] J. Hagedoorn, Organisational Modes of Inter-firm Cooperation and Technology Transfer, *Technovation* (1) (1990).

[6] J. Hagedoorn and J. Schakenraad, Strategic Partnering and Technological Cooperation, in: B. Dankbaar, J. Groenewegen and H. Schenk, *Perspectives in Industrial Economics* (Kluwer, Dordrecht, 1990).

[7] J. Hagedoorn and J. Schakenraad, Inter-firm Partnerships and Cooperative Strategies in Core Technologies, in: C. Freeman and L. Soete, *New Explorations in the Economics of Technical Change* (Pinter, London, 1990).

[8] C.S. Haklisch, Technical Alliances in the Semiconductor Industry, mimeo NYU, 1986.

[9] M. Hergert and D. Morris, Trends in International Collaborative Agreements, INSEAD paper, 1986.

[10] K.J. Hladik, *International Joint Ventures* (Lexington Books, Lexington, 1985).

[11] J.K. Kruskal and M. Wish, *Multidimensional Scaling* (Sage, Beverly Hills, 1978).

[12] OECD, *Technical Cooperation Agreements between Firms: Some Initial Data and Analysis* (OECD, Paris, 1986).

[13] K. Ohmae, *Triad Power* (Free Press, New York, 1985).

[14] M. Porter, *The Competitive Advantage of Nations* (Free Press, New York, 1990).

[15] L. Wilkinson, *Systat: The System of Statistics* (Systat, Evanston, 1986).

[12]

CHOOSING ORGANIZATIONAL MODES OF STRATEGIC TECHNOLOGY PARTNERING: INTERNATIONAL AND SECTORAL DIFFERENCES

John Hagedoorn* and Rajneesh Narula**
University of Limburg

Abstract. This paper focuses on international strategic technology partnerships, and the choice that companies have in terms of the organizational mode, either through complex inter-organizational modes involving equity-sharing, or contractual non-equity alliances. The empirical results show that the choice of particular mode of cooperation varies with the technological characteristics of sectors of industry. Joint ventures are disproportionately represented in relatively mature industries. Contractual alliances dominate strategic technology partnering in so-called high-tech industries.

INTRODUCTION

In this paper, we study international and sectoral determinants of inter-organizational, strategic technology partnering. We define strategic technology partnering as inter-firm cooperation for which a combined innovative technological activity or an exchange of technology is at least part of an agreement [Contractor and Lorange 1988; Hagedoorn 1993]. These strategic technology alliances can be divided in two basic categories. The first concerns inter-organizational modes of technology cooperation that involve equity sharing, in particular joint ventures and jointly owned research corporations. The second category involves contractual alliances that cover a relatively large group of partnerships without equity sharing, such as joint development agreements, joint research pacts, cross licensing, second-sourcing agreements, mutual second-sourcing, and R&D contracts.

---

*John Hagedoorn is Professor of International Business at the University of Limburg at Maastricht, the Netherlands. His research interests center on international competition, strategic alliances, mergers and acquisitions and technological development. His research has appeared in *Strategic Management Journal, Journal of Management Studies, Research Policy, World Development, International Journal of Technology Management, Journal of Common Market Studies*, and *Technovation*.

**Rajneesh Narula is Assistant Professor of International Business at the University of Limburg at Maastricht, the Netherlands. His research interests include foreign direct investment, strategic alliances and technological development. His research has appeared in *Essays in International Business, International Studies of Management and Organization*, and *Management International Review*.

The authors would like to thank Geert Duysters and Marc van Ekert for their assistance and Franz Palm for his valuable comments regarding some methodological topics. Comments by three anonymous referees have also undoubtedly improved the paper. The usual disclaimer applies.

Received: August 1994; Revised: February, June & September 1995; Accepted: September 1995.

Strategic technology partnerships that involve equity sharing tend to possess a governance structure with a new administrative element while contractual alliances do not. The categorization of strategic alliances into these two major modes is supported by recent research that suggests that these modes are fundamentally different: while contractual alliances display quasi-market characteristics, joint ventures are quasi-hierarchical in nature [Osborn and Baughn 1990]. Along similar lines, Hagedoorn [1993] provides an analysis of motives for different modes of strategic technology partnering in a multi-sectoral setting which suggests that inter-organizational modes of technology cooperation characterized by equity sharing are aimed at both market- and technology-mediated objectives. The multidimensionality of the wider set of objectives indicates that equity arrangements are not merely driven by technology sharing, or exclusively aimed at the creation of innovations. In general, even technology-focused joint ventures have a wide range of company objectives. Also, companies do not establish an undertaking such as a joint venture with all its organizational, financial and managerial complexities for a short period of time, and it is generally intended that a joint venture is operational for a substantial number of years. The actual unstable character of many joint ventures [Harrigan 1988; Kogut 1989; Porter 1987], as expressed in the high failure rate, does not contradict the intended long-term perspective of this mode of cooperation nor the multidimensional objectives. Hagedoorn [1993] also establishes that a disproportionate share of non-equity arrangements or contractual strategic technology alliances are primarily R&D- or innovation-driven. Their focus is one-dimensional and their duration is usually relatively short term from the perspective of the alliance itself, although they can have a long-term effect on participating companies.

The international aspect of this phenomenon is studied by concentrating on cross-national alliances. A substantial part of the existing literature focuses on economic, organizational or strategic properties of strategic alliances in general. In recent years a growing body of literature [Dunning 1993; Hladik 1985, 1988; Mowery 1988; Mytelka 1991; Osborn and Baughn 1990] looks in particular at cross-border strategic alliances and joint ventures. In the present contribution we follow the literature mentioned above in stressing the importance of these international alliances for our understanding of changing patterns of international competition. In that context we differentiate between inter-continental and intra-regional alliances. The relevance of this differentiation of nondomestic alliances is supported by Hagedoorn and Schakenraad [1993] who found that international (inter-regional) alliances between firms from Europe, the USA or Japan are particularly aimed at market entry, whereas alliances focusing on joint R&D are, in general, more intra-regional, i.e., aimed at cooperation between partners from one of the major regions of the Triad. In the following, we look in particular at international partnering of four main groupings: U.S.-European Union (EU) alliances, Japanese-EU alliances, U.S.-Japanese alliances, and nondomestic, intra-EU alliances.

The paper presents a description of general trends and patterns in international strategic technology partnering during the 1980s and the early 1990s, followed by a more specific analysis related to a number of propositions derived from the literature. These propositions concern:

- the effect of the level of technological sophistication of sectors of industry on companies' choice of different modes of partnering;
- the degree to which countries differ as to how their companies concentrate their alliances in particular sectors;
- the extent to which there is a disproportional preference of companies to engage in contractual partnerships regardless of the level of technological sophistication of the country of origin of participants.

Following a discussion of the main results, we close with the conclusions that can be drawn from this contribution and some suggestions for future study.

SECTORAL AND INTERNATIONAL ASPECTS OF CROSS-NATIONAL ALLIANCES: SOME PROPOSITIONS

The literature on joint ventures and inter-firm cooperation in general suggests a number of patterns in the sectoral distribution of international strategic technology alliances. Hladik [1985] and Hagedoorn [1993] found a clear association of the technology-intensity of sectors with the occurrence of R&D joint ventures and strategic technology partnering. According to Harrigan [1985], rapidly changing technological development in sectors of industry induces the formation of somewhat more informal forms of cooperation such as non-equity agreements. As industries become mature, more formal modes of cooperation such as joint ventures become the preferred form of collaboration. In a similar vein, Harrigan [1988] states that non-equity agreements are more suited to industrial sectors characterized by an uncertain environment, whereas joint ventures offer better opportunities for partnering companies in stable sectoral environments. Link and Bauer [1989] found that the higher the research intensity of industries, the greater the proportion of R&D funds spent on cooperative research. Osborn and Baughn's [1990] survey of the literature suggests that technological stability of industrial sectors is a crucial factor in explaining different patterns of equity and non-equity partnerships. R&D intensive sectors demand more organizational flexibility, leading to a general preference for contractual agreements, whereas in sectors with low R&D intensity, where organizational flexibility is less crucial, technology partnering agreements tend to be dominated by joint ventures. Dunning [1993] analyses the preference for contractual modes of 'technology-based cross-border alliances' in international high-technology sectors with companies combining their strategic firm-specific advantages with particular advantages of countries. Auster [1992] demonstrates a preference for technical agreements over equity agreements in U.S.–Japanese alliances as the technology involved becomes more sophisticated. In light of the findings reviewed here, we submit that the level of technological sophistication of sectors of industry affects the distribution of equity or non-equity modes of strategic technology partnering. Thus:

> Proposition 1: A larger proportion of contractual agreements (versus equity agreements) is found as the degree of technological intensity of industries increases.

Apart from patterns in the distribution of modes of cooperation and their inter-sectoral divergence, we also intend to evaluate the degree to which companies from different countries deviate in respect to their sectoral preference for international technology partnerships. The literature on foreign direct investment, trade and technology suggests that firms of different nationalities have dissimilar industrial patterns of specialization, and these are associated with their country-specific characteristics, such as technological advantages, market size and so-called national systems of innovation [Narula 1995]. It is well established that the innovative activities of individual countries are largely dominated by their large, multinational companies [Cantwell 1991; Patel and Pavitt 1991]. As such it is not unlikely that a similar pattern of country specialization will prevail in the propensity of these firms to engage in international strategic technology partnering.

The technological sophistication of countries and their industrial patterns of specialization are to a great extent influenced by the national systems of innovation [Nelson 1993; Lundvall 1992]. Several scholars (see for instance contributions to Freeman and Lundvall [1988]) have suggested that, compared to firms from small countries, firms from larger countries tend to have a larger pool of resources, including their national technological infrastructure, from which to draw on, compared to smaller countries. Furthermore, the domestic markets of companies from smaller countries can constitute a disadvantage in the development of new technologies, especially in sectors which require a relatively high minimum efficient scale [Walsh 1988]. In general, given the high cost of R&D associated with most new technologies, companies from small countries are expected to focus on a just a few niche sectors. The national systems of innovations of small countries generally, though not always, tend to further accentuate this 'specialization' of innovative activities by small country firms. Large countries, and firms from these countries, tend to have greater resources at their disposal, and are inclined to be involved in a wider variety of industrial sectors than smaller countries. This line of reasoning seems particularly relevant for the group of very small countries with absolute size disadvantages.

However, country size is not the only determinant of country patterns in international strategic technology partnering. The relative level of technological sophistication of countries has also to be taken into account. For instance, although Spain has a market size that is more than twice that of Belgium (1990 GDP $ 487bn against $ 192bn), it has a lower level of technological sophistication (company R&D/GDP in 1990 was 0.47% against 1.23% – source: OECD).

What we suggest here is that technologically advanced countries are diversified in terms of their sectoral distribution of technology: the level of technological sophistication of these advanced countries is found across a wide array of industries and technologies. On the other hand, technologically less advanced countries also tend to have limited resources that result in the spreading of their resources in fewer sectors. We submit that if this trend in different patterns of specialization of countries extends to the participation of companies in international strategic technology partnering, then:

Proposition 2: The participation in international strategic technology partnering by companies from technologically advanced countries will be evenly allocated across the overall sectoral distribution of partnerships, whereas the participation of companies from technologically less advanced countries will be specialized in a small number of sectors.

Countries differ with respect to their level of technological sophistication, but increased international competition amongst the leading developed economies has led to a technological convergence of sectors into similar levels of technological development. However, it is important to note that even small differences can be decisive in determining the relative international competitiveness of one country over another. In general, technological convergence implies similarity of technologies, but this does not imply that the competitive advantages of countries will be identical [Archibugi and Pianta 1994; Narula 1995]. The argument being made here is simply that various sectors across countries have internationally comparable levels of technological sophistication. Countries can specialize in particular sectors or technologies, but the level of technological development for each field is comparable for most countries in the industrialized world as their companies have to compete in international markets. Therefore, we postulate that the preference for particular modes of international cooperation will be primarily associated with the technological intensity of sectors and not with the level of technological sophistication of countries at large. Thus:

Proposition 3: The assumed disproportional preference of companies to form contractual agreements in high-tech sectors and joint ventures in other sectors is not intermediated by the level of technological sophistication of countries.

METHOD OF DATA COLLECTION

In order to test the propositions and also to discuss some general patterns in international strategic technology partnering, our empirical research employs the MERIT-Cooperative Agreements and Technology Indicators (CATI) data bank. This relational database, with information on over 13,000 technology cooperation agreements involving some 5000 different parent companies, was established in the late eighties. After a pilot project in 1986–1987, systematic collection of inter-firm technology alliances started in 1988. Many sources from earlier years were consulted, which enables us to take a retrospective view. For all sectors of industry or fields of technology in our data bank, we have information on cooperative agreements from at least as early as 1980 up to 1993. In order to collect information on inter-firm alliances and their parents, we consulted various sources, such as informal reports, newspaper and journal articles, books dealing with the subject, and specialized technical journals which also report on business events. Company annual reports, the Financial Times Industrial Companies Yearbooks, and Dun & Bradstreet's 'Who Owns Whom' provide information about dissolved equity ventures and investments, as well as ventures that were not registered when surveying alliances.

This method of information gathering which we might call 'literature-based alliance counting' has its drawbacks and limitations such as inadequacy of certain sources, low profile of certain companies or industries, bias in favour of Anglo-Saxon sources, and underestimation of certain modes of cooperation such as licensing. It also introduces a certain bias in terms of the frequency versus the scale of alliances. Despite these shortcomings, which are largely unsolvable even in a situation of extensive and large-scale data collection, we think we have been able to produce a clear picture of the joint efforts of many companies. This enables us to perform empirical research which goes beyond case studies or general statements. We avoided some of the weaknesses of the database by focusing on the more reliable parts, such as strategic technology alliances, and by ignoring cost-economizing partnerships and licensing agreements.

The databank contains information on each agreement and some information on companies participating in these agreements. The main entity is the inter-firm cooperative agreement. We define cooperative agreements as common interests between independent (industrial) partners that are not connected through (majority) ownership. The CATI database includes only those inter-firm agreements that contain arrangements for transferring technology or joint R&D. Mere production or marketing agreements are excluded. In other words, our analysis is primarily related to technology cooperation. Our focus is on those forms of cooperation and agreements for which a combined innovative activity or an exchange of technology is at least part of the agreement. Consequently, we exclude partnerships that regulate no more than the sharing of production facilities, the setting of standards, collusive behaviour in price-setting and raising entry barriers, although all of these may be side effects of inter-firm cooperation as we define it.

We count as an alliance any agreement made between two or more companies at a particular moment. Subsequent agreements between the same partners are considered as individual and separate agreements. However, if a particular agreement consists of several legal forms of cooperation that are parts of the agreement at large, such as a joint venture with a licensing agreement, we consider this as one agreement. If a partnership is extended with a new partner or a new contract is made between two cooperating firms we view this as a new alliance. Frequencies reported in this paper therefore refer to the number of individual partnerships. Although we do have information on the value of a limited number of alliances (such as the amount of investment) we prefer not to utilize this information for two reasons. First, the coverage of this data excludes contractual agreements for which the value of the agreement is not disclosed, and second, a large percentage of technology partnerships involve interchange of knowledge rather than the exchange of funds or capital.

In this paper, we record for each alliance: the number of companies involved, country of origin of each partner, year of establishment of the agreement, field(s) of technology and/or industry and modes of cooperation. The country of origin of a given company refers to the country where a company or its head office is registered. Therefore, international alliances are partnerships between companies registered in different countries. This obviously has certain limitations in the case of international companies, because a number of their partnerships are between their local subsidiaries and local partners. However, many of these 'artificial' international alliances are monitored from the head-office. Furthermore, there is no other choice in the context of large databases than to follow a strict procedure for categorization. Decisions regarding the possible degree of international subsidiary-level monitoring are impossible to make for a population of thousands of agreements and companies. For the categorization of industries and fields of technology according to their degree of technology intensity, we use the standard OECD classification (see also Hagedoorn [1993]).

We make a distinction between cooperative agreements that are expected to be aimed at the strategic, long-term perspective of the companies involved and those agreements that we think are more associated with the control of either transaction costs or operating costs of companies. In case both general motives appear possible, either because it is not feasible to differentiate between the cost or the strategic argument or because partners often have alternating motives as a consequence of the character of the agreement, we have marked such agreements as being of a mixed character. The procedure is described extensively in Hagedoorn [1993] and Hagedoorn and Schakenraad [1990]. In practice, our decision rules imply that joint ventures with R&D,

research corporations, joint R&D pacts, customer-supplier agreements combined with licensing, cross-licensing, research contracts with licensing, and (mutual) second-sourcing agreements are taken as strategic alliances. Excluded are standard co-makership contracts, co-production agreements, and single licensing agreements for which the cost-economizing argument is thought to be a major motive. The total number of international strategic technology alliances included in the present analysis is 5063.

SOME GENERAL PATTERNS IN INTERNATIONAL STRATEGIC TECHNOLOGY PARTNERSHIPS

The general growth pattern of newly made international strategic technology partnerships and its distribution for the 1980s and the early 1990s is given in Figure 1. This pattern of the annual growth of strategic technology partnering follows the overall growth pattern of domestic and non-domestic alliances as reported in a large number of studies [Chesnais 1988; Contractor and Lorange 1988; Hagedoorn and Schakenraad 1993; Haklisch 1986; Hergert and Morris 1988; Mowery 1988; Mytelka 1991]. Our data suggest that in general the number of strategic technology partnerships for all international combinations has grown rapidly during the first half of the eighties. However, since the mid-eighties there has been some difference in growth patterns. The growth pattern of newly established alliances between companies from Japan and the EU has more or less stabilised. After a steady increase of newly made nondomestic alliances within the EU during the first half of the eighties, there appears to have been a gradual stagnation in the growth of these intra-regional (non-domestic) alliances within the EU towards the end of the period. To some extent this pattern, though at a higher overall growth level, is also visible for alliances made between firms from the U.S. and Japan. Although there are fluctuations in the growth of international alliances between companies from the U.S. and the EU, the pattern in Figure 1 suggests an overall rise for newly established partnerships throughout the period, after some decrease in the growth of newly made alliances at the end of the eighties. Further study of our data reveals that this increase in U.S.-EU alliances is due to the growth of contractual alliances.

Given our interest in both equity and contractual arrangements, we examine the possible differences in preference for each of these modes of governance. An 'equity-contractual arrangement' ratio expresses the number of joint ventures relative to the number of contractual arrangements for intra-EU (nondomestic) alliances, U.S.-EU alliances, Japan-U.S. alliances, and Japan-EU for the period 1980–1993. We found that most of the international alliances are of a non-equity type. However, there are significant differences between the four groupings. For intra-EU and U.S.-EU partnerships, joint ventures are a small minority of the total number of strategic technology alliances, with ratios of 0.37 for both combintions. The higher ratios for Japan-

FIGURE 1
The Growth of International Strategic Technology Alliances (Joint Ventures and Contractual Agreements), 1980–1993

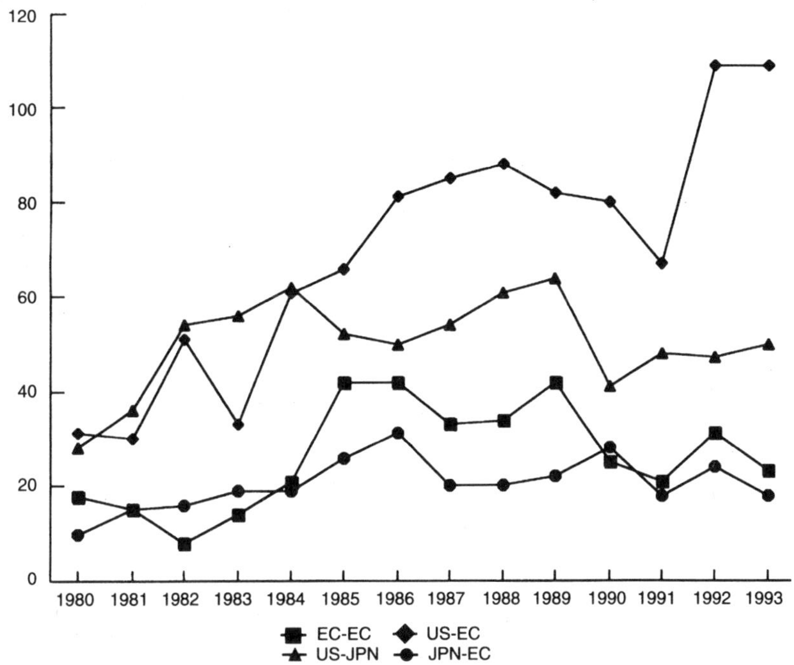

U.S. alliances (0.52) and Japan-EU alliances (0.67) indicate that partnerships with Japanese companies are more frequently governed by equity arrangements. As suggested by internalization theory [Buckley and Casson 1976, 1988; Dunning 1993; Rugman 1980] these equity agreements offer a larger degree of control over technology sharing than non-equity partnerships.

MAJOR FINDINGS

Table 1 shows the sectoral distribution of international equity and contractual technology partnerships for U.S.-Japan, U.S.-EU, Japan-EU and intra-EU groupings. As noted earlier, due to the nature of the MERIT-CATI database, which contains information mainly on technology partnering, a majority of the sectors for which data are available in sufficient detail are high-tech sectors. Furthermore, to test the first proposition, we have to take into account that the majority of international strategic agreements in the MERIT-CATI database – about 70 % – are of a non-equity nature (see Table 1). As the literature attests to, high technology sectors tend to have a higher share of contractual agreements than medium- or low-tech sectors. Therefore, it is reasonable to argue that for high-tech sectors we can expect that 30% or less of the strategic

TABLE 1
Sectoral Distribution of International Equity and Contractual Technology Partnerships, USA, Japan, Intra-EU, 1980–1993

	USA-Japan	USA-EU	Japan-EU	EU-EU	Totals
	JV/Total (%)	JV/Total (%)	JV/Total (%)	JV/Total (%)	JV/Total (%)
Biotechnology ***	16.36	17.53	26.67	13.04	17.69
New Materials Technology ***	47.76	52.81	60.87	38.46	52.80
Computers ***	19.64	20.00	45.45	35.71	22.95
Industrial Automation ***	43.24	30.19	20.00	13.64	30.37
Micro-electronics ***	19.23	17.48	20.45	6.82	16.94
Software ***	20.00	21.43	7.69	20.45	20.81
Telecommunications ***	21.95	29.82	30.77	32.61	28.28
Other I.T. ***	20.00	35.29	38.46	33.33	34.62
Automotive **	70.21	35.71	21.43	33.33	50.51
Aviation/Defense ***	12.50	16.67	25.00	23.53	17.83
Chemicals **	71.83	51.95	72.73	61.54	64.65
Consumer Electronics **	45.45	70.00	37.50	20.00	43.59
Food and Beverage *	66.67	45.45	100.00	100.00	63.16
Heavy Electric/Power ***	7.14	27.78	0.00	14.29	16.22
Instrumentation/Med.Technology ***	75.00	8.33	0.00	25.00	32.61
Other *	75.00	20.00	25.00	33.33	30.00

* low level of technology, ** medium level of technology, *** high level of technology
Classification is based on OECD guidelines (see Hagedoorn [1993]).

technology alliances are joint ventures. For medium- and low-tech industries we assume that more than 30% of the alliances are joint ventures.

In general, the data in Table 1 appear to support the first proposition regarding the association of sectoral technology intensity and the preference for either equity or contractual agreements, as twelve out of sixteen sectors follow the predicted distribution. For high-tech fields such as biotechnology, all information technology sectors (with the exception of 'other I.T.'), the aviation-defense sector, and heavy electrical equipment, we find that the vast majority of partnerships (between 72% and 82%) are of a contractual nature. For twelve of the forty-four international combinations in high-tech sectors, we do not find the expected 70% or more of contractual agreements. However, new materials is the only high technology sector that clearly falsifies the proposition, both across the board and for each of the four individual international combinations. In that sector, the share of equity partnerships is higher than 50%, similar to that expected for medium-tech industries. This similarity is not altogether coincidental. A considerable number of the firms engaged in technology partnering in new materials are in fact from the chemical, steel and engineering industries [Hagedoorn and Schakenraad 1991]. These sectors are characterized by an intermediate level of technological sophistication, and tend to have a higher share of joint ventures. Companies in these industries are engaged in technological development in new materials that frequently replace existing applications of their 'traditional' products. The preference for joint ventures when engaging in joint technology development in their sectors of origin seems to extend to alliances undertaken in new materials. Instrumentation-medical technology also demonstrates a share of joint ventures above 30% but the deviation is rather small and primarily caused by the extremely high share of U.S.-Japanese joint ventures in that sector.

The three sectors with a medium-technology intensity: automotive, chemicals and consumer electronics, follow the expected distribution. However, there is considerable variation for different international combinations. In chemicals, the vast majority of alliances (about 65%) are joint ventures. In the automotive industry, the general distribution is about fifty-fifty, although this is once again largely due to the large number of U.S.-Japanese joint ventures. In consumer electronics over 40% of the partnerships are equity alliances. Finally, the only low-tech sector in our sample, food and beverages, appears to support the assumption that low-tech sectors have a much higher preference for joint ventures than for contractual agreements.

To test Proposition 2 we define 'participation' in terms of the frequency with which companies from a particular country or international region are found in our data. Following some recent contributions to innovation studies [Freeman and Hagedoorn 1995; OECD 1992; Rothwell and Zegveld 1985; Soete and Verspagen 1991] we distinguish between technologically highly developed countries and the somewhat less advanced countries based on an R&D/GDP ratio either above or below 1%.[1] Greece, Ireland, Luxemburg, and Portugal were not included, given the small number of alliances for these countries. Table 2 gives the distribution of the participation of companies from the U.S., Japan and the EU according to nationality and sectoral breakdown. Table 3 shows differences between countries in terms of their degree of specialization in strategic technology partnerships as measured by the coefficient of variation. A higher coefficient indicates that a country is relatively specialized in a few sectors, while a lower coefficient indicates that a country is more diversified in partnering across a larger number of sectors. We see a clear difference in the technological partnering specialization between the technologically less sophisticated countries (Denmark, Spain and Italy) and the technologically advanced countries (Germany, France, U.K., Belgium, Netherlands, Japan, and U.S.).[2]

In general, strategic alliances involving companies from less advanced countries are concentrated in just a few sectors, whereas in the case of the technologically advanced countries they are more evenly distributed over a larger number of sectors. For Spain three sectors – telecommunications, aviation and defense, and microelectronics – account for a share of over 75% of all partnerships. In Denmark two sectors – biotechnology and new materials – account for over 90% of the international agreements. For Italy, however, although the coefficient of variation indicates that the pattern is

TABLE 2
The Sectoral Distribution of the Participation of Companies in International Strategic Technology Alliances, Countries, 1980–1993 (%'s)

	Belgium	Denmark	France	FRG	Italy	Japan	Netherlands	Spain	U.K.	USA	TOTALS
Biotechnology	21.18	68.18	10.40	10.11	9.12	13.49	16.13	10.29	18.97	17.17	16.20
New Materials Technology	7.06	22.73	7.60	9.57	6.42	10.81	6.45	1.47	10.00	9.03	9.29
Computers	1.18	0.00	6.20	3.61	4.73	6.41	4.61	1.47	2.24	5.63	5.30
Industrial Automation	2.35	0.00	4.60	6.32	5.74	6.60	4.38	1.47	6.90	5.24	5.86
Microelectronics	2.35	0.00	14.00	13.90	9.12	21.82	12.67	19.12	8.28	15.94	15.63
Software	10.59	0.00	8.60	9.21	6.76	5.17	15.90	0.00	7.24	8.75	8.55
Telecommunications	14.12	2.27	13.20	13.54	8.45	5.45	9.45	29.41	8.28	9.31	8.60
Other I.T.	1.18	0.00	3.20	2.53	4.73	2.30	4.61	4.41	0.69	1.67	2.26
Automotive	1.18	2.27	2.20	3.43	6.42	5.74	2.07	1.47	2.24	4.18	4.30
Aviation/Defense	4.71	0.00	15.80	13.36	20.95	2.11	2.53	27.94	18.62	6.41	6.82
Chemicals	30.59	2.27	8.20	7.76	11.49	12.44	9.45	1.47	9.31	8.70	9.34
Consumer Electronics	0.00	0.00	1.80	0.54	0.68	2.68	4.61	0.00	0.52	1.17	1.69
Food and Beverage	0.00	0.00	0.60	0.72	0.34	0.57	0.23	0.00	1.55	0.78	0.83
Heavy Electric/Power	1.18	2.27	1.80	1.81	3.04	1.72	0.46	0.00	1.55	2.01	1.61
Instrumentation/Med. Tech.	0.00	0.00	0.80	1.26	0.34	1.82	4.15	0.00	0.86	2.29	2.00
Other	2.35	0.00	1.00	2.35	1.69	0.86	2.30	1.47	2.76	1.73	1.74
Totals	100.00	100.00	100.00	100.00	100.00	100.00	100.00	100.00	100.00	100.00	100.00

TABLE 3
Summary of the Patterns of International Specialization in International Strategic Technology Partnering as Expressed in Coefficients of Variation

Belgium	1.41	Japan	0.92
Denmark	2.79	Netherlands	0.82
France	0.81	Spain	1.61
Germany	0.77	U.K.	0.95
Italy	0.82	USA	0.80

definitely not specialized (Table 3), the industrial distribution of its technology alliances is clearly different from the distribution of the group of technologically more advanced countries (Table 2). For six out of the seven technologically advanced countries, the distribution of alliances is clearly more evenly distributed over all the sectors. The only exception is Belgium, where three sectors – biotechnology, telecommunications and chemicals – account for over 65% of the agreements.

So far, we have been able to establish some preliminary understanding of the relationship between inter-organizational modes of strategic technology partnering, their sectoral setting, and their international environment in a bivariate context for each of the inter-regional combinations. To take our exploration further, we examine the possible association between each of our three basic variables: modes of cooperation, technological sophistication of countries, and technology intensity of sectors.

We have already demonstrated that international strategic technology partnering in high-tech industries appears to be disproportionately organized through contractual arrangements, whereas other sectors with lower levels of technological sophistication have a higher share of joint venturing. We also found some evidence that the level of technological sophistication of countries influences the strategic technology partnering participation of companies from these countries. We now test whether the association between the mode of international technology partnering and the level of technological sophistication of sectors is influenced by the country-level degree of technological intensity (Proposition 3). We distributed the numbers of 'hits', i.e., the number of times that companies from a particular country were engaged in international strategic technology partnering, according to modes of strategic technology cooperation, high-tech or non-high-tech content of sectors, and two groups of countries of different levels of technological sophistication. The differentiation of sectors into levels of technological sophistication and the separation of equity versus non-equity agreements has been discussed above. We again divided the countries into those having a R&D/GDP ratio greater than 1% as technologically sophisticated, and those lower than 1% as less sophisticated. The distribution of the different degrees of participation in

TABLE 4
Participation in Modes of International Strategic Technology Partnering by Different Levels of Technology for Sectors and Countries

	High-Tech Countries		Lower-Tech Countries	
	High-Tech Sectors	Other Sectors	High-Tech Sectors	Other Sectors
Equity	1006	459	106	27
Non-equity	3100	427	244	41

strategic technology partnering with respect to modes of cooperation, technological sophistication of sectors and countries is given in Table 4.

A cursory examination of the results in Table 4 tentatively suggests that within each of the two subgroups of high-tech and lower-tech countries the sub-distributions are very similar. For a more accurate assessment of the possible effect of levels of technological sophistication of countries, we use two statistical procedures to investigate whether the association between modes of partnering and technological characteristics of industries is maintained for the two levels of technological sophistication of countries. The first procedure is known as the partial association of variables in a three-dimensional setting. The other procedure is a modified χ^2 test.

Expanding on the standard procedure for calculating cross products in simple 2×2 tables, the partial association can be measured for three variables. The precise statistical procedure can be found in Moser and Kalton [1971]. Here we will follow accepted practice and only reproduce the measurements for the intermediate and final steps in the analysis. The partial cross product for high-tech countries (the left-hand side of Table 4) is -0.03986, the partial cross product for the lower-tech countries (the other part of Table 4) is -0.0128. The two marginal associations, through which the effect of the third variable (country level of technological sophistication) on each of the two variables are calculated, are -0.00176 and -0.00106. These low levels of marginal association indicate that the association between sectors of industry and modes of cooperation holds. The buildup of both partial associations and the combined marginal association can be expressed in their relative contribution to the association of the variables. The contribution to the partial association for high-tech countries is 97.3%, the contribution for low-tech countries is 2.7%. The contribution of the combined marginal associations is 0%. The extremely low value of the contribution of the combined marginal associations indicates that the gross association of sectors of industry and the preference for equity or non-equity partnerships holds across countries of different levels of technological sophistication.

A second procedure, based on a modified χ^2-test, generates similarly strong results. Our findings of a $\chi^2 = 9.868$ with a significance level of $p<0.025$ also indicate that the preference of companies to form contractual agreements in high-tech sectors and a disproportionate number of joint ventures in other sectors is not significantly intermediated by the level of technological sophistication of countries. Further details of this procedure are provided in Appendix 1.

DISCUSSION AND CONCLUSIONS

It has to be stressed that complexity governs the choice of inter-organizational mode of governance taken by an alliance, and thereby influences the strategic implications for the companies involved. This complexity has internal and external effects on firms involved in strategic technology partnering, which can be described in three dimensions.

The first dimension is the organizational complexity that is inherent when two or more firms engage in an alliance. The mode of governance that is undertaken for a given alliance is affected by the trade-off between minimizing intra-organizational complexity and maximizing control over the alliance by each partner. The more complex inter-organizational mode of technology cooperation, such as a joint venture, raises a number of problems of corporate governance. Its quasi-hierarchical nature not only reaches intermediate levels of corporate control, but also introduces dilemmas related to trust, forbearance and opportunim [Parkhe 1993]. Contractual agreements appear to involve lesser intra-organizational complexity because no separate new administrative element is created. However, although contractual agreements in themselves are less complex, companies often engage in several alliances simultaneously with a variety of partners. This introduces an additional level of complexity namely difficulties associated with both the administration of these partnerships and the need to continuously monitor the net benefits accruing from various contractual alliances [Osborn and Baughn 1990].

The second dimension that determines the organizational complexity (and thus the mode of governance) is the fact that these agreements involve technological development. The process of technological development is both complex and uncertain, and although the return on investment may be high, this reflects the degree of uncertainty that companies face. Technologies for which markets are not well defined and which involve high uncertainty are typically those in which strategic alliances are favoured. If it were not for these characteristics, we probably would not have witnessed the growth of strategic technology alliances over the past decade. In other words, the object of this particular group of alliances – the joint development or sharing of new technologies – is already intrinsically a matter of high complexity, compared with other activities such as manufacturing. However, the level of uncertainty and the resulting complexity differ by sector. Technology development is faster, more complex and uncertain in industries that are high-tech, whereas the value of innovations and their potential benefits are uncertain. By contrast, in mature industries, technology development is relatively slow-paced with incremental changes in technology, and the market and value of the technology are relatively well defined. Technology partnering in high-tech sectors tends to be characterized by a preference for contractual agreements, compared with medium- and low-tech sectors where a disproportionate share of joint ventures is found.

The third dimension is associated with the international context of these alliances, which further exacerbates the already high level of complexity. It is obvious that sharing technology with a partner creates organizational complexity, but this is further increased if the partner is in another country or continent. In terms of control, the agreement has to be monitored from corporate headquarters over long distances or from a local or regional subsidiary. However, even in the latter case, there is often no real reduction in the 'distance of control' but it frequently merely results in the introduction of an additional level of corporate governance. Nonetheless, companies of different national backgrounds are also influenced by their past business experiences as well as by the difficulties of maintaining control over the alliance across borders. Companies that have collaborated on other projects in the past, and are familiar with the business practices and/or regulatory framework in which their partners operate, will perceive less uncertainty in engaging in partnerships that are less complex than with firms with which they have had little experience.

Our results indicate considerable differences between companies of different nationalities in their propensity to engage in strategic technology partnering. These differences do not pertain to differences in preferred modes of governance but more specifically to the sectoral and geographical patterns of the international technology alliances in which companies engage. The participation of companies in strategic technology partnering and the distribution of these alliances over a large number of sectors appear to increase with the increasing levels of technological sophistication of home-countries of these companies. In choosing particular modes of cooperation, the level of technological sophistication of industries plays an important role. In situations of high-tech complexity, in terms of high uncertainties about both the outcomes of the process of technological change as well as the market structural consequences of these changes, the number of alliances are increasing. Under these unstable conditions, when organizational flexibility and speed of information transfer are vital, companies have a disproportionate preference for more informal, contractual and quasi-market oriented agreements. More stable industry structures combined with less dynamic technological changes are parallelled by a relatively higher share of quasi-hierarchical and also more formal modes of cooperation as found in the joint venture mode. In such situations, organizational control over information flows and new technological applications with a long life cycle is more relevant than in the case of technologies that are quickly outdated, when speed of information processing is more important than organizational control itself.

Given the exploratory character of this contribution a discussion of future research topics is appropriate:

- In order to improve both our theoretical and analytical understanding of international strategic technology partnering, the analysis of this phenomenon has to be further integrated in the understanding of corporate internationalization and other aspects of company behaviour, such as technology strategies [Duysters and Hagedoorn 1995; Harrigan 1985; Nohria and Garcia-Pont 1991]. This implies a multivariate setting in which different modes of cooperation are analyzed against a multi-sectoral background with a range of strategic alternatives open to a variety of firms. In that context, organizational and structural characteristics as well as behavioural properties such as entry and exit strategies and a range of technology strategies may be possible explanatory variables.
- A more thorough comprehension of the effects of strategic technology partnering on company performance appears necessary. Measuring these effects has to be evaluated in terms of economic effect, improved or possibly deteriorated technological capabilities for different partners, and the implications of a range of modes of cooperation [Powell and Brantley 1992]. The application of network analysis and the further development of related theoretical contributions can improve our knowledge of the complexities and the inter-organizational context of a large variety of inter-firm relationships that effect companies engaged in strategic technology partnering [Nohria and Eccles 1992]. Also, further analysis of the scale of alliances might increase our insight into the strategic importance of alliances in a particular network configuration.
- Organizational complexity has been mentioned as one of the characteristics of international strategic technology alliances [Parkhe 1993]. The understanding of the interaction of organizational control over technology flows through these various partnerships, each with different degrees of complexity, and the benefits of a partnering strategy appears essential. This is not only crucial for the practical purposes of the participating companies, but a clearer comprehension of the organizational implications and corresponding levels of complexity of control will also be necessary to improve our academic understanding of strategic technology partnering.

APPENDIX

The null hypothesis to be tested for the assumed similarity of the distribution of contractual and equity alliances in high-tech and low-tech industries in both categories of countries in Table 4 is:

$$H_0 : p_{j/1} = p_{j/2} = p_j \qquad (j = 1,2,3,4).$$

The alternative hypothesis H_1 is that for at least one j it holds that $p_{j/1}$ is not equal to $p_{j/2}$. This $p_{j/1}$ refers to the percentage of companies in the population of high-tech countries that belongs to the category j. Schematically this can be shown as:

High-tech		Low-tech	
$p_{1/1}$	$p_{2/1}$	$p_{1/2}$	$p_{2/2}$
$p_{3/1}$	$p_{4/1}$	$p_{3/2}$	$p_{4/2}$

It is obvious that:

$$\sum_{j=1}^{4} p_{j/i} = 1; \; i = 1, 2.$$

Because the p_j's are unknown, we will estimate them from the observations. Under the restrictions of the null hypothesis we estimate the p_j's from the pooled data :

$$\hat{p}_j = \frac{n^{j/1} + n^{j/2}}{N}$$

$$N = n_1 + n_2 \text{ and } n_i = n_{1/i} + n_{2/i} + n_{3/i} + n_{4/i},$$

where $n_{j/i}$ refers to the total number of observations in cell j of population i. n_i refers to the total number of observations in population i.

Now we are able to perform a χ^2-test. The expected number of observations in cell j of population i is:

$$e_{j/i} = n_i \hat{p}_j$$

$$\chi^2 = \sum_{i=1}^{2} \sum_{j=1}^{4} \frac{(n_{j/i} - e_{j/i})^2}{e_{j/i}}.$$

Similar to the statistical procedures we followed with regard to the partial associations our findings of a $\chi^2 = 9.868$ with a significance level of $p<0.025$ indicate that the preference of companies to form contractual agreements in high-tech sectors and a disproportionate number of joint ventures in other sectors is not significantly intermediated by the level of technological sophistication of countries.

NOTES

1. R&D/GDP ratios for 1990 are: Japan 2.12%, Germany 2.02%, USA 1.98%, France 1.48%, U.K. 1.47%, Belgium 1.23%, Netherlands 1.16%, Denmark 0.85%, Italy 0.76%, Spain 0.47% (Source: OECD).

2. These findings have to be interpreted with some caution, given the small sample size associated with the less technologically sophisticated nations.

REFERENCES

Archibugi, Daniele & Maria Pianta. 1994. Aggregate convergence as sectoral specialization in innovation. *Journal of Evolutionary Economics*, 14: 17–33.

Auster, Ellen R. 1992. The relationship of industry evolution to patterns of technological linkages, joint ventures, and direct investment between U.S. and Japan. *Management Science*, 38: 778–92.

Buckley, Peter J. & Marc Casson. 1976. *The future of the multinational enterprise*. London: Macmillan.

——. 1988. A theory of cooperation in international business. In F. J. Contractor & P. Lorange, editors, *Cooperative strategies in international business*. Lexington, Mass.: Lexington Books.

Cantwell, John. 1991. The international agglomoration of R&D. In M. Casson, editor, *Global research strategy and international competitiveness*, 104–33. Oxford, U.K.: Blackwell.

Chesnais, François. 1988. Technical cooperation agreements between firms. *STI Review*, 4: 51–120.

Contractor, Farok J. & Peter Lorange. 1988. *Cooperative strategies in international business*. Lexington, Mass.: Lexington Books.

Dunning, John. 1993. *The globalization of business*. London: Routledge.

Duysters, Geert & John Hagedoorn. 1995. Strategic groups and inter-firm networks in international high-tech industries. *Journal of Management Studies*, 32: 359–81.

Freeman, Chris & John Hagedoorn. 1995. Convergence and divergence in the internationalization of technology. In J. Hagedoorn, editor, *Technical change and the world economy*, 34–57. Aldershot, U.K.: Edward Elgar.

Freeman, Chris & Bengt Lundvall, editors. 1988. *Small countries facing the technological revolution*. London: Pinter.

Hagedoorn, John. 1993. Understanding the rationale of strategic technology partnering: Interorganizational modes of cooperation and sectoral differences. *Strategic Management Journal*, 14: 371–85.

—— & Jos Schakenraad. 1990. Inter-firm partnerships and cooperative agreements in core technologies. In C. Freeman & L. Soete, *New explorations in the economics of technological change*, 3–38. London: Pinter.

——. 1991. Inter-firm partnerships for generic technologies – The case of new materials. *Technovation*, 11: 429–44.

——. 1993. Strategic technology partnering and international corporate strategies. In K. Hughes, editor, *European competitiveness*, 60–87. Cambridge, U.K.: Cambridge University Press.

Haklisch, Carmelia S. 1986. Technical alliances in the semiconductor industry. Mimeo, New York University.

Harrigan, Kathryn R. 1985. *Strategies for joint ventures*. Lexington, Mass.: Lexington Books.

——. 1988. Joint ventures and competitive strategy. *Strategic Management Journal*, 9: 141–58.

Hergert, Michael & Deigan Morris. 1988. Trends in international collaborative agreements. In F. J. Contractor & P. Lorange, editors, *Cooperative strategies in international business*, 99–109. Lexington, Mass.: Lexington Books.

Hladik, Karen J. 1985. *International joint ventures*. Lexington, Mass.: Lexington Books.

——. 1988. R&D and International Joint Ventures. In F. Contractor & P. Lorange, editors, *Cooperative Strategies in international business*, 187–204. Lexington, Mass.: Lexington Books.

Kogut, Bruce. 1989. The stability of joint ventures: Reciprocity and competitive rivalry. *Journal of Industrial Economics*, 38: 183–93.

Link, Albert N. & Laura L. Bauer. 1989. *Cooperative research in U.S. manufacturing: Assessing policy initiatives and corporate strategies.* Lexington, Mass.: Lexington Books.

Lundvall, Bengt A., editor. 1992. *Elements of national systems of innovation.* London: Pinter.

Moser, Carl A. & Gerald Kalton. 1971. *Survey methods in social investigation.* London: Longman.

Mowery, David C., editor, 1988. *International collaborative ventures in U.S. manufacturing.* Cambridge, Mass.: Ballinger.

Mytelka, Lynn K., editor. 1991. *Strategic partnerships and the world economy.* London: Pinter.

Narula, Rajneesh. 1995. *Multinational investment and economic structure.* London: Routledge.

Nelson, Richard R., editor, 1993. *National innovation systems.* New York: Oxford University Press.

Nohria, Nitin & Robert Eccles. 1992. *Networks and organizations.* Boston, Mass.: Harvard Business School Press.

Nohria, Nitin & Carlos Garcia-Pont. 1991. Global strategic linkages and industry structure. *Strategic Management Journal*, 12: 105–24.

OECD. 1992. *Technology and the economy.* Paris: OECD.

Osborn, Richard N. & Chris C. Baughn. 1990. Forms of interorganizational governance for multinational alliances. *Academy of Management Journal*, 33: 503–19.

Parkhe, Arvind. 1993. 'Messy' research, methodological predisposition and theory development in international joint ventures. *Academy of Management Review*, 18: 227–68.

Patel, Pari & Keith Pavitt. 1991. Large firms in the production of the world's technology: An important case of 'non-globalization'. *Journal of International Business Studies*, 22(1): 1–21.

Porter, Michael E. 1987. From competitive advantage to corporate strategy. *Harvard Business Review*, (May-June): 43–59.

Powell, Walter & Peter Brantley. 1992. Competitive cooperation in biotechnology: Learning through networks. In N. Nohria & R. Eccles, *Networks and organizations*, 366–94. Boston, Mass.: Harvard Business School Press.

Rothwell, Roy & Walter Zegveld. 1985. *Reindustrialization and technology.* Harlow, U.K.: Longman.

Rugman, Alan M. 1980. Internalization as a general theory of foreign direct investment: A reappraisal of the literature. *Weltwirtschaftliches Archiv*, 116: 365–79.

Soete, Luc & Bart Verspagen. 1991. Recent comparative trends in technology indicators in the OECD area. In OECD, *Technology and productivity*, 249–74. Paris: OECD.

Walsh, Vivian. 1988. Technology and competitiveness of small countries. In C. Freeman & B. Lundvall, editors, *Small countries facing the technological revolution*, 37–66. London: Pinter.

ELSEVIER

Journal of Economic Behavior & Organization
Vol. 35 (1998) 203–225

JOURNAL OF
Economic Behavior
& Organization

Cooperative agreements and the theory of the firm: Focusing on barriers to change[1]

Frédérique Sachwald

Institut Français des Relations Internationales and University Paris XIII, 27 rue de la Procession, 75740 Paris Cedex 15, France

Received 5 February 1996; received in revised form 20 February 1997

Abstract

Cooperative agreements between firms have become a significant phenomenon in the 1980s and have been analyzed by various approaches, including economics, organizational theory and business strategy. This paper draws on these different strands of the literature to argue that cooperative agreements have led to focus the analysis of firms' organization on the issue of change. As a consequence, cooperative agreements have contributed to recent developments of the theory of the firm in favor of competence-based and evolutionary approaches. The analysis is then extended to the case of multinational firms. © 1998 Elsevier Science B.V.

JEL classification: L22; F23

Keywords: Cooperative agreements; Theory of the firm; Transaction costs; Evolutionary theory; Multinational firms

1. Introduction

Several strands of analysis have contributed to the study of cooperative agreements between firms, including economics, organizational theory and business strategy. This paper shows that these perspectives are complementary and draws on their different insights to examine the influence of cooperative agreements on the evolution of the theory of the firm.

[1] A previous version of this paper has been presented at the EMOT workshop held at Reading University on May 14–16 1995.

PII S0167-2681(98)00054-7

The specific influence of cooperative agreements on the theory of the firm is due to the combination of two types of considerations. Firstly, the analysis of these operations has to consider both the boundaries of the firm and its internal competences, since cooperation aims at pooling resources from different partners. Secondly, numerous empirical studies show that firms resort to alliances in order to manage change.[2] The paper argues that, as a result, the analysis of cooperative agreements leads to a combination of the transaction cost and the evolutionary approaches in order to deal with the relationships between the competitive game and the internal resources of the firm. In its second part, the paper suggests that such an approach is relevant to analyze multinational corporations, which should be better integrated into a general theory of the firm.

2. Evolution of the economic analysis of cooperative agreements

The analysis of cooperative agreements has followed a cycle which is quite typical for a new phenomenon. Indeed, inter-firm alliances were first discovered as a new phenomenon to be defined and measured at the beginning of the 1980s. Cooperative activities between independent firms had existed in the past, but they were more focused on certain types of agreements, such as licences, and less numerous.[3] The second analytical step has mainly consisted of theoretical contributions exploring the reasons for their very existence. Finally, once cooperative agreements became quite common and were recognised as a long-lasting feature of economic organization, the relevant questions moved on to the efficiency of these operations and thus to management matters.

2.1. Identification of a new phenomenon

Cooperative agreements developed quite rapidly in a number of sectors from the end of the 1970s onward. Empirical studies followed this trend with some delay, during the 1980s. Cooperative agreements or alliances were generally defined as the types of transactions undertaken by two or more independent partners and which were an intermediary between spot transactions on a market and mergers or acquisitions, that is, between market and hierarchy. The most striking type of agreements involves competitors, but firms have also developed cooperative schemes with both their suppliers and partners from other sectors.[4] Equity joint ventures are easy to identify, but many cooperative arrangements involve joint activities without the creation of a new corporate entity and even without equity shareholding from the parties. A number of studies thus consider that the important point is the existence of a common project,

[2] *Cooperative agreements* is used here as the most general term, but a number of studies use the term *alliances*, which may be considered as a synonym. Besides, cooperation is implemented through various types of agreements and contracts.

[3] The interwar period is historically specific because of the important role played by cartels in some sectors. These cartels promoted cooperation in various areas, including technological exchanges.

[4] Some cooperative schemes, especially in the field of R&D, involve numerous partners, including public laboratories and various academic institutions.

and hence a need for coordination over time, rather than a specific type of agreement between parties.[5]

Whatever the precise definition which they used, several inter-sectoral studies have shown that the number of cooperative agreements has increased substantially during the 1980s (Marity and Smiley, 1983, Hergert and Morris, 1988, Hagedoorn and Schakenraad, 1991).[6] This general trend has been confirmed by detailed studies on specific sectors or on certain types of agreements such as those organizing cooperation in the fields of research and technology (Jacquemin et al., 1986, Mowery, 1988, Mytelka, 1991, Sachwald, 1994a). A number of studies emphasize the relative intensity of partnering in high-tech industries (GEST, 1986, Hagedoorn and Schakenraad, 1991, Hagedoorn and Schakenraad, 1993). The tendency to cooperate in such sectors is certainly high, but these studies focus on technology partnering, which introduces a bias.

A large number of cooperative agreements has been international in scope. Several studies have focused on the increasing number of international alliances struck by American firms from the 1970s onward (Hladik, 1985, Reich and Mankin, 1986, Mowery, 1988). Others have analyzed intra-European alliances, more particularly in relation with the Single market and the European innovation policy (GEST, 1986, Mytelka, 1993, Kay, 1991). These studies show that international cooperative agreements have multiplied, which does not mean that they have become much more important relative to national ones, since the latter have followed an upward trend too.[7]

2.2. *Market power versus transaction cost interpretations*

Economists had first considered cooperative agreements within the framework of analysis of industrial organization. In this context, two main perspectives competed to explain alliances: that of market power and that of transaction costs. In this respect, the analysis of cooperative agreements was comparable to that of vertical integration, multiple production, or multinational firms.

Whenever markets were supposed to be efficient, cooperation between otherwise competing firms amounted to collusion and tended to reduce competition. In this perspective, cooperation could nevertheless be efficient if it was used to deal with externalities. This argument may be used to explain that benevolent competition policy rules were issued in the 1980s, both in the US and in Europe, to examine cooperative research agreements.[8]

In the aluminum industry, the world capacity to produce bauxite and alumina mostly expanded through jointly owned plants during the 1970s and Stuckey (1983) has noted that these ventures were usually considered as a form of collusion. In the case of the

---

[5] For definitional considerations, see in particular Contractor and Lorange (1988a), Sachwald (1990, 1994a), Garrette and Dussauge (1995). Gaffard (1990) and Rullière and Torre (1995) emphasize the fact that cooperation takes place over time.

[6] One exception is (Ghemawat et al., 1986), but this result may be due to the fact that the data include patent licensing, which in a number of sectors has been a usual phenomenon for a long time. Moreover, it can be argued that licensing is often a rather simple type of transaction and does not lead to substantial cooperation.

[7] An evaluation of the share of international agreements requires precise analyses since it varies according to the type of cooperation; see Hagedoorn and Schakenraad (1993), Sachwald (1994a).

[8] For a discussion of the cost-benefit analysis of R&D cooperative agreements, see Jacquemin et al. (1986).

automobile industry, Bresnahan and Salop (1986) have hypothesized that NUMMI, the Californian production joint venture between GM and Toyota, was a collusion mechanism. They have considered that the sharing of NUMMI's profits generated disincentives for the parent companies to increase production and lower prices. Hence, they constructed 'modified' Herfindahl indexes, which vary according to the legal arrangement with respect to the control of the joint venture, in order to measure the influence of the venture on the distribution of market power. The modified indexes are always lower than the usual Herfindahl index (which corresponds to complete integration) but tend to increase as parent companies exercise tighter control on the joint production unit. This cooperation between Toyota and GM has since then been examined by case studies which have explored the motivations of each partner and assessed the results of the venture. They generally conclude that the agreement has been instrumental to a stronger entry of Toyota in the American market, and thus to an increase in competition.[9]

The transaction cost perspective has challenged the market power interpretation. The argument is similar to that used in the cases of vertical integration or multiproduction.[10] It rests on the observation that technical considerations, as they are embodied in the production function, are not sufficient to determine the way in which activities undertaken within the economic system are divided among firms. The technical potential for economies of scale or economies of scope does not imply that a single firm should undertake the total amount of production necessary to exhaust these economies, because the actual organization of productive units also depends on transaction cost considerations.[11]

Transaction costs occur fundamentally because of bounded rationality in a risky environment: agents cannot devise complete contracts, so that ex post opportunistic behavior is impossible to prevent. Transaction costs tend to increase with market imperfections and uncertainty. Transaction costs economics thus analyzes in detail the sources of market imperfections and their consequences in terms of governance structures.[12] It has explored in particular the characteristics of knowledge to explain that firms often choose to internalize their transfers to counter opportunistic behaviors. Transaction cost analysis shows that specific assets, be they tangible or intangible, cannot be traded on regular markets, so that partners may get locked into bilateral monopoly situations. As a result, firms tend to internalize the transactions of inputs when production processes involve specific assets.

---

[9] See below for further details on this example and the relevant studies.

[10] The case of vertical integration has been extensively analyzed by Williamson (1975, 1985); the case of the multiproduct firm has been dealt with by Teece (1980, 1982).

[11] The transaction cost perspective focuses on them rather than on technical considerations [Williamson (1975): 2].

[12] Transaction cost theory, considering the firm as a cost-minimizing device, may be considered as one contractual approach, along with agency and property rights perspectives (Tirole, 1988, Williamson, 1990, Brousseau, 1993). Williamson (1985) emphasizes the fact that his approach does not seek to deal with all incentive problems ex ante through appropriate contracts. This may be related to his hypothesis of bounded rationality; some authors have nevertheless pointed that the process of minimizing transaction costs may reintroduce strong rationality requirements (Coriat and Weinstein, 1995).

Cooperative agreements can be considered as partial internalization and the analyst has thus to explain why firms should choose this intermediary solution instead of either resorting to the market or fully integrating the transaction (Buckley and Casson, 1988, Hennart, 1988). Firms resort to cooperative agreements because the contracts which would be necessary to use the market are either too expensive to write or insufficiently reliable. The two reasons are of course interrelated since complete contracts would be too costly to negotiate. Simultaneously, complete integration is considered too costly and risky, or even unfeasible in cases where the resources which are pooled together by the cooperative device belong to totally different firms and/or institutions.[13]

In the aluminum industry, bauxite is not a homogenous ore, differing in composition across deposits and the process of refining bauxite into alumina has to be modified in order to suit the different types of ore. The bauxite market for a given processing unit is thus particularly thin. In his study of the aluminum industry, Stuckey (1983) shows that, bauxite and alumina are often not traded at arm's length because the characteristics of the transaction would require very detailed long-term contracts to provide for opportunistic behavior on thin markets. Firms resort to production joint ventures where common assets constitute an incentive to cooperate for each partner by guaranteeing their dedication to a long-term relationship (Stuckey, 1983). Joint ventures enable firms to share intermediate inputs, such as bauxite or alumina, without supporting the full cost of integration, that is, acquiring all the necessary assets. The choice of joint ventures to implement economies of scale or economies of scope in the production of intermediary components results from the same type of considerations. Joint production allows both partners to reach lower unit costs and only support part of the necessary initial investment.

Carmakers have cooperated through a growing number of agreements from the end of the 1970s onward (Sachwald, 1992, 1994b). These agreements have been used to access complementary resources for which market transactions are inefficient, such as distribution channels in foreign countries. Actually, in most cases, the choice was not between a market transaction and a cooperative agreement, but rather between solutions involving various degrees of internalization. Acquisitions have been chosen when firms wanted to benefit from goodwill and a prestigious brand name or when they aimed at extending their production capacities, while controlling the management process. Examples of the first type include the acquisition of Jaguar by Ford, while the acquisition of SEAT by VW corresponds to the second type of objectives. In all cases, the acquisitions have been relatively modest (as compared with the size of the buyer). Equity joint ventures have been preferred to set up joint production units which involve substantial investments. These joint ventures were resorted to in particular when a single firm was not been able to dispose of all the output, as in the case of multipurpose vehicles or upper range engines (Volvo–Renault–Peugeot in the 1970s). Ambitious cooperation covering several functions, such as common distribution, knowledge transfers and exchanges of components has involved minority equity shareholding, as in the case of the

[13] Agreements can organize cooperation between competitors, between vertically related firms, or even between a firm and a public institution, as in the case of R&D cooperations.

Ford–Mazda or Renault–Volvo alliances. Finally, pre-competitive research cooperation, which often pooled several carmakers along with other partners, has been conducted through consortia. The example of the automobile industry thus shows a quite good match between the types of governance structures which firms have chosen and the attributes of the transactions.[14] The higher the risk involved and the larger the assets which are necessary to reach a given objective, the tighter the relationship. In particular, common assets, such as in the case of equity joint ventures, constitute credible commitments which contribute to establishing a long-term relationship. These relationships between the objective of an alliance and its organizational structure have been observed in other industries and may be generalized as principles: the closer cooperation gets to production and distribution and the wider its scope, the more tightly organized it becomes (Rullière and Torre, 1995).

2.3. Knowledge transfers: Competing interpretations

Cooperative agreements are often instrumental to knowledge transfer from one firm to another. In order to examine the choice of these governance structures, transaction cost analysis focuses on opportunism as a major source of market failure in know-how trading, while organizational theory stresses cooperative motivations.

Analyses of knowledge transfers between firms or institutions tend to distinguish between codifiable and tacit knowledge. Codifiable knowledge is costly to exchange because of the fundamental paradox of information and the patent system has been designed to take care of this problem: the inventor describes her discovery and is granted a monopoly on its use, which can then be licensed. By contrast, tacit knowledge cannot be embodied in formulae, designs or other types of specifications. As a result, it cannot be transferred in a codified form such as a patent; tacit knowledge transfers are riddled with uncertainty and hence high transaction costs. Thus firms have to resort to governance structures which economize on transaction costs and facilitate know-how transfers, [Teece (1980, 1982) and Hennart, 1988]. In certain cases, cooperative agreements or joint ventures can be appropriate; in other cases, intra-firm transfers may be the best solution.

The organizational perspective downplays market hazards and focuses on the fact that interactions between parties are necessary to transfer certain types of knowledge, which is sufficient to explain the need for cooperative mechanisms. Kogut and Zander (1992, 1993) claim that the opportunism hypothesis is actually not necessary to explain that firms do not exchange know-how through markets since the most important determinants of the mode of transfer are the attributes of knowledge. Their empirical investigation shows that codifiability and teachability of knowledge favor the transfer from one firm to another through contractual forms rather than complete internalization. On the contrary, complexity increases the probability that a multinational company would transfer knowledge internally, to wholly owned subsidiaries. Such an approach differs from the transaction cost perspective where the integration decision hinges on the risk of opportunistic behavior when the transaction involves specific assets. Knowledge can be

[14] For a more complete treatment, see Sachwald (1990).

quite specific,[15] but from an organizational perspective, the problem does not lie so much in the cost of repeated negotiations and opportunism, but in the effectiveness of the transfer when the capabilities of the parties are too dissimilar (Foss, 1993, Kogut and Zander, 1993). In this perspective, organizational modes for technology transfers should be related to the experience of firms in managing such transfers.

A review of various empirical studies suggests that the determinants of the organizational modes of knowledge transfers between units are a combination of the characteristics of the knowledge itself, as well as both the trust between partners and their experience in managing these operations. These different elements are themselves related to the nationality of origin of the partners [16] and to their cultural background. Parkhe (1991) argues more generally that partners should carefully design their agreement so as to be able to face problems which may result from the confrontation of their diverse backgrounds.

2.4. *Strategy and resources*

Transaction cost and organizational considerations are particularly relevant to exploring governance structures, since they assess the most efficient type of cooperative agreement which a firm should implement in order to use a given set of complementary resources. However, this kind of analysis cannot explain the very reasons why a firm should look for such complementary resources. This explanation requires to relate the firm's internal resources and competences[17] with its strategic needs. In other words, it is necessary to focus on the production process rather than on the exchange devices.

Why did so many firms look for complementary resources from the 1970s onward? In order to answer this question, numerous studies have examined the evolution of firms' strategies. And in sectors where cooperative agreements have been numerous such as semiconductors, computers, pharmaceuticals or automobiles, the answer was to be found in a growing gap between firms' strategies and their internal resources. The relative importance of different types of cooperative agreements has depended upon the resources required by firms' strategies and has thus varied across sectors and countries. However, from the 1980s onward, numerous sectors have had to adapt to two major developments: increasing costs of R&D and the process of globalization.[18]

In a number of sectors, the increasing role of innovation in competition has induced firms to set up specific strategies in order to foster product or process innovations. In such a context, cooperative agreements have been used as one way of complementing internal R&D resources. Firms in high-tech sectors have logically been very active in this respect.

---

[15] Knowledge can have attributes of a public good, but is often costly to acquire and adapt to specific cases.

[16] Hagedoorn and Narula (1994) show that contractual modes are more frequent (relative to equity agreements) in international technology partnering between EC firms or between EC and U.S. firms than between Japanese and U.S. firms or between Japanese and EC firms.

[17] Or capabilities. These two words are used to designate similar notions in recent literature, see, for example, Foss (1993), Teece et al. (1994) and the contribution by Loasby in this issue.

[18] This paper focuses on these two motives for cooperative agreements. In a number of cases, increasing product differentiation, combined with economies of scale or economies of scope, lead to cooperative projects – for example, in the chemical or automobile sectors. But of course, differentiation and scale are not independent of internationalization.

In semiconductors, cooperative agreements have organized technology transfers from American to European firms, as a means of addressing the technological gap (Steinmueller, 1988 and Hobday, 1994). Large pharmaceutical companies have resorted to various types of agreements with small biotechnology firms in order to enter this new research area.[19] Here again, a large share of such cooperations has aimed at transferring technology from American to European firms. A renewed role for innovation in the competitive game has also fostered cooperative agreements in mature sectors such as the automobile industry, where consortia have organized precompetitive research programmes in Europe and in the United States. In a number of cases, these programmes have received public support.

During the 1980s, firms generally increased their degree of internationalization. This was the case in particular for firms which were still relatively little internationalized, such as Japanese or French companies. Traditionally, firms tend to favor control over their foreign units. As a result, when they want to rapidly penetrate a foreign market and if the characteristics of the sector are appropriate, they may choose to acquire foreign companies. During the 1980s, this was very often the case in the chemical industry.[20] Obstacles to this first choice strategy lead firms to choose other means in order to enter foreign markets or to be able to use foreign resources and capabilities. In previous periods, regulations and preferences from host governments used to be one of the major reasons why multinationals resorted to joint ventures.[21] In regulated sectors, such as telecommunications, cooperative ventures may still be considered necessary to enter foreign markets (Mowery, 1988). But firms also often choose a cooperative solution in order to share risks and costs and to speed up their acquaintance with the host country. Japanese firms, which in a number of cases have had to internationalize rapidly during the 1980s, have resorted to joint ventures and cooperative alliances in order to get acquainted with human resource management practices and problems, as well as to assess potential local suppliers. In the automobile industry, this has been one motivation for the joint venture between Toyota and GM or for the set of cooperative agreements between Honda and Rover.[22]

The above examples emphasize the specific perspective of strategic business analyses on cooperative agreements. As opposed to both the market power and transaction costs approaches, it considers firms' capabilities as a fundamental consideration. Lorange and Roos (1992) explicitly propose a two-stage approach to analyzing the motives for alliances between firms. The first step consists in an assessment of the strategic position of firms in order to determine the motives for alliances. The second step classifies alliances by types according to the resources which each partner dedicates to the cooperation. More generally, strategic business analyses tend to first consider firms'

[19] The research tradition of pharmaceutical companies is mainly in organic chemicals; large companies have waited until biotechnology has proved productive in terms of drug research before devoting substantial resources to this new R&D field (Pisano et al., 1988, Sharp, 1994).

[20] Domestic M&As have also been numerous, especially in the US, as part of a vast restructuring movement in the industry (Sachwald, 1994a). A number of operations have concerned only one or a couple of production units from a company.

[21] This was the case in particular in developing countries and Japan. For references on international joint ventures before the 1980s, see Hladik (1985), Porter and Fuller (1986), Dunning (1993).

[22] On these examples, see in particular Kenney and Florida (1993), Mair (1994), Sachwald (1994a).

strategic needs before estimating the costs of different types of transactions designed to meet these needs.[23]

From the strategic management perspective, cooperative agreements explicitly become a way for a firm to match its capabilities to its changing environment. As a result, it focuses much more on firms' resources and their evolution. Moreover, this perspective leads to the consideration of the evolution of cooperations once they have been agreed upon by the partners. Firstly, the need for an alliance can disappear with the need for complementary resources. Secondly, alliances have to be managed in order to yield the expected benefits, otherwise they may well be abandoned. Indeed, as in the case of acquisitions, the rate of failure is far from negligible (Kogut, 1988, Bleeke and Ernst, 1993).

Apart from the cases of failure, an alliance may be terminated either because it has completed its intended objective (at least for one party) or because the wider competitive context has changed. When cooperation aims at learning from one partner, it will be terminated when the latter considers that knowledge transfer has been completed.[24] If a cooperative agreement is used by a firm to get acquainted with the business conditions in a foreign country, it will be abandoned when the multinational will be content with its knowledge of the new environment. In such cases foreign companies may well acquire control over the foreign venture. Changes in the competitive game will of course alter the incentives to cooperate. Let us take two examples of contrasted evolutions. In the computer industry, two major trends have fundamentally modified the competitive game in the 1980s. Firstly, the early gaps between the capabilities and market position of U.S. and non-U.S. firms have narrowed substantially, fostering growing rivalry. Secondly, the rising market share of micro-computers has induced a vertical disintegration of firms. Indeed, many firms have been able to buy microprocessors from Intel and equip their machines with Microsoft software. This has started a standardization move in the industry, which has fostered its restructuring in specialized segments (microprocessors, different types of computers, software...). This new structure, with competing firms on each segment (as opposed to vertical integration) and a high innovation rate, has generated incentives to cooperate between firms (Gomes-Casseres, 1993, Delapierre and Millelli, 1994).

The relationships between specialized biotechnology firms and pharmaceutical companies have followed a reverse pattern: from cooperative agreements to vertical integration. Up to the end of the 1980s, most large pharmaceutical companies were little involved in biotechnology and had often chosen to cooperate with specialized small American companies. Once biotechnology proved to really be a route to discover new drugs and had thus became a core technology, pharmaceutical companies invested much more heavily and acquired their smaller partners in a number of cases.[25] This evolution

23 See, for example, Porter and Fuller (1986), Contractor and Lorange (1988b).

24 Especially if the knowledge transfer is quite self-contained and if the cooperation does not also encompass other objectives. In the automobile industry for example, a number of cooperative agreements have aimed at such transfers (between Japanese and American or European companies), but they often also involved other objectives.

25 Examples include the acquisition of Genentech by Hoffman LaRoche in 1990 and of Genetic Systems by Elf Sanofi (in 1990 after agreements in 1986 and 1988). See Sharp (1994).

has resulted in the termination of several cooperative agreements and in the reduction in the number of new alliances in the field of biotechnology (Hagedoorn and Schakenraad, 1993).

Dynamic and management considerations tend to show that cooperative agreements are now largely integrated into the firms' daily life. This tendency has become more prominent as alliances have multiplied. In a number of cases, firms have progressively built complex networks encompassing different types of alliances with various partners. In the case of innovation, for example, cooperative alliances between competitors may well be integrated into networks along with cooperative schemes involving suppliers and academic laboratories.[26]

As cooperative agreements have become an essential feature of firms' overall organizational structure, management issues have grown in importance and are being explored in more detail.[27] Learning is of course a prominent issue since it is often what cooperation is about. As suggested above, successful learning leads to the termination of cooperation. More generally, effective cooperation entails some amount of learning so that partners are able to overcome the difficulties created by their divergent characteristics. These characteristics may be due to various factors which have already been alluded to: cultural and national background, evolution of strategic goals, corporate culture and management practices. Parkhe (1991) shows that each of these dimensions of diversity between partners constitutes a management problem, threatening the longevity of alliances. As a result, he argues that firms should explicitly provide for means to cope with each of them. Specific learning devices should be thought out in order to enhance understanding between partners from different backgrounds, and the organizational structure of cooperation should be flexible in order to permit adaptation to shifting environments and strategies.

3. Relevance to the theory of the firm

Cooperative agreements have been identified as a specific organizational device, but their underlying motivations have to do with the resources and the inner working of firms. The above analysis has thus argued that a thorough examination of the role of alliances requires relating the firm's competitive position and its strategy with its boundaries and organization. This second part suggests that this is why the analysis of cooperative agreements has contributed to the evolution of the theory of the firm in favor of a combination of transaction cost and evolutionary approaches.

The paper then extends its approach to multinational companies. It firstly examines how they have used cooperative agreements in order to implement global strategies. It then argues that cooperative agreements play a similar role for all types of firms, which leads to an integration of the analysis of multinationals more fully into the general theory of the firm. Such a perspective appears all the more relevant now that multinationals have

[26] Also see the discussion about the organization of multinational firms (Section 3.2).

[27] See, for example, Lorange and Roos (1992), Garrette and Dussauge (1995).

become so banal, and that the focus of analysis shifts away from their existence as transnational entities to their operation and management.

3.1. The firm as an evolving organization

Cooperative agreements are contractual devices which are used to pool together resources from two (more or less) independent firms. They thus contribute to defining the (fluctuating) boundaries of the firm. They also illustrate that firms need to alter their competences and may resort to external resources in order to do so. As a result, cooperative agreements constitute an interesting standpoint to examine the evolution of the theory of the firm.

Analysis of cooperative agreements started as an exploration of the boundaries of the firm: hence the contractual perspective which has been taken at first, using the tools of transaction cost theory. Such a perspective is particularly relevant to addressing the integration strategies that may support competitive advantage, but it does not deal with the constitution and the extension of the latter. The above review (Section 2) has both confirmed the relevance of the transaction cost analysis and shown that it is insufficient; a complete understanding of alliances requires an exploration of firms' strategic needs. This change in the focus of analysis proves particularly significant with respect to the theory of the firm.

3.1.1. From the market perspective to the resource perspective

The fact that strategic considerations are necessary to understand alliances is fundamental as it points to the importance of adopting a resource perspective of the firm. Indeed, "the basic idea behind strategic management is that a firm needs to match its capabilities to its ever-changing environment" (Teece, 1984, p. 87). Strategic management thus aims at establishing coherent relationships between the analysis of the competitive environment and the productive capabilities of the firm. In so doing, strategic management relates the market perspective to the resource perspective of the firm.

The structure–conduct–performance paradigm, as an extension of the analysis of perfect competition, exemplifies the market perspective, where firms' behavior is determined by the characteristics of the market in which they compete. However, thorough analyses of competitive processes and sectoral studies have shown that the conduct of firms cannot be explained by market structures alone. The concept of barriers to entry has been crucial in this demonstration. Barriers to entry are one of the main types of structures in the S–C–P paradigm, but a thorough examination leads to considering them not so much as market structures, but rather as market features on which firms can act. Product differentiation is one of the classical types of barriers to entry. Product differentiation is certainly more or less feasible according to markets, but firms can also act upon the degree of differentiation through marketing strategies.[28] Absolute advantage in favor of incumbent firms is also a category of barriers to entry on which firms can act, by innovating in particular. An even stronger case in favor of manipulation by incumbent

[28] They can propose more differentiated products and alter buyers' perception of the degree of differentiation.

firms can be made with respect to mobility barriers between strategic groups within a given sector.[29]

In a strategic perspective, the influence of firms' conduct on sectoral structures is quite obvious. In his famous analysis of competitive strategies, Porter (1980) draws heavily on the S–C–P paradigm, but considers that firms should exploit market structures rather than comply with them. Typically, in his analysis, incumbent firms should take advantage of barriers to entry as they protect them from price competition. In such a perspective, the relevant question becomes that of the means which firms can use in order to erect or reinforce the different types of barriers which protect them. An exploration of this question shows that barriers to entry and barriers to mobility correspond to the control of rent-earning factors. Wernerfelt (1984) has argued that if these factors cannot be accessed without impediment, they should be considered as a 'resource position barrier.'[30]

The examination of alliances emphasizes the tight interaction between the market perspective and the resource perspective of the firm. Firms have widely resorted to cooperative agreements as a means to lower barriers to entry or mobility barriers. And certainly the objective of alliances was often to ease entry in a market by lowering the cost of the required resources. This approach can be considered all the more general if risk and time considerations are translated into costs. For example, in the case of innovation, the resource barrier is often considered to be the cost of R&D, and average cost is increased by risks of failure. Besides, first mover advantages imply that short innovation and development cycles are fundamental assets in the competitive game. As long as cooperations are used to circumvent barriers to entry, they tend to increase the degree of competition, but they may also be used to build or reinforce barriers. This could be the case in the field of standards or if firms resort to alliances in order to develop 'technological oligopolies.'[31]

3.1.2. Learning

The exploration of cooperative agreements shows that firms do not merely use them in order to exchange resources, but often to learn and enhance their competences. Why should they choose an alliance in such instances? The exploration of this question leads to a focus on the relationships between the technical operations of firms and their organizational structures.

When firms want to enhance their competences, they often have to incorporate tacit knowledge. "A firm's competence is a set of differentiated technological skills, complementary assets, and organizational routines and capacities that provide the basis for a firm's competitive capacities in one or more businesses" (Teece et al., 1994, p. 18). Competences thus include a good proportion of tacit and collective knowledge. They are progressively constructed by firms which then depend on them; in other words, competences are both constituents of competitiveness and constraints which may impede

[29] On the notion of mobility barriers, see Caves and Porter (1977).

[30] If the fact that someone controls a resource affects the costs and/or revenues of later acquirers adversely, the former enjoys the protection of a resource position barrier (Wernerfelt, 1984).

[31] On this notion, see Mytelka (1993), Chesnais (1994). The analysis of cartels during the interwar period suggests that cross-licensing resulted in strengthening barriers to entry into these 'clubs' (Cantwell and Barrera, 1996).

future changes.[32] From this perspective, competences may be considered as a specific kind of barriers to entry, representing sunk costs.[33] As a result, action on a firm's competences requires specific learning devices, which involve an organizational dimension and impose costs. For example, if a firm wants to transfer its work organization, it will have to choose a device through which a number of its employees, and possibly teams, are able to demonstrate and teach their know-how to the new unit. This may be done most efficiently within a firm, but can also be monitored in the context of a cooperative agreement.[34] When American carmakers wanted to learn better production techniques from their Japanese competitors, they could not just read a manual or take a licence and change a couple of isolated steps in their factories. They had to really understand a different approach and review their whole organization in order to implement it, which of course implied extensive training of the personnel (not only production workers). In such a sweeping process, American firms certainly benefitted from their relationships with their Japanese partners, whom they tried to use as 'teachers' in a way.[35]

The exploration of the means firms use in order to transfer different types of knowledge leads to an emphasis on the social nature of firms' competences, which is at the root of the evolutionary theory of the firm.[36] Penrose (1959) argued that the growth of the firm largely depends on the capabilities which belong to teams within the firm and the transfer of which to other units is time-consuming. From this perspective, evolutionary theorists have argued that the firm has to rely on its experience, for better and worse, to act upon its scope, efficiency and ability to innovate. This approach has two important consequences. Firstly, it implies that the firm progressively creates its own productive capabilities and, as a result, competencies are path-dependent. Secondly, internal competences do influence the boundaries of the firm through specific modes of management and transfer which they require.[37] This may be related to management studies which have explored the 'fit' between a firm's strategy, its competences and its organizational structures.[38] The rationale underlying ICI's spectacular de-merger which split the organization into two separate companies in 1992, for example, has been the

[32] On the theme of path-dependency in relation to firms' evolution and innovative capabilities, see in particular Penrose (1959), Nelson and Winter (1982), Teece et al. (1994), Nelson (1995).

[33] This qualification may be illustrated by the productivity dilemma facing firms as explained by Abernathy (1978).

[34] As explained in Section 2, in such cases the reason why markets are inefficient for this type of transfer is not opportunistic behavior by individuals but rather the very technical characteristics of what is to be transfered.

[35] The Big Three have had different types of alliances with Japanese carmakers, which have been more or less efficient with respect to the observation and absorption of Japanese production and product development methods. Moreover, American carmakers have not been equally good at learning (due to the specific history of each firm). On these questions, see Womack (1988), Sachwald (1994b).

[36] See the development of the notion of 'routines' as organizational competences (Nelson and Winter, 1982). This analysis of the collective characteristics of organizations may be compared with the discussion of bounded rationality by Simon (1965).

[37] Innovatory combinations and idiosyncratic activities may generate a need to integrate complementary activities because adequate goods or services actually do not exist in markets (Foss, 1993). Forward integration by large American firms at the end of the 19th century, as explored by Chandler (1977) may be taken as one illustration.

[38] See, for example, Miles and Snow (1984), Prahalad and Hamel (1990). Historical studies have also closely analysed this relationship (Chandler, 1962, Lazonik, 1991).

progressive erosion of this fit between two disparate sets of businesses and an integrated corporate structure (Owen and Harrison, 1995). A firm can successfully manage only a certain degree of diversity within its boundaries.[39]

Since the firm has only a limited ability to manage diversity, the degree to which it integrates a set of knowledge depends, not only on the complexity and codifiability characteristics of the latter,[40] but also on the internal capabilities of the firm. Path dependency in research and knowledge-building results in firm-idiosyncratic technological trajectories. The latter being relatively rigid, cooperative agreements with independent partners may represent an adequate solution to access new knowledge or know-how. Cooperation is more likely when the required competences result from a very distinct set of resources, such as a different science base or research tradition.[41] In such cases, integration is costly, time-consuming and risky, so that firms may choose cooperation as a more efficient solution. This is one reason why cooperation is quite common in the field of research. Innovation often requires pooling different sets of knowledge and having scientists from different horizons cooperating, but the integration solution is not necessary – and may even be counter-productive. However, successful cooperation requires specific investment from both partners in order to be able to learn from one another.[42] Beyond the case of isolated firms, numerous studies of innovation processes have emphasized the role of information exchanges and tacit knowledge transfers in the emergence of national systems of innovation which include various types of cooperative agreements and networks.[43]

The analysis of cooperative agreements points to the need to combine an organizational and a technical perspective to fully understand firms' decisions to enter into alliances. This conclusion certainly fits in with a wider endeavour to combine transaction cost and evolutionary insights. Foss (1994) has argued in favor of such an approach in order to arrive at a 'dynamic neo-institutionalism.' The objective is to deal with the different components of change in firms' activities and structures. The main interest of such an approach, as compared to the contractual perspectives, is to re-integrate the technological and organizational aspects of the productive process.

3.2. *Multinationals are firms*

Theoretical analysis of multinational firms tends to be influenced by the dominant type of foreign production researchers observe. It has been the case with large American FDI flows serving as a model for the product cycle theory and with Japanese FDI in Asia which has been the empirical basis of the 'Japanese model' of foreign production. Since

[39] Teece et al. (1994) discuss this ideas from the symmetrical point of view of coherence, which they propose to measure with indicators of the degree of relatedness between the activities of firms.

[40] As argued by Kogut and Zander (1993) (see in Section 2).

[41] Richardson (1972) contrasted "complementary activities", which required different resources and competences, with "similar activities"

[42] Cantwell and Barrera (1996) suggest that during the interwar period, cooperation in technological development only influenced firms' technological trajectories when it involved 'cooperative learning' and not merely knowledge exchanges through cross-licensing.

[43] Agreements here are very diversified and include academic institutions which cooperate together and with firms.

the 1980s, empirical observations emphasize the process of globalization, with a growing number of multinational enterprises (MNEs) which increasingly tend to organize their value chain on a global scale.[44] As a result, MNEs have become both more common and more complex.

In the context of globalization, international cooperative agreements have influenced studies on MNEs through two channels. From an empirical point of view, transnational cooperative agreements have fully contributed to the process of globalization. From a theoretical point of view, cooperative solutions have played a similar role vis-à-vis multinationals as vis-à-vis firms in general, along the lines explored above. These empirical and theoretical aspects actually interact, in particular, because international production operations have become so humdrum and integrated within global networks.

3.2.1. International cooperative agreements as a factor of globalization

Until the beginning of the 1980s, the MNE was considered as a self-contained and internally controlled administrative system, with a globally optimizing parent supervising a constellation of controlled or fully owned foreign affiliates. The observation of international cooperative agreements has contributed to the growing recognition that multinational companies have become more diverse in their modes of operation. Cooperative agreements have been instrumental to both the extension and the diversification of international production. As such they have been considered as one of the distinctive characteristics of globalization.[45]

Cooperative agreements have been used by a number of firms to reduce risks and costs of going abroad. An alliance with a local firm constitutes an asset to get acquainted with business practices abroad. This may be especially useful with respect to distribution channels and human resource management. As recalled above (Section 2), Japanese firms resorted to alliances to reduce risks and speed up their internationalization process at the beginning of the 1980s.

MNEs have also resorted to cooperative solutions to ease the diversification of their foreign operations. As part of the globalization process, firms try to take full advantage of their global reach, both on the market side and on the resource side. They redeploy their assets and reorganize their operations on a global basis,[46] through both internal restructuring and external growth operations. Depending on circumstances and transaction costs, the latter may take the form of cooperative agreements. This is the case in the field of R&D when firms want to tap specific competences abroad, which may be located in firms or in public institutions. More generally, international cooperative agreements are an organizational tool used by firms to set up their international networks, which require varied types of partners. The internationalization of Japanese companies, the efficiency of which depends on tight and demanding relationships with their suppliers, has induced the latter to follow their clients abroad. The firms being thus co-transplanted

---

[44] These are the aspects of globalization which relate to MNEs and not a description of the whole process.

[45] By authors trying to pin down the specific characteristics of globalization as compared to internationalization, which has been developing since the end of the 19th century. See in particular Dicken (1992), Chesnais (1994), Sachwald (1994b).

[46] This process is being examined in detail by business management studies and the objective here is not to describe these efforts; see, for example, Barlett and Ghoshal (1989), Kogut (1990).

have chosen to ally with local partners in a number of cases. The development of international sub-contracting has also been an aspect of the growing cooperative movement. Indeed, in a number of cases, subcontracting requires extensive communication between the two parties.

3.2.2. Cooperative agreements as an internalization mode across borders

From a theoretical point of view, cooperative agreements have contributed to shifting the focus of analysis from the decision to locate abroad to the organization of mature transnational firms. What is remarkable in this evolution is the fact that MNEs appear much more as one category of firms, which can be studied within the general theory of the firm.

The evolution of the analysis of MNEs and cooperative agreements has been quite similar. At first, both were considered as new and rather paradoxical phenomena, given the hypothesis of efficient markets. In both cases, market power and efficiency considerations have been proposed as competing or complementary explanations.[47] For multinationals, the eclectic paradigm[48] constitutes a synthesis which takes into account the firm's competitive advantage, transaction costs and the characteristics of the host country. In this perspective, 'traditional' MNEs represent an internalization of markets aimed at extending the exploitation of a competitive advantage abroad.

Cooperative agreements can be integrated into the transaction cost theory of MNEs and into the OLI paradigm as partial internalization solutions which constitute alternative channels for the exploitation of firm-specific assets (Hennart, 1991, Dunning, 1993). However, the above analysis of alliances has shown that firms often resort to cooperative solutions in order to access complementary resources, while the OLI paradigm tends to focus on the market seeking a motive for going abroad since ownership advantages are considered to be the roots of the international scope of MNEs. This distinction becomes quite important when firms tend to elaborate global strategies and to progressively create adequate organizational structures and networks. In such a context, each decision to produce abroad does not result from some pre-existent ownership advantage, but may be explained by a specific weakness. Besides, the process of globalization aims at creating or reinforcing competitive advantages through an efficient world organization.[49]

As in the more general case which has been explored above, the analysis of cooperative agreements thus emphasizes the relationship between firms' strategies, their resources and their organization. In such a perspective, the observation of cooperative agreements certainly suggests that MNEs can be analyzed within the general theory of the firm. This may be a way to deal with some of the 'challenges for the theory of MNEs' which Dunning has pointed out. In particular since he considers that "basically... the challenge to scholars is to switch their mode of thinking away from explaining individual cross-

---

[47] The case of cooperative agreements has been examined above; for multinationals on this theme, see Cantwell (1991), Yamin (1991).

[48] Or OLI (Ownership Location Internalization).

[49] Within the ownership advantages, Dunning has progressively expanded the category of those which are based on the capability of the firm to organize internationally (compare, e.g. Dunning, 1981, 1993). This change of focus corresponds to the evolution of MNEs, but is difficult to integrate into the original OLI analysis.

border transactions or acts of foreign production to explaining a system of transactions or acts of foreign production" (Dunning, 1993, p. 93).

3.2.3. Building competitive advantages

MNEs resort to cooperative agreements as one of the means to implement their strategies, along with other types of investments, including acquisitions of foreign companies. Hence, analyses which focus on the study of cooperative agreements are led to deal with the evolution of the sectoral competitive game. From this perspective, MNEs' decisions to produce abroad should be integrated into the sectoral rivalry process and not be considered for each firm in isolation.[50]

The analysis of the internationalization of Japanese carmakers in the 1980s shows the relevance of sectoral analysis, which relates structural impediment to market entry, firms' competitive advantage and their strategy. Each decision to go abroad by a Japanese carmaker may be explained within the OLI paradigm, but sectoral analysis constitutes an important complement by exploring competitive positions.[51] In this case, it shows why and how protectionist pressures have mounted against Japanese imports in North America and Europe. It also enables an understanding of the organizational nature of Japanese carmakers' competitive advantage, which has influenced the means they have chosen once they have been forced to speed up their internationalization. The methods of production which Japanese carmakers had developed relied on specific know-how and routines. The latter being difficult to transfer, Japanese firms had first been prone to set up greenfield sites as a first choice way to produce in the US. Cooperative agreements were used as complements to speed up learning on local practices and reduce risks.[52] The example of Japanese carmakers illustrates the need to integrate market-strategy and resource-competence perspectives. As it has been argued above for firms in general, such a perspective requires the acquisition of new competences, and thus the importance of taking into account the issue of learning.

In the case of the MNE, the issue of learning has been strongly related to the exploration of international technology transfers. Kogut and Zander (1993) have argued in favor of an evolutionary theory of the MNE based on a thorough analysis of knowledge transfers and integration into the firm. Their analysis constitutes an extension of the evolutionary theory of the firm to MNEs. Indeed, the cornerstone of their approach is the treatment of the firm as a social community which builds on its competitive advantage from its productive knowledge – productive knowledge being largely idiosyncratic, the authors consider that firms will internalize its transfer across borders,[53] and become multinational. Such an approach, focusing on the process of innovation itself, has actually led to a shift in the analysis away from international transfers towards the ongoing process of creation within the firm (Cantwell, 1995). As in other areas of management, multinationals have progressively tried to take advantage of their international scope to

[50] Cantwell (1989, 1991) has thus argued in favour of what he labelled 'competitive international industry approaches' to MNEs.

[51] For an application in different industries, see Sachwald (1995).

[52] See above.

[53] On this argument, see Section 1; Section 2 suggest that partial internalization may be a solution in certain cases.

source knowledge from different countries. In doing so, they have increased the complexity of the management of the interactions between knowledge transfers and internal learning throughout the organization and not only among laboratories.[54]

The sectoral and evolutionary approaches on MNEs are complementary, the first one focusing on the process of competition and the second one on the creation and international transfer of knowledge within the firm. The main interest of these approaches is that they adopt a dynamic perspective. Ownership advantages as described by the OLI paradigm result from the firm's specific set of competences. In other words, and in a strategic management perspective, they represent competitive advantages. This is not a mere vocabulary matter, since competitive advantages are built by the firm from its initial resources and given a definite strategy. Besides, as it has been argued above, a firm may need complementary resources to implement its strategy. In this case, internationalization is motivated by a missing resource rather than by a competitive advantage.

Focusing on strategy and change suggests that the theory of the multinational company should be merged into the theory of the firm. In this perspective the focus of the analysis would shift towards the relationships between the firm's competitive position, its competences and its transnational scope. Since the firm is considered as a learning community, such an approach constitutes an adequate framework to analyze the relationships between the MNE and both its country of origin and its various foreign locations. In this respect, it would seem particularly fruitful to build both on the evolutionary theory of the firm and on the notion that countries develop coherent national systems of production and innovation.

The evolutionary perspective shows that change is necessarily progressive and costly. As a result, the process of change is hindered by various sunk costs, which refer to both equipment and competences. This type of approach seems appropriate to discuss such matters as the build up of competitiveness and the relationships between regionalization and globalization in firms' strategies. For example, a firm, the competitive advantage of which is tightly dependent on a specific national production system, will face high risks and costs when it internationalizes.[55] This does not mean that it will refuse to internationalize or fail, but rather that it will have to devise specific means to do it successfully. Actually, this approach shows that the process of change is similar, whether it is motivated by the evolution of the competitive game within a national system or by adaptation to a foreign environment.[56]

[54] On this subject, see Barlett and Ghoshal (1989), Cusumano and Elenkov (1994).

[55] As a consequence, Ruigrok and van Tulder (1995) suggest that in such cases, regionalization may well be a first choice strategy, as opposed to globalization.

[56] The relationships between firms and national systems have been explored through studies on competitiveness and innovation. Recent researches focus more precisely on developing a theory of the firm which would explicitly establish relationships with its wider macroeconomic or structural environment (see work done within the research network GERPISA-Groupe d'Etude et de Recherche Permanent sur l'Industrie et les Salariés de l'Automobile, and Coriat and Weinstein, 1995).

4. Conclusion: Cooperative agreements as accelerators of change

This paper has suggested that the analysis of cooperative agreements has contributed to the evolution of the theory of the firm by focusing on the adaptation of firms' resources and capabilities to change. At first, cooperative agreements have been studied as specific types of transactions and have thus promoted contractual and organizational perspectives. The resource perspective has been promoted by a more complete approach which included an analysis of the reasons why firms need to look for external complementary resources. The evolutionary theory of the firm as well as management studies have then contributed to analyzing the relationship between the competences of a firm, its strategy and its organizational machinery.

The above examination of the empirical and theoretical literature on cooperative agreements leads to an emphasis on their role as accelerators of change within firms' strategies. I want to suggest that, fundamentally, it is through this theme of change that cooperative agreements have influenced the theory of the firm. Penrose (1959) explained long ago that, as she was interested in the process of growth, she could not rely on the theory of the firm which focuses on market equilibrium. A similar argument can be made to explain the development of evolutionary economics in general, which studies the processes of change, taking into consideration both the favorable and the resistance factors. More precisely, the evolutionary perspective shows that firms, as social organizations, are rigid because their effectiveness depends upon learning processes. A firm is not a nexus of contracts or a set of bilateral relationships aimed at optimizing the incentive structure, but a productive organization where collective processes take place.

Since the 1980s, the competitive game has evolved rapidly in a number of sectors, due in particular to the interaction between globalization and innovation tendencies. Cooperative agreements have been one of the means to which firms have resorted to adapt and take advantage of these rapid changes. So, cooperative agreements have contributed to the evolution of the theory of the firm by pointing to the question of change, which ranks so high on firms' agendas. The perceived acceleration of change has quite logically promoted the development of theories which focus on the mechanisms which enable or hinder evolution within firms, as opposed to the contractual approach in particular.[57] Indeed, this paper suggests that the most important motivation underlying the wave of cooperative agreements in the 1980s is not an evolution of transaction costs (that would have become higher or lower for a given operation), but rather the rapid changes required to implement firms' strategies.[58] The main challenge facing firms has often been to increase the diversity of their activities without threatening their internal coherence.

This paper has emphasized the role of cooperative agreements as a way of jumping over barriers to entry, through the provision of complementary resources and learning opportunities. Such a perspective fits well into a more general analysis of the obstacles to change within firms, which should consider the different sources of sunk costs. The

---

[57] As Cantwell (1991) rightly puts it, "...the choice of theoretical issues on which to concentrate is not arbitrary but reflects what each camp believes to be the most important historical driving forces underlying the growth of the modern firm." (p.26)

[58] This does not mean that transaction cost considerations are not relevant; the paper has showed that they have to be considered once the objective of a transaction has been decided upon.

analysis of cooperative agreements has shown that this 'internal' perspective, represented by the evolutionary approach to the firm, has to be combined with an examination of the evolution of the competitive environment. More broadly, the objective would be to tackle the issue of the co-evolution of the environment of the firm and of its internal competences.[59] This type of approach seems well suited to understand firms' evolution, as well as the relationship between firms' competitiveness and their national environment. Here again, the observation that cooperative agreements have been used to speed up changes is relevant. Indeed, public authorities have considered cooperation as one of the means to transfer technologies between countries and to alter some characteristics of national systems.[60]

This paper has also argued that the combination of an evolutionary perspective with an analysis of the competitive game should be particularly fruitful to deal with multinational enterprises. Such a theoretical unification seems particularly relevant since MNEs have expanded so much, now being quite commonplace, but also, in a number of very complex cases involving intricate global networks. Consequently, the most relevant question with respect to international production is not that of its origins, but rather the driving forces underlying its expansion and its management.

The combination of economic and management perspectives has been crucial in analyzing cooperative agreements and in reaching the present conclusions, both with respect to the theory of the firm and with respect to multinational firms. Such interdisciplinary approaches should be further explored. This paper also suggests that the analysis of change in the organization of firms should certainly keep drawing on insights from business history.

Acknowledgements

I want to thank Mo Yamin, whose comments as a discussant of a 1995 version have been very helpful in revising the paper. I am also thankful to John Cantwell and Françoise Nicolas for their support and advice. Comments from two anonymous referees have been useful in preparing this final version of the paper.

References

Abernathy, W., 1978. Productivity Dilemma: Roadblock to Innovation in the Automobile Industry, John Hopkins University Press, Baltimore, MD.

Barlett, C., Ghoshal, S., 1989. Managing Across borders. The Transnational Solution, Century Business, London.

Bleeke, J., Ernst, D., 1993. Collaborating to Compete, Using Strategic Alliances and Acquisitions in the Global Marketplace, Wiley, New York.

---

[59] The evolutionary theory tries to extend the relevant environment to take institutions into account; for two points of view, see Coriat and Weinstein (1995), Nelson (1995).

[60] Innovation policies in Europe and in the United States have supported various cooperative research schemes and network building (involving both firms and academic institutions) (Mytelka, 1991, Sachwald, 1994a, 1994b).

Bresnahan, T., Salop, S., 1986. Quantifying the competitive effects of joint ventures, International Journal of Economic Organization, 4.

Brousseau, E., 1993. L'économie des Contrats, PUF, Paris.

Buckley, P., Casson, M., 1988. A theory of cooperation in international business. In: Contractor F., Lorange P. (Eds.), Cooperative Strategies in International Business, Lexington Books, New York, pp. 31–53.

Cantwell, J., 1989. Technological Innovation and Multinational Corporations, Basil Blackwell, Cambridge.

Cantwell, J., 1991. A survey of theory of international production. In: Pitelis C., Sugden R. (Eds.), The Nature of the Transnational Firm, Routledge, London, pp. 16–64.

Cantwell, J., 1995. Multinational corporations and innovatory activities: Towards a new, evolutionary approach. In: Molero, J. (Ed.), Technological Innovation, Multinational Corporations and New International Competitiveness, Harwood, London.

Cantwell, J., Barrera, P., 1996. The localisation of corporate technological trajectories in the interwar cartels: Co-operative learning versus an exchange of knowledge, Discussion Paper no. 340, University of Reading, Reading.

Caves, R., Porter, M., 1977. From entry barriers to mobility barriers. Quarterly Journal of Economics 9, 241–267.

Chandler, A., 1962. Strategy and Structure: The History of American Industrial Enterprise, MIT Press, Cambridge, MA.

Chandler, A., 1977. The Visible Hand. The Managerial Revolution in American Business, Harvard University Press, Cambridge, MA.

Chesnais, F., 1994. La Mondialisation du Capital, Syros, Paris.

Contractor, F., Lorange, P. (Eds.), 1988a. Cooperative Strategies in International Business, Lexington Books, New York.

Contractor, F., Lorange, P. 1988b. Why should firms cooperate? The strategy and economic basis for cooperative ventures. In: Contractor F., Lorange, P. (Eds.), Cooperative Strategies in International Business, Lexington Books, New York.

Coriat, B., Weinstein, O., 1995. Les Nouvelles Théories de L'entreprise, Le Livre de Poche, Paris.

Cusumano, M., Elenkov, D., 1994. Linking international technology transfer with strategy and management: A literature commentary, Research Policy, March, 195–215.

Delapierre, M., Millelli, C., 1994. Concurrence et transferts de technologie dans l'industrie informatique mondiale. In: Sachwald, F. (Ed.), Les Défis de la Mondialisation: Innovation et Concurrence, Masson, Paris.

Dicken, P., 1992. Global Shift. The Internationalization of Economic Activity, Paul Chapman, London.

Dunning, J., 1981. International Production and the Multinational Enterprise, George Allen and Unwin.

Dunning, J., 1993. Multinational Enterprises and the Global Economy, Addison-Wesley, Wokingham.

Foss, N., 1993. Theories of the firm: Contractual and competence perspectives. Journal of Evolutionary Economics 3, 127–144.

Foss, N., 1994. Why transaction cost economics needs evolutionary economics. Revue d'Economie Industrielle 2e(Trimestre), 7–26.

Gaffard, J-L., 1990. Economie Industrielle et de L'innovation, Dalloz, Paris.

Garette, B., Dussauge, P. 1995. Les Stratégies D'alliance, Les Editions d'Organization, Paris.

GEST, 1986. Grappes Technologiques. Les Nouvelles Stratégies D'entreprises, McGraw-Hill, New York.

Ghemawat, P., Porter M., Rawlinson, R., 1986. Patterns in international coalition activity. In: Porter, M. (Ed.), Competition in Global Industries, Harvard Business School Press, Boston.

Gomes-Casseres, B., 1993. Computers: Alliances and industry evolution. In: Yoffie, D. (Ed.), Beyond Free Trade, Harvard Business School Press, Boston.

Hagedoorn, J., Narula, R., 1994. Choosing modes of governance for strategic technology partnering: International and sectoral differences, MERIT Working Papers Series 2/94-028, Maastricht, MERIT.

Hagedoorn, J., Schakenraad, J., 1993. Strategic technology partnering and international corporate strategies. In: Hughes, K. (Ed.), European Competitiveness, Cambridge University Press, Cambridge, pp. 60–102.

Hagedoorn, J., Schakenraad, J., 1991. The role of interfirm cooperation agreements in the globalization of economy and technology, MONITOR, vol. 8, Commission des Communautés européennes, Bruxelles.

Hennart, J-F., 1991. The transaction cost theory of the multinational enterprise. In: Pitelis C., Sugden, R. (Eds.), The Nature of the Transnational Firm, Routledge, London, pp. 81–116.

Hennart, J-F., 1988. A transaction cost theory of joint ventures, Strategic Management Journal, July–August 361–374.

Hladik, K., 1985. International Joint Ventures: An Economic Analysis Of U.S. Foreign Business Partnerships, Lexington Books, New York.

Hobday, M., 1994. The semiconductor industry. In: Sachwald, F. (Ed.), European Integration and Competitiveness, Edward Elgar, Aldershot, pp. 145–193.

Hergert, M., Morris, D., 1988. Trends in International Collaborative Venture Agreements, Contractor, F., Lorange, P. (Eds.), Cooperative Strategies in International Business, Lexington Books, New York.

Jacquemin, A., Lammerant, M., Spinoit, B., 1986. Compétition européenne et coopération entre entreprises en matière de recherche-développement, Commission des Communautés Européennes, Luxembourg.

Kay, N., 1991. Industrial collaborative activity and the completion of the single market, Journal of Common Market Studies, June.

Kenney, M., Florida, R., 1993. Beyond Mass Production, Oxford University Press, Oxford.

Kogut, B., 1988. A study of the life cycle of joint ventures. In: Contractor F., Lorange, P. (Eds.), Cooperative Strategies in International Business, Lexington Books, New York pp. 169–185.

Kogut, B., 1990. International sequential advantages and network flexibility. In: Bartlett, C., Doz Y., Hedlund, G. (Eds.), Managing the Global Firm, Routledge, London pp. 47–68.

Kogut, B., Zander, U., 1993. Knowledge of the firm and the evolutionary theory of the multinational corporation, Journal of International Business Studies, fourth quarter.

Kogut, B., Zander, U., 1992. The knowledge of the firm in the choice of the mode of technology transaction. Proceedings of the European International Business Association, 18th annual meeting, University of Reading, 13–15 December.

Lazonik, W., 1991. Business Organization and the Myth of the Market Economy, Cambridge University Press, Cambridge.

Lorange, P., Roos, J., 1992. Strategic Alliances, Basil Blackwell, Cambridge.

Mair, A., 1994. Honda's Global Local Corporation, St Martin's Press, New York.

Marity P., Smiley, R., 1983. Cooperative agreements and the organization of industry, Journal of Industrial Economics, October, 437–451.

Miles, R., Snow, C., 1984. Fit, failure and the Hall of Fame. California Management Review XXVI(3), 10–28.

Mowery, D.C. (Ed.), 1988. International Collaborative Ventures in U.S. Manufacturing, Ballinger, Cambridge.

Mytelka, L. (Ed.), 1991. Strategic Partnership. States, Firms and International Competition, Pinter London.

Mytelka, L., 1993. Dancing with wolves: Global oligopolies and strategic partnerships. In: Hagedoorn, J. (Ed.), Technical Change and the World Economy, Edward Elgar, Aldershot, pp. 182–204.

Nelson, R., 1995. Recent evolutionary theorizing about economic change, Journal of Economic Literature, March, 48–90.

Nelson, R., Winter, S., 1982. An Evolutionary Theory of Economic Change, Harvard University Press, Cambridge, MA.

Owen, G., Harrison, T., 1995. Why ICI chose to demerge, Harvard Business Review, March–April, 133–142.

Parkhe, A., 1991. Interfirm diversity, organizational learning, and longevity in global strategic alliances, Journal of International Business Studies, Fourth Quarter, 579–601.

Penrose, E., 1959. The Theory of the Growth of the Firm, Basil Blackwell, Oxford.

Pisano, G., Shan W., Teece, D., 1988. Joint ventures and collaboration in the biotechnology industry. In: Mowery, D. (Ed.), International Collaborative Ventures in US Manufacturing, Ballinger, Cambridge, pp. 183–222.

Porter, M., 1980. Competitive Strategy, The Free Press, New York.

Porter, M., Fuller, M., 1986. Coalitions in global strategies, Porter, M. (Ed.), Competition in Global Industries, Harvard Business School Press, Boston, pp. 325–343.

Prahalad, C., Hamel, G., 1990. The core competence of the corporation, Harvard Business Review, May–June, 79–91.

Reich, R., Mankin, E., 1986. Joint ventures with Japan give away our future, Harvard Business Review, April, 78–86.

Richardson, G., 1972. The organization of industry. Economic Journal 9, 883–896.
Ruigrok, W., van Tulder, R., 1995. The Logic of International Restructuring, Routledge, London.
Rullière, J-L., Torre, A., 1995. Les formes de la coopération inter-entreprises, Revue d'Economie Industrielle, Special issue, Economie industrielle: Développements Récents, 215–246.
Sachwald, F., 1990. Les accords dans l'industrie automobile. Une analyse en termes de coûts de transaction. Economie Prospective Internationale, CEPII/La Documentation Française, 1er trim, 45–65.
Sachwald, F., 1992. Cooperative agreements in the world automobile industry, EIBA, Proceeding of the 18th Annual conference, University of Reading, 13–15 December.
Sachwald, F. (Ed.), 1994a. European Integration and Competitiveness, Edward Elgar, Aldershot.
Sachwald, F. (Ed.), 1994b. Les Défis de la Mondialisation. Innovation et concurrence. Masson, Paris.
Sachwald, F. (Ed.), 1995. Japanese Firms in Europe. Harwood, Luxembourg.
Sharp, M., 1994. Transferts de technologie et politique de l'innovation: le cas des biotechnologies. In: Sachwald, F. (Ed.), Les Défis de la Mondialisation, Masson, Paris, pp. 155–212.
Simon, H., 1965. Administrative Behavior, 2nd edn., Free Press, New York.
Steinmueller, W., 1988. International joint ventures in the integrated circuit industry. In: Mowery, D. (Ed.), International Collaborative Ventures in U.S. Manufacturing, Ballinger, Cambridge.
Stuckey J., 1983. Vertical Integration and Joint Ventures in the Aluminum Industry, Harvard University Press, Cambridge, MA.
Teece, D., 1984, Economic analysis of business strategy, California Management Review, Spring, 87–110.
Teece, D., 1982. Towards and economic theory of the multiproduct firm. Journal of Economic Behavior and Organization 3, 39–63.
Teece, D., 1980. Economies of scope and the scope of the firm. Journal of Economic Behavior and Organization 3, 223–247.
Teece, D., Rumelt, R., Dosi G., Winter, S., 1994. Understanding corporate coherence. Theory and evidence, Journal of Economic Behavior and Organization, 1–30.
Tirole, J., 1988. The Theory of Industrial Organization, The MIT Press, Cambridge, MA.
Wernerfelt, B., 1984. A resource-based view of the firm. Strategic Management Journal 5, 171–180.
Williamson, O.E., 1975. Markets and Hierarchies: Analysis and Antitrust Implications, The Free Press, New York.
Williamson, O.E., 1985. The Economic Institutions of Capitalism, The Free Press, New York.
Williamson, O.E., 1990. The firm as a nexus of treaties: An introduction. In: Aoki, M., Gustafsson B., Williamson, O.E. (Eds.), The Firm as a Nexus of Treaties, Sage London, 1–25 .
Womack, J., 1988. International joint ventures in motor vehicles. In: Mowery, D. (Ed.), International Collaborative Ventures in U.S. Manufacturing, Ballinger, Cambridge, pp. 301–347.
Yamin, M., 1991. A reassessment of Hymer's contribution to the theory of the transnational corporation. In: Pitelis C., Sugden, R. (Eds.), The Nature of the Transnational Firm, Routledge, London, 64–80.

Part III
The Co-evolution of FDI and Technological Development

[14]

Multinationals and Technology Development in Host LDCs

Sanjaya Lall
INSTITUTE OF ECONOMICS AND STATISTICS, OXFORD UNIVERSITY

I INTRODUCTION

The promotion of domestic technological capabilities is becoming a growing concern for the more industrialised of the developing countries (LDCs). Among the many issues that have arisen in this area, some of the most complex and controversial have concerned the proper balance and the nature of the interaction between foreign and indigenous technologies.[1] In an activity such as technological development, where dynamic learning effects and externalities are rife, many analysts have argued that there is a strong need for government intervention: untrammelled market forces by themselves would not lead to the optimal level of technological development in LDCs, and some deliberate exclusion of foreign technologies may be necessary to protect the learning process there.[2]

Multinational corporations (MNCs) from the developed world appear as an important topic in this debate. In many developing countries, MNCs are major producers of advanced industrial products and among the most important conduits for the transfer of modern technologies from abroad. They are often in the forefront of export activity and, in larger countries with relatively developed industrial structures, of local technological activity. It is the very fact of their technological strength, based upon 'frontier' innovation in the industrialised countries, and their competitive prowess in host LDCs which raises fears that they may stifle indigenous capabilities. In the language of the more polemical critics of MNCs, they act as 'agents of technological dependence' of host LDCs.

The case against MNCs in this guise is, however, neither coherent nor well-articulated. There are different version of the criticisms, some of which contradict the others. This chapter is an attempt to separate out some of the strands of the arguments advanced and to evaluate them against some (regrettably scanty) evidence. It will focus on the experience of India, where I have been conducting some research on the process of technological development. It may be said, with some justice, that India is an inappropriate example. It has severely restricted the inflow of foreign direct investment in the last two decades and has sought, behind formidable protective barriers and with widespread inefficiencies, to promote not just self-reliance in production but also self-reliance in technology.[3] Existing foreign investments in the country have been forced to dilute the extent of foreign ownership, and their growth and diversification in response to their competitive strengths in open markeets have been tightly controlled. Nevertheless, most of the world's leading MNCs still operate in India, and are particularly active in high-technology industries.[4] India's attempt to develop its technological capabilities itself makes the role of MNCs an interesting one. Thus, the use of Indian examples is still of relevance to this discussion.

A primary source of confusion in this area is the lack of a clear definition of what 'technology development' means. Section II of this Chapter discusses this problem. Another source of confusion is that the impact of MNCs on different levels of technology development can be assessed in several different contexts: technology development *within* the LDC affiliate of the MNC; technology development of the affiliate *relative to* similar local firms in the host country; and the *direct or indirect effects* of the activities of MNC affiliates upon technology development in related (vertically linked or competitive) firms in the host country. Sections III–V of the chapter deal with these questions in turn. The main conclusions are drawn in Section VI.

Two other introductory points: first, this chapter does not deal with such difficult issues as the 'appropriateness' of products or processes transferred to host countries by MNCs, and, second, it does not define MNC presence *per se* as a bad thing (because it 'denationalises' industry) or as a symptom of indigenous technological 'dependence'. On the first point, it is assumed that these are broader questions of distribution and factor-pricing policy which have to be taken as given, and that foreign and local firms in similar activities do not differ much in their behaviour once the larger parameters are set. On the second, it is assumed that the same technological activity undertaken in a host country yields the same benefit whether it is by a foreign affiliate or a local firm: benefits differ only if behaviour patterns are different.

2 WHAT IS 'TECHNOLOGY DEVELOPMENT'?

This chapter is concerned only with manufacturing industry. In industry, technology development can be defined to encompass an enormous variety of activities.[5] We can simplify these into four categories.

In the sense that technology is simply the knowledge of how to carry out manufacturing activity, even the setting up of a new assembly activity in a developing country can be said to transfer some new knowledge and so contribute to the technology of the host country. From this elementary stage, technology development can progress to the knowledge of operating imported technologies in increasingly sophisticated industries. The gaining of such operating knowledge – for which we use the convenient label of 'know-how' – will include not only the assimilation of imported techniques (which can itself be a lengthy and active learning process) but also quality control (which also involves active technical effort), improved plant layout and production practices, slight modifications to equipment and tooling, trouble-shooting, the use of different raw materials, and so on. It is well accepted now that the acquisition of know-how, even in the context of imported technologies, is a real and significant source of technological progress in LDCs.[6]

The next stage of technology development involves the understanding of the nature of the underlying process and product technologies, and leads to their substantial adaptation, improvement and even replacement by new processes or products. We may call this the development of 'know-why' capabilities. Such technology development arises partly as a natural extension and deepening of know-how capabilities, and partly as a result of conscious efforts to develop design, testing, pilot-plant and similar activities (the 'development' part of R & D in industrial enterprises).

Know-why development may be followed by applied research: the application of given scientific knowledge to the process of commercial innovation. Again, the line between the acquisition of know-why and its extension into genuine innovation is very unclear, but we mention it because of the possibility (real in many LDCs) that the understanding of why technologies work the way they do may not lead on to risky and costly efforts to undertake innovation. The final stage of technology development within industry comprises the ability to undertake basic scientific research, pushing back the frontiers of knowledge without regard to specific commercial application.

The contribution of MNCs to technology development in host LDCs can take place at any or all of these levels. A positive effect at one level may coexist with a negative effect at another. Moreover, at any given level, a positive contribution to technology development within the MNC affiliate may be associated with a negative effect on such development in other enterprises in the host country. The range of potential permutations is enormous, and we clearly cannot analyse them in any detail here. Certainly the lack of sufficient empirical evidence limits the scope for meaningful discussion. For this reason, we will simplify further and refer in the following analysis only to know-how and know-why development (the latter comprising all the advanced stages of technological activity from design and development onwards).

A very important point to bear in mind about the different levels of technology development in LDCs concerns its *net social value*. All technology development entails a cost, and each successive stage probably entails higher costs. The development of indigenous technological capability is not an end in itself: technology is simply one input into the process of industrial production, and the efficiency of the production process which results from the application of technology is the 'bottom line' which the country is concerned with. True, technology development has various cumulative dynamic and external effects, but the social cost entailed must always be compared with the cost of importing it from abroad.

These points are relevant here because India seems to have pushed technological self-reliance almost as hard as inward-looking industrialisation. Of all the industrially advanced LDCs, it is the one with the least relative reliance on foreign technology in any form (MNCs, capital goods imports, licensing and foreign engineering, for example) in the past two decades. As I have argued elsewhere (my paper in Fransman and King (1984)), this policy has been pushed to counter-productive extremes, reinforcing the inefficiencies engendered by industrial policy. Not only is the extent of local technology generation by MNC affiliates and local firms highly conditioned by this framework, the broader economic implications of the efforts also have to be borne in mind.

Let us turn now to the three forms in which the effects of MNCs on technology development are evaluated.

3 TECHNOLOGY DEVELOPMENT WITHIN MNC AFFILIATES

There is a widespread, but mistaken, impression that technologies developed by MNCs in their home countries are transferred costlessly and without adaptation to affiliates in LDCs. Teece (1976) shows in great detail that the transfer of any technology to a new location – developed or less-developed – entails substantial transfer and assimilation costs. These costs arise, of course, whether the recipient is an affiliate or an independent firm, and range from pre-transfer engineering, equipment specification and adaptive R & D (most of these are undertaken in the technology exporting country) to detailed engineering, local design adaptation, start-up, troubleshooting and training (which are mostly undertaken in the recipient country). After a plant is running, subsequent technology creation in the form of know-how (production engineering and minor adaptation) occurs over time, reaching a peak soon after technology introduction and tailing off gradually until a new technology is transferred or new problems encountered (Davies, 1979).

In the know-how phase, there seems no a priori reasons to differentiate between the performance of foreign and local firms. As noted earlier, know-how accumulates as a result of problem-solving, and foreign and local firms encounter identical problems in given locations. Katz's (1978) studies show that foreign affiliates undertake minor innovation-type engineering just as successfully as do local firms, while in terms of adaptation to local factor prices there is no sustained difference between the two groups (Lall, 1978). It is sometimes suggested that MNCs may perform better because of their higher quality technical manpower, or because of their ready access to the accumulated know-how of the parent company. In India, there is little evidence to support or refute this at the level of production engineering.

It is when we come to the know-why and more advanced research phases that the literature leads us to expect MNCs to be relatively backward in local technology development. The economics of R & D location (see Lall, 1979, and Hirschey and Caves, 1981) dictate that the bulk of basic design and development work by MNCs be highly centralised; the need for high level scientific manpower of several kinds and a highly developed technological infrastructure dictates that this centralisation occur in the developed world; and the need for close interaction, especially in engineering industries, between innovation, management and production entails that R & D be placed near the head office or the largest markets.

This does not mean that MNC affiliates do not perform basic design or development work in host LDCs. There has been a growing tendency among MNCs (if US data can be extrapolated to other home countries) to place R & D facilities in *some* LDCs. These are the larger, industrially more advanced and technically better endowed host countries. There are several sound economic reasons why R & D facilities are set up there: first, the transfer of increasingly more complex know-how is itself facilitated by the existence of local know-why capabilities (Teece, 1976); second, in certain products the needs of the local market differ from those of developed countries (for example, motor vehicles, food products, toiletries) and large local markets justify the investment in R & D facilities; third, the use of some LDC facilities for export to the rest of the Third World leads both to long production runs and to the need for product adaptation, strengthening the previous tendency (for example, Volkswagen in Brazil); fourth, the use of local raw materials or the need to test certain products locally induces local investigative R & D activity (for example, pharmaceuticals and food products); and, finally, in exceptional cases LDC R & D facilities can be utilised for the MNCs' global needs of innovation (but this generally occurs *after* local R & D is well established).

Thus, the facts that a relatively small proportion of MNCs' total R & D spending is allocated to the Third World and that most LDCs host no R & D by MNCs are quite compatible with the fact that large countries like Brazil or India today support very active R & D by MNCs. In India there are some 500 officially recognised R & D establishments in private manufacturing industry. Firms with foreign equity (including those which have diluted the foreign share to 40 per cent because of government policy) account for over half these firms; in fact, nearly all foreign affiliates of any size have set up recognised R & D facilities (the next section will present an explicit comparison of R & D by local and foreign firms). Some of these, like the one set up by the Unilever affiliate, have gone in for basic scientific research, contributed major product innovations and led to the utilisation and export of formerly unused local materials.[7] A number of affiliates earn royalties and technical fees from abroad by the export of their know-how. It is often suggested that their access to the parent company enables them to utilise 'frontier' technology to enhance the value of their Indian R & D.

All this suggests that, *up to a certain level*, MNCs do not lag in innovative activity in large LDCs with a pool of skilled manpower. The question then arises: is this level the socially optimal one? Clearly, local R & D in LDCs is much smaller, more application-orientated and less 'deep' than R & D in the home countries of MNCs.[8] To a large extent this is inevitable, given the relative sizes of operations, the level of technological sophistication and the competitive needs of the two locations. However, many host countries (including relatively advanced ones such as Canada) feel that MNC R & D is 'truncated' to an undesirable extent, with the truly innovative functions kept at home even when they could be efficiently located abroad.[9] The allegation is difficult to prove or disprove empirically: the social and private perceptions of the net long-term benefits of establishing R & D in a newly industrialising country can be widely different, depending upon the relative assessments of research capabilities, minimum scale requirements of research, dynamic learning effects and externalities. In their survey of the literature, Kamien and Schwartz (1982) demonstrate clearly how fragile is the empirical basis for generalising about the determinants of innovation even in the highly industrialised countries. In LDCs, where much less is known on the economics of industrial progress, the area of ignorance is far larger. Non-quantifiable and dynamic factors further compound the analytical difficulties.

Since firm conclusions are not feasible, let me round off this section with my own impressions on technology development within MNCs. First, to the extent that MNCs are the most efficient international transmitters of industrial know-how, their presence in LDCs promotes technology development *at the production level* in those sectors where the speed of technical change and the complexity of technology makes other methods of transfer relatively inefficient or costly.[10] By implication, their direct participation may be less necessary in activities with 'lower' or more stable technologies. It may be argued that newly industrialising countries have their comparative advantage in acquiring know-how, rather than spreading their limited technological resources thinly over 'deeper' technical learning. Certainly the most successful NICs, the Gang of Four in East Asia, have benefited enormously from this strategy of efficiently utilising imported technologies without forcing the pace of know-why accumulation. Judged by this criterion, the extent of technology development by MNCs is probably optimal.

Second, to the extent that countries wish to go beyond the know-how stage into know-why and basic research, and have the wherewithal to do so efficiently, MNCs may go to the desired extent only in a few instances. Some government intervention may then be necessary to bring private costs and benefits of local R & D into line with social ones. Two outcomes are possible: the MNC is induced to launch local R & D to the extent desired (and is able to do it very efficiently because of the backing of the parent company), or it refuses to undertake 'deeper' R & D in the given host country (or conducts it inefficiently), and the latter's objectives are only achieved by setting up independent R & D facilities. I suspect the actual outcome lies somewhere between these extremes. A case may then be made that in some cases MNCs do retard 'deep' technology learning. But it is a qualified case, and not as obvious as may appear at first sight.

4 TECHNOLOGY DEVELOPMENT IN MNCs RELATIVE TO LOCAL FIRMS

An earlier survey of the literature (Lall, 1978) had suggested that there was no marked or consistent difference between MNCs and local firms so far as the adaptation of imported technologies was concerned. Given the specific needs of the host country, its technical capabilities and the inherent adaptability of the technologies in question, both sets of firms seemed to respond similarly to the economic environment. The available evidence on the relative efficiency of their responses is very mixed and of uneven quality, so we can deduce little from it.

What, then, of their relative propensities to undertake 'deeper' technological effort? Since no detailed empirical studies exist on the issue of successful innovation by foreign and local firms in LDCs, we may resort to two sets of rather crude indicators: exports of technology by foreign and local firms, and formal R & D spending by them.

Exports of technology have been used as a rough indicator of the level of technological competence reached by a developing country (Lall, 1982). The available evidence suggests that the bulk of technology export activity is undertaken by local firms in LDCs.[11] If this is an indicator of 'deeper' technological learning, we may infer that MNCs lag in the accumulation of know-why in these countries.

We must, however, note two important qualifications. First, many important forms of innovation do not show up in the foreign activities of industrial enterprises, and if they do, they tend to affect the export of products rather than the export of technologies. Secondly, MNC affiliates may transfer their technology abroad via the parent company network (for royalties or in exchange for other technologies[12]) rather than in contractual forms which are caught by the data. In the Indian case, we have already noted that MNC affiliates are actively selling technologies in return for royalties. The Unilever affiliate exchanges technical information freely with its related firms overseas, and also exports its expertise by posting Indian engineers in other LDCs. Many of the innovations made in India are utilised in affiliates in South East Asia and Brazil. Certainly its impressive export effort (Hindustan Lever is the largest single private sector exporter in India) is based partly on the exploitation of its local research effort. Moreover, given the very restricted scope permitted to MNCs in Indian industry, it would be premature to infer from technology export data that foreign affiliates are less capable of know-why accumulation than local firms.

The use of R & D data may appear to be more reliable in this context. There are, however, problems here also. A number of other factors which influence firm-level R & D performance (see Kamien and Schwartz, 1982) must be controlled before the influence of foreign ownership is assessed; R & D expenditures are an indicator of research input, not its output (innovation or learning); different firms may define it differently and the composition of R & D between basic research, applied research and imitation may differ;[13] and a comparison of firms in different industries may be misleading if the inherent research-intensity of the industry is not taken into account. Some data I have analysed on the R & D performance in India, which deal with some of these problems, may be instructive.[14]

These data pertain to 145 firms in only two sectors, engineering and chemicals, minimising the differences in 'technological opportunity' which crucially affect R & D performance. The 100 engineering firms are also assigned a dummy variable distinguishing those engaged in complex (machine building) activities from those in simpler (metal shaping) activities. The two sectors chosen account for 80 per cent of India's foreign collaborations, so far a fair representation of MNC activity there. A number of control variables are included besides the degree of foreign ownership. The shortage of space precludes a detailed analysis of their rationale and expected signs. Some of the results for each of the sectors – we did not aggregate them for this exercise because of the differences in their technological characteristics – are shown in the Appendix.

As far as the impact of foreign ownership (variable x_1) is concerned, the two sectors show different tendencies. Taking a large number of regressions (not shown in the Appendix table) into account, engineering has a positive and consistently significant sign, while chemicals has a negative and only occasionally significant sign.[15] The two sectors also show interesting differences in the impact on R & D of firm size, age, export performance and the proportion of highly paid employees: clearly, generalisations across different sectors cannot be drawn. Foreign ownership has a positive relationship with R & D in one major sector. *There is no support for the case that MNCs in general are less R & D-intensive than are local firms.*

In sum, these two indicators do not, in the final analysis, enable us to conclude that there are significant differences between the relative technological performance of foreign and local firms in India. Given the stringent controls on MNCs in India and the consequent low appropriability of the returns from their research, it is perhaps surprising that the evidence does not show a more negative result for foreign affiliates. A truer test would ideally cover a more open economy with a more competitive industrial structure: but we have to make do with the little information that we do possess.

5 EFFECTS OF MNCs ON RELATED FIRMS' TECHNOLOGY DEVELOPMENT

There are two sets of firms 'related' to MNCs: those that supply inputs to MNC affiliates and those that compete with them in their product markets.

As far as vertically linked firms are concerned, the limited Indian evidence suggests that MNCs are as active in transferring skills, know-how and product design to suppliers as are local firms.[16] The exact pattern of inter-firm linkages depends upon the technical characteristics of the products and the relative sizes of the firms concerned. The highly inward-looking Indian regime has induced all firms, regardless of ownership, to set up extensive local linkages: many of these linkages may, however, be socially inefficient and the resulting production facilities internationally uncompetitive. However, it is likely that the basic propensities to create linkages remains the same in more dynamic and efficient economies.

As far as competing firms are concerned, the debate is still conducted at a very general level. It is argued, on the one hand, that competition from technologically efficient MNCs induces local firms to improve their own technology,[17] and that the presence of MNCs leads to much more rapid diffusion of technology by imitation and 'contagion'.[18] On the other, it is argued that a strong foreign presence inhibits local firms from undertaking risky and costly research activity.[19] The two apparently contradictory views can be reconciled, of course: local firms may well be induced to upgrade their production technology by a greater reliance on licensing when faced with competition from MNCs operating with 'frontier' technologies, and, in the process, their independent know-why development may suffer. A strong MNC presence may, in other words, be associated both with better local know-how and 'shallower' local know-why (needless to say, this is on the assumption that the indigenous industrial sector survives MNC competition on the basis of licensed technology – the phenomenon of 'denationalisation' is assumed to be contained within socially acceptable limits).

We should, however, note some important qualifications to the argument that MNCs inhibit know-why development in competing firms. First, it assumes a specific relationship between licensing and in-house research, treating them as substitutes at 'deeper' levels of R & D; this may not be the case as firms grow more sophisticated and larger. Thus, licensing may become a *necessary input* into in-house technological development;[20] this is certainly the case in Japan, which for decades was the world's largest importer of licensed technology. If local enterprises in the NICs have reached this stage (and I believe many have), increased MNC competition will have net technological benefits. Second, we must be careful not to generalise to all industries: a number of basic sectors (textiles, steel, cement, machine tools for example) are not important stamping grounds for MNCs, and in these the inhibiting effects do not really arise. Third, the Indian case illustrates the dangers of closing the economy too much to foreign competition. Not only has industrial efficiency suffered and damaging technological lags developed (it is becoming increasingly apparent that independent R & D efforts in LDCs simply cannot keep up with world frontiers in many industries, even if they are mere imitators), but a great deal of technological effort has gone into socially unproductive uses such as finding high-cost local substitutes for imported materials and equipment. Finally, some advanced technologies may be available *only* via MNC entry, so the question of inhibition is redundant.

These qualifications still leave a grey area where genuine concern can be voiced about the effects of MNCs on firms in lesser industrialised countries. Moreover, to the extent that know-why development does require large, sheltered markets in theinitial, high-risk phases, a powerful foreign presence may increase the risk to unacceptable levels. The Japanese experience lies behind many arguments that local technology development can proceed very successfully without direct MNC participation. Unfortunately, it is not clear how far Jpan can be emulated by the LDCs of today; and, given Japan's human resources, its actual experience does not prove that it might not have done *even better* with a more liberal policy on MNC entry.

6 CONCLUSIONS

The analysis of the nature and determinants of technology development in LDCs is in its infancy. We know little about its different phases and about the true comparative advantage of LDCs in the scale of technology 'deepening', Not surprisingly, therefore, the relationship between MNCs and local technology development is fraught with confusion and needless polemic.

This chapter has sought to clarify some of the elementary concepts and separate out the different strands of the debate. It has distinguished between various sequential stages of technology development, concentrating on the basic categories of know-how and know-why acquisition. For each of these, it has discussed the impact of MNCs on technology development within the firm, in relation to local firms and in linkage or competition with local firms.

It would appear that the effect of MNCs on know-how development is positive in all its manifestations. Given their undoubted technological superiority in most areas of modern industry, and given also the inherent need to adapt technology to each particular host environment, MNCs contribute to know-how acquisition within their affiliates and, by linkages and competition, to related local firms. As far as know-why development is concerned, the picture is more mixed. MNCs have good economic reasons to keep basic innovative effort centralised in the developed world (though exceptions exist), and generally transfer less to LDCs than many host countries would like. Their presence may also inhibit local firms investing in their own know-why development.

These conclusions need two major qualifications. First, the comparative advantage of LDCs may in fact lie in know-how development, and the correct strategy for them may be to exploit efficiently imported know-why. If know-why develops naturally over time with increased sophistication of production, a reliance on MNCs (in sectors where they are predominant) may well be the best long-term policy. Second, even the development of know-why capabilities in competing local firms may be stimulated by MNC competition once these firms reach a certain size and competence. Thus, MNC entry would hasten the technological maturity of 'infant' industries which may otherwise simply frow into slothful, uncompetitive adolescents.

206

Appendix: Regression analysis (log linear) of determinants of R & D performance in India, 1978–79

	Industry and Equation			*Numbers*	
Independent variables	*Engineering* (n = 100)		*Chemicals* (n = 45)		
	1	2	3	4	5
x_1		0.095^b	-0.120^c	(0.070	
		(1.711)	(1.329)	(0.701)	
x_2	0.903^b		−0.017		0.287
	(2.661)		(0.047)		(0.761)
x_3	0.574	0.694^c			-1.810^b
	(1.164)	(1.359)			(2.456)
x_4	0.900	0.176^c			0.207
	(0.815)	(1.531)			(0.752)
x_5	0.369^a		0.116		
	(2.811)		(0.616)		
x_6	-0.238^b	-0.172^c	0.605^a	0.480^a	
	(2.298)	(1.549)	(3.879)	(3.250)	
x_7	1.322^c	1.596^b		−0.086	
	(1.628)	(1.853)		(0.424)	
x_8				0.222^c	0.300^b
				(1.650)	(2.231)
x_9	1.069^a	1.061^a			
	(4.487)	(4.133)			
x_{10}		−0.718			
		(1.202)			
R^2	0.414	0.356	0.268	0.279	0.208
F	8.79^a	5.95^a	3.168^b	3.407^b	2.821^b

Notes: t statistics in parentheses; significance levels: a = 99%; b = 95%; c = 90%.
Sources: Company balance sheets, Department of Science and Technology, Directorate General of Technical Development.

Symbols and definitions appear at the top of the opposite page

SYMBOLS AND DEFINITIONS

Dependent variable: R & D expenditures in 1978 as a percentage of sales in that year.
Independent variables:
x_1 share of foreign equity in total equity.
x_2 size of firm as measured by sales.
x_3 age of firm in India.
x_4 number of foreign licensing agreements.
x_5 royalties paid as a percentage of sales.
x_6 exports as a percentage of sales, a measure of competitiveness
x_7 percentage of total wages accounted for by highly-paid technical and managerial employees, a measure of top-level skills.
x_8 selling expenses as percentage of sales.
x_9 dummy variable for engineering firms, with 1 = firms engaged in machinery manufacture and other sophisticated activites and 0 = others.
x_{10} average wage level, an index of general production skills.

Notes

1. For concise statements see Katz (1982) and Blumenthal (1979). See also the various papers in Fransman and King (1984) in particular theintroductory chapter by Fransman.
2. See *ibid.*, Dahlman and Westphal (1982), and Lall (1982).
3. See my paper in Fransman and King (1984).
4. For a statistical analysis of the industrial characteristics of MNC investment in India, see Lall and Mohammad (1983).
5. The typology set out below draws upon, but is not identical to, that of Dahlman and Westphal (1982). See also the classification in Lall (1982) pp. 66–8.
6. For a survey of the results of pioneering research in Latin America into plant-level development of know-how capabilities, see Katz (1978).
7. I am grateful to the executives of Hindustan Lever for providing me with detailed information on their R & D in India. This was used for a case study prepared for the ILO on technological change and employment generation by MNCs in developing countries (Lall, 1983a).
8. For a review of the literature see Caves (1982).
9. On Canada, conflicting views on 'truncation' are presented by McFetridge (1977) and Globerman (1976), the former suggesting that truncation presents real problems for Canada, the latter that it is the small scale of protected Canadian operations which restricts the generation of technology and also its efficient application (by licensing).

Globerman concludes that economic nationalism has led to a misallocation of scientific resources in Canada.
10. On this point see Casson (1979).
11. This is the finding of an inter-country study of technology exports by the World Bank and the Inter-American Development Bank in which I am participating.
12. In Australia, for example, Parry and Watson (1979) show that MNC affiliates often have technology-swap arrangements with the parent companies.
13. For an examination of the impact of firm size and market structure on the composition of R & D expenditures, see Mansfield (1981).
14. I am very grateful to Mohammad Saquib for his help in collecting and processing this data. Some of the findings, pertaining to the engineering industry, are discussed in Lall (1983b).
15. In a similar exercise for Australia, Parry and Watson (1979) found a negative impact for the extent of foreign ownership on R & D within a sample of firms which all had some foreign shareholding.
16. See Lall (1980) for an empirical study of the truck manufacturing industry in India.
17. See Caves (1982) ch. 4.
18. The 'contagion' idea is developed by Findlay (1978) who argues, on this assumption, that LDCs' technical change is a direct function of the extent of foreign presence in the local economy.
19. See Lall (1982).
20. See Link (1983).

References

Blumenthal, T. (1979) 'A Note on the Relationship between Domestic Research and Development and Imports of Technology', *Economic Development and Cultural Change*, vol. 27, no. 2.

Casson, M. (1979) *Alternatives to the Multinational Enterprise* (London: Macmillan).

Caves, R. E. (1982) *Multinational Enterprise and Economic Analysis* (Cambridge: Cambridge University Press).

Dahlman, C. and Westphal, L. E. (1982) 'Technological Effort in Industrial Development – an Interpretative Survey of Recent Research', in F. Stewart and J. James (eds) *The Economics of New Technology in Developing Countries* (London: Frances Pinter).

Davies, S. (1979) *The Diffusion of Process Innovations* (Cambridge: Cambridge University Press).

Findlay, R. (1978) 'Relative Backwardness, Direct Foreign Investment, and the Transfer of Technology: A Simple Dynamic Model', *Quarterly Journal of Economics*, vol. XCII, no. 1.

Fransman, M. and King, K. (eds) (1984) *Technological Capability in the Third World* (London: Macmillan).

Globerman, S. (1976) 'Canadian Science Policy and Economic Nationalism', *Minerva*, vol. 14, no. 2.

Hirschey, R. C. and Caves, R. E. (1981) 'Internationalization of Research and Transfer of Technology by Multinational Enterprises', *Oxford Bulletin of Economics and Statistics*, vol. 43, no. 2.

Kamien, M. I. and Schwartz, N. L. (1982) *Market Structure and Innovation* (Cambridge: Cambridge University Press).

Katz, J. (1978) 'Technological Change, Economic Development and Intra and Extra Regional Relations in Latin America', (Buenos Aires: IDB/ECLA) Working Paper no. 30.

Katz, J. (1982) 'Technological Change and Development in Latin America', in R. Ffrench-Davis and E. Tironi (eds) *Latin America and the New International Economic Order* (London: Macmillan).

Lall, S. (1978) 'Transnationals, Domestic Enterprises and Industrial Structure in Host LDCs: A Survey', *Oxford Economic Papers*, vol. 30, no. 2.

Lall, S. (1979) 'The International Allocation of Research Activity by U.S. Multinationals', *Oxford Bulletin of Economics and Statistics*, vol. 41, no. 4.

Lall, S. (1980) 'Vertical Inter-Firm Linkages in LDCs: An Empirical Study'. *Oxford Bulletin of Economics and Statistics*, vol. 42, no. 3.

Lall, S. (1982) *Developing Countries as Exporters of Technology* (London: Macmillan).

Lall, S. (1983a) 'Technological Change, Employment Generation and Multinationals: A Case Study of a Foreign Firm and a Local Multinational in India' (Geneva: International Labour Office, Multinational Enterprises Programme) Working Paper no. 27.

Lall, S. (1983b) 'Determinants of R & D in an LDC: The Indian Engineering Industry', *Economics Letters*, vol 13, no. 4.

Lall, S. and Streeten, P. P. (1977) *Foreign Investment, Transnationals and Developing Countries* (London: Macmillan).

Lall, S. and Mohammad, S. (1983) 'Multinationals in Indian Big Business: Industrial Characteristics of Foreign Investment in a Heavily Regulated Economy', *Journal of Development Economics*, vol. 13, no. 1.

Link, A. N. (1983) 'Inter-Firm Technology Flows and Productivity Growth', *Economics Letters*, vol. 11, nos 1–2.

McFetridge, D. G. (1977) *Government Support for Scientific Research and Development: An Economic Analysis* (Toronto: University of Toronto Press).

Mansfield, E. (1981) 'Composition of R & D Expenditures: Relationship to Size of Firm, Concentration and Innovative Output', *Review of Economics and Statistics*, vol. 63, no. 4.

Parry, T. G. and Watson, J. F. (1979) 'Technology Flows and Foreign Investment in the Australian Manufacturing Sector', *Australian Economic Papers*, vol. 18, no. 32.

Teece, D. J. (1976) *The Multinational Corporation and the Resource Cost of Technology Transfer* (Cambridge, Mass.: Ballinger).

[15]

Japan in a New Phase of Multinationalism and Industrial Upgrading: Functional Integration of Trade, Growth and FDI

Terutomo OZAWA*

Over the past several years Japan's overseas investment has been expanding by leaps and bounds. In 1988 Japan overtook the United States as the world's largest foreign direct investor for the first time, as its outflow of direct investment in that year surged to $33.8 billion while America's slowed to $18.9 billion.[1] In terms of existing stock of direct investments, too, in 1988 Japan surpassed Germany and attained number three position behind the United States and the United Kingdom. In fiscal 1989, Japan's new direct investment continued to expand, recording yet another high of $67.5 billion, up sharply 43.6 per cent over the previous year.[2] If present trends of overseas advance continue, Japan will, within a few years, replace the United Kingdom as the second largest global investor.

Equally as phenomenal as, and more fundamental than, the pace and volume of Japan's foreign direct investment (FDI) in recent years have been changes in its nature and direction. Today's Japanese overseas investments are quite different in motivation and strategies from those of, say, ten or even five years ago. Moreover, the current stage of Japanese foreign direct investment is closely interlinked with the present phase of industrial restructuring at home as was the case with the previous stages. In fact, Japan is currently on the threshold of another new phase of industrial upgrading and multinational advance, each serving, again, as a vital joint-activity for the other.

As this article shows, two major revolutionary changes are underway in Japan, one in the domestic market conditions (i.e. demand side) and the other in the manufacturing sector; they are generating, in synergistic interaction, a *new* dynamo of industrial upgrading—and its accompanying phenomenon of industrial shedding in the form of overseas transplantation of those industrial activities in which Japan can no longer exploit competitive advantages at home.

First of all, Japan is quickly building up its own internal demand and translating it into a huge home market. Japan's *nouveaux riches* consumers have lately been on a shopping spree for a wide array of upscale products, transforming it into one of the

* Professor of Economics, Colorado State University, Fort Collins, CO., U.S.A.; author of *Recycling Japan's Surpluses for Developing Countries* (Paris: OECD, 1989).

[1] JETRO, *Kaigai Chokusetsu Toshi (Foreign Direct Investment) 1990,* Tokyo, 1990, p. 2. Japanese data are based on the balance-of-payments statistics prepared by the Bank of Japan; U.S. data on *Survey of Current Business* (U.S. Commerce Department).

[2] A figure released by the Ministry of Finance as reported in Japan Economic Journal, 12 May 1990, p. 3.

most affluent, upscale markets in the world. The ever-diversifying tastes of Japanese consumers are, furthermore, compelling industry to scourge the world for new ideas for product development. The emergence of a large, sophisticated market in Japan is cementing the basis for *intra-industry* trade and investment in manufactures with the rest of the world. On the supply side, Japan's already efficient factories are being simultaneously metamorphosed into even more advanced ones with intensive applications of computer-aided techniques and robots—and even with the world's first practical use of industrial lasers. These advanced manufacturing systems or "smart factories" are ushering Japan into a new era of flexible manufacturing, capable of meeting the rising sophistication of consumers, both at home and abroad.

BOOMING DOMESTIC MARKETS

The current economic prosperity in Japan is founded on the phenomenal growth of internal demand rather than on the external demand Japan has long been noted for capturing through trade surpluses. The latter's growth rate, in fact, has turned negative in recent years. The domestic markets, especially for high-income products, are booming, with the rising disposable income of Japanese consumers and the ever-greater diversification and sophistication of their tastes. Despite rising oil prices and interest rates and slumped stock markets, the current fiscal year (which will last till the end of March 1991) is expected to register at least a 4 per cent growth rate, having seen a phenomenal 11.4 per cent growth rate in the first quarter of 1990.

The Japanese have long been known for their rather thrifty way of life, but they have recently begun to spend with a vengeance to satisfy their pent-up demand, backed by their huge purchasing power accumulated over the years and augmented by recently swollen liquidity. Their real income is rising not only because of the continuous hefty increases in their wages made possible by economic prosperity and labour shortages at home but also because of the rise in the yen's value, along with import liberalization, that enables them to purchase foreign goods, services and assets at the lowest "bargain" prices ever known to them.

In addition, the Japanese are finicky consumers and quite demanding as to the quality of goods and services they purchase. With rising incomes, lower price stickers, though helpful, are becoming less and less effective inducements to buy. For example, reasonably priced beef imports are on the rise, now that the quota restriction is being gradually dismantled towards complete removal in 1991. But the importers are puzzled to see quickly piling inventories, while the demand for pricey Japanese-grown beef is booming. The reason is: "The Japanese do not like tough, mediocre-tasting beef. They prefer the tender, fat-marbled Japanese-produced varieties, and are willing to pay extremely high prices for it."[3]

[3] Hajima Karatsu, *The "10,000 Visitor Plan"—A Proposal to Revitalize American Industry*, Tokyo Business Today, Vol. 58, No. 8, August 1990, p. 9. Some American ranchers are reportedly attempting to breed the *wagyu* cattle (Japanese breed) in the hope of capturing the lucrative Japanese market by paying as much as $250 a vial of semen (ten times the cost of normal high-grade semen) secured from four bulls smuggled out 14 years ago. Wall Street Journal, 28 September 1990, p. 1.

A recent survey conducted by Japan's Small- and Medium-Enterprise Agency shows that 37 per cent of the responding consumers are "tired of low-quality, mass-produced goods and services", a view matched by 40 per cent of wholesalers who also single out quality as "the biggest worry to overcome", and that 32 per cent of retailers consider goods from abroad not really capable of matching domestic needs.[4] In these days Japanese consumers are demanding something newer, better, and more distinctive.

It is against these backgrounds that Japan's auto makers have gone upscale by introducing luxury cars such as Acura, Lexus, and Infiniti. It is also because of this strong qualitative preference of the Japanese buyers that they are more attracted to high-quality, high-variety, design-attractive European manufactures than American goods which are typically mass-manufactured. More BMWs, Porsches, Jaguars, and Mercedes' are seen in Japan, and Japanese consumers are snapping up well-established, high-fashion, brand-name goods such as Louis Vuitton, Hermès, Pierre Cardin, and Christian Dior. As barriers to manufactured imports are removed in Japan, more sophisticated European imports are capturing a larger market share than American imports. This trend is, indeed, encouraging new European direct investments in Japan's distribution sector—usually in joint ventures with local merchandisers. Suddenly, the Japanese are transforming themselves into strong individualists as consumers, though they may still be group-loyal conformists as workers at their workplaces.

This sudden growth and metamorphosis of the Japanese market has ironically materialized for the most part by external pressure, especially that exerted by the United States. It is the strong U.S. demand that Japan, as an economic superpower, ought to create its own internal demand instead of saving so much and depending on trade surpluses as growth stimuli. In response, Japan began to promote domestic demand along the lines of recommendations as outlined in the so-called Maekawa reports of 1986 and 1987.[5] Most recently the 1990 U.S.-Japan pact concluded under the Structural Impediments Initiative (SII) negotiations stipulated that Japan should expand and open up its domestic market by spending more of its surplus capital on public works, toughening its antitrust policies and streamlining its archaic distribution system. All these U.S. demands are indeed unexpected blessings for Japanese consumers—and industry, since these measures surely contribute to the development of strong domestic demand and home markets. Japan is thus again turning external pressure to its own advantage. A better streamlined, more liberated—hence more efficient—Japan is expected to emerge with greater competitiveness.

The importance of domestic demand as a key determinant of national competitive advantage is emphasized by Michael Porter:

[4] As reported in *Smaller Firms Go Global*, Focus Japan (newsletter), August 1990, pp. 1–2.

[5] These reports were prepared by the Prime Minister's specially appointed Advisory Committee headed by Haruo Maekawa, former Governor of the Bank of Japan.

> "A nation's firms gain competitive advantage if domestic buyers are, or are among, the world's most sophisticated and demanding buyers for the product or service. Such buyers provide a window into the most advanced buyer needs . . . The presence of sophisticated and demanding buyers is as, or more, important to sustaining advantage as to creating it. Local firms are prodded to improve and to move into newer and more advanced segments over time, often upgrading competitive advantage in the process."[6]

Japanese entrepreneurs used to innovate new products with foreign consumers' tastes in mind. In fact, domestic consumption was once regarded as secondary to exporting and was sacrificed for the sake of releasing resources for producing exports and earning foreign exchange. Back in the 1950s and 1960s, for example, the government used to allocate scarce industrial materials to be used specifically for export items alone. Nowadays, on the other hand, Japanese industry's primary interest is in its own home market; it is innovating *inward*—first for domestic consumers, and when a new product has been successfully test-marketed at home, it is then exported overseas.

Japanese stores are now full of brand-new products that have not yet shown up outside the country but that will soon head overseas. For example, DAT (digital audio tape) recorders and players, instant playback camcorders with a color liquid-crystal display, voice-activated cameras, cordless telephones with built-in answering machines, and the like. It is only in Japan so far that fuzzy logic, a very recent engineering principle, has found its commercial applications in trains, elevators, cameras, washing-machines, vacuum-cleaners, rice-cookers and other items, altogether numbering at least 20 that are currently on the market. Fuzzy-oriented vacuum-cleaners, for example, "can automatically sense the amount of dust in a carpet or the conditions of a floor's surface and then set the proper level of suction"; similarly, fuzzy-based washing-machines can "detect the dirtiness of the clothes—and then choose the optimum washing condition".[7]

In the Japanese auto industry, too, the concept of quality is said to be "not just fewer defects anymore", but the Japanese are engineering "unprecedented measures of look, sound, and 'feel' into their cars".[8] They are going above and beyond the stage of defect-free manufacturing which is already taken for granted by the Japanese consumers (*atarimae hinshitsu*) and into the stage of "fine touches" whose cumulative effect will be dazzling and captivating (*miryokuteki hinshitsu*)—all in response to the more sophisticated demands of the home consumers.

The upshot of these developments is that more and more innovations are trickling down from Japan into overseas markets. Japan is thus becoming a major initiator of the product life cycle of many new products.

Indeed, the emergence of this affluent and ever-more sophisticated home market is dramatically altering the nature of Japan's overseas investment and industrial

[6] Michael Porter, *The Competitive Advantage of Nations*, New York: Free Press, 1990, p. 89.
[7] Takashi Ono, *Everything's Fuzzy Among Household Appliances*, Japan Times, weekly international edition, 2 September 1990, p. 13.
[8] *A New Era for Auto Quality*, Business Week, 22 October 1990, pp. 84–96.

upgrading. This is a godsent development for growth-conscious Japan, since it can continue to develop and sharpen its competitive edge in high-end products by capitalizing on its own home-market—instead of the Western (notably American)—markets on which it has long been dependent to secure the basis for scale economies for new products but whose relative affluence and sophistication have recently declined considerably. Japanese industry can also continue to shed low-end manufacturing through direct overseas operations without worrying about a possible hollowing-out of its home industrial base. FDI is thus created basically as a by-product of the Schumpeterian process of "creative destruction".

Toward flexible manufacturing

Throughout the post-World War II period, the Japanese have striven—and significantly succeeded—in "perfecting" mass-production techniques for standardized, low-cost, non-luxury, yet quality-reliable goods, for which price and durability were the key determinants of competitiveness. Their strategy was initially focused on limited lines of goods, thereby avoiding the costs of change-overs for short model cycles and cultivating economies of scale. Thus the manufacturing system of *limited variety, large-lot mass production* has established itself as the main strength of Japanese industry. Japan's car makers, for instance, built their initial competitive edge by concentrating on low-end, small cars and their limited variety. To attain economies of scale, however, they had to export aggressively to overcome the small size of the domestic market.

At the plant level, early on, Japanese firms also vied vigorously with each other in installing the latest machinery and equipment, benefiting from the state-of-the-art production techniques therein embodied. Their young capital stock in modern manufacturing industries contributed enormously to productivity growth. In addition to keeping their physical facilities technologically up-to-date, they introduced numerous organizational innovations of their own by streamlining work and material flows, minimizing inventories on the factory floor through the "just-in-time" (JIT) system, and training and retraining workers for changing manufacturing requirements. Above all, they succeeded in building quality control into the production process itself by making workers quality-conscious with spot-inspecting rather than being dependent on independent quality inspectors at the end of the line or at the receiving end—their customers.

In addition, Japanese industry built a reputation for its extensive use of a subcontracting system, an arrangement unique in its vertical depth and in its use of *keiretsu* (or "a vertical link of enterprises"). This pyramidal supply-support system is designed and co-ordinated in such a way that a very fine vertical division of labor is attained along the lines of different factor- and skill-intensity requirements in producing myriad parts, components, and accessories—all creating a staggering cumulative improvement in labor productivity.

Moreover, the practices of the JIT inventory management and total quality control (TQC) have come to permeate the entire layers of the vertical system from final assemblers to subcontractors in a vigorous and continuous search for technological improvements. The greater the number of parts and components to be produced separately and assembled into a final good, the more efficient this manufacturing system has become. This feature is succinctly captured in the following observation: "The Japanese labor productivity advantage is enormous in high-volume assembly processes where hundreds, even thousands, of interdependent steps must be co-ordinated".[9] Organizational innovations and co-ordinations are a key to the success of any assembly-based industry.

Yet this manufacturing system, so far perfected for the age of mass production of standard products, is fast becoming outdated in the light of the new age of consumerism in Japan described above—and in other advanced countries. In this coming age, a greater variety of goods and more custom-tailored goods are demanded and need to be manufactured in small lots and in a timely manner at an ever-faster rate of turnover to meet the incessantly changing needs and tastes of customers.

Harald B. Malmgren aptly describes the quiddity of the new age of manufacturing:

> ". . . the most significant effect of the transformation of industrial activity is the ability to adjust products continuously, tailoring them to continuously changing requirements of individual users. The flexibility and agility of the new industrial processes allow a much closer interaction between user and producer. It allows, and even requires, producers to work directly with users in the design and application of products, interweaving engineering and other services with the provision of hard goods.
>
> What this means, in essence, is that modern industry will be able to come back full circle to its beginnings in the last century, when individual suppliers and cottage industries produced one-off, one-of-a-kind consumer goods like suits, or producer goods like single-purpose machinery. In other words, *the long experience of producing standardized items for off-the-shelf sale may be coming into a period of decreasing relevance, while provision of tailored or specially designed products may be an entirely new mode of competition, based upon large-scale systems that are highly flexible.* Such a transformation would also tend to shorten or compress product life cycles, giving the competitive edge to the suppliers with the fastest response time rather than to those with the cheapest off-the-shelf, standardized items.
>
> In the context of the new industrial revolution, competitiveness will increasingly depend upon speed of response and character of product-related services." (emphasis added).[10]

Indeed, Japanese industry is in the midst of striving to set up a computer-integrated, highly automated manufacturing system. In fact, their efforts to rationalize the factory floor of mass production, the efforts exerted up until very recently, have turned out to be an effective *preparatory* step to a more advanced system of manufacturing, the "smart factory". For five crucial steps are, in essence, required to set up the "smart factory" that could run itself, producing multi-variety goods in short model cycles and in accordance with electronically relayed specifications:[11]

---

[9] James C. Abegglen and George Stalk, Jr., *Kaisha. The Japanese Corporation,* New York: Basic Books, 1985, p. 61. They make an excellent analysis of the production costs of limited varieties and explain how the Japanese rationalized work-flows at the factory level.

[10] Harald B. Malmgren, *Technology and the Economy,* in William Brock and Robert Hormats, eds., *The Global Economy: America's Role in the Decade Ahead,* New York: Norton, 1990, pp. 98–99.

[11] These five steps are described in *Smart Factories: America's Turn?* Business Week, 8 May 1989, pp. 142–148.

Step 1: Pre-automation-stage simplification and reorganization of the shop floor for optimum efficiency (i.e. rationalization of the conventional factory floor for optimum work-flows).

Step 2: Creation of "islands of automation" with robots, flexible manufacturing cells, computer controls, and other advanced shop-floor systems (i.e. partial application of automation hardware and software to certain segments of a production flow).

Step 3: Linking of the islands of automation to each other and to computer-aided design (CAD) through a shared data-base (i.e. computer link-up of automated work segments).

Step 4: Application of artificial intelligence (AI) and object-oriented software to integrate some operations with CAD and to automate such complex steps as production scheduling and diagnostics (i.e. the birth of a "smart factory").

Step 5: Establishment of a "computer-integrated enterprise (CIE)" by extending AI and object-oriented systems to all decision-making steps, from product planning to customer service (i.e. the birth of a "smart enterprise").

The last step is the completion of building a "smart enterprise" in which the company's "smart factories" are in turn all interlinked with "smart service departments" and managed comprehensively by its "smart headquarters".

Observing a global race toward the Factory of the Future, "Business Week" makes the following assessments: Japan is way ahead of the United States in Step 1; Step 2 is "widely implemented" in both the U.S. and Japan, but much more effectively in Japan; Japan is "slightly ahead" in Step 3; Step 4 represents the "cutting edge among U.S. companies" but is done "more effectively" in Japan, though still rare; and Step 5 is "underway at only a couple of U.S. companies" but "so far not implemented in Japan, but many Japanese companies could do so rapidly". On the other hand, Europe is judged "to lag behind Japan and the United States" in Step 1; "just beginning" in Step 2; "far behind" in Step 3; and "almost non-existent" in Steps 4 and 5.[12]

Japanese industry, indeed, has stumbled upon this five-stage sequence of building the smart factory, since to compete, it first had to emphasize the work-flow rationalization and computer-aided robotization of its large-scale assembly-based industries during the passing age of limited-variety mass production. It was thus led to prepare itself serendipitously through Steps 1, 2 and 3, and perhaps well into Step 4. These sequential steps involved a move from factory automation (FA) with the installation of robots, to the computer-integrated-manufacturing (CIM). It is estimated that in 1990, "one-third of Japanese manufacturers will have completed their introduction of CIM, with the remaining two-thirds in the process of putting in such systems or studying the feasibility of doing so."[13]

Japanese achievement in this new endeavor is well represented by Fanuc of Japan, one of the leading builders of the Factory of the Future:

[12] Ibid., pp. 142–143.

[13] *CIM—The Manufacturing/Information Wave of the Future,* Tokyo Business Today, Vol. 58, No. 8, August 1990, p. 48.

"Even more prophetic is an emerging series of factories that Fanuc is establishing in the remote regions of Japan. The items produced in the first of these factories will be products for electric spark machinery and electric motors. Fanuc has chosen to call these 'village blacksmith' factories. Each of these factories is on a site of only about 54,000 square feet. Each one of these factories has a floor space of approximately 1,800 square feet. Each one of these factories has four CNC (computer numerically controlled) machine tools and three robots. Each facility has only two employees, both engineers. The total facility cost of each one of these approximately $370,000. The goal of Fanuc's President, Seiuemon Inaba, is a 'future network of small-scale automated factories'. Note that he has left out a significant word: *global*. For once you achieve this kind of manufacturing plant network in Japan, there is no reason why these factories could not be duplicated at will around the world."[14]

A decisive spur to Japan's drive to factory automation has been severe labor shortages, shortages that Japan endeavors to solve by introducing labor-saving technology instead of importing workers from abroad. In 1989, workers in the auto industry, for example, averaged 2,275 hours because of overtime, more than any other manufacturing sector (2,150 hours on average)—and certainly much higher than the average of 1,900 hours in the United States and 1,613 hours in West Germany.[15] Japan's auto industry is already drawing on temporary workers for about 30 per cent of factory employment to meet the booming demand, as more skilled permanent employees are shifted to the production of luxury models. More and more robots are thus used wherever possible to save labor, and production of low-end models is, therefore, more willingly than ever before, transferred overseas, both to solve the problem of a severe labor shortage at home and simultaneously to give their overseas transplants an additional basis for economies of scale.

The components-intensive electronics industry is another sector in which factory automation has progressed very rapidly:

"In a factory where Matsushita Electric makes Panasonic VCRs, a robot winds wire a little thinner than a human hair 16 times through a pinhole in the video head, and then solders it. There are 350 of these robots in the factory and they wind, and then wind some more, 24 hours a day. They do it five times faster and much more reliably than the 3,000 housewives who, until recently, did the same job with microscopes on a subcontract basis in Japan's countryside. The robots even inspect their own work.

A U.S. company can't get this technology—even if there were an American consumer electronics industry to take advantage of it. Matsushita invented and custom-made all 530 wire-winders to gain a competitive edge."[16]

The zeal for in-house robotization is widespread throughout Japanese industry, from manufacturing to services, especially now in the construction industry whose booming business encounters serious labor shortages. Advances in automation are thus firm-specific, giving the individual Japanese companies their own unique competitive edge. And it is these labor-saving innovations that enable Japanese companies to produce more cost-effectively than local competitors even in high-wage Western countries where they set up factories.

---

[14] Thomas G. Gunn, *Manufacturing for Competitive Advantage*, Cambridge, Mass.: Ballinger, 1987, p. 205.
[15] Figures are from Japan's Minsitry of Labor as reported in Wall Street Journal, 17 September 1990, p. A10.
[16] Andrew Tanzer and Ruth Simon, *Why Japan Loves Robots and We Don't*, Forbes, 16 April 1990, p. 148.

Indeed, Japanese companies, notably large ones, are endeavoring to go beyond the "smart factory" and moving toward the "smart enterprise"—and even toward the "smart industry" by interlocking business ties of their *keiretsu* groups by way of on-line computer-based communication channels, especially between final assemblers, parts/components suppliers, and distributors:

> "At Toyota Motor Corporation, the international standard manufacturing automation protocol (MAP) has been adopted as a basic telecommunications network, with robots and other FA equipment controlled by computer. This network directly links not only Toyota plants, but its affiliated parts makers and sales sector as well, effectively constituting a new 'electronic bulletin board system'."[17]

In early 1990, moreover, Japan launched a $1 billion ten-year research project called the Intelligent Manufacturing System Project (or more popularly called "Global Factory Research Project") with the aim of collaborating with other nations in revolutionizing the world's factories through international standardization of manufacturing technologies. According to the original plan, Japan's Ministry of International Trade and Industry (MITI) and its private sector are to contribute $600 million, with $200 million each expected from the United States and Europe.

So far, more than 80 Japanese companies and several foreign companies such as United Technologies and Rockwell International have already signed up as research participants. The project, however, has become somewhat controversial, since cynics in the West warn that it may be a Japanese Trojan Horse to absorb advanced Western technology.[18] The participating Japanese companies, on the other hand, bristle at such charges because they consider themselves already state-of-the-art manufacturers who are willing to share their expertise in the interest only of enhancing global manufacturing efficiency.

Be that as it may, Japanese industry is steadily gearing up to the new age of manufacturing. Japan is on the threshold of a new phase of industrialization, leaving the era of *limited variety, large-lot mass manufacturing* and entering the new era of *multi-variety, small-lot flexible manufacturing*.

A NEW ROUND OF INDUSTRIAL SHEDDING AND GROWTH

Overseas direct investment has been serving as a house-cleaning-and-renovating vehicle for Japan as it has repeatedly metamorphosed from one phase of industrialization to another. To understand the implications of the current revolutions in Japan's home market and manufacturing system described above, it is necessary to review Japan's postwar efforts for industrial upgrading.

Since the end of World War II, Japan has gone through three major phases of industrialization, though overlapping to some extent in phases, and is currently in the midst of another new round, whose unfolding characteristics have just been seen

[17] See footnote 13, *op.cit.*, p. 48.

[18] Jacob M. Schlesinger, *Japan's Global-Research Project is Seen by Some in West as Trojan Horse,* Wall Street Journal, 1 May 1990, p. A17.

above. Each structural metamorphosis at home has, moreover, been accompanied by a phase-specific outward migration of business operations. As presented elsewhere, the nature and direction of Japan's FDI can be best explained in terms of the "industrial restructuring" theory,[19] within whose framework the Japanese experience is described below.

Phase 1: Labor-Intensive Industrialization and the "Elementary" Stage of Overseas Investment

In the early postwar period (approximately from the start of the 1950s to the latter half of the 1960s) Japan emphasized labor-intensive industrialization as a way of creating local employment and exporting price-competitive manufactures to earn badly needed foreign exchange. In those days, Japan's major exports came from light industries such as textiles, household electric appliances, sundries, and whatever low-wage labor was then capable of producing. It was in the mid-1960s, when shortages of low-cost workers were encountered for the first time and wages began to rise rapidly, that labor-intensive manufacturing activities were transplanted from Japan into its neighboring Asian countries, such as Taiwan, Hong Kong and South Korea, where low-wage labor was then still in abundant supply. And this low-wage-seeking type of FDI, which may be identified as the "elementary" stage of overseas investment, continued to be made throughout the 1970s, accentuated by the first two rounds of appreciation of the yen in 1971 and 1973 that destroyed Japan's labor-intensive exports whose marketability had depended on the price—rather than quality—factor.

Phase 2: Heavy and Chemical Industrialization and the "Resource-Seeking" Stage of Multinationalism

The rapid growth of heavy and chemical industries (namely, steel, shipbuilding, heavy machinery, petrochemicals, and synthetic fibers) began in the 1960s, replacing light industries as Japan's primary exporters throughout the 1960s and until the early 1970s. As Japan emerged as the world's leading producer and exporter of resource and energy-intensive goods such as steel, aluminum, ships, heavy machinery and chemicals, the security of resource supplies from overseas became the priority policy consideration for natural-resource-scarce Japan.

Early on, Japan relied on the trade route for importing vital resources but soon began to secure the supply sources by foreign investment. It basically took two forms: (1) the "invest-and-import" approach in which equity ownership, mostly partial, of foreign resources is sought in order to ensure import supplies; and (2) the "loan-and-import" approach in which direct loans are extended in exchange for long-term supply contracts.[20]

[19] Terutomo Ozawa, *Europe 1992 and Japanese Multinationals: Transplanting a Subcontracting System in the Expanded Market*, in B. Burgenmeier and J. L. Mucchielli (eds.), *Multinationals and Europe 1992*, London: Routledge, 1990.

[20] For this distinction between the two types of resource-seeking investment, see Terutomo Ozawa, *A Newer Type of Foreign Investment in Third World Resource Development*, *Rivista Internazionale Scienze Econmiche e Commerciali*, Vol. 29, No. 12, December 1982.

These resource-seeking investments were made in large amounts for minerals and energy resources such as iron and copper ores, bauxite, oil and natural gas—especially in the early 1970s and immediately after the 1974 oil crisis, though Japan had initiated such investments back in the 1950s. Toward the end of the 1960s, however, Japan's heavy and chemical industrialization reached its peak, bumping against the limits of the domestic environment; it was physically no longer possible to keep expanding heavy and chemical industries in a space-constrained island economy. And the first oil crisis of 1974 was a decisive additional factor that finally compelled Japan to shift gears from resource-intensive industrialization to less-pollution-prone, knowledge-intensive segments; heavy and chemical industries (such as aluminum, paper, and crude steel) also began to be shed overseas through direct investment. This stage of Japan's overseas investment was thus as fully compatible with and supportive of industrial upgrading at home as its previous stage.

Phase 3: Subcontracting-Dependent, Assembly-Based Industrialization and the "Assembly-Transplanting" Stage of Multinationalism

The third phase of industrialization was characterized by a shift toward less resource-based, environmentally more compatible, higher value-added, assembly-based industries, notably automobiles and electronics products. Japan actually had begun to foster these more knowledge-intensive pursuits as infant industries as early as the late 1950s. In this third phase of development, Japanese industry adroitly capitalized on its dualistic industrial structure, in which numerous small- and medium-sized enterprises coexisted alongside a limited number of large-scale firms, a remnant of Japan's prewar pattern of economic development as a late-starter.

This dualistic (or multi-layered) structure has proved to be instrumental in producing the *networks* of co-operative subcontracting arrangements necessary for the vertically multilayered assembly operations of modern machinery industries, which require numerous subassemblies, parts, components and accessories. In fact, the multilayered industrial structure—composed of manufacturers of different sizes and different levels of technological sophistication ranging from large, modern corporations down to very small, family-managed shops—is an ideal structure for assembly-based industries.

But the very success of building up the efficient, large-scale (hence exploitative of scale economies) hierarchies of assembly operations in automobiles and consumer electronics, along with stepped-up R&D activity and technological accumulation, resulted in Japan's export drive and its ever-expanding trade surpluses. These developments in turn quickly led to trade conflicts and the sharp appreciation of the yen.

To skirt protectionism, Japanese producers of color TV sets and microwave ovens were the first group of manufacturers to begin, in the mid-1970s, to replace their exports with local assembly operations in the Western markets, initially in North America and then in Europe. The next group to follow suit, in the 1980s, comprised manufacturers of photocopiers, electronic typewriters, printers, microchips and automobiles.

This type of investment was replacing existing exports from Japan; hence it was feared that Japan would soon end up exporting comparative advantage, thereby weakening its trade-based economy. In fact, a talk of the "hollowing-out" effect of export-substituting investment became widespread among scholars, policymakers, and industrialists. This fear was heightened all the more as assembly-based manufacturers of autos and electronic goods were further compelled to transplant abroad more and more upstream activities (i.e. production of parts and components) under the pressure of local content rules and demand for local management. The "screwdriver" factories initially set up by Japanese corporations to assemble imported parts soon had to be transformed into more and more localized operations. Even some segments of R&D, particularly those suitable for local research resources and those aimed at redesigning products to local tastes, began to be performed in major Western countries.

It is against this background that in the current "assembly-transplanting" stage of foreign investment Japan is apparently eroding its own competitive advantages in autos and electronic products.

In a recent paper most appropriately entitled *Japanese Multinationals in Europe and the United States: Exporting Cars, Computers, and Comparative Advantage through Foreign Direct Investment,* Michelle Gittelman and John Dunning raised a critically important question that was essentially at the back of many Japanese policymakers' minds only a few years ago:

> "The story of Japanese FDI in the United States and Europe is one of the remarkable strategic dexterity in response to the rapidly changing O-L-I (ownership, location, internalization) configuration of the three locations in the 1980s. In moving swiftly from exports to defensive FDI strategies (the early 1980s) to offensive FDI strategies (the late 1980s/early 1990s), Japanese multinationals have managed to maintain their O-advantages in increasingly competitive markets. The question that begs to be answered at this point is whether, and for how long, the process can continue: *can Japan continue to upgrade its industrial infrastructure if many of the necessary instruments of the upgrading (especially R&D assets; technical and managerial personnel; inter-firm supplier networks) are geographically dispersed outside of Japan?* If not, will the resultant decline in the L (location-specific) advantages of Japan undermine the O (ownership-specific) advantages of its firms?" (emphasis added).[21]

True, the phenomenal success of Japan's economic development and the dramatic appreciation of the yen are now changing Japan's industrial landscape as its FDI replaces exporting and as Japan now finds it imperative to sharpen its competitive edge by increasingly employing foreign corporate assets and technical resources located outside the country. But the spread of Japan's industrial infrastructure beyond

[21] Michelle Gittelman and John H. Dunning, *Japanese Multinationals in Europe and the United States: Exporting Cars, Computers, and Comparative Advantage Through Foreign Direct Investment.* Paper presented at a conference at the American Institute for Contemporary German Studies, Johns Hopkins University, Washington, D.C., September 1990, and to be published in Paul J. J. Welfens, ed., *Multinationals in Europe and Global Trade in the 1990s*, forthcoming.

its borders will not necessarily erode its effort to continuously upgrade home industries. Indeed, acquisitions of foreign corporate assets add to domestic assets, the former augmenting the latter.

The question of whether FDI becomes a mechanism through which the home country loses part of its industrial base (i.e. the "hollowing-out" effect) or FDI still serves its house-cleaning-and-renovating function (i.e. the "upgrading" effect) depends on that country's ability to induce continuous innovations domestically for creating new growth markets and industries, and simultaneously to employ overseas ventures as a catalyst to stimulate technological progress at home.

Given the favorable changes in both the home market and the manufacturing sector examined earlier, Japan is well on its way to another new phase of industrialism, a phase to be detailed below. In other words, Japan's assembly-based overseas ventures initially compelled to be set up willy-nilly in North America and Europe as an inferior (namely "second-best") alternative to exports have turned out to be the "first-best" choice—that is, they have become far more desirable, after all, than exporting from home. The present "assembly-transplanting" stage of Japan's foreign direct investment is being transformed into another round of industrial shedding that allows Japanese industry to concentrate further on higher value-added, more knowledge-intensive, and higher income, upscale products.

Stated differently, FDI has again become protrade and restructuring-supportive, but this time changing the nature of Japan's trade from the mainly home-based inter-industry type to the *intra-industry* or *intra-company* type.

So far, two types of intra-industry—and intra-company—division of labor have come into existence. One involves an international division of labor in intermediate goods; production of fairly standardized (hence relatively low value-added) parts and components or those that can be cost-effectively produced locally is now being transferred both to low-wage developing countries and to high-wage Western countries—in the latter, with the installation of labor-cost-reducing and labor-quality-augmenting automation equipment mostly shipped out of Japan.

The establishment of these operations abroad has thus been accompanied by the rising exports for Japan of plant machinery, equipment, and other capital goods (including those firm-specific, custom-tailored robots). In addition, more and more sophisticated components are produced at home and supplied to the overseas manufacturing outposts. This development, therefore, is resulting in a *vertical* division of labor under which Japanese industry endeavors to retain—and further develop—higher value-added, critical components at home. Here can be discerned a *more refined* or *more sharply delineated and specialized* form of trade within an industry (that is, intra-industry trade) and even within a company (that is, intra-company trade). And this partially explains why the sharp appreciation of the yen has so far failed to curb Japanese exports while it has drastically encouraged overseas investments.

A second type of a newly emerging division of labor is seen in the transfer overseas of standardized, low-end models of a product which Japanese industry used to mass-produce at home for export, a point briefly touched on earlier. This constitutes a *horizontal* pattern of specialization. For example, standardized models (mostly low-end, low-profit-margin) of Japanese automobiles are increasingly shifted to overseas manufacturing. Honda is transferring the production of the Civic and Accord models to the U.S. and Europe, while at home it intends to concentrate mainly on the Prelude and the Acura models (including the latest $60,000 sports car, NSX). So are Toyota and Nissan, who are retaining the production of luxury models such as Lexus and Infiniti only in Japan. Thus Japanese car makers' efforts to operate their U.S. and European factories successfully may contribute to the revitalization of the latter's auto industry, but Japan is giving up a comparative advantage merely in standardized compact to mid-size models, while it creates a new comparative advantage in upscale, higher value-added ones at home.

Viewed in the above light, the current stage of transferring those comparative advantages that are destined soon to be superannuated is no different from the previous stages of industrial transfers. Separate papers could easily be written similar to the Gittelman/Dunning paper; *Japanese Multinationals in Asia: Exporting Textiles, Sundries and Comparative Advantage through FDI,* and *Japanese Multinationals in Resource-Abundant Countries: Exporting Steel, Petrochemicals and Comparative Advantage through FDI,* each describing the Phase 1 and the Phase 2 experiences of Japan's FDI, respectively.

It is in the context of the above development that Japanese industry is now ready to meet the demand for local content and full localization of its manufacturing operations in the West, especially in the European Community which imposes such requirements more strictly than North America:

> ". . . Europeans grasp at the straw of 'reciprocity', 'local content', and genuine local manufacturing (as opposed to 'screwdriver' assembly). All of these are now becoming heated subjects of EC discussion in Brussels. But *as the old saying goes, you must be careful what you wish for, because you might get it.* In France, the government has long excoriated Japanese companies for exporting finished products to the French market—sometimes even dumping them below cost in order to build market share—without manufacturing *in* France and generating local employment. At the moment, however, Japanese auto manufacturers now say they *want* to establish French plants. They claim, however, that their efforts to do so are being rebuffed by government policies that seek to protect the existing setup of the French auto industry from Japanese competition. They make the same complaint about Italy." (emphasis added).[22]

Indeed, the European Community is about to get what it wished for in local content and indigenized manufacturing—complete localization but only in those product lines that are already standardized and rather low-end.

---

[22] Daniel Burstein, *Yen! Japan's New Financial Empire and Its Threat to America,* New York: Fawcett Columbine, 1990, p. 318.

Phase 4: Flexible Manufacturing and the "Strategic Localization" Stage of Global Operations

As the result of the revolutions in consumerism and manufacturing, the Japanese have just *begun* to savor the fruits of economic success as consumers with their rapidly rising incomes and to be liberated from the grind of tedious "brawn" jobs as factory workers. The Middle East crisis, though it is no doubt a disruptive factor to Japan's economic growth, cannot stop and reverse the momentum and dynamism of technological progress that underlie its current investment boom. In 1989, in fact, Japan overtook the United States to become the world's biggest investor in plant and equipment despite its size, it being much smaller than the latter, marking the first time that any country had out-invested the United States in the post-World-War-II period.[23] A drive to the Factory of the Future necessitates drastic renovations of the existing manufacturing facilities, an imperative demonstrated in the current investment boom. Japan is basking in a virtuous circle of economic expansion, since investment expenditures in turn create more incomes—hence a faster growth of the home market justifying such a capacity expansion.

Given the rapid upgrading of Japanese home market and industry, what would be the major characteristics of Japanese FDI in the emergent age of new consumerism and manufacturing? Although Japan is still in the midst of the "assembly-transplanting" stage of multinationalism and its next stage is inchoate, there are a number of harbingers that hint at some key features of Japan's new global advances. Four broad types of next-generation activities are discernible:

(1) *M&As (mergers and acquisitions) in search of foreign corporate assets*

In recent years, the M&A transactions carried out by Japanese corporations abroad have been sharply on the rise. In 1986, there were a total of 206 such transactions, but in 1989 the total reached 405, and during the first half of 1990 alone it already amounted to 226.[24] Until recently, well over a half of these purchases of foreign firms had taken place in the United States, but the proportion of European deals rose from 13.7 per cent in 1986 to 25.7 per cent in 1990. Europe 1992 is obviously a strong inducement. Yet in terms of value of M&As the United States is expected to remain the major market for Japanese mega-purchases. Indeed, Japan may soon be surpassing Britain as the leading corporate buyer in the United States. And this new form of Japan's overseas investment is expected to grow steadily in the future.

The nature of these overseas M&A activities is *strategic*, in the sense that they are aimed at securing new technology, human resources, production bases, marketing channels, and other foreign corporate assets—all in one bundle, an activity that is quite distinct from the conventional type of FDI. Traditional direct investment (i.e. setting up shop overseas) has been designed to transplant the investor's *own* corporate assets

[23] Wall Street Journal, 26 July 1990, p. A14.

[24] These statistics are compiled by Yamaichi Securities Co., as cited in *M&A Activity Shifting to Europe*, Japan Economic Journal, 28 July 1990, p. 1.

abroad so as to produce locally in the host economy. In contrast, M&A is the mechanism through which to acquire directly and control the ownership of foreign corporate assets that are strategically complementary with the buyer's own assets. From the multinational's point of view, it may be characterized as an *asset-capturing* activity in contrast to the *asset-disseminating* nature of the conventional form of direct investment.

The strategic approach to M&A adopted by Japanese corporations is best illustrated by Sony's $3.4 billion takeover of Columbia Pictures and Matsushita's $6.6 billion acquisition of MCA, owner of Universal Studios. Both are aimed at acquiring foreign corporate assets ("entertainment software" in both cases) that complement the "electronics hardware" of these giant Japanese manufacturers. Fujitsu's purchase of Britain's International Computers Ltd (ICL) is another prime example of the asset-capturing role of M&A exploited by Japan's multinationals.

(2) *Establishment of overseas R&D, design/engineering/testing centers in search of foreign human resources*

Overseas transplanted production, especially when local content and localization of management are demanded, requires adaptive development efforts; intermediate goods need to be re-designed and re-engineered, depending upon the productive capacities of local support industries. Product designs are often modified to meet the local market conditions. These tasks can be better performed by local personnel who are familiar with their own localities.

In addition to adaptive efforts, the research centers devoted more toward original R&D activities are increasingly arranged in the host economy by Japanese multinationals who are in a better position than the local firms to organize and finance such an endeavor because of their stronger financial and commercialization capacities. They are intended to capitalize on unique indigenous technological resources in the host economies. For example, many new car models are designed and engineered at Japanese-owned R&D/design/engineering centers in the United States and Europe not only for local markets but for Japanese market as well.

This type of R&D takes the form of wholly owned operations or joint ventures with local interests. A 1989 survey conducted by JETRO (Japan's External Trade Organization) indicates that 150 of the responding 485 Japanese ventures in the United States (or 30.9 per cent of the respondents) conduct some R&D-related functions locally, and that 34 of them (or 7 per cent) operate separate local research centers, while 19 (or 4 per cent) are engaged in joint R&D projects with local firms—thus about 40 per cent of them, taken together, are carrying out research activities in one way or another.[25] Another survey by JETRO for Europe also reveals that only 41 of the 227 responding Japanese ventures in Europe (or 18 per cent) are keeping R&D/

[25] JETRO, *Zaibei Nikkei Seizogyo Keiei no Jittai* (Business Conditions of Japanese-affiliated Manufacturing in the United States), Tokyo, 1990. pp. 34–35.

design activities only in Japan; the rest (that is, the majority) are delegating at least some segments of research activities locally.[26]

(3) *Setting up of regional headquarters in search of an optimal integration of intra-regional operations*

As the operations of Japanese corporations spread world-wide, a centralized, home-based way of managing and running diverse activities in different markets becomes cumbersome and inefficient. The host economies' pressure for localization of multinationals' activities also compels them to become "insiders". The emergence of a tripolar world economy with the centers in North America, Europe, and Pacific Asia is an additional factor in inducing the multinationals to localize their overseas activities as a separate unit in each of these three regions. Hence, Japanese multinationals, especially large ones with globally spread investments, are setting up regional headquarters, say, one in Europe, one in North America, and one in Asia (outside Japan), each in charge of running its regionally co-ordinated network of overseas business operations.

The trends for regionalization or "becoming an insider" are most pronounced in Europe, especially among Japan's auto and electronics makers. For example, Nissan Motors, perhaps the most Europeanized among Japanese multinationals, operates both an assembly plant for passenger cars and a design center in England, an assembly plant for commercial vehicles in Spain, a continental distribution center in Belgium (through which its cars made both in England and Spain are distributed to the rest of Europe), and its regional headquarters, along with its parts, maintenance and financial centers, in the Netherlands. This type of intra-regional integration will be more frequently observed in the future, though it may be still limited to a handful of large corporations.

(4) *Intra-company imports into Japan from overseas investments*

As the assembly-transplanting stage of FDI nears its completion, more and more products assembled overseas by Japanese transplants will be exported back to Japan. This development is partly in response to the export requirements often imposed by the host economies and, more importantly, because this "assemble-abroad, then-import-back" operation of low-to-medium-end products makes perfect economic sense for Japanese multinationals; as they concentrate on manufacturing higher-end products or varieties at home, their lower-end counterparts that have been transplanted overseas need to be imported back in order for them to offer complete lines of products in their home markets.

Honda, which operates two assembly plants for Accord and Civic in the United

[26] JETRO, *Zaiou Seizogyo Keiei no Jittai* (Business Conditions of Japanese Manufacturing in Europe), Tokyo, 1990, pp. 49–51.

States (in Ohio), expects to raise the output to 510,000 units in the mid-1990s, out of which it plans to export 70,000 outside North America, including 50 000 back to Japan. Similarly, standardized electronic consumer goods such as calculators, radios with cassette-recorders, low-end VCRs, and video-players are now produced in the Asian NICs (newly industrializing countries) and shipped back to Japan.

Overseas investments—and intra-company trade—in the food and beverage industry are also rising, as Japanese firms acquire ranches and vineyards with an eye to importing beef and wine back to the home markets. As Japan's home market expands, many more of this type of investment and trade are expected to emerge. The growing home market is thus increasingly captured by Japan's own multinationals through both FDI and intra-company trade.

Concluding observations

In the existing literature, theoretical as well as descriptive, on multinationals' investment activities, the role of FDI as a critical agent of industrial restructuring is little emphasized.[27] Yet the Japanese experience clearly demonstrates how a country can make use of its overseas investment as a vital catalyst of industrial upgrading, serving both as a *transfer* mechanism through which Japan's comparatively disadvantaged industrial activities are passed on to other countries and as a *supportive* system for moving up to higher value-added sectors.

Back in 1975, I made the following observation: "the trend of Japan's FDI . . . reflects more the adaptive behavior of the *entire* Japanese economy to changing world economic conditions than the random market behavior of its individual firms."[28] This description is still valid today. There has been a strong measure of harmony between the private and national interests in Japan's FDI. This is largely because of the dynamics of rapid industrialization that generates powerful economic incentives for the private sector to advance overseas in such a manner to escape from the newly emerged constraints of growth at home and simultaneously to capitalize on the fresh opportunities created in both the home and overseas markets, a phenomenon illustrated in the "industrial restructuring" paradigm of FDI. Overseas investment is a *joint* macroeconomic apparatus, along with industrial, trade and technology policies, that functions as the *élan* vital for Japan's drive to continuous industrial upgrading.

[27] Only a few economists have lately begun to look into the relationship between structural changes and international business. John H. Dunning, for example, observes: "One of the lacunae in the literature on international business is a dynamic approach to its role on economic development." (p. 21) in *Multinationals, Technology and Competitiveness*, London: Unwin Hyman, 1988. Also, John Cantwell identifies an evolutionary process of technological accumulation as the common denominator for economic development and FDI in *Technological Innovation and Multinational Corporations*, Oxford: Basil Blackwell, 1989. See also Kiyoshi Kojima and Terutomo Ozawa, *Micro- and Macro-economic Models of Direct Foreign Investment: Toward a Synthesis*, Hitotsubashi Journal of Economics, Vol. 25, No. 1, June 1984.

[28] Terutomo Ozawa, *Peculiarities of Japan's Multinationalism: Facts and Theories*, Banca Nazionale del Lavoro Quarterly Review, No. 115, December 1975, p. 425.

[16]

Foreign direct investment and economic development

Terutomo Ozawa*

Although transnational corporations play the crucial role as transplanters of technology, skills and access to the world market, how thy facilitate structural upgrading and economic growth in developing countries has not been adequately conceptualized in terms of a theory of economic development. This article develops a dynamic paradigm of TNC-assisted development by recognizing five key structural characteristics of the global economy as underlying determinants. The phenomena of trade augmentation through foreign direct investment, increasing factor incongruity, and localized (but increasingly transnationalized) learning and technological accumulation are identified as three principles that govern the process of rapid growth in the labour-driven stage of economic development and, eventually, the emergence of TNCs from the developing countries themselves also plays a role in this process.

The theory of foreign direct investment (FDI) has so far been built most extensively around industrial organization economics, the theory of the firm and economics of internalization, with a particular focus on either market structure or the firm as a unit of analysis. Surprisingly, it has not been conceptualized in terms of a theory of economic development (how FDI facilitates structural upgrading and economic growth), despite the crucial role TNCs play as generators and transplanters of technology, skills and linkages to the world market.[1] As John Dunning [1988a, p. 21] phrased it, "One of the lacunae in the literature on international business is a dynamic approach to its role in economic development"; there is little systematic exposition of "the impact of (TNC) activity on dynamic comparative advantage".

*Professor of Economics, Colorado State University, Fort Collins, Colorado.

Many developing countries, especially the Asian newly industrializing countries and the emerging newly industrializing countries (Indonesia, Malaysia and Thailand), and more recently some Latin American countries (especially Brazil, Chile and Mexico), are successfully developing by opening up their economies under outward-oriented policies, albeit in varying degrees. Although outward orientation alone is not a sufficient condition for rapid growth, it does create a climate favourable for the transfer by TNCs—and the absorption by local enterprises—of modern managerial, production and marketing technologies which are the *sine qua non* of industrial modernization. Undoubtedly, TNCs are the prime mover behind the industrial dynamism of those rapidly developing countries. Indeed, they are now increasingly counted upon to duplicate their role as development agents elsewhere in the world, especially in Eastern Europe, which is turning outward by adopting pro-market policies. Moreover, TNCs now originate not only from the advanced countries but also more and more from rapidly growing developing countries, notably the newly industrializing countries (NICs), and those TNCs, especially in manufacturing, are more active in other developing countries than in the advanced countries.

There is thus a definite need to incorporate the recognized developmental functions of TNCs into a theory of open-economy development so as to explain, in one integrated theoretical paradigm, what type of investment activities of TNCs can facilitate a process of industrial upgrading and growth in a developing host country and in what ways; and how the developing country itself, once it gathers momentum for industrialization, is, in turn, induced to create its own TNCs in the wake of such TNC-facilitated economic transformation. The aim of the present article is to move in that direction by

[1] There are several important works that touch on some key developmental aspects of FDI by treating it either as an agent of economic growth or as a function of such growth [Dunning, 1981a, 1988a; Dunning and Cantwell, 1990; Tolentino, forthcoming; Cantwell and Tolentino, 1990; Kojima, 1975; Kojima and Ozawa, 1985; Ozawa, 1979, 1991, 1991a, 1991b]. But those works are still tangential to the dynamic (intertemporal) process of structural transformation in developing host countries and do not yet constitute a well-structured *sui generis* theory of TNC-facilitated economic development.

integrating the relevant ideas and analytical models into a synthesized framework so as to delineate the causal relationship between the operations and economic development of TNCs, particularly in the early critical stages of transformation in a labour-surplus developing country.

Structural characteristics of the world economy

Theories of international economic activities are built on some perceived realities of the world economy. Some basic structural characteristics (SC) need to be analysed for the purpose of constructing a dynamic theory of TNC-facilitated structural upgrading. Five of them are described below:

- SC No. 1 *Inter-economy divergences in supply and demand conditions*

The individual economies are divergent in their factor endowment and level of technological competence on the supply side, and in consumers' needs and tastes on the demand side.

- SC No. 2 *Firms as creators and traders of intangible assets*

Individual firms are the major generators, procurers and disseminators of technology, skills and marketing channels.

- SC No. 3 *A hierarchy of economies*

There is a hierarchy (and sub-hierarchy) of economies, globally as well as regionally, with respect to economic development; "leader" economies serve as growth centres for a cohort of "follower" economies. In other words, the individual economies in the world are at various stages of industrial upgrading and per capita income (that is, at different stages of dynamic comparative advantage in terms of levels of technological competence and factor proportions).

- SC No. 4 *Natural (stage-compatible) sequencing of structural upgrading and development*

The currently advanced economies gradually went through multiple phases of industrial upgrading, each phase compatible with its corresponding factor endowment (capital-labour ratio) and technological capacity, albeit over a long period of time by today's standards. This evolutionary path fits the notion of an optimal sequencing of development starting from the initial stage of labour-intensive, low-skill manufacturing (or from the initial stage of natural resource extraction) and moving on to the subsequent stage of relatively physical capital-intensive industrial activities and finally to the more advanced stage of human capital-intensive growth. When that sequencing is allowed to develop, evolutionary forces are set in motion to improve steadily the country's industrial structure, step-by-step, in the most orderly manner.

- SC No. 5 *A strong trend away from inward-looking and towards outward-looking orientation in trade and investment policy, but simultaneously an increasing recognition that Government, though it should deregulate and privatize economic activities so as to unleash the vitality of the private sector, can play a positive role in augmenting the market system.*

There are basically two types of trade and investment regimes: an outward-looking, export-oriented (OL-EO) type, and inward-looking, import substituting (IL-IS) type. But their features are often combined to produce a hybrid regime, though it still can be identified as either more strongly or weakly characterized by one approach. It is now widely recognized that OL-EO is more effective than IL-IS in achieving a faster growth and structural upgrading in the developing countries. Many of the developing countries, once enamoured of IL-IS under economic nationalism, are switching to OL-EO. This new *zeitgeist* is a powerful force for rapid changes in international economic relations. At the same time, the crucial role that Governments can play in supporting the competitive market to assist the development of export industries and structural upgrading is increasingly recognized.

Conventional theories of international business and structural characteristics

From time immemorial, different regions have been exchanging goods with each other, enriching themselves in the process. The predominant basis for trade has been interregional divergences in supplies of primary factors, technological and climatic conditions and patterns of demand (SC No. 1). It was indeed against the supply side of this structural feature alone that first Adam Smith and then David Ricardo set out to theorize the phenomenon of trade and its benefits. Smith introduced the doctrine of absolute advantage, which is built on the economies of scale realizable from an extended market through exports, while Ricardo constructed the doctrine of comparative advantage, which is enhanced by specialization-induced allocative efficiency. The Heckscher-Ohlin theory, too, was founded on the *uneven* distribution of factors of production among countries. All of those trade theories assume the international immobility of factors (hence, the permanence of SC No. 1), partly because of the informational and socio-psychological costs of transactions involved in transferring them[2] and, perhaps more importantly, for the sake of analysing trade alone.

The conventional trade theories thus capture international exchanges of commodities basically as *arbitrage* operations that exploit and profit from any differentials in price and availability of commodities between different locations, the differentials being determined by productivity and factor endowments. Moreover, all of those conventional trade theories treat firms as if they were non-existent because of the usual assumption of perfect competition; hence no firm-specific advantage is permissible. What matters is the country-specific features that bring about discrepancies in pre-trade commodity prices among countries. That is to say, those conventional theories are constructed against the background of SC No. 1 alone.

The theory of FDI as it has derived from Stephen Hymer's seminal work (1976) explicitly recognizes and, indeed, places in the spotlight the existence of firm-specific assets (SC No. 2). In other words, FDI draws on the role of firms as creators and exploiters of intangi-

[2] For example, Smith [1776/1908, p. 291] observed: "[the fortune] of the trader who is obliged frequently to commit it, not only to the winds and the waves, but to the more uncertain elements of human folly and injustice, by giving great credits in distant countries to men, with whose character and situation he can seldom be thoroughly acquainted." Ricardo [1817/1888, p. 77] also stated: "Experience . . . shows that the fancied or real insecurity of capital, when not under the immediate control of its owner, together with the natural disinclination which every man has to quit the country of his birth and connections, and intrust himself, with all his habits fixed, to a strange government and new laws, check the emigration of capital."

ble corporate assets. Hence it is no longer logical to assume perfect competition. FDI is viewed to take place only in imperfect markets. It is, moreover, the firm, not the country, that is the *real actor* motivated to trade intangible assets across national borders.

But the basis for FDI resides not only at the level of individual firms, but also at the level of regions or countries, that is, SC No. 2 should be related back to SC No. 1. Paul Krugman [1990, p. 83] for example, relates firm-level variables to country-level variables:

> ". . . multinational enterprise occurs whenever there exist related activities for which the following is true: There are *simultaneously transaction cost incentives to integrate these activities within a single firm and factor cost or other incentives to separate the activities geographically.*
>
> "Suppose, for example, there is a two-stage product process consisting of a capital-intensive upstream activity and a labour-intensive downstream activity and that (for any of the usual causes) there are compelling reasons to combine these activities inside vertically integrated firms. Suppose further that countries are sufficiently different in factor endowments that, unless these activities are geographically separated, there will be unequal factor prices. Then the result will clearly be the emergence of firms that extend across national boundaries." (Emphasis added.)

As is well known, this way of explaining the emergence of TNCs has long been stated in Dunning's eclectic paradigm [1981b, 1988b], in which the necessary triumvirate conditions for international production are specified: (a) the firm must possess some ownership-specific advantages; (b) to exploit those advantages, internalization (local production under equity ownership) is more beneficial to the firm than arm's-length transactions (exporting and licensing, for example); and (c) overseas locational factors are more favourable than domestic ones. Dunning was among the first who started to combine SC No. 2 and SC No. 1 in theorizing the direct investment of TNCs.

Raymond Vernon's product-life-cycle theory of trade and investment (1966) explicitly brings in SC No. 3, namely a hierarchy of economies, as an additional factor. New products are initially introduced in a high-income country, notably in the United States during the 1950s and 1960s, but will eventually spread to the world, first to other advanced countries, but later to the developing countries, in a trickle-down fashion as the products mature and become technologically standardized. It is a dynamic model of changing comparative advantage and technology transfers. Yet how the transfer of standardized production affects the economic development and structure of the developing host countries is outside the explanatory property of the model. It is a model constructed from the perspective of an innovating advanced country.

Towards a dynamic paradigm of development facilitated by TNCs

Any developing country, if it is serious about raising its standard of living, must open its economy so as to avail itself of opportunities to trade, interact with and learn from the already advanced. In fact, ever since the industrial revolution in England, industrialization in the rest of the world, wherever successful, has been a "derived" phenomenon. Continental Europe succeeded by following the United Kingdom's footsteps through commercial contracts and conscious efforts for learning and emulation [Landes, 1969]. So did the United States: "America started off as a copier" [Thurow, 1985].[3] Likewise, Japan's economic miracles in both the pre- and post- Second World War periods have been based on that mechanism of learning and emulation.

A process of learning and emulation necessarily means the existence of a hierarchy of economies in terms of stages of economic development and national wealth (SC No. 3), *a structure that creates opportunities for the less developed to emulate the advanced—and for the advanced to transfer their knowledge and skills down the hierarchy.* The advanced countries are the rich reservoirs of industrial technology, information and experiences which the followers can tap. They also provide the promising export markets from

[3] Thurow [1985, p. 1] continues: "New England's textile mills hired craftsmen (we would now call them engineers) who had worked in, or toured, the British textile mills and had memorized or written down enough of what they had seen to copy those mills in the New World. American history school books remember this copying as a good example of Yankee ingenuity. British history school books see it in another light: as theft. America stole British technology."

which the less developed can earn precious hard currencies. As Adam Smith observed [1776/1908, p. 378] "A nation that would enrich itself by foreign trade is certainly most likely to do so when its neighbours are all *rich and industrious*" (emphasis added). Nowadays, those trading and learning opportunities are all the more enhanced, because the advanced countries themselves take the initiative to make use of hierarchical relations by way of the activities of their TNCs and their official economic and technical assistance.

But to exploit those *hierarchical externalities* fully (positive externalities emanating from SC No. 3), a developing country must *align its pattern of comparative advantage and its stage of development with advanced countries*. That is to say, a stages-compatible order of sequencing structural upgrading (SC No. 4) needs to be achieved. For the sequence of development should not be jumbled as is the case with the IL-IS approach; the process needs to be organized step by step, and stage by stage, in an evolutionary fashion along the lines demonstrated in the past by the currently advanced countries. This means that, when a country is still scarce in human and physical capital but abundant in labour (unskilled and semi-skilled), any attempt to build a capital-intensive, skill-requiring industry is ineffective and unachievable. Such a country must first focus on introducing labour-intensive industries to employ its most abundant factor. After all, economic development is characterized by the steady accumulation of physical and human capital, which produces the sequential pattern of dynamic comparative advantage in any market-based, growing economy. Such a path assures *compatibility* between the factor intensities and technological requirements of the economic activities pursued at a particular stage and the domestic availability of appropriate factors and technological capacity.

Indeed, it was Adam Smith (1776/1908) who emphasized the importance of "a natural order of things" in the developmental sequence of an economy (the "natural progress of opulence"). He envisaged intersectoral complementarities (first, between "town" and "country", between agriculture and manufacturing, and then, later, between manufacturing and foreign trade) and the accumulation of capital and the development of political/technological/organizational capacities ("order and good government" and the natural progression of manufactures from the "necessities" to the "conveniences" and to the "elegancies of life") [pp. 290-314].

In that regard, the "stages theory of competitive development" recently introduced by Michael Porter (1990) deserves special attention. He maintains that, "despite the diversity of most economies, we can identify a predominant or emergent pattern in the nature of competitive advantage in a nation's firms at a particular time" (p. 545) by way of four distinct stages: (i) factor-driven, (ii) investment-driven, (iii) innovation-driven, and (iv) wealth-driven. But [Porter, 1990, p. 546] "the first three stages involve successive upgrading of a nation's competitive advantages and will normally be associated with progressively rising economic prosperity", while "the fourth stage is one of drift and ultimately decline".

The factor-driven stage (akin to Smith's agriculture-first phase) is characterized either by natural-resource-based activities (primary extraction) or by labour-intensive manufacturing. The investment-driven stage, on the other hand, is associated with the manufacturing of intermediate and capital goods (heavy and chemical industrialization) and infrastructural building (housing, transportation, communications and public works construction). The innovation-driven stage arrives when a country is human-capital-abundant and active in research and development (R&D). Most developing countries are in the factor-driven stage, and some are already well on their way to the investment-driven stage.

It is important to keep in mind that *a particular stage of competitive development is associated with a particular pattern of export competitiveness.* The factor-driven stage is related to factor-based trade advantages (in either primary goods or labour-intensive goods), the investment-driven stage to scale-based advantages (in large-scale, capital-intensive goods), and the innovation-driven stage to R&D-based advantages (in high-tech manufactures). Thus economic growth and transformation is accompanied by the changing patterns of dynamic comparative advantage.

Interestingly enough, moreover, *the pattern (nature and direction) of FDI, both inward and outward, changes* pari passu *with the structural transformation of the economy.* For example, the beginning of the factor-driven stage attracts resource-seeking or labour-seeking inward FDI. The transition from the labour-driven to the investment-driven stage generates outward investments towards lower-wage countries in labour-intensive manufacturing (and in resource extraction abroad, particularly if the economy happens to be natural-resource-scarce) and attracts inward investments in capital and intermediate goods industries. Similarly, the transition from the investment-driven to the innovation-driven stage brings about *simultaneously* inward investments in technology-intensive industries and outward investments in intermediate goods industries. This orderly flow of investment tends to occur as long as the country pursues an unjumbled "progress of opulence".

As shown in figure I, those stages can be distinguished in terms of the changing endowment proportions of three major factors used in a nation's overall industrial activity: resources (both natural and labour), physical capital and human capital. In other words, the nation grows by upgrading its structure as its factor and technological endowments change with the accumulation of physical and human capital (relative to resources): *the higher the per capita gross national product (GNP), the larger the per capita stock of physical and human capital relative to that of resources and the larger the per capita stock of human capital relative to that of physical capital.* The economy continuously evolves to develop new comparative advantages by shifting from technologically less sophisticated, low-productivity products to more sophisticated, higher-productivity industrial activities (analogous to Smith's sequence of "necessities—conveniences—elegancies"). For example, this changing pattern of dynamic comparative advantage may be measured in terms of an index of the revealed comparative advantage in unskilled labour-intensive goods and an index of the revealed comparative advantage in physical/human capital-intensive goods as schematically illustrated in the figure. At the same time, different stages-based patterns of FDI, both inward and outward, appear over a certain span of time, a span determined by how rapidly a given economy industrializes, as outlined.

Indeed, the stages-based evolutionary progression of FDI has been most clearly demonstrated by Japan's rapid post-war structural transformation and its concomitant FDI. As elaborated elsewhere [Ozawa, 1991a, 1991b, 1991c], Japan's industrial structure

Figure I. Stages of development, changing factor proportions, and dynamic comparative advantage

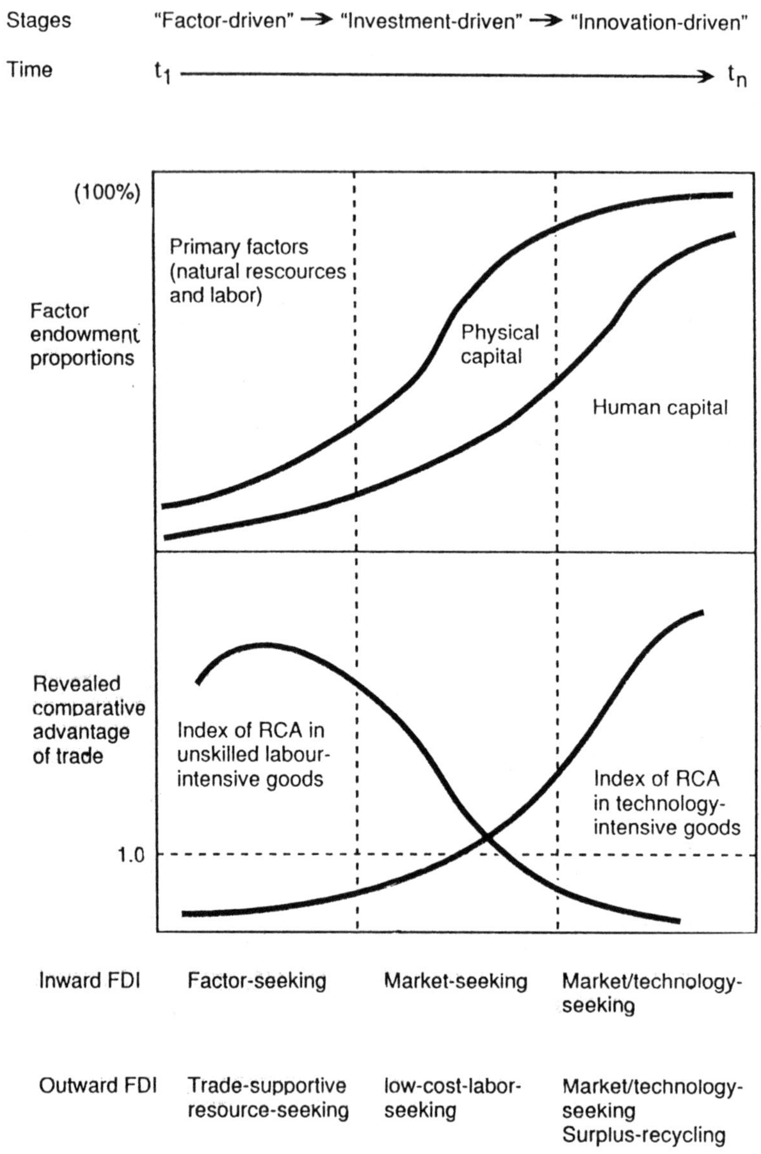

has so far metamorphosed in three continuous stages, accompanied by different stages-specific patterns of FDI—and is currently undergoing the latest, fourth, stage:

Phase I: Expansion of labour-intensive manufacturing in textiles, sundries and other low-wage goods ("what may be called Heckscher-Ohlin industries) (1950 to the mid-1960s).

Accompanied by selective imports of capital goods (embodied technology) and purchase of licensing agreements (disembodied technology), and followed by the "elementary (low-wage labour-seeking)" type of FDI in standardized, labour-intensive industries in the neighbouring Asian countries (starting in the late 1950s, but notably after the mid-1960s and growing strongly, especially after the revaluations of the yen in the early 1970s).

Phase II: Scale economies-based modernization of heavy and chemical industries, such as steel, shipbuilding, petrochemicals and synthetic fibres ("non-differentiated Smithian industries") (the late 1950s to the early 1970s).

Accompanied by the incessant adoption and adaptation of advanced Western technology, mostly through licensing but also via joint ventures (inward FDI) in petrochemicals, machinery and chemicals, and the "resource-seeking" (later followed by the "house-cleaning") type of outward FDI in resource-extractive and processing industries (most actively throughout the 1960s and 1970s).

Phase III: Assembly-based mass production of consumer durables, such as automobiles and electric/electronics goods" ("differentiated industries") (the late 1960s to the present).

Continuous adaptation of, and improvement on, imported technology and stepped-up domestic R&D, and accompanied by the "export-substituting-cum-surplus-recycling" type of outward FDI in overseas assembly operations in automobiles, electronics and related parts and components, as well as banking and finance (especially after the Plaza Accord of 1985), most actively in the United States and Europe.

Phase IV: Mechatronics-based flexible manufacturing, small-lot, multi-variety production, along with innovations in HDTV, new materials, fine chemicals, advanced micro-chips and opto-electronics ("Schumpeterian industries") (the early 1980s onwards).

Hosting inward FDI in chemicals, pharmaceuticals and machinery, and generating the "triadization" type of outward FDI in which Japanese TNCs are establishing the intraregional clusters of integrated operations and strategic alliances in the Americas, Europe and the Pacific Rim.

The Japanese experience thus delineates the *close inter-relatedness* between structural upgrading, dynamic comparative advantage and FDI, along the paths of its physical/human-capital-intensive factor endowment and technological progress. Because of the rapidity of Japan's economic transformation, its structural progression (from the "H-O industries" to the "non-differentiated Smithian industries", to the "differentiated Smithian industries", and finally to the "Schumpeterian industries") actually retraces, in a highly time-compressed fashion, the evolutionary changes previously experienced by the advanced Western countries over a much longer span of time. And international business is *the main cause* of this structural growth acceleration.

Indeed, there have been a number of empirical studies made on the relationship between a nation's changing pattern of comparative advantage in manufactured goods and the accumulation of physical and human capital.

Bela Balassa's study [1979/1989, p. 26], for example, concluded that:

> "The empirical estimates show that inter-country differences in the structure of exports are in a large part explained by differences in physical and human capital endowments. The results lend support to the 'stages' approach to comparative advantage,

according to which the structure of exports changes with the accumulation of physical and human capital. The approach is also supported by inter-temporal comparison for Japan, which indicates that Japanese exports have become increasingly physical capital and human-capital-intensive over time."

He then extracted (p. 26) some important policy implications:

"To begin with, [those findings] warn against distorting the system of incentives in favour of products in which the country has a comparative disadvantage. The large differences shown among product categories in terms of their capital intensity point to the fact that there is a substantial penalty for such distortions in the form of the misallocation of product factors. This will be the case in particular when the system of incentives is biased in favour of import substitution in capital-intensive products and against exports in labour-intensive products."

These types of distortions are naturally reduced when a developing country pursues an OL-EO strategy of development rather than an IL-IS one (SC No. 5). And it has been empirically and amply proved, and now accepted as an axiom, that the OL-EO approach is far superior to its IL-IS counterpart in growth performance and industrial upgrading in the developing countries, for the former allows efficacious market-guided evolutionary forces to operate at maximum efficiency.

In short, those structural characteristics of the global economy (SC Nos. 3, 4 and 5) that have not been explicitly reckoned with as determinants in the conventional theories of FDI constitute a new global environment highly conducive to an expansion of trade, FDI, knowledge transfer and economic growth, a climate significantly favourable for TNCs. Those characteristics have to be brought into analysis as explanatory factors in an effort to construct a theory of FDI in the field of economic development.

Comparative-advantage-augmenting type of foreign direct investment

Throughout the present article, the term FDI is used in a very broad sense: it is meant to describe the transfer of *development ingredients* of all sorts by TNCs (that is, intangible corporate assets, finance and links with global markets) in *both* the traditional form of investment (whole or majority equity ownership) and "new" forms (minority ownership and a variety of non-equity contractual arrangements, such as licensing, managerial contracts and turnkey operations [Oman, 1984]).

What is crucial for outward-looking, export-based development is that FDI needs to be approached in such a manner that the developing countries' existing or potential comparative advantage can be fostered and fully maximized, that is, in a comparative-advantage-augmenting fashion. Here, the notion of pro-trade FDI (and that of anti-trade FDI) introduced by Kojima (1975) is quite relevant. Kojima's macro-ecnomic theory of FDI [Kojima, 1975; Kojima and Ozawa, 1984] is built on two key propositions. One is the Ricardian doctrine of comparative advantage, the other his own proposition as a supplement:

Proposition I: Countries gain from trade and maximize economic welfare when they export comparatively advantaged goods and import comparatively disadvantaged goods.

Proposition II: Countries gain *even more* from an expanded basis for trade when intangible assets are transplanted from the home countries' comparatively disadvantaged industries onto the host countries' comparatively advantaged ones (both current and potential).

The second proposition indicates a mutually beneficial type of FDI that parallels the mutually gainful type of trade posited in the first proposition. Ricardo built his trade model on the assumed immobility of factors (hence the persistence of SC No. 1). Furthermore, he thought that, even if factor movement occurred, it would be of the anti-trade type—and for that matter, indeed, of the hollowing-out type [Ricardo, 1817/1888, p. 77]:

> "It would undoubtedly be advantageous to the capitalists of England, and to the consumers in both countries, that under [the circumstances of higher labour and productivity in Portugal in absolute terms] the wine and the cloth should *both* be made in Portugal, and therefore that the capital and labour of England employed in making cloth should be removed to Portugal for that purpose." (Emphasis added.)

Thus, once factor mobility is admitted, Portugal alone ends up producing *both* goods; hence there would be no basis for trade. The end result is that Portugal flourishes, whereas England is hollowed out and languishes. Surprisingly enough, however, it did not occur to Ricardo that if, *instead of* moving to Portugal "the capital and labour of England employed in making cloth", Portugal's secret (technology) of higher productivity in cloth is transferred to England through FDI, not only is England saved from hollowing out, but also both nations can prosper *even more*, since England's comparative advantage in cloth is now *enhanced* by such technology transfers. More surprisingly, Ricardo resorted to the doctrine of absolute advantage when it came to factor movement.

Kojima's distinction between pro-trade FDI and anti-trade FDI is criticized as "normative" [Gray, 1985], but it can be converted into "positive" analysis when it is combined with the structural characteristics of Nos. 3, 4 and 5, since the comparative-advantage-oriented policy of market-friendly OL-EO economies is such that pro-trade FDI rather than anti-trade FDI will be automatically attracted to such a regime. The IL-IS regimes set up in labour-surplus, capital-scarce developing countries are the ones that attract anti-trade FDI into their highly protected, stage-incompatible, import-substituting industries. The phenomenon of a comparative-advantage expansion made possible by FDI in the OL-EO regimes will hereafter be identified as trade augmentation through FDI.

Outward orientation and supergrowth in the initially labour-surplus developing countries

The income growth effect envisaged in the static Ricardian model of trade (which allows no transfer of factors and no structural change) can be summarized as follows:

Inter-economy discrepancies in supply ⟶ Static comparative advantage ⟶ Trade ⟶ Income growth

According to Ricardo, as gains from trade are realized through specialization and reallocation of resources into comparatively advantaged industries (that is, attainment of allocative efficiency), national income rises, but no outward shift of the country's production possibility curve occurs because there is *neither* factor growth *nor* technological progress. The whole purpose of the Ricardian model is to show the basis for, the direction of, and gains from, trade and nothing more.

Yet this *static* model can be transformed into a *dynamic* structural upgrading model once the role of TNCs as a facilitator of industrial transformation and the additional set of structural characteristics are explicitly brought into analysis as causative factors.

Economic development under OL-EO policy is a function of dynamic comparative advantage, but here the magnitude of comparative advantage is not only determined by SC No. 1 alone, but also enhanced—and upgraded—by the confluence of SC Nos. 2, 3, 4 and 5. This is *a dynamic paradigm of FDI-facilitated development* (hereafter called the dynamic paradigm) which depicts the *magnified* power of trade as an engine for growth. The power of trade is magnified because (i) a strong outward orientation reduces market distortions and encourages competition; (ii) the basis for trade is expanded by comparative-advantage-augmenting (CAA) FDI (inward as well as outward), hence a greater expansion of trade; and (iii) self-propelling market forces are set in motion to upgrade the country's industrial composition towards higher value-added sectors in an "unjumbled" manner, given a market-friendly environment (including incentives and supportive infrastructure) by the Government. The net result is a supergrowth of OL-EO economies accompanied by rapid structural upgrading, the type of "economic miracles" exhibited by the Asian newly industrialized countries (NICs). A high correlation between outward orientation and supergrowth has been empirically confirmed in many situations [Balassa, 1989; World Bank, 1991]. But the causative links between those two phenomena are still left unexplained. One possible growth mechanism is suggested for this black box:

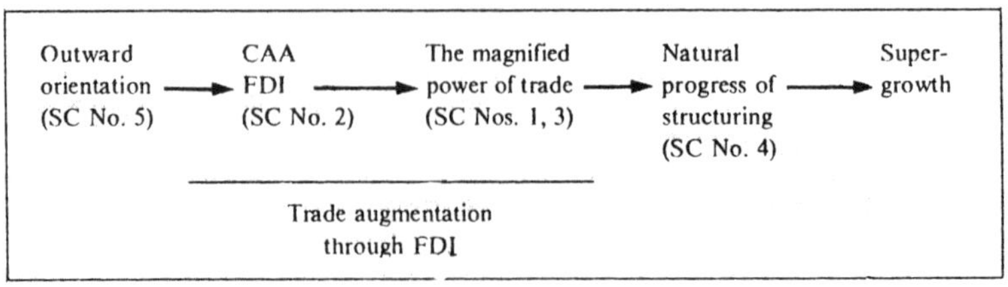

More specifically, how this scenario unfolds itself over a time period, say from t1 to t5, can be explained in terms of the dynamic paradigm. Consider a typical developing country that is initially plagued with high unemployment and low income (figure II).

(t1) A labour-abundant developing country (with high unemployment and underemployment) adopts an outward-looking, export-focused path of development (SC No. 5), thereby reducing market distortions and setting free the market forces for stage-compatible economic activities. A market-friendly economic system thus emerges with a strong outward orientation towards trade and investment interactions with advanced countries. The country thus stands to benefit from the *hierarchical externalities* (SC No. 3).

(t2) This initial condition attracts inward FDI in standardized labour-intensive manufacturing activities, that is, in a comparative-advantage-augmenting fashion (in a direction conducive to trade augmentation through FDI). At the start, the developing host country usually supplies labour from its unemployed (mostly urban) and underemployed (mostly rural) labour pool at low wage rates (a process well postulated by Arthur Lewis (1954) in his model of unlimited labour supply). TNCs, often in joint ventures with local interests, are able to develop export-oriented manufacturing without generating upward pressure on local wages, although they usually pay a real wage rate higher than the local average, especially in fringe benefits (for example, lunches, basic health care and training). Low wages are the basis for price competitiveness. Exports are mostly of the intra-firm type, either purchased as intermediate goods by the investing TNCs or marketed through the distribution channels of TNCs (on the theoretical level, that phase of export growth corresponds to the Smithian "vent-for-surplus" phenomenon [Myint, 1958]).

(t3) But the more labour-seeking FDI flows in, the sooner the disappearance of unemployment, if not underemployment. Upward pressures on wages appear, and local wages begin to rise (that is, the Lewisian phase of unlimited labour supply ends along with the Smithian "vent-for-surplus" phenomenon, and the "factor-price-

Figure II. Outward orientation, FDI-magnified trade competitiveness and structural upgrading in a developing country

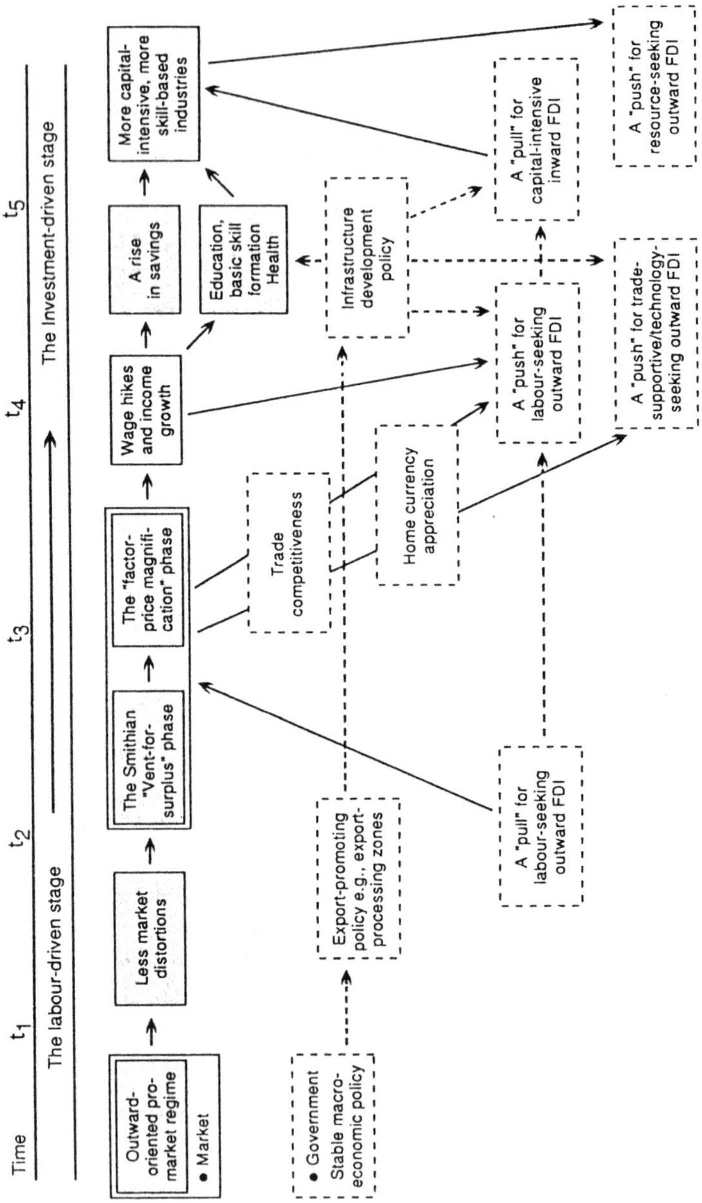

magnification" effect [Stolper and Samuelson, 1941] now takes over, as long as labour utilization continues to be pursued under OL-EO policy).

(t4) The rising wages and the increasing availability of jobs, though still in low-productivity (hence low-wage by international standards) sectors, will result in a rapid rise in national income and an expansion of domestic markets. Wage rates are still low, but all family members (not only the traditional breadwinner, the father, but also the mother and their adolescent children) who seek employment can find jobs; hence a rise in family income.[4]

Given the appropriate incentives (for example, attractive interest rates on deposits, tax exemptions and hard-currency-denominated deposits), this increased household income can generate domestic savings.[5] Employment also leads to basic skill formation via training and a stronger desire to upgrade skills, a higher motivation for learning. The country's factor endowment begins to change from low-skilled labour abundance to a relatively more capital abundance and a greater skill accumulation.

As a consequence, a *factor incongruity* emerges soon between the hitherto promoted export industries (whose competitiveness depends on a high input ratio of low-wage, unskilled labour to capital) and the country's changing factor endowment (which now becomes less and less unskilled-labour-abundant). This is an ineluctable outcome of the labour-driven process of development and may be called *the principle of increasing factor incongruity*. Those grass-roots developments (improved job opportunities, rising wages, increased household savings and the desire to seek education and upgrade one's skills) are crucial in shifting the country's factor endowment towards more human- and physical-capital abundance, endogenously paving the way to the next phase of investment-driven industrialization.

[4] The rapid build-up of the garment industry in Bangladesh provides an interesting example: "Women who never worked before are earning between $40 and $55 a month--far above the projected 1991 national per capita income of $202 a year. Many of these women's husbands, meanwhile, can get work only as day labourers or rickshaw drivers, earning around $1 a day." *The Wall Street Journal*, 6 August 1991, p. A8.

[5] Balassa (1989) points out a strong correlation between export growth and domestic savings. What is argued in this article suggests one causal link.

As wages rise, the country's comparative advantage in labour-intensive industries naturally wanes. Given the macroeconomic stability that ensures a competitive exchange rate for the country's currency at the start of economic development, the more successful labour-intensive exports are (that is, the greater the improvement in the country's balance-of-payments conditions), the greater the chances that the currency will begin to appreciate. Higher wages and the appreciated currency thus render low-productivity, labour-intensive industries less competitive, compelling domestic industry to move into higher value-added, more capital-intensive sectors. At that point, in order to protect their hard-won export markets, local entrepreneurs, as well as TNCs, begin to transplant their existing operations to other developing countries where they can still find a supply of low-wage labour. Of course, those local firms that succeed in moving upmarket (via an introduction of labour-saving measures and new higher-value-added varieties, or moving into a more capital-intensive industry) may remain at home.

This is an interesting—and seemingly paradoxical—turn of events: *a once highly labour abundant (hence low-wage) developing country itself has metamorphosed, over a relatively short span of time, into an active foreign investor, seeking lower-wage labour in other developing countries.*[6] Its firms now possess their own advantages and experiences freshly gained over the period of from t1 to t4 (SC No. 2), mostly through the "apprenticeship" of firms to foreign technical assistance [Amsden, 1989, p. 20]. The Japanese, back in the 1950s, 1960s and 1970s, and more recently the Asian NICs have amply demonstrated the capability of followers to come up with their situation-specific improvements on imported technology as well as their own innovations. This may be called *localized learning and technological accumulation* [Lall, 1983; Stiglitz, 1987; Cantwell and Tolentino, 1990] and has been emphasized by Cantwell and Tolentino (1990) as the key determinant of FDI from developing countries. That becomes even more relevant in the later stages of development that are increasingly characterized by a growing abundance of human capital.

[6] For recent stories on TNCs from developing countries, see, for example, Fujita (1990), Cantwell and Tolentino (1990), and Tolentino (forthcoming).

What is more, the momentum of learning in both production and marketing skills continues and even takes the form of technology-seeking, information-gathering and trade-supportive outward investment in advanced countries. Thus "localized learning" now actually turns to more and more "internationalized learning".

Increasing factor incongruity is also observed as primary resource-seeking outward investment and is induced in search of secure supplies of natural resources abroad. In particular, as a natural-resource-indigent "Ricardian" economy experiences rapid industrialization, it inevitably becomes more and more dependent on overseas resources [Ozawa, 1982]. This incongruity between the resource requirements of the economy and its own endowments (and hence the rising vulnerability of dependence on imported industrial resources) can only be dealt with by means of FDI in resource-extractive industries abroad so as to secure long-term supply bases. All this outward FDI is fully compatible with and supportive of the evolving pattern of comparative advantage, enhancing the basis for trade (that is, trade augmentation through FDI is in operation).

(t5) By now, the country is on the threshold of the investment-driven stage in which more physical-capital- and more skill-intensive industries are to be built; it may not be long before the country is classified as a newly industrialized country. To build higher value-added, more scale-dependent industries, the role of the Government becomes even more crucial in providing physical infrastructural facilities, education, public health, R&D incentives and other "public goods". As compared to the labour-driven stage, which can be left largely to the dictates of the market with the minimum selective involvement of Government, the development of more advanced industries entails an appropriate management of externalities (market imperfections), since education and training (which are the causes of positive externalities) are now the key industrial inputs, along with the minimum scale of operation required by indivisible capital investments and the accompanying economies of scale. The Government is thus called upon to provide the stage-appropriate institutions and infrastructural investments to support dynamic comparative advantage [Ozawa, 1987; Dunning, 1988a and 1991; World Bank, 1991].

Although the above scenario is presented in a highly stylized fashion, it nevertheless fits very nicely with the recent experience of Asian newly industrialized countries, which have been skilfully taking advantage of *both* inward and outward FDI as catalysts of industrial transformation. In fact, they are already in the midst of investment-driven industrialization or even at the beginning of innovation-driven growth in some sectors.

In this stages-theoretic framework of analysis, it is often said that Asian NICs are following the footsteps of Japan, and that emerging NICs (Indonesia, Malaysia, the Philippines and Thailand) are, in turn, duplicating the experiences of NICs and that others (Bangladesh, China, India, Pakistan and Sri Lanka) are about to repeat the same labour-driven episode. That sequential pattern of development in the Asian-Pacific region is often referred to as "a wild-geese flying formation" (after the phrase introduced by Kaname Akamatsu in 1935[7]). But Japan itself has been closely emulating the West in its effort to catch up and take over—and has very quickly succeeded in going through the first three stages described by Porter, thereby emerging as a high advanced economy. Japan's rapid industrial transformation in the period following the Second World War discussed earlier illustrates the efficacy of a stages-based "progress of opulence" and may serve as an important paradigm for developing countries.

The wild-geese flying formation or a stage-based development can alleviate the "fallacy of composition" problem of OL-EO industrialization: if the developing countries adopt the OL-EO approach all at once, will there be enough export markets? As Bela Balassa [1989, p. 28] succinctly phrased it:

> "The stages approach to comparative advantage . . . permits one to dispel certain misapprehensions as regards the foreign demand constraint for manufactured exports under which developing countries are said to operate. With countries progressing on the comparative advantage scale, their exports can supplant the exports of countries that graduate to a high level."

[7] See Kaname (1961). The interaction between TNCs and the flying-geese formation is explored in Ozawa (1990).

TNCs are no doubt the facilitators of that type of *market recycling* as they are able to shift manufacturing activities from one location to another, in a fashion compatible with the prevailing factor and technological endowments of the economies of host countries, within a hierarchy of economies (SC No. 3).

Conclusions

Although the functions of TNCs as providers of technology, management expertise, finance and links with the world market have often been mentioned, the role of their FDI in industrial upgrading and growth has not been incorporated into a theory of outward-oriented economic development. The dynamic paradigm presented here recognizes five crucial structural characteristics of the world economy as explanatory factors. Those structural features fall under the rubric of the environment-system-policy frame of analysis proposed by John Dunning (1988a). Hence, the dynamic paradigm may be construed as one detailed structural model of this genre.

Focusing on the case of a representative developing country, that is, an initially labour-surplus economy, the dynamic paradigm traces out the *intertemporal* sequence of interactions between FDI (first inward and, later, outward) and structural transformation, a sequence that results in an evanescent pattern of comparative advantage during the course of labour-driven industrialization. That is the stage from which the Asian NICs have just graduated, and which many other developing countries are expected to go through in their early phase of development.

The operational mechanism of the dynamic paradigm so far rests on three basic principles: those of trade augmentation through FDI (according to Kojima, 1975), of increasing factor incongruity and of localized (but increasingly transnationalized, at least in part) learning and technological accumulation (according to Cantwell and Tolentino, 1990). Some more conceptual work is needed to explore the investment-driven and the innovation-driven stages in order to complete the stages theory of TNC-facilitated growth in its entirety. No doubt the demand-side factors play a greater and greater role as development proceeds to later stages. This evolutionary aspect of demand needs to be explicitly conceptualized. Perhaps what may be tentatively called the principle of catch-up in income and increasing diversity in tastes is a relevant consideration. In that regard, the Japanese experience, which encompasses all the developmental stages from the factor-driven to the innovation-driven in a highly condensed time-capsule, may serve as an important guidepost. Yet such a comprehensive analysis requires far more space than is available in the present article. Only the first dimension has been elaborated, which is none the less the most relevant to the role of TNCs as "jump-starters" for economic development. ■

References

Akamatsu, Kaname (1950). A theory of unbalanced growth in the world economy. *Weltwirtschaftliches Archiv*, 86, pp. 196-217.

Amsden, Alice H. (1989). *Asia's Next Giant: South Korea and Late Industrialization.* Oxford: Oxford University Press.

Balassa, Bela (1979/1989). The changing pattern of comparative advantage in manufactured goods. *Review of Economics and Statistics*, 61 (May), pp. 259-266. As reproduced in *Comparative Advantage, Trade Policy and Economic Development.* New York: New York University Press.

________ (1989). *New Directions in the World Economy.* New York: New York University Press.

Cantwell, John A. (1989). *Techological Innovation and Multinational Corporations.* Oxford: Basil Blackwell.

________ and Paz Estrella E. Tolentino (1990). *Technological Accumulation and Third World Multinationals.* Discussion Paper No. 139, Department of Economics, University of Reading, Series B, III.

Dunning, John H. (1981a). Explaining the international direct investment position of countries: Toward a dynamic or developmental approach. *Weltwirtschaftliches Archiv*, 11, pp. 30-64.

________ (1981b). *International Production and the Multinational Enterprise.* London: George Allen & Unwin.

———— (1988a). *Multinationals, Technology and Competitiveness.* London: Unwin Hyman.

———— (1989b). *Explaining International Production.* London: Unwin Hyman.

———— (1991). Governments—markets—firms: towards a new balance? *The CTC Reporter*, 31 (Spring), pp. 2-7.

———— and John A. Cantwell (1990). The changing role of multinational enterprises in the international creation, transfer and diffusion of technology. In *Technology Diffusion and Economic Growth: International and National Policy Perspectives*, F. Arcangeli, P. A. David and G. Dosi, eds. Oxford: Oxford University Press.

Fujita, Masataka (1990). TNCs from developing countries. *The CTC Reporter*, 30 (Autumn), pp. 42-45.

Gray, Peter H. (1985). Multinational corporations and global welfare: an extension of Kojima and Ozawa. *Hitotsubashi Journal of Economics*, 26 (December), pp. 125-133.

Hymer, Stephen (1976). *The International Operations of National Firms.* Cambridge, Massachusetts: The MIT Press.

Kojima, Kiyoshi (1975). International trade and foreign investment: substitutes or complements. *Hitotsubashi Journal of Economics*, 16 (June), pp. 1-12.

———— and Terutomo Ozawa (1984). Micro- and macro-economic models of direct foreign investment: toward a synthesis. *Hitotsubashi Journal of Economics*, 25 (June), pp. 1-20.

———— (1985). Toward a theory of industrial restructuring and dynamic comparative advantage. *Hitotsubashi Journal of Economics*, 26 (December), pp. 135-145.

Krugman, Paul (1990). *Rethinking International Trade.* Cambridge, Massachusetts: The MIT Press.

Lall, Sanjaya (1983). Multinationals from India. In *The New Multinationals*, Sanjaya Lall, ed. Chichester: John Wiley & Sons, pp. 21-87.

Landes, David S. (1969). *The Unbound Prometheus.* Cambridge: Cambridge University Press.

Lewis, W. Arthur (1954). Economic development with unlimited supplies of labour. *The Manchester School of Economic and Social Studies*, 12 (May), pp. 139-191.

Myint, Hla (1958). The "classical theory" of international trade and the underdeveloped countries. *Economic Journal*, 68 (June), pp. 317-337.

Oman, Charles (1984). *New Forms of International Investment in Developing Countries*. Paris: OECD.

Ozawa, Terutomo (1979). A newer type of foreign investment in third world resource development. *Rivista Internazionale di Scienze Economiche e Commerciali*, 29 (December), pp. 1,133-1,151.

________ (1987). Can the market alone manage structural upgrading? A challenge posed by interdependence. In *Structural Change, Economic Interdependence and World Development*, 1, J. H. Dunning and M. Usui, eds. Basingstoke, United Kingdom: Macmillan, pp. 45-61.

________ (1990). Multinational corporations and the "flying-geese" paradigm of economic development in the Asian Pacific region, paper presented at the 20th Anniversary World Conference on Multinational Enterprises and 21st Century Scenarios, Tokyo (July).

________ (1991a). Japanese multinationals and 1992. In *Multinationals and Europe 1992*, B. Burgenmeier and J. L. Mucchielli, eds. London: Routledge, pp. 135-154.

________ (1991b). The dynamics of Pacific Rim industrialization: how Mexico can join the Asian flock of "flying geese". In *Mexico's External Relations in the 1990s*, Riordan Roett, ed. Boulder and London: Rienner, pp. 129-154.

________ (1991c). Japan in a new phase of multinationalism and industrial upgrading: functional integration of trade, growth and FDI. *Journal of World Trade*, 25 (February), pp. 43-60.

Porter, Michael (1990). *The Competitive Advantage of Nations*. New York: Free Press.

Ricardo, David (1817/1888). *Principles of Political Economy and Taxation.* In *The Works of David Ricardo*, J. R. McCulloch, ed. London: John Murray, pp. 1-584.

Smith, Adam (1776/1908). *An Inquiry into the Nature and Causes of the Wealth of Nations.* London: Routledge; New York: E. P. Dutton.

Stolper, W. F., and P. A. Samuelson (1941). Protection and real wages. *Review of Economic Studies*, 9 (November), pp. 58-73.

Stiglitz, J. E. (1987). Learning to learn, localised learning and technological progress. In *Economic Policy and Technological Performance*, P. Dasgupta and P. Stoneman, eds. Cambridge: Cambridge University Press, pp. 125-153.

Tolentino, Paz Estrella E. (forthcoming). *Technological Innovation and Third World Multinationals.* London: Routledge.

Thurow, Lester C. (1985). *The Management Challenge: Japanese Views.* Cambridge, Massachusetts: The MIT Press.

Vernon, Raymond (1966). International investment and international trade in the product cycle. *Quarterly Journal of Economics*, 80 (May), pp. 190-207.

World Bank (1991). *World Development Report.* Washington, D.C.: World Bank Publications.

[17]

The investment development path revisited
Some emerging issues

John H. Dunning and Rajneesh Narula

PART 1: THE THEORY

The nature of the investment development path

The notion that the outward and inward direct investment position of a country is systematically related to its economic development, relative to the rest of the world, was first put forward by John Dunning in 1979, at a conference on 'Multinational Enterprises from Developing Countries' which took place at the East–West Center at Honolulu.[1]

Since then the concept of the investment development path (IDP)[2] has been revised and extended in several papers and books (Dunning 1981, 1986, 1988a, 1993; Narula 1993, 1995; Dunning and Narula 1994). The following paragraphs summarize the state of thinking – prior to this volume – on the nature and characteristics of the IDP.

The IDP suggests that countries tend to go through five main stages of development and that these stages can be usefully classified according to the propensity of those countries to be outward and/or inward direct investors. In turn, this propensity will rest on the extent and pattern of the competitive or ownership specific (O) advantages of the indigenous firms of the countries concerned, relative to those of firms of other countries; the competitiveness of the location-bound resources and capabilities of that country, relative to those of other countries (the L specific advantages of that country); and the extent to which indigenous and foreign firms choose to utilize their O specific advantages jointly with the location-bound endowments of home or foreign countries through internalizing the cross-border market for these advantages,[3] rather than by some other organizational route (i.e. their perceived I advantages).

2 The investment development path revisited

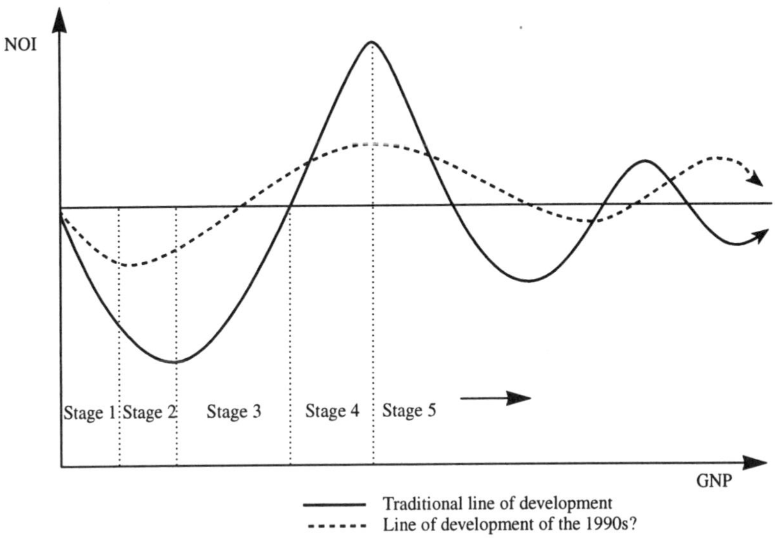

Figure 1.1 The pattern of the investment development path
Note: Not drawn to scale – for illustrative purposes only

A diagrammatic representation of the IDP, which relates the net outward investment (NOI) position of countries (i.e. the gross outward direct investment stock less the gross inward direct investment stock) – as a continuous line – is presented in Figure 1.1. We shall briefly summarize the main features of these stages, but pay particular attention to Stage 5, which we did not consider in our earlier writings.

Stage 1

During the first stage of the IDP path, the L specific advantages of a country are presumed to be insufficient to attract inward direct investment, with the exception of those arising from its possession of *natural* assets. Its deficiency in location-bound *created* assets[4] may reflect limited domestic markets – demand levels are minimal because of the low per capita income – inappropriate economic systems or government policies; inadequate infrastructure such as transportation and communication facilities; and, perhaps most important of all, a poorly educated, trained or motivated labour force. At this stage of the IDP, there is likely to be very little outward direct investment. *Ceteris paribus*, foreign firms will prefer to export to and import from this market, or conclude cooperative non-equity arrangements with indigenous firms. This is because the O specific advantages of domestic firms are few and far between, as there is little or no indigenous technology accumulation and hence few created assets. Those that exist will be in labour-intensive manufacturing and the primary product sector (such as mining and agriculture), and may be government influenced through infant industry protection such as import controls.

Government intervention during Stage 1 will normally take two forms. First it may be the main means of providing basic infrastructure, and the upgrading of human capital via education and training. Governments will attempt to reduce some of the endemic market failure holding back development. Second, they engage in a variety of economic and social policies, which, for good or bad, will affect the structure of markets. Import protection, domestic content policies and export subsidies are examples of such intervention at this stage of development. At this stage, however, there is likely to be only limited government involvement in the upgrading of the country's created assets, e.g. innovatory capacity and human resources.

Stage 2

In Stage 2, inward direct investment starts to rise, while outward investment remains low or negligible. Domestic markets may have grown either in size or in purchasing power, making some local production by foreign firms a viable proposition. Initially this is likely to take the form of import substituting manufacturing investment – based upon their possession of intangible assets, e.g. technology, trademarks, managerial skills, etc. Frequently such inbound foreign direct investment (FDI) is stimulated by host governments imposing tariff and non-tariff barriers. In the case of export-oriented industries (at this stage of development, such inward direct investment will still be in natural resources and primary commodities with some forward vertical integration into labour-intensive low technology and light manufactures) the extent to which the host country is able to offer the necessary infrastructure (transportation, communications facilities and supplies of skilled and unskilled labour) will be a decisive factor. In short, a country must possess some desirable L characteristics to attract inward direct investment, although the extent to which foreign firms are able to exploit these will depend on its development strategy and the extent to which it prefers to develop technological capabilities of domestic firms.

The O advantages of domestic firms will have increased from the previous stage, wherever national government policies have generated a virtuous circle of created asset accumulation. These O advantages will exist due to the development of support industries clustered around primary industries, and production will move towards semi-skilled and moderately knowledge-intensive consumer goods. Outward direct investment emerges at this stage. This may be either of a market seeking or trade-related type in adjacent territories, or of a strategic asset seeking type in developed countries. The former will be characteristically undertaken in countries that are either lower in their IDP than the home country, or, when the acquisition of created assets is the prime motive, these are likely to be directed towards countries higher up in the path.

The extent to which outward direct investment is undertaken will be influenced by the home country government-induced 'push' factors such as subsidies for exports, and technology development or acquisition (which influence the I advantages of domestic firms), as well as the changing (non-government-induced) L advantages such as relative production costs. However, the rate of outward direct investment growth is likely to be insufficient to offset the rising rate of growth of inward direct investment. As a consequence, during the second stage of development, countries will increase their net inward investment (i.e. their NOI position will worsen), although towards the latter part of the second stage, the growth rates of outward direct investment and inward direct investment will begin to converge.

Stage 3

Countries in Stage 3 are marked by a gradual decrease in the rate of growth of inward direct investment, and an increase in the rate of growth of outward direct investment that results in increasing NOI. The technological capabilities of the country are increasingly geared towards the production of standardized goods. With rising incomes, consumers begin to demand higher quality goods, fuelled in part by the growing competitiveness among the supplying firms. Comparative advantages in labour-intensive activities will deteriorate, domestic wages will rise, and outward direct investment will be directed more to countries at lower stages in their IDP. The original

O advantages of foreign firms also begin to be eroded, as domestic firms acquire their own competitive advantages and compete with them in the same sectors.

The initial O advantages of foreign firms will also begin to change, as the domestic firms compete directly with them in these sectors. This is supported by the growing stock of created assets of the host country due to increased expenditure on education, vocational training and innovatory activities. These will be replaced by new technological, managerial or marketing innovations in order to compete with domestic firms. These O advantages are likely to be based on the possession of intangible knowledge, and the public good nature of such assets will mean that foreign firms will increasingly prefer to exploit them through cross-border hierarchies. Growing L advantages such as an enlarged market and improved domestic innovatory capacity will make for economies of scale, and, with rising wage costs, will encourage more technology-intensive manufacturing as well as higher value added locally. The motives of inward direct investment will shift towards efficiency seeking production and away from import substituting production. In industries where domestic firms have a competitive advantage, there may be some inward direct investment directed towards strategic asset acquiring activities.

Domestic firms' O advantages will have changed too, and will be based less on government-induced action. Partly due to the increase in their multinationality, the character of the O advantages of foreign firms will increasingly reflect their ability to manage and coordinate geographically dispersed assets. At this stage of development, their O advantages based on possession of proprietary assets will be similar to those of firms from developed countries in all except the most technology-intensive sectors. There will be increased outward direct investment directed to Stage 1 and 2 countries, both as market seeking investment and as export platforms, as prior domestic L advantages in resource-intensive production are eroded. Outward direct investment will also occur in Stage 3 and 4 countries, partly as a market seeking strategy, but also to acquire strategic assets to protect or upgrade the O advantages of the investing firms.

The role of government-induced O advantages is likely to be less significant in Stage 3, as those of FDI-induced O advantages take on more importance. Although the significance of location-bound created assets will rise relative to those of natural assets, government policies will continue to be directed to reducing structural market imperfections in resource-intensive industries. Thus governments may attempt to attract inward direct investment in those sectors in which the comparative O advantages of enterprises are the weakest, but the comparative advantages of location-bound assets are the strongest. At the same time, they might seek to encourage their country's own enterprises to invest abroad in those sectors in which the O advantages are the strongest, and the comparative L advantages are the weakest. Structural adjustment will be required if the country is to move to the next stage of development, with declining industries (such as labour-intensive ones) undertaking direct investment abroad.

Stage 4

Stage 4 is reached when a country's outward direct investment stock exceeds or equals the inward investment stock from foreign-owned firms, and the rate of growth of outward FDI is still rising faster than that of inward FDI. At this stage, domestic firms can now not only effectively compete with foreign-owned firms in domestic sectors in which the home country has developed a competitive advantage, but they are able to penetrate foreign markets as well. Production processes and products will be state of the art, using capital-intensive production techniques as the cost of capital will be lower than that of labour. In other words, the L advantages will be based almost completely on created assets. Inward direct investment into Stage 4 countries is increasingly sequential and directed towards rationalized and asset seeking investment by firms from other Stage 4 countries. The O specific advantages of these firms tend to be more 'transaction' than 'asset' related (Dunning 1993), and to be derived from their multinationality *per se*. Some inward direct investment will originate from countries at lower stages of development, and is likely to be of a market seeking, trade-related and asset seeking nature.

Outward direct investment will continue to grow, as firms seek to maintain their competitive advantage by moving operations which are losing their competitiveness to offshore locations (in countries at lower stages), as well as responding to trade barriers installed by both countries at Stages 4 and 5, as well as countries at lower stages. Firms will have an increasing propensity to internalize the market for their O advantages by engaging in FDI rather than exports. Since the O advantages of countries at this stage are broadly similar, intra-industry production will become relatively more important, and generally follows prior growth in intra-industry trade. However, both intra-industry trade and production will tend to be increasingly conducted *within* multinational enterprises (MNEs).

The role of government is also likely to change in Stage 4. While continuing its supervisory and regulatory function, to reduce market imperfections and maintain competition, it will give more attention to the structural adjustment of its location-bound assets and technological capabilities, e.g. by fostering asset upgrading in infant industries (i.e. promoting a virtuous circle) and phasing out declining industries (i.e. promoting a vicious circle). Put another way, the role of government is now moving towards reducing transaction costs of economic activity and facilitating markets to operate efficiently. At this stage too, because of the increasing competition between countries with similar structures of resources and capabilities, governments begin taking a more strategic posture in their policy formation. Direct intervention is likely to be replaced by measures designed to aid the upgrading of domestic resources and capabilities, and to curb the market distorting behaviour of private economic agents.

Stage 5

As illustrated in Figure 1.1, in Stage 5, the NOI position of a country first falls and later fluctuates around the zero level. At the same time, both inward and outward FDI are likely to continue to increase. This is the scenario which advanced industrial nations are now approaching as the century draws to a close, and it possesses two key features. First, there is an increasing propensity for cross-border transactions to be conducted not through the market but internalized by and within MNEs. Second, as countries converge in the structure of their location-bound assets, their international direct investment positions are likely to become more evenly balanced. It has been suggested elsewhere (Dunning 1993) that these phenomena represent a natural and predictable progress of the internationalization of firms and economies. Thus the nature and scope of activity gradually shifts from arm's length trade between nations producing very different goods and services (Hecksher–Ohlin trade) to trade within hierarchies (or cooperative ventures) between countries producing very similar products.

Unlike previous stages, Stage 5 of the IDP represents a situation in which no single country has an absolute hegemony of created assets. Moreover, the O advantages of MNEs will be less dependent on their country's natural resources but more on their ability to acquire assets and on the ability of firms to organize their advantages efficiently and to exploit the gains of cross-border common governance. Another feature of Stage 5 is that as firms become globalized their nationalities become blurred. As MNEs bridge geographical and political divides and practise a policy of transnational integration, they no longer operate principally with the interests of their home nation in mind, as they trade, source and manufacture in various locations, exploiting created and natural assets wherever it is in their best interests to do so. Increasingly, MNEs, through their arbitraging functions, are behaving like mini-markets. However, the ownership and territorial boundaries of firms becomes obscured[5] as they engage in an increasingly complex web of trans-border cooperative agreements.[6]

The tendency for income levels to converge among the Triad countries has been noted, among others, by Abramovitz (1986), Baumol (1986), Dowrick and Gemmell (1991), and Alam (1992). Indeed, during the 1970s and 1980s, Japan, the EC and EFTA countries have experienced a 'catching up' in their productivity and growth relative to the United States (the 'lead' country); while a range of the newly industrializing countries began to move from Stage 2 to Stage 3 in their IDP.

As a result of these developments, the economic structures of many industrial economies have become increasingly similar. Countries which were once the lead countries in Stage 4 now find themselves joined by others. This tends to reduce their NOI position and pushes them into Stage 5 of the IDP. At the same time, there has also been a 'catching-up' effect among MNEs since the 1970s. Firms that have had relatively low levels of international operations have been internationalizing at faster rates than their more geographically diversified counterparts. These two effects are not unrelated; firms have had to compensate for slowing economic growth in their home country by seeking new markets overseas. Given the similarity in income levels, the factors of production are broadly similar, and, as Cantwell and Randaccio (1990) have shown, firms that are trying to catch up seek to imitate competitors and develop similar O advantages as their competitors in the same industry, *but not necessarily in the same country*.

To take this argument a step further, as income levels, economic structures and patterns of international production among the Triad countries converge, the relative attractions of a particular location will depend less on the availability, quality and price of their natural assets and more on those of their created assets. It has been noted elsewhere that the prosperity of modern industrial economies is increasingly dependent on their capacity to upgrade continually, or make better use of their technological capacity and human resources (Cantwell and Dunning 1991). Since many of these advantages are transferable across national boundaries, it may be predicted that, in the long run, this will lead to a more balanced international investment position, and to an increasing convergence of created asset L advantages.

However, the ability of a country to upgrade its technological and human capabilities is a function of its own location-bound endowments and, in particular, of its natural assets, the characteristics of its markets and the macro-organizational strategies of its government. We believe the role of government in affecting dynamic economic restructuring cannot be overstated. In a myriad of ways, governments can promote new trajectories of economic growth which some countries are better able to cope with than others. This has been amply illustrated by the evolution of Japan's economy compared to that of the United States, especially in the 1980s.

In terms of their gross inward and outward direct investment positions, Stage 5 countries, after an initial burst of new inward direct investment (e.g. as occurred in the United States in the 1980s), may be expected to settle down to a fluctuating equilibrium around a roughly equal amount of inward and outward investment. Inward investment will be of two kinds. The first will come from countries at lower stages of the IDP and will be essentially of the market seeking and knowledge seeking type. The second will be from Stage 4 (or Stage 5) countries whose firms will continue to indulge in rationalized investment among themselves, as well as making outward direct investments in less developed countries, especially in the natural-resource-intensive sectors. In other words, truly rationalized or efficiency seeking MNE activity will occur as plant and product specialization is encouraged in sectors where economies of scale and scope are important.

As the world economy begins to resemble a global village, strategic asset seeking investments may also be expected to rise, and this, too, will lead to increasing convergence among countries as firms seek to improve their O advantages by cross-border mergers and acquisitions (M&As) or strategic alliances. Therefore, in the shorter time frame, inward and outward investment will fluctuate depending on relative innovatory and organizational strength of the participating countries. However as Cantwell has noted,

> The sectoral pattern of innovative activity gradually changes as new industries develop and new technical linkages are forged between sectors. Yet this is a slow process which in general only slightly disturbed the pattern of technological advantages held by firms of the major industrialized countries in the 20 years between the early 1960s and the early 1980s.
>
> (Cantwell 1989, p. 45).

Thus, *pro tem*, at least, it is possible for one country to be a net outward investor compared to another. But over time, according to the extent and speed at which created assets are transferable, the investment gap will again close, leading to a fluctuating investment position around an equilibrium level. It is within this context that the fifth stage will exist.

In other words, an equilibrium of sorts will be perpetuated, but it will not be a stable equilibrium as the relative comparative and competitive advantages of countries and firms are likely to be continually shifting. Hence, along with these fluctuations in relative comparative advantages, when combined with external and internal changes in the domestic economy, gradually the number of countries at Stage 5 will fluctuate.

The acquisition, diffusion and transfer of O advantages will be influenced by the cumulative causation in trade, production and technology, and whether the industry or sector in each of the countries at Stage 5 experiences a 'vicious' or a 'virtuous' circle (Dunning 1988b, Cantwell 1989). In the former case, it may serve to increase technological divergences between countries; in the latter, it may strengthen the technological linkages between them.

In summary, Stage 5 is marked by a gradual convergence of industrial structures among countries and a change in the character of international transactions. MNE activity, in particular, will be directed to efficiency seeking investment with greater emphasis on cross-border alliances, mergers and acquisitions; and the governance and equity position of MNEs will become increasingly *transnational*. The success of countries in accumulating technology, as well as inducing continued economic growth, will depend increasingly on the ability of their firms to coordinate their resources and capabilities at a regional and global level. The economic convergence of industrialized countries on one hand, and high rate of intra-Triad FDI growth on the other, may be expected to foster regional and/or global integration as well as lessening the role of natural assets as a country-specific determinant of FDI. In Stage 5, governments will increasingly assume the role of strategic oligopolists, taking into account the behaviour of other governments in the formation and execution of their own macro-organizational strategies. In this stage too, governments are likely to play a more pro-active role in the fostering of efficient markets, and cooperating with business enterprises to reduce structural adjustment and other transaction costs.

We conclude. Beyond a certain point in the IDP, the absolute size of GNP is no longer a reliable guide of a country's competitiveness; neither, indeed, is its NOI position. This is for two reasons. First the competitiveness of a country is better measured by the rate and character of growth of GNP *vis-à-vis* that of its major competitors. Second, as the motivation of FDI evolves away from being primarily geared to the exploitation of existing O advantages to the acquisition of new O advantages, countries which offer the appropriate location-bound resources for the creation of such advantages may increase their attractiveness to inbound FDI. Investments made to acquire or exploit indigenous competitive advantage, far from representing a weakness of the recipient country, could represent a strength. Certainly, recent evidence seems to suggest that in the Triad at least, inbound and outbound FDI are increasingly complementary to each other, especially at a sectoral level (UNCTAD 1993).

Most of the empirical testing of the basic proposition of the IDP, viz. that there is a systematic relationship between a country's inward and outward investment and its GNP per capita, has used cross-sectional data, and is generally supportive of the proposition.[7] However, new cross-sectional and time series data – some of which are set out in this and later chapters – seem to be pointing to two things. The first is that the shape and position of the IDP probably varies much more between individual countries than it was originally thought. In particular, the economic structure of countries, and the development strategies and macro-organizational policies of governments, appear to be critical in influencing both the role of

MNEs in a country's economy at a given moment of time, and how inward and outward direct investment may help fashion the growth and structure of the economy over time (Dunning 1993). The second is that the underlying nature of the IDP for all countries appears to be undergoing some change, due to a series of events of the global economy, which leads us to revise some of our hypotheses about its trajectory. This issue is taken up in greater detail in Part 2 of this chapter.

Country-specific factors and the IDP

As most of the contributions to this volume will be considering the nature of the interaction between inbound and outbound FDI and the level, composition and growth of the GNP of countries, we will need to write comparatively little at this point about country-specific factors. It may, however, be appropriate to remind ourselves that the IDP was first put forward to illustrate the relevance of the eclectic paradigm of international production in explaining the NOI position of countries, and, indeed, this is one of the main purposes of this volume. It follows, then, that any predictions about the IDP must rest on the contents of the paradigm itself.

Now, as stated earlier, the paradigm avers that a country will attract inbound FDI when (i) foreign firms possess certain O specific advantages over and above those of indigenous firms; (ii) its L bound resources and capabilities favour the deployment of these competitive advantages, relative to those offered by other countries; and (iii) foreign firms perceive that it is to their benefit to internalize the intermediate product markets for these advantages, rather than selling them via the external market, or by a cooperative arrangement, to domestic firms in the host country. Similarly the paradigm hypothesizes that the propensity of a country to be an outward direct investor will rest on the strength and character of the O advantages of its indigenous firms, and the extent to which these might best be exploited by adding value to them in a foreign location, and organized through an MNE hierarchy rather than through a non-equity relationship with a foreign firm.

The eclectic paradigm further suggests some of the ways in which, over time, inbound and outbound investment may affect the trajectory of a country's development path. This it might do by its impact both on the composition and productivity of domestic economic activity, and the ease or difficulty with which a country is able to restructure its resources and capabilities to meet the needs of endogenous and exogenous change. The critical role played by inbound FDI in the upgrading of Singaporean indigenous endowments, and that of outbound FDI in the dynamic restructuring of Japan's post-war development, are two cases in point, although other examples show that FDI does not always have such salutary effects on economic welfare.[8]

Over the past 30 years, there have been a large number of studies on the impact of both outbound and inbound MNE activity on the development and economic restructuring of the countries in which they operate. The overwhelming consensus of these studies is that, for good or bad, this is critically dependent on three main variables, viz. (i) the type of FDI undertaken, (ii) the structure of the indigenous resources and capabilities of the countries concerned, and (iii) the macro-economic and organizational policies pursued by governments.[9] We would then expect the shape and position of the IDP of countries – which traces the interaction between inbound and outbound FDI and advances in the prosperity of those countries – to be determined by the same variables.

We have suggested that one of the characteristics of economic development identified by several writers (e.g. Porter 1990, Ozawa 1992, Narula 1995, Dunning and Narula 1994) is that, as development proceeds, the significance of indigenous assets relative to *natural* assets as a locational attraction to inbound FDI increases. Ozawa (1992), in describing the post-war development of Japan, identifies four distinct stages, viz. those of labour-intensive manufacturing, scale-economies-based production of heavy and chemical industries, assembly-based mass production of consumer durables, and mechatronics-based flexible manufacturing. For his part, Porter (1990) writes about the nature of competitive advantages of a country according to whether they are factor driven, investment driven, innovation driven or wealth driven. While the prosperity of most poor countries is largely resource driven, that of the richest is largely innovation or wealth driven. Naturally, the precise balance of a country's natural and created assets will vary depending on the extent to which it is endowed with the former (cf. Canada with Japan, and Kuwait with Singapore) but as Ohmae (1987) has powerfully shown, even the value of natural resources such as minerals and agricultural products can be dramatically increased by secondary processing and astute marketing, both of which require the input of created assets.

In view of the fact that an increasing number of countries are now at the innovation stage of their IDP – indeed, as we shall suggest later, innovation-led production is changing the trajectory of global economic development - it is not surprising that both outward and inward FDI are being increasingly evaluated by national governments in terms of their perceived contribution to technological capacity and human resource development (Dunning 1994). Governments, too, are becoming aware that, if their FDI is properly to achieve their objectives, they need to provide the location-bound resources and capabilities essential for the efficient creation and deployment of the O specific advantages of both foreign investors and their own MNEs. An appropriate combination of the competitive advantages of firms and countries is likely to make for a *virtuous* cycle of upgrading economic development, with each advantage fostering the other. An inappropriate combination of such advantages – or the lack of one or both – is likely to lead to a vicious cycle to the detriment of economic development.

Structural changes and the IDP

Let us now turn to the main focus of Part 1 of this chapter, which is to consider the ways in which recent technological and organizational changes, as they have impinged upon the governance of both firms and national economies, have affected our thinking about the shape and character of the IDP.

(a) Some shifts in the rationale for FDI

Most of the received literature on MNE-related activity tends to assume that firms engage in FDI in order to exploit best, or organize more efficiently, their existing competitive advantages. Sometimes, these resources and capabilities are combined with foreign location-bound assets to supply domestic or adjacent markets, and sometimes to service more distant markets. In some instances, too, inbound FDI may be used to restructure the existing portfolio of foreign value-added activities by MNEs. Such sequential investment is best thought of as efficiency seeking trans-border activity, as contrasted with market or resource seeking trans-border activity.

In the last decade or more, however, MNE activity has been increasingly motivated by the desire to acquire new competitive advantages, or protect existing advantages. Such strategic asset acquiring FDI has been particularly pronounced within the Triad of advanced industrial countries, and is most dramatically shown by the spate of cross-border mergers and acquisitions (M&As) which have occurred since the mid- and late 1980s.[10] Essentially, such M&As have been (and still are) undertaken by firms for five main reasons: (i) the rising costs of innovation and of entry into unfamiliar markets, (ii) competitive pressures for firms to be more cost effective, (iii) the growing need to tap into complementary technologies and to capture the economies of scale and scope expected from the merger or acquisition, (iv) a desire to protect or advance their global markets, *vis à vis* oligopolistic competitors, and (v) the need to encapsulate the time of the innovating or market entry process.

Such strategic asset acquiring FDI implies that firms may engage in outward FDI from a position of weakness, and that countries may attract inbound FDI because their resources and capabilities offer competitive advantages to foreign MNEs. Thus, part of the contemporary outbound MNE activity directed to the United States is designed to gain access to the technological capabilities of US firms, and their privileged access to US or adjacent (e.g. NAFTA) markets. Such FDI is likely to be determined by a different configuration of O and L advantages than those facing traditional market or resource seeking MNEs.

The effect of strategic asset seeking investment on the IDP is that it is likely to increase the level of outward investment of all countries, but particularly that from medium income and fast growing industrializing nations, as they seek to establish a speedy presence in the most innovatory and dynamic markets of the world. Frequently, firms from developing countries do not have the full range of resources and capabilities to promote a fully blown 'stand-alone' competitive strategy, and certainly not one which would help them penetrate unfamiliar global markets. Depending on their particular strengths and weaknesses, their liquidity position and the type of assets to which they need access, the mode of foreign involvement by firms is likely to vary between an FDI, a minority joint venture and some form of cooperative alliance. However, *ceteris paribus*, the first of these routes is most likely to be preferred whenever the assets sought are perceived to be critical to protect or advance the core competencies of the investing firm.

As yet, there has been little systematic research into the significance of strategic asset acquiring FDI, relative to that of other kinds.

But, taking inbound FDI in the United States as an example, it is generally agreed by scholars that, although the resurgence of activity by European firms in the 1970s and much of the greenfield investment by Japanese firms in the 1980s reflected the growing O advantages of these firms, relative to those of US firms, many of the transatlantic European M&As in the late 1980s and early 1990s have been geared towards strengthening the O advantages of the investing firms (or diminishing those of their competitors) by buying into US resources and capabilities, and/or markets. One suspects that had it been feasible, there would also have been a substantial amount of M&A activity by US and European firms in Japan during this period.

The presence of strategic asset acquiring FDI is then tending to raise the level of inward investment of industrialized countries – and particularly those which are the leading repositories of advanced created assets. It is also tending to increase the outward investment of these countries – but not to the same extent as inward investment – while raising the outbound FDI by industrializing developing countries, as they seek to aid and accelerate the entry of their firms into global markets. In short, the presence and growth of asset acquiring MNE activity is leading to a flattening out of the NOI position of countries, as compared to that suggested by the traditional version of IDP, viz. at lower levels of GNP the net *inward* investment position will be less, and at high levels of income the NOI position will be lower, than in the absence of such investment.[11] The suggested reshaping of the IDP is portrayed by the dotted line in Figure 1.1.

(b) The emergence of alliance capitalism

Another feature of the last decade has been the growth in non-equity collaborative arrangements of one kind or another. Sometimes, these are being pursued as alternatives to FDI, but, for the most part, they are complementary to it. Increasingly, cross-border *intra*-firm FDI and *inter*-firm cooperative schemes are being perceived as part of a holistic and multi-modal strategy of the leading global players.

It is possible to identify many different kinds of collaborative schemes, but the vast majority fall into two categories. The first take the form of strategic alliances which are specifically intended to gain access (or preclude a competitor from gaining access) to foreign assets or markets. The second embrace a galaxy of international subcontracting relationships, in which inter-firm cooperation goes beyond the production of materials, parts and components, to the design and development of new materials, parts and components. In each case, too, it would seem that the terms of any collaboration are contained less in a formal contract and more by a sense of agreed mission and mutual commitment.

It is the latter characteristic of inter-firm relationships which has led to the term 'alliance' capitalism being coined (Gerlach 1992).[12] Alliance capitalism differs from hierarchical capitalism in that, whereas in the case of the former, the coordination of economic activity is determined primarily by arm's length markets and inter-firm cooperation, in the latter it is decided primarily by arm's length markets and hierarchical *intra*-firm fiat.[13]

While alliance capitalism has long since been a prominent feature of many East Asian economies – most noticeably Japan and South Korea – in the mid-1990s it is spreading – albeit in a modified form – to other parts of the world. This is for three main reasons. The first reflects the lack of experience of hierarchical capitalism by the transition economies of Central and Eastern Europe and China. As these countries struggle to embrace the discipline of free markets, they are finding that the speediest and most effective way to upgrade their natural and created assets is for their newly privatized firms to form cooperative alliances with other domestic, or foreign, firms, rather than to pursue the route of internal economic growth. Second, it reflects the increasing inability of firms to pursue 'stand-alone' strategies in situations in which their core competencies need to be efficiently combined with those of other firms if they – the former – are to be fully effective.

Third, one of the features of the emerging techno-economic paradigm of micro-economic activity, viz. flexible and innovation-led production, is that it requires a symbiotic and continuing relationship between the various participants in the production process, and that this is likely to be most effective if it is based on mutually agreed upon goals and on active and purposeful cooperation, rather than on administrative fiat. Thus, although many firms are currently downsizing the range of their activities in order to concentrate on those central to their core competencies, i.e. becoming *less* hierarchical, they are also concluding new strategic alliances with their critical suppliers and industrial customers along the value chain, and with their competitors across value chains. This they are doing both to leverage more effectively their own special capabilities, and to ensure, by appropriate control procedures, that the goods and services they transact with other firms, and which critically affect these capabilities, are provided at the highest quality and/or the lowest cost.[14]

What are the implications of the advent of alliance capitalism on the international direct investment position of countries and their IDP? Perhaps, the main implication is that such non-equity forms of cross-border production are becoming too important to be ignored in discussing the export and import of resources and capabilities, and the way in which their use is influenced either by hierarchical or by cooperative arrangements involving foreign firms. Here what scant evidence we have[15] suggests that, apart from situations in which inbound and/or outbound FDI is disallowed or regulated by governments, cross-border alliance formation to gain access to new technologies and markets, or to exploit the economies of scale and synergy, tends to involve a two-way exchange of resources and capabilities between firms from advanced industrial countries. By contrast, other kinds of alliances, and especially those which involve firms from both developed and developing countries, are primarily concluded in order to facilitate a transfer of resources and capabilities from the former to the latter countries.[16]

Incorporating such alliances into the IDP would then suggest that inbound transfer MNE activity[17] of the poor or middle income countries would increase, but in the case of the richer countries, one might predict an increase of both inbound and outbound resource transfer. Unfortunately, apart from some data on cross-border inter-firm royalties and fees, it is extremely difficult to quantify either the extent to which alliances *do* transfer resources and capabilities, or the consequence of such transfers on the welfare of the exporting and importing countries. As a subject for further research, the relationships between alliance formation, economic structure and development surely demand some degree of priority.

(c) The role of non-market country-specific differences in explaining the investment development path

In our earlier testing of the hypothesis that a country's outward and inward FDI is systematically related to its stage of development (Dunning 1981, 1986, 1988a), we identified a number of contextual variables which could explain why the shape and position of the IDP differs between countries. *Inter alia*, our research showed that industrial or industrializing countries were likely to generate more outward direct investment at any given level of GNP per head than the natural resource-based economies.[18] At the same time, we made the general assumption that countries – or, more specifically, the governments of countries – pursued market friendly economic strategies, and intervened as little as possible in the organization and allocation of resources within their jurisdiction.

In retrospect, it is clear that, throughout the last three decades, by a bevy of macro-economic and organizational policies, national governments have considerably affected the structure of the IDP of their countries. This they have done by both specific actions to influence the level and composition of inbound and outbound FDI, and by their general economic and social policies, which impact the attractiveness of their location-bound resources and markets to foreign investors. Moreover, notwithstanding the liberalization and deregulation of many markets over the past decade, national governments continue to exercise a powerful influence on a country's international investment position and the profile of its IDP.

Several of the country case studies of this volume confirm this proposition. It is most obviously seen in the case of centrally planned and East Asian economies. But, as revealed by some quite dramatic shifts in the outward and inward FDI position of particular countries – which have often occurred as a direct result of a reorientation in government economic policy – it is no less evident in economies such as the UK, France, Sweden, Greece and Portugal in Europe; Chile, Columbia, Argentina, Jamaica, Mexico and Venezuela in Latin America; China, India, the Philippines, Vietnam and South Korea in Asia; Canada in North America; Morocco, Egypt, Nigeria and South Africa in Africa; and in Iran in the Middle East.

There is already a good deal of evidence that the liberalization and privatization of markets, and the attempts by many governments to increase inbound investment, have led to an increased flow and restructuring of inward investment into many countries, and, *pari passu*, an increased flow of outward investment from other countries.[19] It is also apparent that the role of national governments in affecting the price and quality of location-bound resources within their jurisdictions, and the motivation and capabilities of their own firms to be outward investors, is becoming increasingly significant – and especially in so far as inter-Triad strategic asset and efficiency seeking investment is concerned. Thus, for example, government policies which aim to upgrade the quality of indigenous resources and capabilities to meet the demands of the international marketplace are likely to engage in more cross-border transactions (e.g. FDI, trade and cooperative alliances) than those which are designed to promote economic self-reliance.[20]

It follows from the above paragraphs that the relationship between FDI and income levels using cross-sectional country data may well be expected to vary at different points of time because of changes in the role of governments in affecting inward and outward investment. It is also likely to fluctuate according to the direction and character of technological advances and the country from which they originate. Such advances, by their impact on the competitive advantages of firms of a particular nationality, on the cost-effectiveness of the location-bound resources and capabilities of countries, and on the way in which economic activity is organized, are likely to lead to a repositioning of the IDPs of countries. At the same time, longitudinal data show that variations in the trajectory of a country's IDP may occur because of changes in the actions of national governments. In short, then, although we have some ideas about the interaction of the behaviour of governments and inward and outward investment, we need to explore this in a more systematic and rigorous way.

The theory: some concluding remarks

In this chapter, we have suggested that some of the propositions initially put forward to relate the inward and outward direct investment position of countries to their stages of development need reconsideration. Partly, we have argued this is because the *raison d'être* and character of FDI has undergone some important changes. Partly it is because other forms of cross-border involvement – and notably cooperative arrangements – by MNEs need to be incorporated in the analysis; and that the interaction between these other forms and the stages of development may well be different from those of FDI. And partly we have suggested this is because differences in national government policies need to be more explicitly identified as an explanatory variable of the international direct investment of countries, before any satisfactory relationship between outbound and inward investment and economic development can be established.

Lastly, we believe that further attention needs to be given to the form and characteristics of the fifth stage of the IDP, i.e. that in which the outward and inward investment positions – like those of exports and imports – fluctuate around the same level. Here the hypothesis is that the structure of the most advanced industrial economies are both similar and inextricably linked with each other. Although, for a period of time, one nation, through a series of path breaking technological or organizational advances and/or superior macro-economic or macro-organizational policies, might gain a major competitive advantage over other nations, any marked increase in outbound direct investment may well be tempered by a corresponding rise in inbound strategic asset seeking investment and alliance formation. Moreover, due to the increasing ease at which knowledge, information and even organizational techniques can move across national boundaries, any lead by one country is likely to be quite quickly eroded by its competitors, and sometimes, the catching-up process, itself, may be aided and abetted by inbound and outbound FDI and alliance formation.

However, little is known about the mechanism by which this is achieved. Some hints about the dynamic interplay between the O specific advantages of firms and the L specific advantages of countries have been given by Tolentino (1993), Dunning (1993) and Narula (1995); while Ozawa (1992) has explored some of the ways in which inbound and outbound Japanese FDI has affected the structure of Japanese economic development. The conditions under which inward and outward FDI[21] can promote the upgrading of a country's resources and capabilities and advantageously restructure its resource allocation have also been explored by several writers,[22] including Cantwell (1989) and Cantwell and Dunning (1991), who used the concepts of virtuous and vicious cycles to explore the interplay between inward FDI and the competitiveness of a particular industrial sector.

While some of the contributions to this volume take this discussion a little further, much remains to be done. Why, indeed, has inbound FDI promoted advantageous structural economic development in some countries, e.g. Singapore and Thailand, but not in others, e.g. Chile and Nigeria? What determines whether outward direct investment helps a country to upgrade the quality of its indigenous assets, and to promote its dynamic competitive advantage, or, instead, to erode its technological strengths and human resource development? What is the impact of alliance capitalism on both the optimal mode of resource transfer and usage, and on the ways in which these may affect the structure and pace of economic development? It is questions such as these which need the attention of scholars in the years to come.

PART 2: A STATISTICAL EVALUATION

Structural change and the IDP

At the outset it is important to point out that any statistical evaluation of the IDP must necessarily be a tentative one. Any attempt to conduct a thorough empirical analysis of a complex and changing relationship has severe limitations. Given this fact, it is not our intention to develop a rigorous statistical specification and test of the IDP. In fact, our aim is almost exactly the opposite as we wish to demonstrate that a statistical evaluation of the relationship between FDI and economic development cannot be conducted on an aggregate basis across countries, as the IDP represents a paradigm which is idiosyncratic and country specific, and therefore best analysed on a country-by-country basis. We intend merely to argue that the basic relationships postulated by the IDP are still applicable, and how the lacunae regarding the extent and evolution of natural and created assets as well as the changes in the world economy affect the relationships suggested by the IDP.

Structural changes in the world economy

There have been two major developments in the world economy which have impacted the character of the IDP. The first is the introduction of Stage 5, which reflects the catching-up and convergence process of the industrialized economies. As we have discussed in Part 1 of this chapter and elsewhere (Narula 1993, 1995, Dunning and Narula 1994), as countries reach Stage 4 and begin to enter Stage 5, the activities and growth of their MNEs are no longer a function of just the economic conditions of their home country, but the various host countries in which they have subsidiaries. The more globalized the operations of a firm, the greater the extent to which its O advantages are likely to be firm specific, rather than determined by the economic, political and cultural conditions of its home country. Moreover, the O advantages of firms will increasingly be dependent on their ability to acquire and develop *created* assets and their ability to organize these assets efficiently in order to exploit the advantages due to common governance, making the MNE less dependent on its home country's natural resources. As such, O advantages become increasingly firm specific as MNEs become more internationalized. The consequence of this is that the outward direct investment position of a country's firms at this stage is no longer entirely dependent on the economic status and competitiveness of their home country, and increasingly affected by the conditions in the various other countries in which they operate, and therefore, after reaching a certain NOI position, a country's investment position will *not* necessarily be proportional to *its income level or relative stage of development*. To put it another way, we hypothesize that, *ceteris paribus*, a Stage 5 country will continue to experience change in its FDI position regardless of whether its relative stage of development or income levels change. This is readily apparent when examining the NOI position and GDPs of countries like the United States and the UK (see also Chapters 2 and 3), which have either remained at the same relative stage of development or fallen, but continue to experience high growth in both their inward and outward position. Indeed, not only has the share of total worldwide inward investment to industrialized countries increased, but a greater extent of outward investment from these countries is being directed towards other industrialized countries.

Furthermore, the use of GDP as a proxy for development does not take into account the profound changes in the economic structure of the industrialized countries, which have shown a clear trend towards tertiary (i.e. service) sectors. In other words, while their overall economic growth has slowed over the past two decades, there has been considerable structural adjustment between sectors. This has also had an effect on the composition of their inbound and outbound FDI and, because of this, its geographical composition. Since 1980, for example, much of the growth in inward and outward FDI has been directed to the tertiary sector, and has been between industrialized countries (Narula 1995). These changes make a statistical evaluation of the relationship between NOI and GDP of Stage 4 and 5 countries an increasingly difficult exercise through an aggregate, cross-sectional test.

The growth of alternative forms of overseas value-added activity such as strategic alliances needs also to be taken into account, especially in high technology sectors. As suggested in Part 1 of this chapter, strategic alliances have become an important means by which MNEs from industrialized countries have begun to engage in cross-border activities since the 1980s. The evidence suggests that there is an increasing preference for Triad-based MNEs to utilize non-equity-based cooperative agreements in preference to equity-based agreements (such as joint ventures) in their intra-Triad partnering activities in these sectors[23] (Hagedoorn and Narula 1994). Such agreements are naturally not reflected in the FDI data, and since well over 95 per cent of all strategic alliances[24] are intra-Triad in scope (Freeman and Hagedoorn 1994) the use of FDI data for industrialized countries without allowing for the growth of strategic alliances may make the results questionable.

The second consequence of changes in the world economy has been the growing divergence of at least some of the developing economies away from the industrialized economies. The catching-up process described earlier has not occurred among the poorer countries, who have diverged as a group from the wealthier countries and are not exhibiting a tendency to converge in relation to the world leaders (Dowrick and Gemmel 1991, Dowrick 1992, Verspagen 1993), and are in fact 'falling behind'. The effect on the FDI activities of developing countries is that less inward direct investment is from industrialized countries, and those developing countries that are outward investors prefer to invest in the industrialized countries wherever possible to acquire created assets. However, the 'falling-behind' effect is associated primarily with Stage 1 and 2 countries, while a handful of developing countries that are regarded as newly industrializing economies (Stage 3) have been shown to be 'catching up' with the industrialized economies. Indeed, data on FDI flows indicate that the four Asian NICs account for 57.6 per cent of total outward flows from non-oil exporting developing countries between 1980 and 1990, and 83 per cent over 1988–90.

The idiosyncratic nature of countries

Most previous tests of the validity of the IDP have used a cross-sectional study across countries as a surrogate for longitudinal analysis. As the various country studies in this volume will later show, the exact circumstance of each country is unique, and while there are some general similarities between groups of countries, the explanatory power of the 'ideal' IDP based on cross-sectional analysis of a large group of countries is severely limited. This aggregation of countries for a given time period assumes that countries follow a broadly similar investment development path, whereas, in fact, each country follows its own particular path which is determined by three main variables: (i) the extent and nature of its created and natural assets, (ii) its strategy of economic development, and (iii) the role of government. These factors essentially determine the nature and extent of the firm-specific assets of both foreign MNEs and domestic firms operating within its borders.

The character of a country's resource endowments

The extent and nature of a country's natural and created assets are determined by two main issues: (i) its resource structure and (ii) its size.

(i) Resource structure

A country may possess a significant comparative advantage, or an absolute advantage, in primary commodities. Such a country is likely to spawn domestic firms that possess O advantages in the exploitation of such assets. However, especially if such an advantage is a near absolute one, it is likely to be the recipient of considerable inward investment from MNEs that wish to internalize the supply of primary products to their upstream activities located in other countries, and the extent of this inward investment will almost continue to rise as the other L advantages associated with the host country develop. These L advantages include the availability of skilled labour and other infrastructural facilities, and may lead to sequential vertical investment in upstream activities by both domestic firms and MNEs. As a result, a comparative advantage in a natural-resource-based industry may be sustained even when the income levels rise to developed country standards. Such a scenario would result in an NOI position that continues to be negative even when its economy is developed, as for example in Australia. Any outward investment would also tend to be in industries that are either in or related to the primary sector, but would be dwarfed by the increasing extent of inward investment. Such countries would tend to have a much lower (i.e. negative) level of NOI at considerably advanced stages of development.

The *lack* of a natural resource base (i.e. a comparative disadvantage in primary commodities) would, *ceteris paribus*, result in the opposite effect. Inward investment at earlier stages would be muted, and outward investment might begin at an earlier stage to secure the availability of necessary natural resources. Such a country is also more likely to begin strategic asset seeking investment at an earlier stage (e.g. Japan). Overall, these countries would become net outward investors at a considerably earlier stage of development than those well endowed with natural resources.

(ii) Market size

Countries that possess small domestic market size, such as Hong Kong, Singapore and Switzerland, are likely to have not just limited natural resources such as primary commodities, but limited attraction in terms of market size. Thus the lack of economies of scale will inhibit inward foreign investment in earlier stages. As their human capital and infrastructure improves, some inward investment may occur for export processing purposes. The small populations may mean not just small aggregate consumption, but that domestic firms would need to seek overseas markets in order to achieve economies of scale. This not only would result in outward direct investment at earlier stages of development, but also suggests that as income levels rise, domestic investors that were involved in export-oriented production will seek overseas locations to compensate for the shortage of low wage human capital for labour-intensive production. Such countries will reach (and remain at) a positive NOI position at a considerably earlier stage of their development. The opposite scenario would apply for large countries, which would attract larger amounts of inward investment due to the attractions of their large markets, and domestic firms may not have as much incentive to seek overseas markets since economies of scale can be achieved at home.

The dynamics of the natural/created asset evolution are primarily determined by those associated with the economic, social and political environment issues that are generally a direct result of the actions of governments. A statistical analysis of the IDP cannot, given the static nature of a cross-sectional test, capture the dynamic development of created assets. The role of government is even more idiosyncratic and peculiar to each country, and it is exceedingly difficult to translate this into a general variable, or to group countries into distinct groups according to the role of governments in influencing the created/natural asset balance. However, since these issues are dealt with in considerable detail in the country studies throughout the rest of this volume, we will briefly discuss the two main issues that primarily influence the dynamics of created asset development.

Economic system

The economic orientation of a country may either be outward looking, export oriented (OL–EO) or inward looking, import substituting (IL–IS) (Ozawa 1992). Depending on the orientation of an economy, the use of either (or a hybrid of the two) will substantially affect both economic development and the extent and pattern of FDI, and hence the nature of the path taken by a particular country. An OL–EO regime is likely to achieve faster growth and structural upgrading. Ozawa (1992) argues that an OL–EO regime is a necessary condition for FDI facilitated development. We suggest here that although it is not a necessary condition for growth in the first two stages, the greater the extent of OL–EO policy orientation, the faster the process of structural adjustment and economic growth and the quicker a country's progress through the stages of the IDP. Our earlier discussion of the various stages assumes an OL–EO type of policy regime beyond the second stage, but not for the first two. IL-IS countries would tend to have relatively little inward and outward FDI activity.

The failure of countries to proceed beyond the second stage is associated with the vicious cycle of poverty (VCP). This, when applied in the traditional sense, is explained as follows: low income levels in less developed countries are associated with low savings rates which, in turn, result in low capital investment, thereby keeping income levels low. In the parlance of the eclectic paradigm there is a lack of ownership advantages of domestic firms and location advantages of the country, as well as an inability to develop or acquire these. The O advantages referred to here include financial assets as well as the Oa and Ot types of advantages, whereas the L advantages are those of infrastructure. This cycle can be broken, *inter alia*, through the infusion of capital through FDI, which allows for technological spillovers and financial capital.

Governments and organizations of economic activity

Although the kind of economic system associated with a country broadly determines the path taken by a country, the nature of government policy associated with a particular system can vary between countries with the same economic system and at the same stage of development. There are two main areas of government strategy which directly impinge on the nature of the IDP of a country: macro-economic strategy and macro-organizational strategy (Dunning 1992). The role of governments in determining macro-economic policy is relatively well defined, and is often associated with the economic system. On the other hand, there is considerable variance amongst countries in the role of governments in determining macro-organizational strategy. Macro-organizational strategy primarily influences the structure and organization of economic activity, and the nature of the policies most appropriate at a particular stage should, in an 'ideal' situation, change as the economy evolves, reflecting the nature of market imperfections that the policy is designed to circumvent (Hamalainen 1993). Essentially, in such a best-world scenario, government plays a market facilitating role in which its macro-organizational policy dynamically evolves over time. Increasing economic specialization associated with economic development leads to a growth in market failures and increases the potential benefits of government macro-organizational policy (Durkheim 1964). However, as Hamalainen (1993) points out, governments may also fail, and society is often faced with a choice between imperfect markets and imperfect governments. Given that macro-organizational policy embraces a wide variety of issues,[25] and the fact that there is little agreement on what the optimal involvement of government should be, the macro-organizational policy stance varies widely among countries. The differences between the macro-organizational strategy of countries at the same stage of development influence both the structure of markets and the extent to which economic activity is efficiently conducted, thereby affecting the specialization and economic structure of the country, as well as the extent of FDI activity associated with it.

Evaluating the IDP

As we have earlier indicated, it is not our intention to develop a rigorous specification of the IDP, but merely to examine whether a causal relationship exists between FDI and economic development for 1992, and to illustrate the deviation from the 'ideal' path due to the extent of natural and created assets, as well as those due to structural changes in the world economy. We shall utilize data on

FDI stocks published in UNCTAD (1994). GDP and population data are derived from World Bank (1994). All data are in nominal US dollars. All FDI and GDP figures are normalized by population. Inward FDI per capita is denoted as IWK, outward FDI as OWK, NOI per capita as NOIK, and GDP per capita is denoted as GDPK. In conducting the analysis, we shall also attempt to illustrate that the nature of the relationship varies with the extent of natural and created assets associated with a country.

The extent of natural/created assets is based on a two-fold criteria to allow for differences due to resources intensity as well as differences in country size. High natural asset countries are defined as those countries whose primary exports as a percentage of total exports (PRX) are greater than or equal to 50 per cent or whose area is greater than 1.9 million km^2.[26] Since most developing countries tend to have a comparative advantage in primary commodities, the sample tends to consist largely of developing countries. Although it would be more appropriate to include only those countries which have an absolute advantage in natural resources, rather than those with a comparative advantage, it is exceedingly difficult to find such a measure. Low natural asset countries are assumed to represent countries with a high created asset base, or have a potential to become economies with a created asset base. This group is defined as countries for which PRX $\leq$ 20 per cent or area is less than 5000 km^2.

Net outward investment

We examine the relationship between NOI and GDP utilizing a quadratic specification. This allows for the fact that the dependent variable changes over time and stages, but it also assumes that the rate of change is more or less constant. Apart from running regressions for the entire sample (ALL) we also do so for a smaller subsample that excludes the most industrialized countries. This has two purposes. First, since we have not developed a specification for the fifth stage of the IDP, by excluding the countries that are most likely to be in Stage 5, we are able to test whether in fact the J-curve initially proposed by Dunning (1981) is still valid for the pre-Stage 5 countries. Second, by excluding the Stage 5 countries we are able to avoid 'stretching' of the IDP due to the cluster of a large number of developing countries at the origin and the spread of the industrialized countries around the *X*-axis, due to the process of

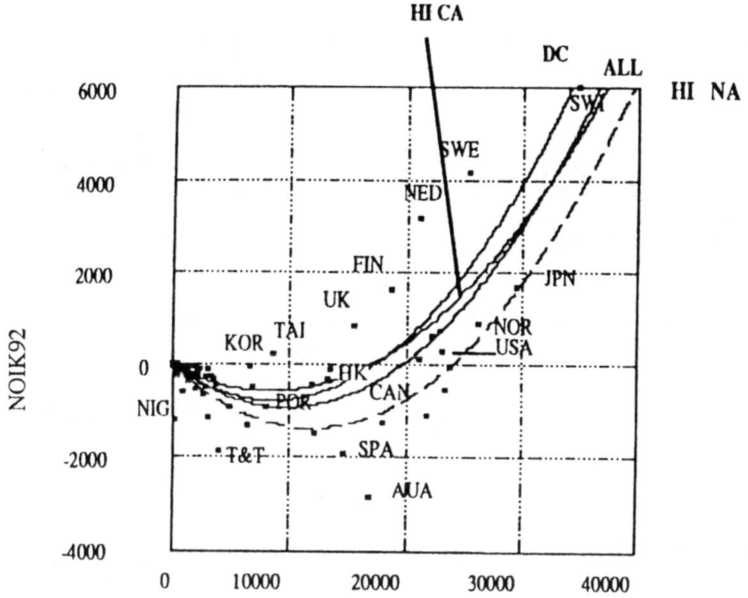

Figure 1.2 NOI and GDP for 1992

convergence and divergence. This would make the second NOI = 0 point further to the right than might actually be the case. The cut-off point for this sample is taken to be the first industrialized country with an NOI > 0. Figure 1.2 shows the plot of NOI per capita against GDP per capita for 1992. We also run regressions for two other subsamples, the created asset countries (HI CA) and the natural asset countries (HI NA). These four curves have been superimposed on Figure 1.2 while the results of these regressions are given in Table 1.1. We have not included the results for 1980 here since these are broadly similar to those for 1992. Nonetheless, there are broad differences in the distribution of the observations for the developing countries and the industrialized countries that are due to the catching-up and falling-behind scenarios due to convergence and divergence which have been evaluated using some simple measures in Table 1.2. In the case of the industrialized countries, which dominate much of the graph, there has been an increasing trend towards a wider distribution along the Y-axis since 1980. The

Table 1.1 Linear regression equations for NOI with GDP based on quadratic relationship

Sample	*GDPK*	*GDPK*^{2}	*ADJ.R*^{2}	*F-value*	*N*
ALL	−0.1872***	0.957×10^{-5}***	0.542	51.69	88
DC	−0.1767***	0.102×10^{-4}***	0.418	26.35	73
NA	−0.2292***	0.962×10^{-5}***	0.582	36.64	53
CA	−0.1329**	0.789×10^{-5}***	0.579	16.33	24

*** Significant at the 1% level. ** Significant at the 2.5% level.

Table 1.2 FDI activity and GDP, 1980 and 1992

	1980			**1992**				
	Mean	*Standard deviation*	*Ratio*	*Mean*	*Standard deviation*	*Ratio*	*Ratio of means*	*Ratio of standard deviations*
	(i)	(ii)	ii/i	(iii)	(iv)	iv/iii	iii/i	iv/ii
All countries								
Inward FDI	246	386	1.6	783	1263	1.6	3.2	3.3
Outward FDI	164	533	3.2	721	1783	2.5	4.4	3.3
NOI	−84	402	−4.8	−62	1076	−17.4	0.7	2.7
GDP	3453	4200	1.2	6231	8717	1.4	1.8	2.1
Industrialized countries								
Inward FDI	749	525	0.7	2671	1746	0.7	3.6	3.3
Outward FDI	837	985	1.2	3562	2717	0.8	4.3	2.8
NOI	87	804	9.2	890	2157	2.4	10.2	2.7
GDP	10919	2167	0.2	22816	4804	0.2	2.1	2.2
Non-industrial countries								
Inward FDI	128	223	1.7	363	574	1.6	2.8	2.6
Outward FDI	6	26	4.2	90	338	3.8	14.8	13.2
NOI	−124	217	−1.7	−273	429	−1.6	2.2	2.0
GDP	1696	2088	1.2	2545	3533	1.4	1.5	1.7
Stage 3 countries								
Inward FDI	556	789	1.4	2694	4414	1.6	4.8	5.6
Outward FDI	40	84	2.1	727	954	1.3	18.2	11.4
NOI	−515	711	−1.4	−1966	3815	−1.9	3.8	5.4
GDP	5402	2975	0.6	11118	3671	0.3	2.1	1.2
Stage 1 and 2 countries								
Inward FDI	97	183	1.9	221	347	1.6	2.3	1.9
Outward FDI	5	26	5.5	12	35	2.8	2.6	1.4
NOI	−94	177	−1.9	−208	335	−1.6	2.2	1.9
GDP	1110	1095	1.0	1285	1182	0.9	1.2	1.1

Note: All values are normalized by population.

mean NOI has become more positive, increasing by a factor of 10.2, while the mean GDP has increased by a factor of just 2.1. Furthermore, the ratio of the standard deviation to the mean of NOI has fallen from 9.2 in 1980 to 2.4 in 1992 for industrialized countries, while this ratio for GDP has remained constant at 0.2. This suggests that convergence phenomena regarding GDP have halted, while the NOI positions of these countries have become increasing similar.

The high growth rate of NOI relative to GDP for the industrialized countries can be contrasted with that of the non-industrial countries (Table 1.2). The mean NOI levels for the non-industrial countries have become more negative but only by a factor of 2.2, whereas their mean GDP levels have increased by a factor of 1.5. The ratio of the standard deviation to the mean of NOI has decreased only slightly, while that of GDP has increased from 1.2 to 1.4 between 1980 and 1992. This suggests there is an increasing variation in the income levels of these countries as a whole, while their NOI positions have remained at the same level of dispersion.

However, as Figure 1.2 illustrates, there seem to be two groups of pre-Stage 4 countries. The majority of developing countries seem to be clustered at the origin, while just a handful of countries are more widely distributed, and roughly correspond to the newly industrializing countries (NICs). If we extract this group, we are able to distinguish between the Stage 1 and 2 countries, and the Stage 3 countries, identified separately in Table 1.2. It is readily apparent that much of the growth associated with the entire sample of non-industrialized countries was primarily associated with the NICs. The mean level of GDP for Stage 1 and 2 countries between 1980 and 1992 has shown only a marginal increase even in nominal terms by a factor of just 1.2, implying that there may even have been a decline in real terms. As for the Stage 3 countries, their GDP growth rate was equivalent to that of the industrial countries. The mean NOI level for Stage 1 and 2 countries doubled over the same period, becoming more negative, while NOI for Stage 3 became more positive, growing by a factor of 3.8.

The change in distribution over time lends support to our earlier comments regarding the changes in the world economy. Nonetheless, there are only minor differences in the regressions between 1980 and 1992, and therefore we shall only present those for the most recent period.

The results of the regressions, set out in Table 1.1 and plotted on Figure 1.2, confirm our hypotheses. By excluding the industrialized countries from our analysis, the results of the regressions seem to be weaker. The curve (labelled as 'DC') provides a better estimation of the true relationship between NOI and GDP for non-industrialized countries.

As Figure 1.2 shows, the results of our regressions also confirm our hypotheses regarding the differences in the 'idealized' IDP due to the differences in the extent of natural and created assets. Countries with above average natural assets tend to demonstrate a lower level of NOI for any given value of GDP relative to the average expected path. Countries with above average created assets, on the other hand, demonstrate a much higher value of NOI relative to the average expected path, and to the natural-asset-type countries. It is interesting to note, however, that although there are differences in NOI for any given level of GDP between the two groups, the difference narrows considerably at higher levels of GDP.

Inward and outward FDI

The effects noted above regarding changes to the extent of NOI are more apparent when examining the two components of NOI separately. As Table 1.2 shows, the mean outward FDI for the industrialized countries increased by a factor of 4.3 between 1980 and 1992 – twice that of the growth of GDP. The ratio of the standard deviation to the mean has also fallen, implying that the level of outward FDI has tended to converge among this group of countries. The mean inward FDI has also increased by a factor of 3.6, but the level of disparity has remained constant.

For pre-Stage 4 countries, outward FDI grew faster than inward FDI between 1980 and 1992, but even in 1992 the extent of outward investment remained at very low levels. It is interesting to note that the ratio of the standard deviation to the mean for outward FDI for developing countries has fallen from 5.5 to 2.8, implying that a larger number of developing countries have begun to engage in outward FDI since 1980. The Stage 3 countries, on the other hand, have showed an 18-fold increase in their outward investment levels, and a fall in the ratio of the standard deviation to the mean from 2.1 to 1.3. Inward investment into the NICs also grew twice as rapidly as that into other developing countries. Thus, as expected, the Stage 3 countries have demonstrated growth of both inward and outward FDI at a much higher pace than both the developing countries as well as the industrialized ones. Their GDP has grown at the same rate as that of the industrialized countries, and twice as

Table 1.3 Log–linear regression equations for inward and outward FDI against GDP

Independent variable	*Sample*	*Constant*	*LOGGDPK*	*ADJ.R^2*	*F-value*	*N*
LOGOWK	ALL	−11.866***	1.9487***	0.866	342.58	54
LOGOWK	NA	−10.812***	1.8199***	0.809	115.63	28
LOGOWK	CA	−14.572***	2.2457***	0.768	57.35	18
LOGIWK	ALL	−3.7024***	1.1626***	0.746	256.85	88
LOGIWK	NA	−4.2996***	1.2751***	0.705	127.39	54
LOGIWK	CA	−4.166***	1.176***	0.755	68.87	23

*** Significant at the 1% level. ** Significant at the 2.5% level.

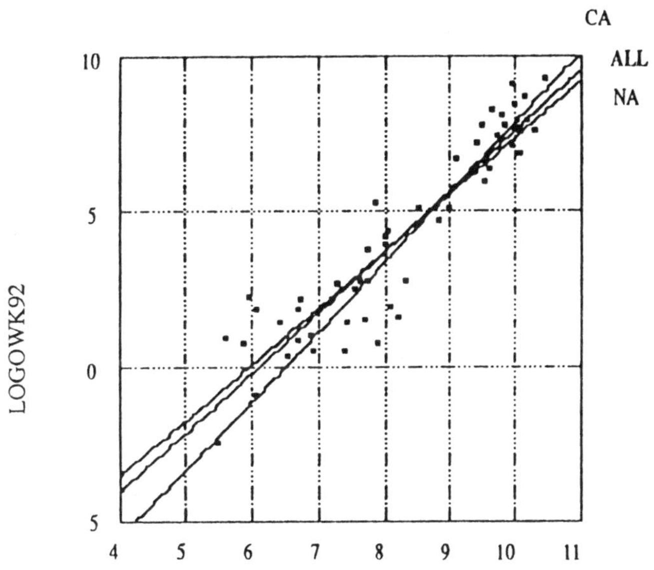

Figure 1.3 Outward FDI and GDP, 1992 (log–linear)

fast as the developing countries. The ratio of the standard deviation to the mean of GDP for the NICs is almost the same as that for the industrialized countries, implying that, as a group, their levels of GDP have converged.

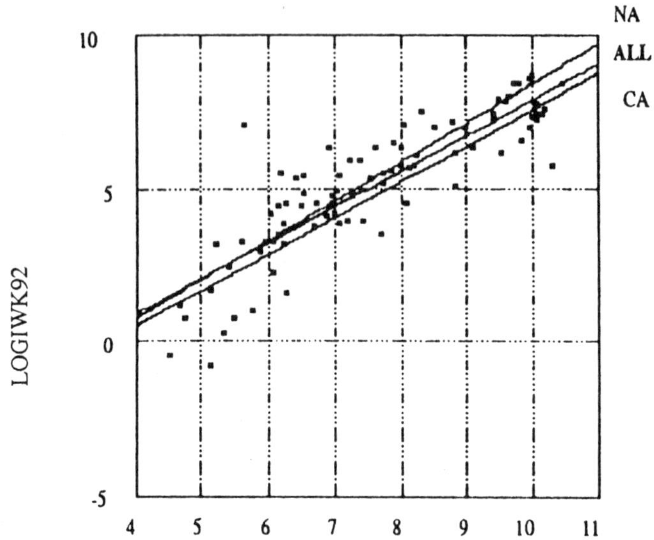

Figure 1.4 Inward FDI and GDP, 1992 (log–linear)

In running regressions for gross inward and outward direct investment, we have utilized a log–linear specification as originally suggested by Dunning (1981). We have done so for three samples – all countries (ALL), the natural asset countries (NA) and the created asset countries (CA), and the results are set out in Table 1.3 and graphed against the data in log form on Figures 1.3 and 1.4. Since a large number of developing countries have no outward investment, there is a considerable loss of sample size when outward direct investment per capita (OWK) is logged. As a result of this, the estimated value of the intercept is inaccurate. In the case of inward direct investment per capita (IWK), although theoretically a constant term is not required, since there is no country for which GDP = 0, it is necessary to include one. It is therefore not meaningful to reconvert the data back into linear form. Nonetheless, it is significant to note that the intercept terms for IWK equations are considerably lower than for OWK, which confirms that in fact inward FDI tends to precede outward FDI.

As Table 1.3 shows, the results are highly significant for all six regressions at the 1 per cent level, with R^2 values greater than 0.7 in all cases. In the case of inward FDI, as hypothesized, created asset countries demonstrate a lower rate of growth of inward FDI than do natural asset countries, while created asset countries have a higher slope for outward FDI than do natural asset countries.

CONCLUSIONS

The current version of the IDP, in introducing dynamic aspects to its framework, represents a paradigm that encapsulates complex phenomena which are exceedingly averse to aggregation. The relationship between FDI and economic development requires the comparison of two phenomena at different levels of economic analysis. While FDI is primarily a micro-economic or firm-specific activity, economic development is a macro-economic or country-specific phenomenon (Gray 1982). The examination of FDI as a country-specific variable requires the assumption that the activities of domestic and foreign MNEs can be aggregated in terms of their motivation, both within industrial sectors and across industrial sectors. Such aggregation can only be justified in countries where the nature, mode and motivation of MNE activity are relatively homogeneous, and the extent of their value adding activities remains at relatively low levels, such as in the lesser developed countries. However, as MNEs become more globalized and engage in more complex investment activity the importance of firm-specific factors in determining the FDI profile of a country becomes increasingly significant. This increasing complexity, together with the differences in measuring FDI between countries, makes any such analysis a hazardous one.

In addition, some of the data reviewed confirm that there have been profound changes in the world economy as a whole, as well as among particular groups of countries and within these countries. The process of catching up and falling behind has resulted in a polarization of countries into three distinct groups. The first group consists of the industrialized countries, which have been shown to have a convergence within the group of income levels, but considerable growth in income levels in the past 15 years. On average, their NOI positions have become increasingly positive, while their levels of gross inward and outward FDI are tending to converge. The second group consists of a handful of economies that can be regarded as being in Stage 3, which are exhibiting high growth in income levels and have become larger net outward investors since the 1980s. Their income levels also show signs of converging as a group, as well as with the industrialized countries. More significantly from our point of view, however, is the fact that their inward and outward FDI has been growing at a rate that outstrips even that of the industrialized countries. The third group consists of the Stage 1 and 2 countries which have experienced a divergence of income levels away from those of the industrialized countries as well as from the Stage 3 countries. Income levels have grown only marginally even in nominal terms, although the extent of FDI activity associated with these countries continues to grow faster than their domestic economies. Nonetheless, their NOI position has become increasingly negative, and both their inward and outward FDI positions have grown at a slower pace than that of either of the other two groups.

Nonetheless, the data reviewed in this chapter indicate that, despite the limitations associated with a static analysis of a dynamic phenomena, changes in the world economy as well as the large differences between countries, there continues to be a relatively strong causal relationship between FDI activity and economic development. Furthermore, we have also been able to examine, albeit simplistically, the hypotheses regarding how the extent of natural/created assets determines the shape of the IDP. Countries that are relatively well endowed in natural assets have a higher growth rate of inward FDI, but a much lower growth rate of outward FDI, than countries that do not have a strong natural asset position as well as those countries with a strong created asset position. In terms of the traditional J-relationship between NOI and development, the J-curve continues to be valid for pre-Stage 5 countries. Natural-asset-type countries tend to have a considerably higher NOI position relative to the created asset countries at any given income level.

Both Parts 1 and 2 of this chapter have sought to show that cross-sectional analysis is not an appropriate tool to capture the dynamic character of the IDP. This requires an oversimplification of complex economic activity into a few general and aggregated variables, an exercise which cannot be undertaken without great caution. For instance, if we restrict ourselves to using GDP as a single indicator of development, the process of economic restructuring as well as the growth of the technological competitiveness of countries are not taken into account. Likewise, the growth of strategic alliances as an alternative mode of international value-added activity and the growth of the activities of governments in the organization of economic activity cannot be usefully included in an aggregate analysis across countries.

For this and other reasons the following chapters examine some of the critical country-specific variables likely to determine both the gross inward and outward investment positions at a given level of income, and the shape of their IDP as they move through the various stages of economic development. A final chapter seeks to relate these differences to the more general statistical relationship which we have shown to exist between a country's international investment position and its gross domestic product per capita.

NOTES

1 See Kumar and McLeod (1981).
2 Earlier called 'cycle'.
3 Or the right to their use.
4 *Natural assets* consists of the 'fruits of the earth' and the stock of unskilled labour. *Created assets* are those derived from the upgrading of natural assets. The latter assets may be tangible or intangible, and include capital and technology as well as those pertaining to skilled labour, such as technological, managerial and organizational expertise. For further details, see Dunning (1994) and Narula (1995).
5 See the article by Reich (1990) for a succinct discussion on this issue.
6 See Gugler (1991).
7 For a review of the various empirical studies, see Narula (1993).
8 Cited, for example, in Dunning (1993), Chapters 11–19.
9 As reviewed, for example, in Dunning (1993).
10 Acquisitions accounted for 83.2 per cent of the outlays by foreign direct investors in the United States between 1986 and 1992 (Fahim-Nader and Bargas 1993).
11 Put another way, in their search for created assets and markets which they perceive necessary to advance their objectives, both firms and countries are acquiring these resources and capabilities by buying out (or investing in) foreign firms, in addition to being bought out (or invested in) by foreign firms. The extent to which these routes are complements to, or substitutes for, each other has not been explored in the literature. For an examination of the consequences of alliance capitalism for theorizing about the determinants of international production, see Dunning (1995).
12 Also called 'collective' capitalism (Lazonick 1992) and the 'new' capitalism (Best 1990).
13 What one writer (Maister 1993) has referred to as a 'farmer' rather than a 'hunter' organizational management style.
14 Other motives for strategic outsourcing include the need to capture the

specialized professional capabilities of suppliers, to shorten cycle times, and to respond better to customers. Examples of firms engaging in 'close' control procurement strategies include Marks and Spencer, Nike and Honda (Quinn and Hilmer 1994).

15 Notably from the work of John Hagedoorn and his colleagues at MERIT. For recent contributions see Hagedoorn and Schakenraad (1991, 1992, 1994) and Hagedoorn (1992, 1993).

16 For example, host governments may compel foreign firms to conclude cooperative arrangements with indigenous firms if they wish to produce within their boundaries, and/or force domestic firms to conclude such arrangements with foreign firms if they wish to transfer their O advantages abroad.

17 It is also worth recalling that the greater part of non-equity technology and organizational transference between countries is undertaken by MNEs.

18 Primarily because the *created* to *natural* asset ratio of the former countries was higher.

19 See especially Contractor (1990).

20 This proposition is explored in more detail in Dunning (1992, 1994).

21 Thus to quite a large extent the US net inward investment position in the 1960s has been eroded by the catching up of many European countries; while the seemingly invincible economic prowess of Japan in several sectors is now being challenged by the United States and Europe.

22 For a review of these, see Dunning (1993), especially the writers mentioned in Chapter 10. See also Dunning (1994).

23 Freeman and Hagedoorn (1994) suggest that between 1980 and 1989, almost 70 per cent of strategic technology partnering agreements between Triad firms and developing country firms were equity based, while for intra-Triad agreements, it was less than 50 per cent.

24 These data only cover strategic technology partnering for the period 1980-9.

25 See Dunning (1992, 1993).

26 These data are both derived from the World Bank (1994).

REFERENCES

Abramovitz, M. (1986) 'Catching up, forging ahead, and falling behind', *Journal of Economic History*, *XLVI*, 385–406.

Alam, M. S. (1992) 'Convergence in developed countries: an empirical investigation', *Weltwirtshaftliches Archiv*, *128*, 189–200.

Baumol, W. (1986) 'Productivity growth, convergence, and welfare: what the long-run data show', *American Economic Review*, 76, 1072–85.

Best, M. (1990) *The New Competition: Institutions of Restructuring*, Cambridge, MA: Harvard University Press.

Cantwell, J. (1989) *Technological Innovation and Multinational Corporations*, Oxford: Basil Blackwell.

Cantwell, J. A. and Dunning, J. H. (1991) 'Multinationals, technology and the competitiveness of European industries', *Aussenwirtschaft*, *46*, 45–65.

Cantwell, J. and Randaccio, F. (1990) 'The growth of multinationals and the catching up effect', *Economic Notes*, *19*, 1–23.

Contractor, F. (1990) 'Ownership patterns of US joint ventures abroad and the liberalization of foreign government regulation in the 1980s: evidence from the benchmark surveys', *Journal of International Business Studies*, *21*, 55–73.

Dowrick, S. (1992) 'Technological catch up and diverging incomes: patterns of economic growth 1960–88', *The Economic Journal*, *102*, 600–10.

Dowrick, S and Gemmell, N. (1991) 'Industrialization, catching up and economic growth: a comparative study across the world's capitalist economies', *Economic Journal*, *101*, 263–75.

Dunning, J. H. (1981) 'Explaining the international direct investment position of countries: towards a dynamic or developmental approach', *Weltwirtschaftliches Archiv*, *119*, 30–64.

——— (1986) 'The investment development cycle revisited', *Weltwirtshaftliches Archiv*, *122*, 667–77.

——— (1988a) *Explaining International Production*, London: Unwin Hyman.

——— (1988b) *Multinationals, Technology and Competitiveness*, London: Unwin Hyman.

——— (1992) 'The global economy, domestic governance, strategies and transnational corporations: interactions and policy recommendations', *Transnational Corporations 1*, December, 7–45.

——— (1993) *Multinational Enterprises and the Global Economy*, Wokingham, England, and Reading, MA: Addison-Wesley.

——— (1994) 'Reevaluating the benefits of foreign direct investment', *Transnational Corporations 3*, February, 23–52.

——— (1995) 'Reappraising the eclectic paradigm in an age of alliance capitalism', *Journal of International Business Studies* 26(3).

Dunning, J. H. and Narula, R. (1994) 'Transpacific FDI and the investment development path: the record assessed', *University of South Carolina Essays in International Business*, No. 10, May.

Durkheim, E. (1964) *The Division of Labour in Society*, New York: The Free Press.

Fahim-Nader, M. and Bargas, S. E. (1993) 'US business enterprises acquired or established by foreign direct investors in 1992', *Survey of Current Business*, May, 113–23.

Freeman, C. and Hagedoorn, J. (1994) 'Catching up or falling behind: patterns in international interfirm technology partnering', *World Development*, *22*, 771–780.

Gerlach, M. L. (1992) *Alliance Capitalism: The Social Organization of Japanese Business*, Oxford: Oxford University Press.

Gray, H. P. (1982) 'Macroeconomic theories of foreign direct investment: an assessment,' in A. Rugman (ed.) *New Theories of the Multinational Enterprise*, London: Croom Helm.

Gugler, P. (1991) *Les Alliances Strategiques Transnationales*, Fribourg: Editions Universitaires.

Hagedoorn, J. (1992) 'Organizational modes of inter-firm cooperation and technology transfer', *Technovation*, *10*, 17–30.

——— (1993) 'Understanding the rationale of strategic technology partnering: inter-organizational modes of cooperation and sectoral differences', *Strategic Management Journal, 14*, 371–85.

Hagedoorn, J. and Narula, R. (1994) *Choosing Modes of Governance for Strategic Technology Partnering: International and Sectoral Differences*, MERIT Working Paper Series No. 94-025.

Hagedoorn, J. and Schakenraad, J. (1991) 'The internationalization of the economy, global strategies and strategic technology alliances', *Nouvelles de la Science et des Technologies, 9*, 29–41.

——— (1992) 'Leading companies in networks of strategic alliances in information technologies', *Research Policy, 21*, 163–90.

——— (1994) 'The effects of strategic technology alliances on company performance', *Strategic Management Journal, 15*, 291–309.

Hamalainen, T. (1993) *The Evolving Role of Government in Economic Organization*, mimeo, Newark: Rutgers University.

Kumar, K. and McLeod, M. (eds) (1981) *Multinationals from Developing Countries*, Lexington, MA: D. C. Heath.

Lazonick, W. (1992) 'Business organization and competitive advantage: capitalist transformation in the twentieth century', in Dosi, G., Gianetti, R. and Toninelli, P. A. (eds) *Technology and Enterprise in a Historical Perspective*, Oxford: Clarendon Press, 119–63.

Maister, D. H. (1993) *Managing the Professional Service Firm*, New York: Free Press/Macmillan.

Narula, R. (1993) *An Examination of the Evolution and Interdependence of Foreign Direct Investment and Economic Structure: The Case of Industrialized Countries*, Ph.D. Thesis, Newark: Rutgers University.

——— (1995) *Multinational Investment and Economic Structure*, London: Routledge.

Ohmae, K. (1987) *The Borderless World*, New York: The Free Press.

Ozawa, T. (1992) 'FDI and economic development', *Transnational Corporations, 1*, 27–54.

Porter, M. E. (1990) *The Competitive Advantage of Nations*, New York: The Free Press.

Quinn, J. B. and Hilmer, F. G. (1994) 'Strategic outsourcing', *Sloan Management Review, 35*, Summer, 43–55.

Reich, R. (1990) 'Who is us?', *Harvard Business Review*, January-February, 53–64.

Tolentino, P. (1993) *Technological Innovation and Third World Multinationals*, London: Routledge.

UNCTAD (1993) *World Investment Report 1993*, New York: United Nations.

UNCTAD (1994) *World Investment Report 1994*, Geneva: United Nations.

Verspagen, B. (1993) *Uneven Growth Between Interdependent Economies*, Avebury: Aldershot.

World Bank (1994) *World Development Report 1994*, Oxford: Oxford University Press.

Part IV
Geographical Localization in Multinational Corporations and Technology Spillovers

[18]

TECHNOLOGICAL CAPABILITIES AND JAPANESE FOREIGN DIRECT INVESTMENT IN THE UNITED STATES

Bruce Kogut and Sea Jin Chang*

Abstract—This article examines the effect of relative technological capabilities on Japanese direct investment into the United States by looking simultaneously at industry conditions in the two markets. A negative binomial regression model is specified to estimate the effects of R & D capability and industry structure on a count measure of Japanese entries across 297 industries. The results indicate that Japanese direct investment in the United States is drawn to industries intensive in R & D expenditures summed across both countries; voluntary restraints on Japanese exports encourage direct investment. When the entries are disaggregated by mode (e.g., new plant or acquisition), there is a significant indication that joint ventures are used for the sourcing and sharing of U.S. technological capabilities.

THERE is an on-going policy concern in the United States over the loss of the resident technological capabilities due to foreign direct investment. Popular attention has focused particularly on Japanese investments in high technology industries. Yet, little analysis has been given to whether these investments reflect the extension of Japanese firm advantages through foreign direct investment as opposed to the targeting of American technology.

These issues are also of interest to the broadening of theoretical and empirical studies on the relationship of R & D expenditures to foreign direct investment. Home technological capability has been found in numerous empirical studies to correlate significantly with outward exports and foreign direct investment.[1] The explanation for this finding has conventionally emphasized that foreign direct investment and exporting are jointly determined by a combination of factors reflecting dynamic comparative advantage and industry rivalry. R & D expenditures lead to the creation of valuable rent-yielding and intangible assets, such as the knowledge involving new product and process technologies. Home oligopolistic rivalry not only promotes the creation of these technologies, but leads to eventual investments in overseas markets.[2]

Frequently neglected in the empirical studies on the determinants of foreign direct investment are the conditions of the foreign market. This neglect is particularly important to identifying whether foreign direct investment is motivated by the home technological advantage or by the desire to source technology in the foreign market. Because of the high correlation of the industry distribution of R & D expenditures among countries, the positive relationship between home R & D and outward direct investment cannot itself distinguish between these two distinct motives.

For example, due to the paucity of non-American data on R & D expenditures, most empirical studies on foreign investment into the United States have relied on American industry data, assuming a high correlation of R & D expenditures among countries. But the positive relationship between U.S. R & D expenditures and foreign entry raises, in fact, a paradox that foreign investments into the United States favor industries intensive in domestic R & D expenditures.[3] Without observations on R & D intensity of the investing country, it has been impossible to sort out whether foreign direct investment in the United States is motivated by the exploitation of an initial technological advantage or by the seeking of new technologies resident in the United States.

Received for publication December 27, 1989. Revision accepted for publication August 9, 1990.

* Wharton School of the University of Pennsylvania.

This paper is supported under a project grant from the German Marshall Fund; additional funding was given by AT & T under the auspices of the Reginald H. Jones Center. We acknowledge the research assistance of James Jurney and the comments of Erin Anderson, Paul Rosenbaum, Louis Wells, and the anonymous referees. The data for the study were collected by the International Trade Administration at the Department of Commerce under the guidance of the late Mr. Richard Apcar, to whom we would like to express our gratitude and appreciation of his dedicated public service.

[1] See the summaries by Deardorff (1984) for exports and by Caves (1982) for foreign direct investment.

[2] Caves (1971) and Magee (1976) note the relationship between foreign direct investment and oligopoly may be spurious, as both arise from the possession of entry deterring and intangible assets by a firm.

[3] See the reviews by Pugel (1981) and McClain (1983).

The importance of distinguishing between the relative technological capabilities of the source and recipient countries has recently been raised by the studies of Audretsch and Yamawaki (1988) and Yamawaki and Audretsch (1988) on the determinants of Japanese exports to the American market. In their work, Japanese exports have been shown to be sensitive to competitive conditions in the home and foreign market. Relative Japanese advantage in R & D is a significant determinant of exports to the American market.

The following article addresses the question of the effects of Japanese and U.S. technological capability on Japanese foreign direct investment into the United States. Two industry measures of technological capability are examined: R & D expenditures for each country and innovation frequency (though data for the latter exist only for the United States). These variables, along with other determinants of foreign direct investment, are then regressed against a count measure of the number of Japanese entries in 297 United States industries. Because the dependent variable is a count of entries, a negative binomial regression is specified, with heterogeneity being captured by an error term. The count measure allows for a lower level of aggregation (4-digit SIC) than previous studies and for the elimination of censoring bias if an industry has not evidenced an entry.

I. Industry Influences on Foreign Entry

There is little theoretical controversy over the importance of the role of the competitive conditions of the foreign market in the theory of foreign direct investment. The incorporation of the effects of the competitive nature of the foreign market on inward direct investment poses, however, complications to the traditional explanation stressing intangible assets of the firm and oligopolistic rivalry of the home market. These complications are of two sorts. One is simply whether the foreign country is advantaged in location.[4] Lower factor costs of the foreign relative to the home country should increase the attractiveness of direct investment. There is some empirical evidence to support this commonsensical relationship, especially in the findings regarding the positive effect of tariffs on inward investment flows.[5]

Common sense notwithstanding, the high degree of flows of foreign direct investment among developed countries suggest that location advantages in factor costs are not pivotal. In fact, Swedenborg (1979) found that the country distribution of Swedish outward investment flows are correlated with high wages. One possible explanation for this anomaly derives from Caves' (1971) explanation that foreign direct investment stems from the public good character of the rent-yielding intangible asset and the necessity of local production to adapt products to the foreign market. Alternatively, Swedenborg's result could represent a confounding of horizontal and vertical investment. Vertical direct investment need not be restricted to raw material and low-cost labor sites, but also can occur for the sourcing of the technical capabilities (as embodied in skilled and professional workers) from foreign countries. Both Dunning (1988) and Cantwell (1989) hypothesize, with empirical support, that countries leading in technology will draw foreign direct investment.

The second complication posed by incorporating competitive conditions of the target country in the theory of foreign direct investment is that the foreign market may itself be oligopolistic. In the 1960s, when the overwhelming proportion of the world's foreign direct investment flows was American, the interaction between national oligopolies did not seem very consequential. Hence, the emphasis by Vernon (1966) solely on the influence of home rivalry in new product and process technologies on foreign direct investment was reasonable and supported by considerable empirical evidence.[6] Subsequent empirical work, especially of Flowers (1976) and Graham (1978), and the theoretical model of cross-hauling investments of Brander and Krugman (1983) indicate oligopolistic rivalry at the international, rather than national, level as influencing the foreign direct investment decision. This perspective coincides with the recent treatment of exports and international R & D rivalry, where export compe-

[4] The effect of location is best summarized in Dunning's "eclectic" approach. See Dunning (1981).

[5] See Horst (1972) and Caves (1982).

[6] See the volume of empirical studies edited by Wells (1972), as well as Knickerbocker (1973).

TABLE 1.—JAPANESE ENTRY INTO 2-DIGIT SIC INDUSTRIES

SIC Code	Industry Group	Entry Count	Value (million $)
20	Food and kindred products	73	1298
21	Tobacco products	0	0
22	Textile mill products	13	130
23	Apparel and related products	7	40
24	Lumber and related products	10	0
25	Furniture and fixture	8	37
26	Paper and allied products	9	174
27	Printing and publishing	9	166
28	Chemical and allied products	100	2382
29	Petroleum and coal products	2	− 18
30	Rubber and plastics, n.e.c.	29	804
31	Leather and leather products	2	0
32	Stone, clay, and glass products	22	205
33	Primary metal industries	72	2195
34	Fabricated metal products	48	434
35	Machinery, except electrical	133	3210
36	Electric and electronic equipment	209	2334
37	Transportation equipment	114	3111
38	Instruments and related products	49	1389
39	Miscellaneous manufacturing	29	335
	Total	938	18,226

Note: The amount of Japanese investment in the United States is calculated for 1980–88. The entry count is for 1976–87.

tition (sometimes abetted by governments) is driven by investments in R & D.[7]

Though R & D influences jointly exports and outward direct investment, there is a subtle difference. By Vernon's product cycle argument, foreign direct investment should occur in industries that are maturing and, hence, declining in R & D expenditures. Similar implications are to be found in Dornbusch, Fischer, and Samuelson's (1977) Ricardian analysis of trade, where differences in relative productivity among two countries result in each country specializing in the industries where it maintains a relative technical advantage. Productivity gains localized to a few export industries force the marginal industries to withdraw from the market. Two implicit alternatives are for home firms to transfer production to the foreign market if they are at a home location disadvantage but otherwise maintain a competitive position in the foreign market, or to increase investments in R & D expenditures in order to renew the technological base of export or import-competing industries.

In summary, there exists three issues of theoretical and policy importance that are currently unsettled: whether foreign direct investments into the United States respond to conditions of the home or foreign market structure; whether they are motivated by the exploitation of home technological advantage or by the sourcing of the U.S. locational advantages in technology; and whether exports is the preferred mode in those industries where home technological capabilities are relatively advantaged. The first two of these issues are examined below by looking jointly at the influence of Japanese and U.S. industry factors on Japanese direct investments in the United States; the last issue is only addressed by joint inference from our and other studies' results.

[7] See Brander and Spencer (1983, 1985), as well as the empirical treatment by Scherer (1988).

II. Empirical Model

Measurement of Foreign Entry

Sorting out explanations for these issues has always been difficult because foreign direct investment figures are prone to substantial measurement errors. (R & D measurement has also been a stumbling block, as discussed below.) Balance of payment records neglect the investment portion derived from local debt or the capitalized contribution of intangibles, such as technology or goodwill. Foreign sales in a country have been a preferred measurement, but are also vulnerable to error by neglecting local content in the value-added. Like R & D expenditures, most available data on foreign direct investment or foreign sales tend to be aggregated to the 3 or 2 digit SIC levels.

We employ a novel measure in the analysis below, that is, a count of the number of entries in each 4-digit industry made by Japanese firms between 1976 and 1987. (See the appendix on data sources.) An entry count has the methodological advantage in providing a disaggregated measurement of foreign direct investment, thus allowing for a more refined determination of the influence of industry variables. Table 1 provides the industry breakdown of Japanese entries into the United States and the value of these investments as provided by balance of payments statistics. (We give only the entry count and value for manufacturing industries; minority investment positions are excluded.[8]) The correlation between these two series is 0.89.

Variable Measurement

A major problem in analyzing the relative influence of industry and country variations in R & D expenditures on foreign entry is the paucity of data. When available, the data are highly aggregated. In addition, problems of collinearity are unavoidable, because industry R & D expenditures are correlated across countries.

Audretsch and Yamawaki (1988) and Yamawaki and Audretsch (1988) address these problems by using multiple measures of technological intensity, in addition to looking at the raw differences in industry R & D expenditures between the two countries. We follow this approach by gathering information on R & D expenditures as a percentage of sales and (for the United States only) on innovation frequencies. This latter measure is a count of the number of innovations reported at the four-digit industry level.[9] These measures have been used successfully in previous applied work.[10] All three variables (*Japan R & D*, *U.S. R & D*, and *Innovation Frequency*) are expected to be positively related to Japanese entry.

The drawback of the R & D expenditures is the high correlation between the U.S. and Japanese series. We addressed this problem by both summing and subtracting the two countries' expenditures.[11] The R & D sum provides a measure of the overall effect of R & D on entry; it captures the dimension of international rivalry in R & D, as well as its properties as an intangible and, thus, internationally transferable asset. The R & D difference serves to indicate whether entry is motivated by a home (i.e., Japanese) technological capability or by the desire to source U.S. technology. The two variables are weakly correlated with each other.[12]

The remaining variables are designed to capture other conditions of rivalry in Japan and the United States, as well as barriers to entry into the United States. Following Hymer (1976), Knickerbocker (1973), and others, we use the industry concentration ratio as an index of rivalry (*U.S. 8-Firm Concentration* and *Japan 8-Firm Concentration*).[13] Because intermediate levels of concentration have been found empirically to lead to oligopolistic instability, 8-firm concentration indices for Japan and the United States were used. (Regression results with 4-firm concentration rates are similar, as to be expected given the high correlation between the two indices.)

As foreign entry shares many of the characteristics of entry into new product markets, we can rely on the results of the recent literature on entry to identify additional variables that influence foreign investment. One variable characterizing the attractiveness of an industry for entry is shipment growth (*Shipment Growth*).[14] A growing industry permits entry without necessarily displacing encumbents. Industry advertising and marketing expenditures (*US Advertising*) have, however,

[8] Because a few industries (especially auto-related) experienced an inordinately large number of entries, the regressions reported below were reestimated without them; no change in the results were found.

[9] See Acs and Audretsch (1987) for a description.

[10] See Audretsch and Yamawaki (1988) and Yamawaki and Audretsch (1988). An alternative strategy of using company data is provided by Goto and Suzuki (1989), but this information is available only for a fraction of the cases.

[11] We thank Paul Rosenbaum for this suggestion.

[12] Since the U.S. and Japanese R & D series are not identical in their aggregation, we also built the same sum and difference measures using the National Science Foundation R & D data, which aggregated similarly to the Japanese series. None of the results reported below were changed.

[13] Similar findings for the degree of concentration and outward U.S. direct investment were found by Baldwin (1979) and Pugel (1978) and by Caves, Porter, and Spence (1980) for U.S. investment in Canada.

[14] See Gorecki (1975), Duetsch (1984), and Khemani and Shapiro (1986). Shapiro (1983) also confirms this finding for foreign entry into Canada.

TABLE 2.—DATA DESCRIPTIONS AND PREDICTED SIGNS FOR INDEPENDENT VARIABLES

Variable	Definition	Predicted Sign
Japan R & D	Japanese R & D expenditure/sales, (%) eleven years average for 1976–85	+
U.S. R & D	U.S. R & D expenditure/sales (%) 1977	+
R & D Sum	Japanese R & D expenditure + U.S. R & D expenditure	+
R & D Difference	Japanese R & D expenditure − U.S. R & D expenditure	−
Innovation Frequency	Number of new innovations, collected by Small Business Administration, 1982	+
Japan R & D Growth	Average growth rate of Japanese R & D expenditure/sales, 1975–85	−
Japan 8-Firm Concentration	8-firm concentration ratio (%), 1982	+
U.S. 8-Firm Concentration	8-firm concentration ratio (%), 1982	−
Shipment	Eleven years average value of industry shipment, 1975–85	+
Shipment Growth	Average growth rate of industry shipment, 1975–85	+
Import	Import/Shipment (%), eleven years average 1975–85	+
U.S. Advertising	Advertising expense/sales (%), 1977	−
Export Restriction	Dummy variable noting the existence of quota and voluntary restraints for Japanese export	+

been usually found to deter entry.[15] While advertising levels have been found to be correlated with outward flows of foreign direct investment, they should deter inward entry. To these two variables, we add the degree of import penetration (*Import*); industries with high penetration of imports should also be easier to enter by local operations. To control for industry size effects, we include shipments volume (*Shipment*) as a right-hand variable. Finally, in line with the findings of Horst (1972) that import tariffs encourage inward investment, a dummy variable indicating quotas (*Export Restriction*) is constructed.

The above arguments and variable definitions are given in table 2. (Sources for the data are given in the appendix.) Descriptive statistics are given in table 3. The correlation matrix of the variables (not reported here) indicates a substantial degree of collinearity exists for the Japanese and U.S. R & D expenditure variables. A low correlation between innovation frequency and Japanese R & D expenditures suggests one avenue to compare the relative effects of Japanese and U.S. technological investments without degrading the standard errors. Some caution in interpretation is required, however, for, as discussed below, innovation frequency and R & D expenditures are imperfect substitutes for each other.

[15] An exception is Highfield and Smiley (1987) who found no relationship between advertising and entry rates, whereas Kessides (1986) finds an overall positive impact of advertising on entry.

Sample Size

The Department of Commerce listing of foreign entries indicates 938 Japanese entries occurred in 218 industries in the period of 1976 through 1987; the remaining 245 industries, as identified by the U.S. Standard Industrial Classification, did not experience a Japanese entry during the period. Entries only into wholesale and retail distribution are not classified in the manufacturing industries and are not included in the analysis; industries experiencing no entries are included. Unfortunately, matching Japanese industry concentration data to the U.S. industries left 122 U.S. industries without a match; an additional 34 industries were lost due to missing values of other variables. We eliminated these industries in the runs. (Including these industries and eliminating the Japanese concentration variable does not change the results reported below.) This elimination left 825 entries in 297 industries, of which 165 industries experienced entries and 132 industries experienced none. These 297 in-

TABLE 3.—DESCRIPTIVE STATISTICS AND CORRELATION MATRIX

Variable	Mean	Standard Deviation	Lowest	Highest
(1) *Entry Count*	2.78	6.74	0.00	72.00
(2) *Japan R & D*	1.72	1.14	0.20	5.70
(3) *U.S. R & D*	1.35	1.46	0.00	10.20
(4) *R & D Sum*	3.07	2.29	0.40	15.90
(5) *R & D Difference*	0.36	1.25	−6.76	3.98
(6) *Innovation Frequency*	11.44	56.78	0.00	860.00
(7) *Japan R & D Growth*	0.05	0.02	−0.01	0.12
(8) *Japan 8-Firm Concentration*	76.04	19.26	17.50	100.00
(9) *U.S. 8-Firm Concentration*	54.27	22.70	5.00	100.00
(10) *Import*	0.12	0.27	0.00	2.98
(11) *Shipment*	5,503.90	14,689.00	65.34	209,924.27
(12) *Shipment Growth*	0.02	0.05	−0.15	0.41
(13) *U.S. Advertising*	1.62	2.39	0.00	20.20
(14) *Export Restriction*	0.03	0.17	0.00	1.00

Note: $N = 297$.

dustries accounted for 88% of all manufacturing entries.

Model Specification

The use of a count of foreign entries presents econometric and measurement issues similar in the study of patents. We follow the approach of Hausman, Hall, and Griliches (1984) on patent counts by specifying a Poisson regression to model the probability that the number of entries will occur n times (with $n = 0, 1, 2, \ldots$) as follows:

$$\text{Prob}(Y = yj) = e^{-\lambda j}\lambda_j^{Yj}/Y_j! \qquad (1)$$

with Y_j being the count of relationships for the entries of the jth industry. To incorporate exogenous variables, lambda can be made a function of the covariates:

$$\lambda_j = \exp(\Sigma B_i X_{ij}), \qquad (2)$$

where B's are the coefficients, X's are the covariates (with $X1$ set to one), i indicates the ith variable, and j is the jth industry. The exponential function ensures non-negativity.

The Poisson distribution contains the strong assumption that the mean and variance are equal to lambda. Based on an OLS regression, an estimate of the Breusch-Pagan statistic for heteroscedasticity was found to be highly significant. Since some of the (heteroscedastic) variation can be attributed to the large differences in shipment volume across industries, we calculated the Goldfeld-Quandt statistic, which indicated significant differences in variances due to industry size.

To eliminate some of this variation, the initial Poisson regression (using the same variables given in column 3 of table 4) was weighted by shipment volume (*Shipment*). As a heuristic diagnostic (described in Cameron and Trivedi (1986)), the squared residual (i.e. $(y_j - \lambda_j)^2$) was calculated from this initial regression and then regressed on the predicted count (λ_j). The coefficient of 0.02 is significant at less than 0.001, indicating substantial overdispersion.

To address the persisting problem of heteroscedasticity, we incorporate explicitly the disturbance of industry heterogeneity by specifying a compound distribution through an addition of an error term. Equation (2) now becomes

$$\lambda_j = \exp(\Sigma B_i X_{ij})\exp(u_j); \qquad (3)$$

λ_j is no longer determined but is itself a random variable. As u_j is unobserved, it is integrated out of the expression by specifying a gamma distribution for the error term, whereupon the now compound Poisson reduces to the negative binomial model (Johnson and Kotz, 1970). Only the scale of the distribution is permitted to vary as a function of the covariates, with the variance of Yj parameterized to equal $(1 + \alpha)E(Yj)$, that is, a

TABLE 4.—NEGATIVE BINOMIAL REGRESSION ESTIMATES ON ENTRY COUNTS

Variable Name	Equations (1)	(2)	(3)	(4)	(5)
Intercept	0.36 (0.90)	1.22 (3.12)[a]	0.57 (1.41)	0.58 (1.44)	0.58 (1.43)
Japan R & D	0.47 (6.04)[a]		0.26 (2.67)[a]		
U.S. R & D		0.29 (6.39)[a]	0.20 (3.69)[a]		
R & D Sum				0.23 (5.78)[a]	0.22 (5.35)[a]
R & D Difference				0.02 (0.36)	0.04 (0.60)
Innovation Frequency					0.48×10^{-3} (0.17)
Japan R & D Growth	−14.27 (−2.30)[b]	−23.06 (−4.06)[a]	−16.03 (−2.70)[a]	−16.17 (−2.72)[a]	−16.21 (−2.74)[a]
Japan 8-Firm Concentration	0.02 (3.34)[a]	0.02 (4.39)[a]	0.02 (4.55)[a]	0.02 (4.56)[a]	0.02 (4.50)[a]
U.S. 8-Firm Concentration	−0.01 (−3.30)[a]	−0.02 (−4.37)[a]	−0.02 (−4.37)[a]	−0.02 (−4.39)[a]	−0.02 (−4.31)[a]
Import	0.16 (0.25)	0.41 (0.62)	0.25 (0.40)	0.25 (0.40)	0.25 (0.40)
Shipment	-0.54×10^{-5} (−1.04)	-0.47×10^{-5} (−0.89)	-0.49×10^{-5} (−1.01)	-0.49×10^{-5} (−1.01)	-0.49×10^{-5} (−0.99)
Shipment Growth	4.56 (3.11)[a]	1.18 (0.39)	1.90 (1.18)	1.86 (1.15)	1.24 (0.66)
U.S. Advertising	−0.03 (−0.72)	−0.04 (−1.17)	−0.04 (−0.39)	−0.04 (−0.86)	−0.03 (−0.82)
Export Restriction	2.58 (6.99)[a]	3.02 (8.99)[a]	2.84 (8.43)[a]	2.85 (8.46)[a]	2.83 (8.26)[a]
α	0.72 (7.73)[a]	0.70 (7.58)[a]	0.65 (7.47)[a]	0.65 (7.48)[a]	0.65 (7.11)[a]
Log-Likelihood	−714.84	−712.10	−707.08	−707.12	−707.01

Note: t-statistics are in parentheses.
[a] $p < 0.01$.
[b] $p < 0.05$.

constant variance-mean ratio.[16] Given this specification, it becomes straightforward to account for overdispersion.[17] The parameter estimate of α provides a measure of contribution of intrinsic randomness or (as is more probable) heterogeneity caused by omitted variable bias. One likely source of heterogeneity is the omission of the effects of time. Given the availability of a single panel for most of the variables, the data were pooled over time and analyzed by cross-section. Using 1981 and 1985 as breakpoints (which closely correspond to obvious disturbances caused by changes in the value of the dollar/yen exchange rate), the data were partitioned into three samples. No important differences were found in the estimates of the coefficients.

III. Statistical Results

In the results described below, we report only the negative binomial regression estimates. The Poisson estimations tend to give substantially higher t-statistics; there are few cases where a variable that is significant under the negative binomial specification is not also significant under the Poisson. All tests are two-tailed. We discuss first the R & D results across all the regressions and then turn to the other estimates.

In table 4, the first two columns report the results when R & D expenditures for Japanese and the United States are separately entered. The findings indicate strongly that the number of

[16] We verified the specification against an alternative by running the diagnostic recommended by Cameron and Trivedi (1986) to regress the squared residual (divided by the predicted count) on a constant and the predicted count. The fit is positive and only significant for the intercept, supporting the constant variance-mean parameterization.

[17] Greene's statistical package (LIMDEP) provides this test as a standard feature.

entries is related to Japanese and U.S. R & D expenditures, at least when they are separately entered. Interestingly, growth in Japanese R & D expenditures depresses the tendency to enter the United States. As explained further below, this result implies that in industries where Japanese corporations are increasing their technological capabilities, foreign direct investment in the United States is relatively less prevalent. The next regression, reported in column 3, enters U.S. and Japanese R & D expenditures simultaneously. Despite their collinearity, both variables are estimated to be positively and significantly related to entry.

Because of possible contamination of the standard errors due to collinearity, other specifications were explored. As explained earlier, we summed the Japanese and U.S. R & D expenditures (*R & D Sum*), as well as subtracted the U.S. expenditure from the Japanese (*R & D Difference*). The results, given in column 4, indicate that entry is more likely in high R & D expenditure industries; the relative technological advantage of either country does not itself appear to influence the likelihood. The finding points to the importance of the intangibility and transferability of R & D assets, as well as to international rivalry in high technology industries.

In column 5, *Innovation Frequency* is added to the regression. Though positive, its coefficient is not significant. This result is somewhat surprising given a relatively large bivariate correlation (0.4) between *Innovation Frequency* and entry. (Inspection of the correlation table suggested that this correlation is potentially spurious due to the multicollinearity of *Innovation Frequency*, *Shipment Growth*, and *U.S. R & D*.)

The coefficient signs and significance levels of the other variables are fairly robust over the estimated regressions. The coefficients to the 8-firm concentration measures indicate that market structure in both countries exerts a very significant effect on entry. High home concentration clearly encourages direct investment in the United States; high U.S. concentration deters Japanese entry. These results reinforce the traditional findings of the importance of oligopolistic conditions in the home market, but they also show that direct investment is sensitive to the market structure of the target country.

The only other variable that is consistently significant is *Export Restriction*. Because the auto-related industries were outliers in entries and also were characterized by export restrictions, we reestimated the regression in their absence. The results remained largely the same. The imposition of voluntary export restraints promotes Japanese entry into the United States through foreign direct investment. (A coefficient of 2.8 implies that entry should increase by 16, a significant amount!)

The estimates to the remaining variables are not impressive. High important penetration (*Import*) and domestic growth in shipments (*Shipment Growth*) are very weakly related to foreign entry. The coefficient to U.S. advertising and marketing expenditures is negative, indicating that entry is deterred, but the result is not significant. The absolute volume of shipments is also not significantly related to entry.

In all of the regressions, the estimates for unexplained heterogeneity (α) are highly significant. Clearly, there is a significant proportion of inter-industry variation in entry that is not being explained by the included variables. One possibility is that investment motivations differ by the mode of entries (i.e., acquisition, new plant, or joint venture), which vary in frequency across industries. We address this issue below.

Discussion

The regression analysis strongly confirms the traditional findings that foreign direct investment is related to the intensity of industry R & D expenditures. In fact, the strong correlation to both U.S. and Japanese R & D expenditures indicate substantial competition in technological intensive industries. These results are consistent with the recent findings of Scherer (1988), Cantwell (1989), and Yamawaki and Audretsch (1988).

The results regarding the negative effects of the growth of Japanese expenditures suggest that the relationship is more complicated than is frequently depicted. Part of the explanation may lie in a specification bias due to the omission of the export alternative. As Horst (1972) pointed out, exports and foreign direct investment are jointly determined and are likely to be influenced by many of the same factors. It appears that exports remain favored in those industries where

Japanese firms are proportionately expanding most rapidly their technological capabilities.

When analyzed in conjunction with Yamawaki and Audretsch's (1988) positive finding between exports and Japanese expenditures relative to the United States, the results on R & D point to the speculation that growth in Japanese R & D capabilities discourages foreign direct investment because exports persist as the optimal mode. Both modes require a domestic technological capability to be viable. But the switch point from exports to foreign direct investments appears to be driven by dynamic changes in comparative advantage in maintaining and developing the technological capability of export industries.

An examination of the data provides further insight into the analysis. Among the fast growing R & D industries are sectors which are known to be facing considerable import competition in Japan and which are difficult to transplant, e.g., metals (zinc, aluminum, iron and steel) and petroleum refining. The slowest R & D growth industries include television and radio, transportation (exclusive of aircraft), and chemicals. This pattern suggests the immobility of production in import-competing industries stimulates R & D expenditures; the causal path between the feasibility and desireability to shift production offshore and R & D expenditures is likely to flow in both directions. The data are consistent with the broad arguments of Vernon (1966) and Dornbusch, Fischer, and Samuelson (1977), but they also give the important reminder of the endogeneity of R & D and productivity expenditures.

Though not significant, *U.S. Advertising* is negatively signed, consistent with the literature on entry.[18] As the usual finding is that domestic advertising is associated with outward foreign direct investment flows, this negative relationship between advertising and inward entries suggests a potential qualification to past empirical work. In particular, one can speculate that advertising advantages are harder to transport across borders because it is not product embodied and is often tailored to national cultures (Caves, 1982). In addition, entry is deterred by the presence of encumbent brands.

It is of some considerable policy interest that export restrictions promote foreign entry, with a significance of less than 0.001. In light of Yamawaki's (1986) finding that quotas decrease Japanese exports, the effect of export restrictions is to discourage exports and increase foreign entry into the United States. This result is not surprising, but whether it is the intention of policymakers is dubious.

In summary, the statistical analysis generally confirmed the expected relationships between the independent variables and the count frequency of Japanese investment entry into U.S. manufacturing industries. The surprising findings that entries are less likely from Japanese industries showing growing technological capabilities can be reconciled with recent theories of export and production hysteresis, where production from marginally-viable industries are shifted to foreign sites (Kogut and Kulatilaka (1990)). Yet, the alternative explanation that Japanese investments are promoted to source the technological advantages resident in the United States is equally consistent with the results.

IV. Entry by New Plants, Acquisition, or Joint Venture

Some insight into the relative merits of these arguments can be gleaned by partitioning the data by mode of entry.[19] In table 5, the regression results are provided for each entry mode. Columns 1 to 3 are the negative binomial regression estimates for entry with both Japan and U.S. R & D expenditures included; columns 4 to 6 give the estimates with the R & D expenditures summed and subtracted.

The results given in columns 1 to 3 indicate important differences from the results in table 4 and substantial variation across mode of entry. Of most importance, Japanese R & D expenditures have no significant influence on entry into the United States by acquisition or joint venture. For new plants, the effect of Japanese R & D investment is to increase the likelihood of entry. U.S. R & D expenditures are associated with higher entry rates by all three modes, but especially by

[18] See Shapiro and Khemani (1987).

[19] Because of its small sample size, entry by plant expansion is not examined.

TABLE 5.—NEGATIVE BINOMIAL REGRESSION ESTIMATES BY TYPE OF ENTRY

	Acquisition	Joint Venture	New Plant	Acquisition	Joint Venture	New Plant
Intercept	−0.14	−0.94	−0.50	−0.10	−0.91	−0.50
	(−0.20)	(−1.50)	(−0.88)	(−0.15)	(−1.44)	(−0.88)
Japan R & D	0.05	−0.11	0.41			
	(0.44)	(−0.82)	(3.00)[a]			
U.S. R & D	0.18	0.34	0.20			
	(2.56)[b]	(3.92)[a]	(2.87)[a]			
R & D Sum				0.10	0.11	0.31
				(1.95)[c]	(2.12)[b]	(5.87)[a]
R & D Difference				0.03	−0.25	0.08
				(0.35)	(−2.34)[b]	(0.81)
Innovation				0.003	−0.001	−0.001
Frequency				(1.04)	(−0.24)	(−0.26)
Japan R & D	−14.80	−2.08	−21.99	−13.50	−2.26	−22.25
Growth	(−1.89)[c]	(−0.27)	(−2.43)[b]	(−1.91)[c]	(−0.29)	(−2.48)[b]
Japan 8-Firm	0.02	0.02	0.02	0.02	0.02	0.02
Concentration	(3.13)[a]	(2.69)[a]	(3.20)[a]	(2.94)[a]	(2.70)[a]	(3.28)[a]
U.S. 8-Firm	−0.03	−0.01	−0.02	−0.03	−0.01	−0.02
Concentration	(−5.18)[a]	(−2.54)[b]	(−2.77)[a]	(−4.85)[a]	(−2.58)[a]	(−2.90)[a]
Import	0.57	−0.88	−0.19	0.71	−0.87	−0.20
	(0.61)	(−0.96)	(−0.24)	(0.85)	(−0.95)	(−0.25)
Shipment	-0.28×10^{-5}	-0.94×10^{-5}	0.12×10^{-5}	-0.30×10^{-5}	-0.95×10^{-5}	0.12×10^{-5}
	(−0.46)	(−1.33)	(0.20)	(−0.47)	(−1.25)	(0.20)
Shipment	2.75	−3.64	1.52	−0.69	−3.19	2.51
Growth	(1.54)	(−1.75)[c]	(0.80)	(−0.28)	(−1.35)	(1.06)
U.S. Advertising	0.02	−0.21	−0.05	0.03	−0.22	−0.05
	(0.47)	(−2.75)[a]	(−0.77)	(0.74)	(−2.78)[a]	(−0.82)
Export	1.67	3.33	2.69	1.53	3.37	2.71
Restriction	(4.74)[a]	(8.40)[a]	(7.56)[a]	(4.09)[a]	(7.79)[a]	(7.53)[a]
α	0.62	0.32	0.83	0.56	0.33	0.82
	(3.25)[a]	(2.98)[a]	(5.54)[a]	(3.22)[a]	(2.73)[a]	(5.35)[a]
Log-Likelihood	−394.46	−330.94	−518.71	−391.36	−330.60	−518.61

Note: *t*-statistics are in parentheses.
[a] $p < 0.01$.
[b] $p < 0.05$.
[c] $p < 0.10$.

joint ventures. These results suggest the use of cooperative ventures to source, or benefit from, U.S. technological capabilities in industries where the United States, but not Japan, are intensive in R & D expenditures. But it should be noted that the effect of the growth in Japanese R & D expenditures is to dissuade entry, no matter by what mode.

The remaining variables present some interesting contrasts. Acquisitions are strongly favored when the investment is from a concentrated Japanese industry but into a low concentrated U.S. industry. This relationship suggests that home rivalry might spill over into a rush to invest by acquisition. Sparsity of American targets appears to discourage acquisitions, presumably because the likelihood of a takeover is captured in the valuation of share prices.

Joint ventures favor industries where advertising expenditures are low. This finding reinforces the explanation that joint ventures are oriented towards technological sourcing. All three modes are stimulated by the existence of export restraints.

Columns 4 to 6 report the estimates by replacing each country's R & D expenditure by the variables *R & D Sum*, *R & D Difference*, and *Innovation Frequency*. The results point to a striking difference. Unlike the other entry modes, joint ventures clearly favor industries where the United States has greater R & D expenditures. The results for entry by new plants resemble the findings for the overall sample, as given in table 4. Acquisitions are insensitive to differences in R & D expenditures and are only weakly related to the total expenditures in both countries.

The difference between new plants and the other two modes is echoed in the other results. Joint ventures are shown again to be less likely in high advertising expenditure industries. A positive but modest effect is evident for acquisitions; the effect is insignificant but it raises the speculation that acquisitions are also used for brand label and distribution channel access. The importance of home (i.e., Japanese) market concentration is confirmed for all three entry modes. The highly significant and negative coefficient to the measure of U.S. concentration indicates, however, that foreign market structure (and, possibly, the lack of acquisition targets) is an important deterrent to entry.

In summary, the results on entry mode choice provide an interesting blend of support for a Ricardian investment model of the product cycle and technological sourcing explanations.[20] Entries by new plants occur in industries where the total R & D expenditures are high and where Japanese R & D expenditures are not growing quickly. As predicted by Vernon's product cycle explanation or by recent arguments of direct investment under shifts in relative technical efficiency, industries with declining relative shares of R & D are eventually forced to withdraw from exports or to shift overseas. However, the industry distribution of entries by joint ventures points to the motivation to source U.S. technological capabilities.

V. Conclusions

The above analysis of Japanese entries into 297 U.S. industries throws light on a number of areas of theoretical and policy importance. In large part, the findings are consistent with recent work on entry into U.S. industries, no matter if the entrant is U.S. or foreign. Industry growth and R & D expenditures attract entry; high levels of seller concentration deter entry.[21]

But unlike previous studies on domestic entry or on foreign direct investment, our findings look not only at the conditions of the target industry, but also permit an analysis of the effects of the industry structure of entering firms on the probability of entry. The implications of the findings are straightforward and appealing. Japanese direct investments in new plants predominate in high R & D expenditure industries, but are not sensitive to a relative technological advantage in either market.

These issues are important to the on-going debate on the benefits of foreign direct investment in the United States. It has long been a contention in the popular press that foreign investments drain American technology. With the recent strengthening of the oversight provisions of the interagency Committee on Foreign Investment in the United States, the executive branch has the authority to block foreign acquisitions in violation of national security interests. Many of these interests have been defined as the safeguarding of U.S. technological strengths.

The results of this paper do not indicate that Japanese acquisitions are more frequent in high-technology industries. In fact, the entry count of acquisitions is related only with weak significance to any of the R & D measures. The results indicate, however, that Japanese–U.S. joint ventures appear to be motivated by the sourcing of U.S. technology. As recent work has shown a strong tendency for manufacturing joint ventures to terminate by acquisition (Kogut, 1991), attention should be directed—to the extent that policy makers should be concerned at all over direct investment—in areas of joint U.S. and Japanese cooperation.

There is an important and unobserved institutional explanation for the relationship of R & D to Japanese foreign entry, for it has been the policy of MITI to encourage the investment of both physical and technological capital in targeted export and import-competing industries. The strong effect of the variable *Japan R & D Growth* may be capturing this institutional effect. In part, this interpretation raises questions of the robustness of the findings to other countries. It also raises, more intriguingly, important issues of the relationship of government policy to the composition of trade and investment and the long-term evolution of a country's comparative advantage.

[20] The significance levels to the estimates of α are also lower compared to those in table 4, suggesting that some of the earlier heterogeneity stemmed from the pooling of counts across entry modes.

[21] See the review of Acs and Audretsch (1989). They qualify the R & D results by noting that high R & D expenditure industries deter small firm entry.

REFERENCES

Acs, Zoltan, and David Audretsch, "Innovation, Market Structure, and Firm Size," this REVIEW 69 (Nov. 1987), 567–574.

______, "Small-firm Entry in US Manufacturing," *Economica* 56 (May 1989), 255–265.

Audretsch, David, and Hideki Yamawaki, "R & D Rivalry, Industrial Policy and U.S. Japanese Trade," this REVIEW 70 (Aug. 1988), 438–447.
Baldwin, Robert E., "Determinants of Trade and Foreign Investment: Further Evidence," this REVIEW 61 (Feb. 1979), 40–48.
Brander, James A., and Paul R. Krugman, "A 'Reciprocal Dumping' Model of International Trade," *Journal of International Economics* 15 (1983), 313–321.
Brander, James A., and Barbara Spencer, "International R & D Rivalry and Industrial Strategy," *Review of Economic Studies* 50 (Oct. 1983), 707–722.
_____, "Export Subsidies and International Market Share Rivalry," *Journal of International Economics* 18 (Feb. 1985), 83–100.
Cameron, A. Colin, and Pravin Trivedi, "Econometric Models Based on Count Data: Comparisons and Applications of Some Estimators," *Journal of Applied Econometrics* 1 (Jan. 1986), 29–53.
Cantwell, John, *Technical Innovations in Multinational Corporations* (London: Basil Blackwell Inc., 1989).
Caves, Richard, "International Corporations: The Industrial Economics of Foreign Investment," *Economica* (Feb. 1971).
_____, *Multinational Enterprise and Economic Analysis* (New York: Cambridge University Press, 1982).
Caves, Richard, Michael E. Porter, and A. Michael Spence, with John T. Scott, *Competition in the Open Economy: A Model Applied to Canada* (Cambridge: Harvard University Press, 1980).
Deardorff, A., "Testing Trade Theories and Predicting Trade Flows," in R. W. Jones and P. B. Kenen (eds.), *Handbook of International Economics* (Amsterdam: Elsevier Science Publishers, B.V., 1984).
Dornbusch, Rudiger, Stanley Fischer, and Paul Samuelson, "Comparative Advantage, Trade, and Payments in a Ricardian Model," *American Economic Review* 67 (1977), 823–839.
Deutsch, Larry, "Entry and the Extent of Multiplant Operations," *Journal of Industrial Economics* 32 (1984) 477–487.
Dunning, John H., "Explaining the International Direct Investment Position of Countries," *Weltwirtschaftlisches Archiv* (1981), 30–64.
_____, *Multinationals, Technology, and Competitiveness* (Boston: Unwin Hyman, 1988).
Flowers, Edward, "Oligopolistic Reactions in European and Canadian Direct Investment in the United States," *Journal of International Business Studies* 7 (1976), 43–55.
Gorecki, Paul, "The Determinants of Entry by New and Diversifying Enterprises in the UK Manufacturing Sector, 1958–1963: Some Tentative Results," *Applied Economics* 7 (1975), 139–147.
_____, "The Determinants of Entry by Domestic and Foreign Enterprises in Canadian Manufacturing Industries," this REVIEW 58 (Nov. 1976), 485–488.
Goto, Akira, and Kazuyuki Suzuki, "R & D Capital, Rate of Return on R & D Investment and Spillover of R & D in Japanese Manufacturing Industries," this REVIEW 71 (Nov. 1989), 555–564.
Graham, Edward, "Transatlantic Investment by Multinational Firms: A Rivalristic Phenomenon?" *Journal of Post-Keynesian Economics* 1 (1978), 82–99.
Greene, William H., *LIMDEP* (Philadelphia: Social Science Data Center).
Hausman, Jerry, Bronwyn Hall, and Zvi Griliches, "Econometric Models for Count Data with an Application to the Patents–R & D Relationship," *Econometrica* 52 (July 1984), 909–938.
Highfield, Robert, and Robert Smiley, "New Business Starts and Economic Activity: An Empirical Investigation," *International Journal of Industrial Organizations* 5 (1987), 51–66.
Horst, Thomas, "The Industrial Composition of U.S. Exports and Subsidiary Sales to the Canadian Market," *American Economic Review* (Mar. 1972), 37–45.
Hymer, Stephen, *International Operations of National Firms: A Study of Direct Foreign Investment* (Cambridge: MIT Press, 1976).
Johnson, Norman, and Samuel Kotz, *Discrete Distributions* (Boston: Houghton Mifflin, 1970).
Kessides, Ioannis, "Advertising, Sunk Costs, and Barriers to Entry," this REVIEW 68 (Feb. 1986), 84–95.
Khemani, R. S., and D. Shapiro, "The Determinants of New Plant Entry in Canada," *Applied Economics* 18 (1986), 1243–1257.
Kindleberger, Charles P., *American Business Abroad: Six Lectures on Direct Investment* (New Haven: Yale University Press, 1969).
Kogut, Bruce, "Joint Ventures and the Option to Expand and Acquire," *Management Science* 37 (1991), 19–33.
Kogut, Bruce, and Nalin Kulatilaka, "Direct Investment, Hysteresis, and Real Exchange Volatility," Working Paper, Reginald H. Jones Center for Corporate Strategy, The Wharton School (1990).
Knickerbocker, Federick, *Oligopolistic Reaction and Multinational Enterprise* (Boston: Division of Research, Graduate School of Business Administration, Harvard University, 1973).
Magee, Stephen, "Information and the Multinational Corporation: An Appropriability Theory of Direct Foreign Investment," in Jagdish Bhagwuti (ed.), *The New International Economic Order* (Cambridge: MIT Press, 1976), 317–340.
McClain, David, "Foreign Direct Investment in the United States: Old Currents, 'New Waves,' and the Theory of Direct Investment," in C. Kindleberger and D. Audretsch (eds.), *The Multinational Corporation of the 1980's* (Cambridge, MA: MIT Press, 1983).
Pugel, Thomas, *The Industry Determinants of Foreign Direct Investment into the United States* (New York: Division of Research, Graduate School of Business Administration, New York University, July 1986).
_____, "The Determinants of Foreign Direct Investment: An Analysis of U.S. Manufacturing Industries," *Managerial and Decision Economics* 2 (1981), 220–228.
_____, *International Market Linkages and U.S. Manufacturing: Prices, Profits and Patterns* (Cambridge, MA: Ballinger, 1978).
Scherer, Frederick, "International R & D Races: Theory and Evidence," The Prince Bertil Symposium on "Corporate and Industry Strategies for Europe," Stockholm School of Economics, Nov. 9–11, 1988.
Shapiro, Daniel, "Entry, Exit and the Theory of the Multinational Corporation," in C. Kindleberger and D. Audretch (eds.), *The Multinational Corporation in the 1980's* (Cambridge, MA: MIT Press, 1983).
Shapiro, Daniel, and R. S. Khemani, "The Determinants of Entry and Exit Reconsidered," *International Journal of Industrial Organization* 2 (1987), 15–26.
Swedenborg, Birgitta, *Multinational Operations by Swedish Corporations* (Stockholm: Almqvist and Wiksell, 1979).
Vernon, Raymond, "International Investment and International Trade in the Product Life Cycle," *Quarterly Journal of Economics* 80 (May 1966), 190–207.

Wells, Louis (ed.), *The Product Life Cycle and International Trade* (Graduate School of Business, Harvard University, 1972).

Yamawaki, Hideki, "Exports, Foreign Market Structure and Profitability in Japanese and U.S. Manufacturing," this REVIEW 68 (Nov. 1986), 618–627.

Yamawaki, Hideki, and David B. Audretsch, "Import Share Under International Oligopoly with Differentiated Products: Japanese Imports in U.S. Manufacturing," this REVIEW 70 (Nov. 1988), 569–579.

DATA APPENDIX

Since the entries are counted at the 4-digit SIC level, most of the independent variables were collected also at the 4-digit level, except for Japanese R & D expenditures and its growth rate. *ENTRY COUNT* was constructed by calculating total numbers of Japanese firms' entries into the U.S. manufacturing industries during the period 1976–87. The raw data on the Japanese entries into the United States were collected by the International Trade Administration at the Department of Commerce, as published annually in *Foreign Direct Investment in the United States*, Department of Commerce, ITA. *JAPAN R & D* was constructed from the *Kagaku Gizyuitsu Kenkyou Chousha Houkou* (Research Report on Science and Technology), based on the survey by Statistics Bureau, Office of Prime Minister, Japan. *JAPAN R & D* is for 1975–85 and is SIC aggregated to the equivalent 2 or 3 digit level. *JAPAN R & D GROWTH* is computed as the average of annual growth rates of *JAPAN R & D*. *JAPAN CONCENTRATION* are 8-firm concentration ratios, constructed from *Syuyou Sangyoni Okeru Seisan Syouchyoudo to Herfindahl Index no Syui* (Trends in Production-based Concentration Ratios and Herfindahl Indices for Major Industries), Report #12-87-001, 262-00-A, Fair Trade Commission, Japan, 1987. The matching of Japanese industry concentration data to the U.S. industries left 122 out of 454 4-digit SIC manufacturing industries without a match.

U.S. R & D and *U.S. ADVERTISING* are constructed from the Federal Trade Commission Line of Business Report for 1977. *SHIPMENT*, eleven years' average value of industry shipment, 1975–85, was obtained from unpublished reports from the Department of Commerce. *SHIPMENT GROWTH* is the average of annual growth rates of *SHIPMENT*. *IMPORT* is import value divided by industry shipment, eleven year average 1975–85, from the same source as *SHIPMENT*. *US CONCENTRATION* are 8-firm concentration ratios for 1982, as listed in the *1982 Census of Manufacturing*, Department of Commerce. *INNOVATION FREQUENCY* measures the numbers of new innovations for each industry, which are collected by Small Business Administration for 1982. *EXPORT RESTRICTION* is a dummy variable noting the existence of U.S. quota or voluntary export restraints by Japanese companies. Those industries include steel industries (3312, 3313, 3315, 3316, and 3317), television and radios (3651), semiconductors (3674), automobiles and parts (3711 and 3714). Textile industries were not included in the sample because of the missing data in other variables.

Journal of Development Economics 43 (1994) 279–293. North-Holland

Technology, market characteristics, and spillovers

Ari Kokko*

Stockholm School of Economics, Stockholm, Sweden

Received June 1992, final version received January 1993

Using detailed (unpublished) industry data from the Mexican manufacturing industry, this paper examines the relation between observed technology spillovers from foreign direct investment and proxies for the complexity of MNC technology. The results suggest that advanced MNC technologies or large technology gaps *alone* do not constitute unsurmountable obstacles to spillovers, but that spillovers are less likely in industries with 'enclave' characteristics, i.e. where large technology gaps and high foreign shares coincide.

Key words: Foreign direct investment; Multinational corporations; Technology transfer; Spillovers

JEL classification: F23; O30

1. Introduction

There are numerous case studies to suggest that technology spillovers from foreign direct investment (FDI) may provide important benefits for the host countries of multinational corporations (MNCs) [see Kokko (1992) for a review]. The technology and productivity of local firms may improve as foreign firms enter the market and demonstrate new technologies, provide technical assistance to their local suppliers and customers, and train workers and managers who may later be employed by local firms. The competitive pressure exerted by the foreign affiliates may also force local firms to operate more efficiently and introduce new technologies earlier than what would otherwise have been the case.

Studies of aggregate manufacturing in several countries – Caves (1974) on Australia, Globerman (1979) on Canada, and Blomström and Persson (1983) on Mexico are the best known examples – have also found that foreign presence has a positive impact on the productivity of local firms, and

Correspondence to: Dr. A. Kokko, Department of International Economics and Geography, Stockholm School of Economics, P.O. Box 6501, S-113 83 Stockholm, Sweden.

*I would like to thank Svenska Bankforskningsinstitutet and SAREC for financial support, and Magnus Blomström, Robert E. Lipsey, Mario Zejan, and two anonymous referees for helpful comments and suggestions on earlier drafts of the paper. All remaining errors and omissions are my own.

0304-3878/94/$07.00
SSDI 0304-3878(93)E0046-Z

therefore concluded that spillovers are typically significant. However, there are cases where spillovers have apparently not taken place at a large scale, as suggested by studies of FDI in various European countries [Cantwell (1989)], in Moroccan manufacturing industries [Haddad and Harrison (1991)], and in Venezuelan manufacturing [Aitken and Harrison (1991)].

A possible reason for the apparently contradictory findings from the country studies is that various host industry and host country characteristics may influence the incidence of spillovers. For instance, the technology imports of MNC affiliates seem to be larger in countries and industries where the educational level of the local labor force is higher, where local competition is tougher, and where the host country imposes fewer formal requirements on the affiliates' operations [Blomström et al. (1992), Kokko (1992)]. The affiliates' levels of technology or technology imports, in turn, probably have some impact on how large the spillovers to local firms will be.

Yet, the available empirical evidence does not reveal how the spillovers we can observe are related to the level of technology in MNC affiliates, and it is not entirely obvious from a theoretical perspective what the relation should be. In fact, two opposing arguments can be found in the literature on foreign investment and technology transfer. On the one hand, it is sometimes argued in discussions of the 'appropriateness' of MNC technology that MNC affiliates may be too advanced to leave any mark on local host country firms. The reason is that 'technical advances [and technologies] applicable to the factor-proportions of capital-rich developed countries are hardly of any use in improving techniques of low capital-intensity in less developed countries' [Lapan and Bardhan (1973, p. 585)]. This suggests the hypothesis that spillovers are negatively related to the complexity of MNC technology or the size of the technology gap between affiliates and local firms. On the other hand, it is obvious that a certain technology gap is necessary for those spillovers that occur as local firms copy MNC technology or benefit from the MNCs' training of local employees, and it has been hypothesized that spillovers grow with the size of the technology gap [Findlay (1978), Wang and Blomström (1992)].

The few empirical studies that more or less indirectly discuss the relation between technology and spillovers reflect this confusion. Blomström and Wolff (forthcoming) analyze data for 20 two-digit industries in Mexico in 1970 and 1975, and examine some factors influencing the rates of growth of gross output per employee in locally-owned Mexican firms and the rates of convergence of gross output per employee between locally-owned firms and foreign affiliates. They find that both measures are positively related to the foreign share of industry employment and the initial labor productivity gap between locals and multinationals. This is consistent with the hypothesis that spillovers increase with the size of the technology gap. By contrast, Haddad and Harrison (1991), who study the effects of foreign presence on the relative productivity of local firms in Moroccan manufacturing, imply that large technology gaps or advanced MNC technology inhibit spillovers. They find that higher foreign presence leads to smaller deviations from best practice technologies in 'low tech' industries, but that there is no such effect in 'high tech' sectors, that are defined to include machinery, transport equipment, electronics, scientific instruments, and chemicals. Their interpretation is that competition from foreign firms may push local industry toward best practice, e.g. by forcing the least efficient firms out of business, but that there are no spillovers of advanced technology. Similarly, Cantwell (1989), seems to suggest that spillovers are most important in the industries where the technology gap is small. He examines the responses of local firms to the entry and presence of U.S. multinationals in the European markets 1955–1975, and argues that the most positive impact on local technological capability can be seen in the industries where the local firms had a strong technological tradition and were able to challenge the invading U.S. affiliates.

The purpose of this paper is to investigate how spillovers observed in static cross-section analyses are related to various proxies for the complexity of MNC technology and the technology gap between locally-owned host country firms and MNC affiliates.[1] In the course of the study, we will also find reason to briefly examine the relation between spillovers and the market shares of foreign firms. The next section describes the data set and the statistical model used for the empirical tests, section 3 presents and discusses the regression results, and section 4 concludes the paper.

2. Data and variables

The data used here to examine the effects of technology differences on spillovers refer to the Mexican manufacturing industry 1970, and are from unpublished worksheets provided by *Dirección de Estadísticas de la Secretaría de Industria y Comercio* in Mexico, collected for the Mexican Census of Manufactures 1971.[2] The information is gathered at the plant level and covers the entire manufacturing sector, which is divided into 230 four-digit industries with a further break-down according to three ownership categories: domestic private, foreign, and state ownership. Plants with at least 15 percent of shares owned by foreigners are defined as 'foreign', and those where the Mexican state owns more than 49 percent are defined as state-owned, irrespective of the share of foreign ownership. The state-owned plants are excluded from the present sample, since they may operate under soft

[1] It would be preferable to examine spillovers over time rather than in a static context, but the data used for the empirical analysis are only available for one year.

[2] This is essentially the same data base as that used by Blomström and Persson (1983). Estados Unidos Mexicanos (1973) summarizes much of the information, but the published data are not presented separately for local and foreign firms.

budget constraints, or have other goals (related to e.g. employment creation or geographical localization of operations) than profit maximization, as discussed in Blomström and Persson (1983).

Because of missing information, 14 of the 230 industries had to be discarded from the sample. For the remaining 216 industries, the data set includes information on employment, assets, value added, industry concentration, payments of patents, and the division of employees between blue-collar and white-collar workers. For the 156 industries that hosted foreign MNCs in 1970, the data on employment, assets, and value added are available separately for foreign and private local firms. For the other variables, information is only available for industry totals, without the breakdown according to ownership.

The statistical models used to examine spillovers in most earlier empirical studies of aggregate manufacturing [Caves (1974), Globerman (1979), Blomström and Persson (1983)] have been based on linear estimations of the labor productivity of locally-owned firms as a function of the foreign firms' market share and various other industry characteristics, such as capital–labor ratios, labor quality, scale economies, and concentration. If foreign presence has been found to have a significant positive effect on local labor productivity (after the effects of other variables have been accounted for), it has been concluded that spillovers take place.[3] The main difference between the earlier models is that only Blomström and Persson (1983) use the capital-intensity of local firms as an exogenous variable. Hence, they focus on spillovers that affect the efficiency in the use of given factor inputs, whereas Caves (1974) and Globerman (1979) also capture the effect of foreign presence on the local firms' capital–labor ratios. We adopt the former model here, because it is likely to provide a more conservative estimate of the extent of spillovers. However, the data allow us to check the results by performing tests of the latter type as well.

Thus, we hypothesize here that the labor productivity of local firms can be estimated by the function

$$VA/L^{d} = f(K/L^{d}, LQ, HERF, FOR), \tag{1}$$

where variables are defined as follows: The dependent variable is VA/L^{d}, the average labor productivity in domestic firms, and it is measured as the ratio of value added to total employment in locally-owned plants. The domestic firms' capital–labor ratio, K/L^{d}, is the ratio of total assets to total employment in locally-owned plants. The labor quality measure, LQ, is based on the

[3]The models presented by Aitken and Harrison (1991) and Haddad and Harrison (1991) differ from the earlier studies on this point. They use plant level data for several years, and examine the determinants of total factor productivity growth and deviations from best practice technology. We are not able to replicate their tests because of the lack of suitable data.

ratio of white-collar workers to blue-collar workers (W/B) in each industry's total employment.[4] However, we know *a priori* that foreign firms are likely to employ a higher share of white-collar workers, and since the W/B ratio refers to each industry as a whole, there is also a troublesome correlation between that and FOR, our measure of foreign presence. We have therefore constructed our labor quality proxy LQ as the sum '$\alpha+\epsilon$' from the regression

$$W/B=\alpha+\beta\ FOR+\epsilon. \tag{2}$$

In other words, LQ intends to reflect the share of W/B that is not explained by the degree of foreign presence in the industry.[5] The variable $HERF$ is the Herfindahl index, and measures the level of concentration in each industry.[6] It is included to account for the expected effect of market structure on the value of output: more concentrated industries are supposedly better able to engage in monopoly pricing and should therefore display higher labor productivity, ceteris paribus. FOR, finally, is the ratio of the foreign plants' employment to total employment in each industry, and measures the degree of foreign presence: if spillovers take place, it is expected to have a significant positive effect on local labor productivity.[7]

In addition to the variables appearing in function (1), we will use three proxies for the technological characteristics of industries to divide the sample in 'high' and 'low' technology groups, and to construct interaction terms for the statistical tests. The first one is labelled PAT, and measures the average payments of patent fees per employee in each industry. PAT reflects the formal technology payments of all actors taken together (private local, foreign, and state-owned) but it is likely to be particularly highly correlated with those of foreign firms (although that information is not available separately for 1970): MNC affiliates account for most of the world's formal

[4]Labor quality is often proxied by various wage measures, but these cannot be used here because of simultaneity problems: wages are determined by labor productivity, which is what we are trying to explain.

[5]Blomström and Persson (1983) construct a similar labor quality proxy, but define it as only the error term 'ϵ' in the regression estimating W/B.

[6]The Herfindahl index is calculated from plant level data, and it is defined as: $HERF=\sum^{n}_{i}(x_i/X)^2$, $i=1,\ldots,n$, where x_i is the employment of the n individual plants, and X is the total employment of the industry.

[7]We also used a proxy for advertizing expenditures in some estimations, but its effect was not significant, and it was dropped from the model. Similarly, a proxy for effective rates of protection was included in a sub-set of the sample, but it was dropped because the estimated effect was not significant, and because it was only available for 93 industries. Some additional explanatory variables have appeared in earlier studies. Blomström and Persson (1983) include proxies for economies of scale and average working time, but their effects are not significant: we were not able to replicate either of them because of lack of data. Globerman (1979) uses the average number of hours worked and proxies for plant-scale and product-scale economies to capture the same effects, and finds a significant positive impact of plant-scale economies, but in most cases no significance for the other two variables.

technology payments.[8] Hence, we treat *PAT* as a proxy for the technology level of the foreign affiliates in each industry, and the higher the patent payments, the more advanced are the affiliates' technologies assumed to be.[9]

The average capital intensity of the foreign affiliates, K/L^f, is another industry characteristic that may be positively related to the level of MNC technology. However, the simple correlation between *PAT* and K/L^f is only 0.11, and it is apparent that both cannot measure technological complexity accurately. With perfect data, we would have preferred the *PAT* measure, but because of its possible shortcomings – MNC parents' technology pricing practices may not be cohesive, all technologies may not be patented, transfer pricing may occur, et cetera – we have included also K/L^f as a proxy for technology.

Our third proxy, *PGAP*, reflects the labor productivity gap between local and foreign firms. It is defined as the ratio of value added per employee in foreign plants to value added per employee in private locally-owned plants. This is an indirect measure of the technology gap discussed by Findlay (1978) and Wang and Blomström (1992).

Table A.1 summarizes the variables, table A.2 provides some descriptive statistics for the data set, and table A.3 presents a correlation matrix.

To examine the effect of the size of the technology gap and the complexity of MNC technology on spillovers, we go on to make two types of tests. First, we divide the sample into six sub-samples characterized by 'high' or 'low' values for the variables *PAT*, K/L^f, and *PGAP*, estimate eq. (1) for each of these, and then compare coefficient estimates across each pair of sub-samples. This allows the coefficients of all right-hand variables to vary across sub-samples, and we are particularly interested to see if foreign presence has the same apparent effect on local productivity in industries with 'low' and 'high' technology (or 'small' and 'large' technology gaps). Comparisons of the broad functional relationships between sub-samples are based on Chow-tests, whereas statements comparing the coefficients of any specific variables are based on *F*-tests of the null hypothesis that there are no differences. Second, we include some interaction variables into eq. (1) to see whether the

[8]Fairchild (1977) surveys 25 pairs of manufacturing firms with direct U.S. equity investment and 100 percent Mexican ownership, respectively, operating in Monterrey, Mexico. Comparing the two types of firms, she concludes that four times as many U.S. firms (or 'joint ventures') than local firms used U.S. patents or licenses in 1973. Since her sample of firms was closely matched in terms of the firms' age, assets, net sales, and employment, and the goods produced, it is possible that the foreign affiliates' share of patent payments is even higher for our more heterogeneous sample.

[9]We also used data on license payments by U.S. manufacturing industries in 1977 as an alternative for *PAT*. However, the U.S. data were only available at the two-digit level, and the Mexican four-digit industries had to be aggregated into 16 broad groups for the construction of the variable [see Blomström and Wolff (forthcoming) for a concordance scheme between U.S. two-digit and Mexican four-digit SIC codes]. The results were similar to those for PAT, and are therefore not shown separately.

Table 1

Results of OLS estimations: Labor productivity in private locally-owned manufacturing firms in Mexico 1970.[a]

Equation	Constant	K/L^d	$HERF$	LQ	FOR	Adj.R^2	F	N
All industries								
(1.1)	0.265	0.464	−0.003	0.161	0.112	0.53	61.84	216
	(2.47)**	(11.77)***	(0.04)	(2.58)**	(3.49)***			
Industries with MNCs								
(1.2)	0.256	0.443	0.030	0.130	0.141	0.43	29.62	156
	(1.99)**	(8.47)***	(0.30)	(1.88)*	(2.38)**			

[a]The dependent variable is VA/L^d. All variables are normalized by division with the sample means. Estimated coefficients are shown together with the absolute value of the t-statistic in parentheses. *, **, and *** denote significance at the 10, 5, and 1 percent level of significance (two-tailed tests). For definitions of variables, see table A.1.

coincidence of advanced technology (or large technology gaps) and large foreign shares has any effects on the results.

3. Regression results

Regression equations (1.1) and (1.2) in table 1 present the results of OLS estimations of function (1) for the entire sample of 216 industries, and for the 156 industries where foreign MNCs were present in 1970.[10] Eq. (1.1) is included to allow a comparison with the tests in Blomström and Persson (1983), and the results here are very similar to theirs. Capital-intensity, labor quality, and foreign presence all have highly significant positive effects (in other words, spillovers, as we have defined them, are present), but the impact of producer concentration on local labor productivity is negligible. The main difference from the Blomström and Persson (1983) estimation is that we have no scale variable, but its absence does not seem to have any large effect on the coefficients of other variables: similarly, dropping the $HERF$ variable leaves the results unchanged.[11]

Turning to eq. (1.2), where only the 156 industries with foreign firms present are included, we see some small changes in estimated coefficients, and a fall in the significance levels for the impacts of labor quality and foreign presence. However, the spillover effect is still significant. There is also a decline in the explanatory power of the equation – both adjusted R^2 and the F-value fall – which may be caused by the smaller number of observations.

[10]All results reported in the tables refer to linear estimations of the model. The estimations have also been made with logarithmic transformations of the variables: typically, adjusted R^2 increases, but the signs and significance levels of the coefficients remain unchanged. We retain the linear model throughout this paper in order to be able to compare the results with earlier studies.

[11]More detailed comparisons with the results in Blomström and Persson (1983) are not possible, since it is not known how their variables were normalized.

The estimations discussed below include only industries where foreign firms were present in 1970, so eq. (1.2) may serve as a point of reference.

3.1. *Impact of industry characteristics on spillovers*

Table 2 presents estimation results for six equations covering sub-samples of industries grouped according to their levels of *PAT*, K/L^f, and *PGAP*, and for three equations where interaction terms have been added to the model.

The first two equations refer to groups defined according to the values of the variable *PAT*. The average payments for patents, trade marks, and other fees per employee are lower than 600 pesos for the industries included in eq. (2.1), and higher than 600 pesos for those in eq. (2.2).[12] The results show that the estimated coefficient of *FOR* (i.e. the spillover effect) is positive and significant at the five percent level in the group with low average patent payments. The coefficient estimate in the group with high patent payments is comparable in size, but not significant. In isolation, this implies that spillovers do not generally occur in the industries with the most complex technologies. However, *F*-tests suggest that the null hypothesis of equal effects of *FOR* in both groups cannot be rejected. A possible explanation for this partly conflicting finding is that there is a large variance in the group with higher patent payments, and that we cannot, at reasonable levels of confidence, say much about whether spillovers will occur or not. Hence, high technology seems to make spillovers less likely, but there is no evidence to show that spillovers cannot occur in industries with complex technologies.

Eq. (2.3), for industries where the average capital-intensity of foreign affiliates is below 200,000 pesos per employee, and eq. (2.4), for industries with higher capital-intensities in affiliates, provide a similar picture of spillovers. The spillover effect is positive and highly significant in the industries with less capital-intensive foreign affiliates, but not significant when affiliates are more capital-intensive. Yet, the hypothesis that the coefficients of *FOR* are equal in the two sub-samples cannot be rejected, judging by *F*-tests. In other words, spillovers seem less likely when foreign technologies are capital-intensive, but may nevertheless occur.[13]

[12]The cut-off points between 'high' and 'low' technology groups have been chosen on the basis of analyses of residuals from regression equation (1.2). The samples have been split approximately where changes in the pattern of residuals occur. The estimated coefficients in each pair of sub-samples become more similar if lower cut-off points are chosen (since the sub-samples become more similar), whereas higher cut-off points do not change the results significantly (as long as the 'high' groups contain at least 30 observations).

[13]Alternative proxies for technology levels, based on the labor quality measures and the average labor productivity of foreign affiliates were also used for the grouping of industries, with results very similar to those presented in table 2. The impact of *FOR* was consistently significant and positive in the 'low' technology group, but not significantly different from zero in the 'high' technology group; yet, *F*-tests failed to reject the null hypothesis that the coefficients of *FOR* were equal.

Table 2

Results of OLS estimations: Labor productivity in private locally-owned manufacturing firms in Mexico 1970. Impact of industry characteristics.[a]

Equation	Low *PAT* (2.1)	High *PAT* (2.2)	Low K/L^f (2.3)	High K/L^f (2.4)	Small *PGAP* (2.5)	Large *PGAP* (2.6)	(2.7)	(2.8)	(2.9)
Constant	0.140	0.218	0.390	0.253	0.106	0.433	0.254	0.265	0.262
	(0.74)	(1.01)	(2.65)***	(1.09)	(0.72)	(3.38)***	(1.97)*	(1.99)**	(2.12)**
K/L^d	0.433	0.449	0.425	0.426	0.375	0.441	0.447	0.438	0.426
	(5.99)***	(5.36)***	(7.24)***	(3.54)***	(5.41)***	(9.99)***	(8.44)***	(7.94)***	(8.46)***
HERF	−0.059	0.155	−0.136	0.254	0.261	−0.343	0.023	0.030	0.024
	(0.15)	(0.98)	(1.20)	(1.26)	(2.05)**	(3.67)***	(0.22)	(0.29)	(0.25)
LQ	0.318	0.065	0.113	0.054	0.075	0.342	0.130	0.126	0.119
	(2.62)**	(0.70)	(1.41)	(0.66)	(1.13)	(4.38)***	(1.88)*	(1.77)*	(1.78)*
FOR	0.168	0.114	0.208	0.014	0.182	0.127	0.151	0.128	0.274
	(2.41)**	(0.85)	(3.34)***	(0.11)	(2.47)**	(2.32)**	(2.42)**	(1.72)*	(4.09)***
*FOR*PAT*	-	-	-	-	-	-	−0.006	-	-
	-	-	-	-	-	-	(0.51)	-	-
$FOR*K/L^f$	-	-	-	-	-	-	-	0.014	-
	-	-	-	-	-	-	-	(0.28)	-
*FOR*PGAP*	-	-	-	-	-	-	-	-	−0.104
	-	-	-	-	-	-	-	-	(3.75)***
Adj. R^2	0.41	0.40	0.46	0.22	0.43	0.74	0.42	0.42	0.47
F	17.85	10.55	30.40	4.46	27.36	37.12	23.64	23.57	28.56
N	97	59	107	49	105	51	156	156	156

[a]The dependent variable is VA/L^d. All variables are normalized by division with the sample means. Estimated coefficients are shown together with the absolute value of the t-statistics in parentheses. *, **, and *** denote significance at the 10, 5, and 1 per cent levels of significance (two-tailed tests). For definitions of variables, see table A.1.

The differences in the incidence of spillovers are smaller for the sub-samples grouped according to the size of the productivity gap between affiliates and local firms. The coefficient of FOR is positive and significant both in eq. (2.5), where the value added per employee in foreign plants is at most twice as high as that in private locally-owned plants, and in eq. (2.6), where affiliates are more than twice as productive as private locally-owned plants.[14] Productivity differences per se are apparently not serious obstacles for spillovers.[15]

Although the impact of *FOR* may be equal in all groups, there are some other significant differences between sub-samples. A Chow-test suggests that local labor productivity is not determined by the same factors in the two *PAT* groups. Labor quality has a significantly larger impact when patent payments are low, judging by an *F*-test. This is surprising, since it would be more reasonable to expect a strong impact of labor quality in industries with complex technologies that demand high labor skills. The most likely explanation is simply that *LQ* is an imperfect proxy for labor quality, but the data needed for more accurate measures are not available. Labor quality also has a larger positive coefficient when the labor productivity gap is larger, but it is impossible to draw any firm conclusions from this finding, for the same reason.

Another significant difference between the two *PGAP* groups is that the coefficient of the variable *HERF* is significant in both sub-samples, but with opposite signs. This is notable, since concentration did not have any significant impact in the full sample. More specifically, concentration seems to enhance local productivity when the gap between affiliates and local firms is small, but depress it when the gap is large. One possible explanation, suggested by Blomström (1986), is that *HERF* reflects the existence of scale economies: local firms may benefit from scale economies when they are relatively productive compared to foreign affiliates, but not when they are lagging far behind the affiliates.

It is also possible that high levels of concentration and large technology gaps together signal situations where spillovers are less likely. In fact, the foreign affiliates in industries with these characteristics may operate in 'enclaves', in isolation from local firms. High levels of concentration and large productivity gaps may, for instance, coincide in industries where the affiliates have been strong enough to take over the bulk of the market, and forced local firms into narrow niches where the products and technologies of

[14]It should be noted that there is a possible selection bias in the grouping of industries according to *PGAP* if spillovers reduce the productivity gap. The industries with the smallest spillovers may automatically end up in the group with large *PGAP*s, whatever the reason for the lack of spillovers. However, it seems that the bias does not have a large impact on the results, since we observe positive spillovers also in that group.

[15]It would also be interesting to examine if wholly- or majority-owned foreign affiliates and joint ventures have identical impacts on local technology and productivity, but the data do not distinguish between these two kinds of 'foreign' firms.

MNCs are not profitable. In these cases, there is not much reason to expect any spillovers. In other industries, where MNCs are less dominant and where local firms are in more direct competition with affiliates, spillovers may well occur in spite of large technology gaps or advanced MNC technologies.

In eqs. (2.7) to (2.9), we have added some interaction terms to the model we have used so far, in order to examine if the combination of advanced technology (or large technology gaps) and foreign domination inhibit spillovers. Eq. (2.7) adds the term *FOR*PAT* to the estimation, but its coefficient is very small and not significantly different from zero. Comparing eq. (2.7) with eq. (1.2), it can be seen that the interaction term does not have much effect on the estimated coefficients of other variables either. Hence, it seems reasonable to conclude that complex technology (as proxied by the variable *PAT*) does not impede spillovers even when it is accompanied by large foreign market shares. Eq. (2.8), where the interaction term is defined as FOR^*K/L^f, gives similar results.

Eq. (2.9), by contrast, suggests that the simultaneous presence of large productivity gaps and large foreign shares has a notable impact on the outcome. The estimated coefficient of the interaction term *FOR*PGAP* is negative and highly significant. This appears to support the hypothesis that spillovers are smaller in industries with 'enclave' characteristics. Moreover, the coefficient of *FOR* is markedly larger in eq. (2.9) than in the base case, i.e. eq. (1.2), which means that spillovers may be even more important in industries where affiliates and local firms are in direct competition with each other than what Blomström and Persson (1983) concluded.

Before these results can be presented as conclusions, there is one more issue to consider. The statistical model used by Globerman (1979) differs significantly from ours on one point, as noted earlier. The model we have estimated treats local capital-intensity as given, and assumes that the spillovers we can measure are related to 'disembodied' technology. Globerman (1979) uses K/L^f (instead of K/L^d) to control for industry differences in productivity and technology, which allows him to capture also the impact of foreign presence on the average capital-intensity of local firms (or on the average deviation between local and foreign firms' capital-intensities). To make sure that this distinction does not seriously affect the findings from table 2, it is necessary to comment on the results of 'Globerman-type' estimations of the equation

$$VA/L^d = f(K/L^f, LQ, HERF, FOR). \tag{3}$$

The estimation results are very similar to those presented in eqs. (2.1)–(2.9), and will therefore not be shown separately. The estimated spillover effects are not significantly different in the 'high' and 'low' technology groups, the interaction term *FOR*PGAP* has a significant negative impact on local productivity, but the other interaction terms are not significant. Hence, the conclusion that advanced technology alone is probably not an obstacle to spillovers, but that large technology gaps and high foreign shares together may impede spillovers, survives.

Finally, a brief look at what industries exhibit both high foreign shares and large productivity gaps may be useful to confirm that it is reasonable to talk about 'enclaves' and 'dual markets'. Most of the industries with the highest values for the interaction term *FOR*PGAP* are such where superior marketing abilities and proprietary brand names and labels are likely to make up the competitive assets of foreign firms. Examples of the goods produced are perfumes and cosmetics, alcoholic beverages, cigars and cigarettes, chewing gum, instant coffee, prepared food products, clocks and watches, and film. The monopolistic nature of these industries may mean that the products of foreign and local firms are not directly comparable, that foreign and local firms aim at different segments of the market, and that there may not be much scope for spillovers of production technology.[16]

Many of the other industries are such where the market leadership of the foreign affiliates is likely to be based on economies of scale. Examples here are office and computing machines, synthetic fibres, light bulbs and fluorescent tubes, some chemical industries, and primary metal industries. The affiliates operating here may have 'crowded out' the local firms so that there is nobody to absorb the potential spillovers. It is also possible that the small local entrepreneurs in these industries are not able to benefit much from learning about machines or management practices that are optimal for a plant that is ten, fifty, or a hundred times larger than their own firm. Hence, it does not seem implausible to use the term 'enclave' for the industries where the combined impact of large productivity gaps and foreign domination is strongest, although the dual character of the markets can be caused by either highly differentiated products or scale economies.

4. Concluding comments

This paper has examined technology spillovers in various groups of Mexican manufacturing industries 1970, and has attempted to determine whether differences in the technology gap between local and foreign firms or differences in the level of technological complexity have any impact on the observed spillovers. A conclusion from the regression analysis is that factors related to technology *alone* do not seem to inhibit spillovers, but that large productivity gaps and large foreign market shares *together* appear to make up significant obstacles.

The industries where large productivity gaps and large foreign shares occur simultaneously seem to be characterized by differentiated products and/or

[16]However, Caves (1971) suggests that marketing technology may also spill over, so that foreign presence has an impact even when production technologies are not comparable.

significant economies of scale, which allow the foreign affiliates to crowd out local competitors from important segments of the market. As a result, the affiliates may operate in 'enclaves', i.e. isolated segments of the market where technologies, products, and plant sizes are very different from those used by local firms. The large differences between foreign and local firms in these industries may explain why spillovers do not occur.

The finding that spillovers may differ across industries could explain some of the conflicting results from earlier studies of spillovers in aggregate manufacturing. Whether or not enclaves will occur is determined by various host country characteristics, such as history, market size, domestic technological capability, economic policies, and so forth. These characteristics differ between host countries and make enclaves more common in some countries than in others – the possibility to observe spillovers in aggregate manufacturing may vary accordingly.

The analysis also offers some obvious policy conclusions for host country governments that wish to encourage foreign investment in order to benefit from technology spillovers. Efforts to promote FDI should perhaps focus on industries where local technological capability is already relatively strong, or where product differentiation and scale economies are not so significant that foreign firms can easily take over the whole market.

Appendix A

Table A.1

Definitions of variables.

	Dependent variable
VA/L^d	The ratio of value added to the total number of employees in private locally-owned manufacturing plants in each industry. Proxy for labor productivity.
	Explanatory variables
K/L^d	The ratio of total assets at book value to the total number of employees in private locally-owned plants in each industry. Proxy for capital intensity.
HERF	A Herfindahl index for each industry, including also foreign and state-owned plants. Measure of producer concentration.
W/B	The ratio of white-collar workers to blue-collar workers in each industry, including also foreign and state-owned plants. Proxy for labor quality.
LQ	The intercept and error term, $\alpha + \epsilon$, from the regression $W/B = \alpha + \beta FOR + \epsilon$, to correct for the a priori known correlation between W/B and *FOR*. Proxy for labor quality.
FOR	Share of each industry's total employment accounted for by foreign plants.
	Selection criteria for grouping of observations
PGAP	Ratio of value added per employee in foreign plants to value added per employee in private locally-owned plants in each industry. Proxy for the technology gap.
PAT	Ratio of the value of payments of patent fees and trade marks to total number of employees in each industry, including foreign and state-owned plants. Proxy for technological complexity.
K/L^f	Ratio of total assets at book value to total number of employees in foreign plants in each industry. Proxy for capital intensity.

Table A.2

Descriptive statistics for 156 Mexican manufacturing industries with foreign firms, 1970.[a]

Variable	Unit	Mean	St. Dev	Minimum	Maximum
VA/L^d	'000 pesos	54.77 (49.10)	35.56 (34.34)	12.97 (2.63)	259.89 (259.89)
K/L^d	'000 pesos	112.57 (98.47)	92.06 (91.28)	14.54 (4.55)	568.63 (568.63)
HERF		0.43 (0.44)	0.19 (0.19)	0.19 (0.19)	1.00 (1.00)
W/B		0.33 (0.29)	0.23 (0.21)	0.06 (0)	1.63 (1.63)
LQ		0.01	0.21	−0.35	1.39
PAT	'000 pesos	1.84 (1.36)	6.03 (5.19)	0 (0)	51.14 (51.14)
FOR		0.31	0.25	0	0.99
PGAP		2.03	1.62	0.41	10.19
K/L^f	'000 pesos	196.98	207.80	0	2296.17

[a]Figures within parentheses include also the 60 industries where foreign plants were not present in 1970. For definitions of variables, see table A.1.

Table A.3.

Simple correlation coefficients for exogenous variables, 156 Mexican manufacturing industries with foreign firms, 1970.[a]

	K/L^d	*HERF*	*LQ*	*FOR*	*PAT*	*PGAP*	K/L^f
K/L^d	1.00						
HERF	0.23 (0.19)	1.00					
LQ	0.11 (0.19)	−0.15 (0.11)	1.00				
FOR	0.34 (0.38)	0.48 (0.23)	−0.08 (0.00)	1.00			
PAT	0.19 (0.20)	0.04 (0.14)	0.13 (0.14)	0.28 (0.31)	1.00		
PGAP	−0.02	0.14	−0.04	0.13	0.53	1.00	
K/L^f	0.38	−0.01	0.22	0.05	0.11	0.28	1.00

[a]Figures in parentheses include also the 60 industries where foreign firms were not present in 1970. For definitions of variables, see table A.1.

References

Aitken, B. and A. Harrison, 1991, Are there spillovers from foreign direct investment? Evidence from panel data for Venezuela, Mimeo. (MIT, Cambridge, MA and World Bank, Washington, DC).

Blomström, M., 1986, Foreign investment and productive efficiency: The case of Mexico, Journal of Industrial Economics 35, 97–110.

Blomström, M. and H. Persson, 1983, Foreign investment and spillover efficiency in an underdeveloped economy: Evidence from the Mexican manufacturing industry, World Development 11, 493–501.

Blomström, M. and E. Wolff (forthcoming), Multinational corporations and productivity convergence in Mexico, in: W. Baumol, R. Nelson and E. Wolff, eds., International convergence of productivity, with some evidence from history (Oxford University Press, Oxford).

Blomström, M., A. Kokko and M. Zejan, 1992, Host country competition and technology transfer by multinationals, Working paper no. 4131 (NBER, Cambridge, MA).

Cantwell, J., 1989, Technological innovation and multinational corporations (Basil Blackwell, Oxford).

Caves, R.E., 1971, International corporations: The industrial economics of foreign investment, Economica 38, 1–27.

Caves, R.E., 1974, Multinational firms, competition, and productivity in host country markets, Economica 41, 176–193.

Estados Unidos Mexicanos, 1973, IX Censo Industrial 1971: Tomo 1 (Secretaría de Industria y Comercio, Mexico City).

Fairchild, L.G., 1977, Performance and technology of United States and national firms in Mexico, Journal of Development Studies 14, 14–34.

Findlay, R., 1978, Relative backwardness, direct foreign investment, and the transfer of technology: A simple dynamic model, Quarterly Journal of Economics 92, 1–16.

Globerman, S., 1979, Foreign direct investment and 'spillover' efficiency benefits in Canadian manufacturing industries, Canadian Journal of Economics 12, 42–56.

Haddad, M. and A. Harrison, 1991, Are there positive spillovers from direct foreign investment? Evidence from panel data for Morocco, Mimeo. (Harvard University, Cambridge, MA and World Bank, Washington, DC).

Kokko, A., 1992, Foreign direct investment, host country characteristics, and spillovers, Ph.D. thesis (EFI/Stockholm School of Economics, Stockholm).

Lapan. H. and P. Bardhan, 1973, Localized technical progress and transfer of technology and economic development, Journal of Economic Theory 6, 585–595.

Wang, J.-Y. and M. Blomström, 1992, Foreign investment and technology transfer: A simple model, European Economic Review 36, 137–155.

[20]

Strategic Management Journal, Vol. 17(Winter Special Issue), 155–165 (1996)

KNOWLEDGE SOURCING BY FOREIGN MULTINATIONALS: PATENT CITATION ANALYSIS IN THE U.S. SEMICONDUCTOR INDUSTRY

PAUL ALMEIDA
School of Business, Georgetown University, Washington, DC, U.S.A.

Do multinationals go abroad to acquire technological knowledge? Do they also contribute knowledge locally? We investigate the learning and contribution patterns of multinational firms in the U.S. semiconductor industry through the analysis of citations to their patents and through field interviews. We find that the knowledge used in innovation by foreign subsidiaries in U.S. regions is predominantly local (at the regional and country level). In fact, foreign firms use regional knowledge significantly more than similar domestic firms. In the case of European and Korean firms, foreign investment is directed towards offsetting home country technological weaknesses. The study finds that foreign firms also contribute to local technological progress—a significant proportion of the citations to their patents are local. Local learning without contributing may not be possible.

INTRODUCTION

Do multinational firms go abroad to acquire technological knowledge? Do they also contribute to local technological progress? The answer to these questions, central to the theory of foreign direct investment (FDI), depends in part on the geographic patterns of knowledge diffusion. If knowledge is sticky and remains confined within narrow spatial boundaries, plant location serves as an important source of competitive advantage to firms. Firms located in innovative regions have greater access to new technological knowledge than their spatially distant counterparts. To remedy this disadvantage, firms from other countries set up plants in knowledge-intensive regions. Innovative regions therefore serve as magnets to new investment. In fact, Alfred Marshall, in his comparative analysis of nations published in *Industry and Trade* (1920), noted that economic activity was drawn to regions rich in the 'atmosphere' of knowledge.

Knowledge-intensive districts or regional clusters (firms in related industries clustered in geographic space) are as pervasive today as they were in Marshall's nineteenth-century Britain (Krugman, 1991). Several of these regional clusters, like Silicon Valley, continue to grow by attracting new investment, not just of domestic firms but of foreign-based multinationals as well (Teece, 1992). The increasing presence of foreign firms in U.S. regions raises questions regarding the knowledge-acquiring and -sharing patterns of these multinationals, especially in critical high-technology industries.

By following the paper trail left by patent citations, this paper tracks interfirm knowledge flows in the semiconductor industry. Compared to earlier work which relies primarily on descriptive evidence, this paper investigates empirically the phenomenon of knowledge sourcing by foreign firms located in U.S. regions. By examining the citation data of fabrication and design patents,

Key words: multinational; knowledge; patents; semiconductor; learning

CCC 0143–2095/96/S20155–11

the study reveals the geographic patterns of the technological learning of these firms. The study also examines the contribution that the foreign firms make to local technological progress.

MULTINATIONALS AND TECHNOLOGY

One of the main directions of research in international business has been towards understanding the motivation for FDI and therefore multinationality. The seminal ideas first put forth by Hymer (1957) and extended by Kindleberger (1969) and Caves (1971) explained FDI as the outcome of firm-specific monopolistic advantages. These advantages (economies of scale, technology or even managerial and marketing know-how) were transferred by firms internally across international borders, thus allowing them to compete with host country firms. Thus it was the possession and exploitation of firm-specific advantages which drove FDI. Recent research in international management looks upon multinationals not just as exploiters of home country knowledge but as vast networks which access technology from various locations and share it with other parts of the organization (Bartlett and Ghoshal, 1989; Hedlund, 1986). In fact, this ability to learn and use knowledge elsewhere in the firm is one of the primary advantages of multinationality (Kogut, 1989).

Learning-oriented foreign investment, however, is only necessary if knowledge remains within geographic boundaries. There is good reason to believe that the production and flow of knowledge may be prescribed within spatial boundaries. The history of music, art and commerce is characterized by invention and discovery within specific regions. In a recent study, Jaffe, Trajtenberg, and Henderson (1993) analyzed patent citation data pertaining to domestic university and corporate patents to test the extent of localization of knowledge. They found evidence that patents tend to be cited in the same area where the originating patents are located, even after controlling for the existing concentration of patenting activity. Similarly, Almeida and Kogut (1994) studied the diffusion of technological knowledge in the semiconductor industry and found that knowledge remains localized in the U.S.A. and particularly in Silicon Valley.

Empirical research tracking the capabilities of multinationals to access, transfer and use knowledge is fairly sparse, though there is some evidence pointing to learning-oriented FDI. For instance Cantwell (1993) found that foreign subsidiaries in Britain invested in areas of host country technological expertise. Kogut and Chang (1991) showed that the entry of Japanese firms into the U.S.A. is linked to sectoral R&D intensity in the host country. Shan and Song (1995) found that American biotechnology firms with high levels of patenting activity are more likely to be the targets of takeover bids by foreign firms. Though technology-seeking investment appears to be on the rise, we have little direct evidence that multinationals actually succeed in learning from local sources.

A question of equal import is whether MNCs contribute to local technological progress. Again there is scant empirical evidence of the extent to which multinationals share technology locally. Kokko (1992) reviews studies that show that the operations of foreign subsidiaries of European firms in LDCs have resulted in productivity gains to the host countries. He points out, however, that these productivity gains could have resulted from the diffusion of knowledge and expertise from the multinational firm to host firms or from the effects of greater competition and better resource allocation (due to the presence of MNCs) in the host economy. This stream of research fails to separate out these sources of productivity gains by the host country economy and does not offer any direct evidence of knowledge flows to local firms.

In fact, research thus far has provided us with a quite limited picture and inadequate understanding of the overall patterns of knowledge acquisition and contribution of multinational enterprises. This paper seeks to track both the learning and sharing aspects of multinational activities by analyzing the citations to patents belonging to multinational subsidiaries in the U.S.A.

RESEARCH SETTING

The semiconductor industry originated from the invention of the first solid state transistor at the laboratories of AT&T (Bell Labs) in New Jersey in 1947. Over the next four decades process and product technology in the semiconductor industry advanced at a rapid pace while the industry grew increasingly international.

Formal technology transfer arrangements, like alliances, served as mechanisms to diffuse technology internationally. In a study of technology flows in the semiconductor industry, Griffin (1989) found that most technological agreements resulted in an outflow of technology from the U.S.A. In addition to the formal transfer of technology, there is also evidence (primarily ethnographic) of the spread of knowledge through informal channels. Since its inception, the American semiconductor industry has been characterized by interfirm mobility of scientists and engineers (Rogers and Larsen, 1984). Technology diffusion within the industry has also been facilitated by the culture of networking and information exchange between engineers (Saxenian, 1991). In their history of the semiconductor industry, Braun and MacDonald (1982) provide the example of a bar located close to Intel, Fairchild, Raytheon and other semiconductor companies, which served as a place where engineers 'drank, exchanged information and hired employees.' It is this culture of intraregional communication and interfirm mobility that causes local knowledge sharing between firms in the semiconductor industry.

A recent flood of FDI into high-technology regions, like Silicon Valley, suggests that multinationals in the semiconductor industry are using FDI to gain access to local information channels and thus source location-specific knowledge (Yoffie, 1987). Between, 1985 and 1989 Japanese equity investments in high-technology firms (primarily biotechnology and semiconductors) increased from $44 million to $320 million. Similarly R&D investment by the U.S. affiliates of foreign electronic firms quadrupled between 1982 and 1992; for U.S. firms investment remained static (Kogut and Gittleman, 1994). Informal evidence suggests that these investments are geared towards accessing and using local technology which would be difficult to obtain from a distance (Teece, 1992). Interviews in Silicon Valley reveal that foreign firms buy into the local knowledge networks through hiring employees locally and using local suppliers. Over 90 percent of the executives interviewed by the author in these foreign firms had previously worked for another local firm. These arguments lead us to the following hypotheses related to the sources of knowledge used in innovation.

> *Learning Hypothesis A: Foreign subsidiaries in the U.S.A. use local (regional and country) knowledge to a* greater *extent than would be expected given the geographic distribution of innovative activity in the industry.*

If the primary reason for the location of the foreign subsidiaries in the U.S.A. is to source local knowledge, foreign firms can be expected to locate plants in regions where this knowledge is prevalent. Thus, through plant location in knowledge-intensive regions, foreign firms are able to target and use regional knowledge in their innovative processes. Domestic firms are guided by an array of factors in their choice of location, only one of which is access to technology (Malecki, 1991). Foreign firms on the other hand seek primarily to gain information (Mariotti and Piscitello, 1995). Confirming these studies, interviews held with executives of several foreign firms revealed that the primary reason for their plant's location in Silicon Valley was to keep track of advanced technological developments and expand the firm's knowledge base beyond that existing in their home country. Foreign firms can, therefore, be expected to source local knowledge even more than host country firms located in the same region.

> *Learning Hypothesis B: Foreign subsidiaries in the U.S.A. use local knowledge to a* greater *extent than similar domestic firms.*

Local presence increases a firm's access to knowledge, but it also leads to the leakage of firm-specific knowlege to the host region. Firms located in a region and conducting research become a part of the regional knowledge network (Cohen and Levinthal, 1989). This network which facilitates absorption of external knowledge also allows leakage of firm knowledge to others in the region through the same channels. The high turnover and extensive informal exchanges which characterize knowledge-intensive regions like Silicon Valley facilitate a two-way flow of knowledge (Almeida and Kogut, 1994; Saxenian, 1991). These arguments lead us to the following hypothesis.

> *Contributing Hypothesis: Foreign subsidi-*

aries in the U.S.A. contribute to local (regional and country) knowledge to a greater *extent than would be expected given the geographic distribution of innovative activity in the industry.*

We test these hypotheses through the analyses of citation data relating to patents of foreign and domestic semiconductor firms located in the U.S.A.

DATA AND METHODS

The geographic characteristics of knowledge flows are analyzed (a) through the analysis of patent citations relating to patented innovations and (b) through field research.[1] This patent citation analysis is carried out using the case-control method.

Patent data have commonly been used to throw light on the process of innovation. Patent data are systematically compiled, have detailed information, and are available continuously across time. Patent documents allow us to pinpoint the geographic location, the time, and the technology of an invention. In addition, a patent document contains a list of citations to other patents which represent the technological antecedents to the particular innovation. This citation list is compiled through a uniform and rigorous process applied by the patent examiner as a representative of the patent office. These citations enable researchers to observe interfirm technology building.

Appendix 1 explains why patent data are useful for analyzing the patterns of knowledge flows in the semiconductor industry. Appendix 2 explains the appropriateness of the use of U.S. patent data to compare the patterns of knowledge influences of foreign and domestic firms.

The sample

The patenting activity of subsidiaries of foreign firms in the U.S. regions formed the basis for our analysis. We first identified every foreign (non-U.S.) semiconductor firm with subsidiaries engaged in either design or fabrication activity in the U.S.A.[2] Twenty two semiconductor firms from Britain, Canada, France, Germany, Italy, Japan, Korea and Taiwan had U.S.-based plants which had patented semiconductor innovations.[3] The firms analyzed include NEC, Toshiba, Hitachi, Samsung, Philips and Siemens, all of which are among the top patent holders on the semiconductor industry. We randomly selected a maximum of three design and three fabrication patents filed between 1980 and 1990 for each foreign subsidiary. (These 114 patents belonging to foreign subsidiaries in the U.S.A. comprised Sample A.) Most of the patents selected came from Silicon Valley (52%) reflecting the concentration of foreign firms and patenting activity in that region.[4]

For each patent in Sample A we selected a matching patent belonging to a U.S. firm. The patents were matched by (a) technology,[5] (b) time, (c) geographic region and (d) firm size to those in Sample A. These 114 matched patents comprised Sample B. We thus created matched samples of patents belonging to foreign and domestic firms and analyzed the learning and contributing patterns of these innovations by looking at (a) the patents they cited and (b) patents that cited them (Figure 1). The geographic location of *cited* patents *vis-à-vis* the sample patent offers an insight into the spatial learning patterns of the firms being studied while the geographic location of *citing* patents offers an insight into the contributing patterns of these firms.

Localization is indicated by the joint condition that the citing patent (or cited patent) and the major patent belong to the same geographic

---

[1] Semistructured interviews of 28 design and fabrication engineers of semiconductor firms were conducted. The firms at which interviews were held belonged to Silicon Valley, Texas and New York-New Jersey areas. Engineers employed by non-American firms, such as Philips, Siemens, Samsung, Fujitsu and NEC, were interviewed at plants located within the U.S.A.

[2] The list was compiled through the examination of the volumes of FDI in the U.S.A.: transactions (1974–1993) and from industry sources. Status of firms was confirmed through telephone calls.

[3] We identified the regional location of the innovative activity through the inventor's home location available on the front page of the patent document.

[4] Major regions of semiconductor activity in the U.S.A. were identified by plotting actual plant locations of over 450 fabrication units and 300 design facilities and demarcating the corresponding regional clusters. Major regions identified included Santa Clara county in California and the NY-NJ-PA belt in the northeast.

[5] Patents are classified by technology into 400 classes and nearly 100,000 subclasses. We matched patents by the identical technology subclass.

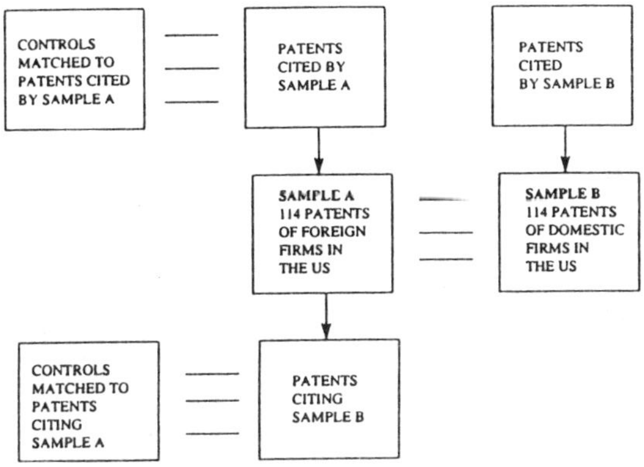

Figure 1. Research design. Note: arrows indicate the direction of technological build-up

region. To measure the frequency of localization we geographically matched each set of patents from foreign subsidiaries (Sample A) with the cited patents. However, the observed frequency of geographic coincidence of the sample patent and the cited patents may reflect the distribution of patenting activity rather than the localization of knowledge. To adjust for any bias due to this existing distribution of technological activity, we followed Jaffe *et al.* (1993) in the construction of a control sample. For each patent cited by Sample A, we identified a corresponding control patent. This patent was identified such that the patent (technology) class was identical to that of the cited patent and the application date was as near as possible to the cited patent. The control patent thus closely resembles the cited patent in terms of technology and time of innovation. Since the control patent, however, does not cite the sample patent, the frequency of a geographic matches between the two reflects the existing concentration of patenting activity for a particular region. This frequency of geographic matches between the sample patent and the control patent sets the baseline against which we compare the frequency of major patent-cited patent matches.

Table 1 provides the descriptive statistics for the patent citations in the learning and contributing studies. Each patent in the major samples cited approximately six patents and was cited by nearly three patents each. Interestingly, both Sample A and Sample B had approximately the same number of citing patents after adjusting for self-cites. The number of citations to a patent are an indicator of the technological and economic importance of the innovation it represents (Albert *et al.*, 1991; Narin, Noma, and Perry, 1987). It appears that the matched samples of U.S. and foreign samples contain patents of equal importance.

Table 1. Descriptive statistics

Learning Study

Sample	Number of major patents	Number of cited patents	Mean cited patents	% Self citations
Sample A	114	677	5.94	8.12
Sample B	114	756	6.63	10.85

Contributing study

Sample	Number of major patents	Number of citing patents	Mean citing patents	% Self-citations
Sample A	114	338	2.96	10.95

Note: Self-citations are not used in statistical tests.
Sample A consists of patents belonging to foreign firms in the U.S.A.
Sample B consists of matched patents belonging to U.S. firms.

Statistical test for localization of knowledge

Figure 2(A) illustrates the design of the statistical test for Learning Hypothesis A. In this analysis,

P_A is the frequency probability that the patent from Sample A and the cited patent are geographically matched, and P_{Acon} is the corresponding frequency probability that the patent from Sample A and the control patent are geographically matched. The null hypothesis is

$$H_{1Ao}: P_A = P_{Acon}$$

and the alternate hypothesis is

$$H_{1Aa}: P_A > P_{Acon}$$

The t-statistic is calculated as follows:

$$t_A = (P_A - P_{Acon})/[(P_A(1-P_A) + P_{Acon}(1-P_{Acon}))/n]^{0.5}$$

The t-statistic tests the difference between two independently drawn binomial proportions. A positive significant value of Student's t indicates support of the proposition that knowledge flows (represented by patents cited by Sample A) are geographically more localized than expected. The test was similarly constructed to test the Contributing Hypothesis. For the contributing study, we geographically matched the patents in Sample A with the patents that cited these sample patents (as opposed to those cited by a sample patent as was the case in the learning study) and calculated the t-statistic.

Figure 2(B) illustrates the design of the statistical test of Learning Hypothesis B. Let P_A be the frequency probability that the patents in Sample A and the cited patents are geographically matched, and P_B be the corresponding frequency probability that the patents in Sample B and their cited patents are geographically matched. Assuming binomial distributions, the null hypothesis is

$$H_{1Bo}: P_A = P_B$$

and the alternate hypothesis is

$$H_{1Ba}: P_A > P_B$$

The t statistic for Hypothesis 2A is calculated as follows

$$t_{AB} = (P_A - P_B)/[(P_A(1-P_A)/n_A + P_B(1-P_B)/n_B]^{0.5}$$

Again a positive value of Student's t indicates support of the proposition that knowledge flows are geographically more localized for Sample A than for the matched Sample B. The tests were carried out at two different geographic levels (country and U.S. regions).

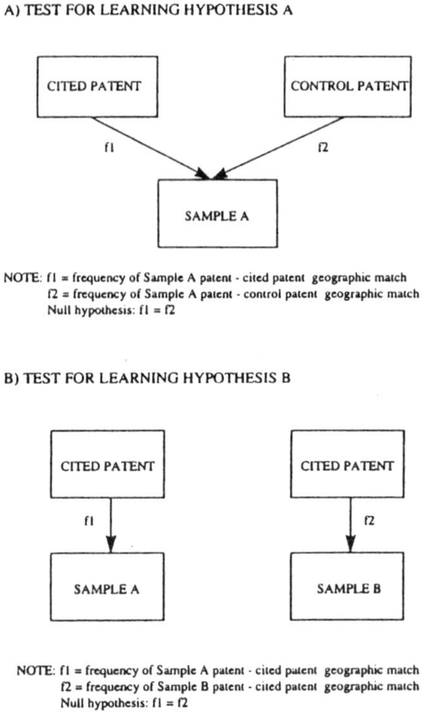

Figure 2. Test of localization of patent citations.

FINDINGS

The results of the case–control tests for the learning study are provided in Table 2 for both the regional and country level of analysis. 'Number of citations' corresponds to the total number of patents cited by the major sample less self-cites (when the major patent and the cited patent have the same assignee). 'Citation matching %' and 'Control matching %' are the percentage of cited patents and controls respectively that belong to the same geographic location as the major patent. The t-statistic tests the equality of the cited or control proportions as described previously.

Table 2. Learning study: Means test of localization of knowledge

(I) Regional-level analysis

Hypothesis A		Hypothesis B	Sample A	Sample B
Number of citations	622	Number of citations	622	674
Citation matching %	13.8	Citation matching %	13.8	10.5
Control matching %	9	Difference	3.3	
Difference	4.8	*t*-Statistic	1.81	
t-statistic	2.68			

(II) Country-level analysis

Hypothesis A		Hypothesis B	Sample A	Sample B
Number of citations	622	Number of citations	622	674
Citation matching %	65.9	Citation matching %	65.9	64.5
Control matching %	55.3	Difference	1.4	
Difference	10.6	*t*-statistic	0.51	
t-Statistic	3.85			

Learning Hypothesis A is supported at two geographic levels (region and country); the *t*-statistic is significant at the 0.01 level in both cases. There is a significantly higher proportion of citation matches than control matches indicating localization effects. As the patents cited by foreign firms (in the U.S.A.) are more likely to be local than expected, we can conclude that the knowledge building patterns of these firms are localized.

The tests for Learning Hypothesis B (the comparison with similar domestic firms) have mixed results. While the percentage of citation matches for Sample A (foreign firms) are greater than that for Sample B (domestic firms) at both geographic levels of analyses, we have a significant *t*-statistic (0.05 level for a single-tailed test) for only the regional test. The results suggest that within geographic regions foreign firms not only learn more from local sources than expected, they do so to a greater extent than domestic firms.

The results in Table 3 show that there is support for the Contributing Hypothesis; patents belonging to foreign firms are cited locally more than would be expected given the spatial distribution of patenting activity. This result is significant (0.01 level) for U.S. regions but weaker at the country level.

Table 3. Contributing study: Means test of localization of knowledge

(I) Regional-level analysis

Contributing hypothesis	
Number of citations	301
Citation matching %	14.6
Control matching %	7.3
Difference	7.3
t-Statistic	2.89

(II) Country-level analysis

Contributing hypothesis	
Number of citations	301
Citation matching %	56.8
Control matching %	50.8
Difference	6
t-Statistic	1.47

Country technological strength

A multinational can be expected to seek knowledge abroad when its home country is relatively weak in a particular technological field. The firm would choose a host country with strength in the desired technology. We investigated the relative strengths of the home countries and host country (in this case the U.S.A.) with respect to the technologies patented by the foreign firms in the U.S.A. To measure relative technological stengths of countries we followed Cantwell (1989) in the computation of the indicator Revealed Technological Advantage (RTA).[6] The indicator is calculated here as shown below.

$$RTA_{A1} = X/Y$$

where RTA_{A1} is the Revealed Technological Advantage of Country A in Technology 1. X is Country A's share of total patents granted in Technology 1 in a given year and Y is Country A's share of total patents granted in all semiconductor technologies in the same year. Technology 1 is the technological subclass (one of the approximately 100 semiconductor subclases) assigned by the U.S. Patent Office to each patent in the sample. An RTA of greater than 1 would indicate that the country is relatively strong in the technology being considered (compared to other semiconductor technologies) while an RTA of less than 1 would indicate a relative weakness. The RTA was calculated for the home country of the firm owning the patent and for the U.S.A. (host country) for each of the 114 patents in our study.

Table 4 shows the results of the mean RTA for the technologies under consideration for both the home country of the multinational and for the host country (U.S.A.). It can be observed that the European countries (including Germany, France, Italy and Britain) and Korea had RTAs of much less than 1, revealing home country weaknesses in the technologies being developed in the U.S A. The RTA indicator of approximately 1 for Japan shows neither home country strength nor weakness in the patented technologies. Similarly, the U.S.A. is not particularly strong in the technologies in which U.S.-based Japanese firms are patenting. However, the U.S.A. does show relative strength for the technologies in which Korean and European firms patent (RTA > 1).

Table 4. Revealed technological advantage

Country	Home country RTA			Host country (U.S.) RTA		
	X	*Y*	RTA	*X*	*Y*	RTA
Japan	36.8	36.7	1.01	50.3	49.4	1.02
Europe	2.9	3.9	0.73	57.2	51.1	1.12
Korea	1.7	2.8	0.61	60.1	49.7	1.21

Note: X = country share (in percent) of total patents granted in technologies under consideration in the given year.
Y = country share (in percent) of total patents granted in all semiconductor technologies in the given year.
RTA = X/Y.

DISCUSSION AND CONCLUSION

The results of this study confirm the local character of both learning and contributing. The findings of the Learning Study suggest that foreign firms are aware of the difficulty of learning from afar, and use local plants to upgrade the technological ability in fields which may be weak in their home countries. Furthermore, though both domestic firms and foreign firms use regional knowledge, foreign firms use regional knowledge to a significantly greater extent. This finding contradicts the conventional view that foreign firms are burdened with the 'liability of foreignness' (Hymer, 1957) and hence should find it more difficult to tap into regional networks. Any disadvantages foreign firms possess are outweighed by the motivation to learn locally. A review of the citation patterns of patents of foreign firms reveals not just a regional concentration of citations but specific firm-to-firm linkages. SGS-Thomson in the Dallas area of Texas commonly builds on patents by Texas Instruments. Similarly, the Siemens plant in New Jersey overwhelmingly cites patents from nearby Bell Labs and AT&T. These patterns suggest that foreign firms may not be targeting just regions but specific firms in their learning efforts.

The results of the Learning Study present an incomplete picture unless they are viewed in the context of the findings of the Contributing Study. There is evidence that foreign firms do not just

[6] Cantwell used the RTA indicator to assess the comparative strengths of industries for the U.S.A. and European countries at different periods in time. We use the indicator to assess a country's strength in a particular technology within the semiconductor industry.

learn locally but also contribute locally (at the regional level). An explanation for this local two-way knowledge flow lies in the sociological aspects of knowledge exchange. In a detailed ethnography of Silicon Valley and Route 128, Saxenian (1990) concluded that interfirm knowledge flows are facilitated by the interaction of employees. Almeida and Kogut (1994) show that the intraregional mobility of engineers is an important mechanism for the flow of technology across firms within regions. In order to learn, firms tie into the local social networks and hire engineers locally. These regional connections, however, also serve to diffuse knowledge from the firm to others in the region. Local learning (and contribution) is, therefore, as much an aspect of a region's sociology as it is of a firm's motivation. Learning without contributing may not be possible.

The country comparisons of the RTA indicators for Japan and Europe reveal interesting differences. The European and Korean firms appear to target technologies in which they are weak and the U.S.A. is particularly strong. However, it is Japan which is often viewed as the main rival to the U.S.A. in the battle for dominance in the semiconductor industry. There is no evidence to suggest that Japanese firms are attempting to narrow the gap in technologies where the country is weak. The combined evidence of the (approximately) unity RTA indicators and the substantial innovative presence of Japanese firms in the U.S.A. suggests that these firms play the role of participants in U.S.-based innovative activity rather than the role of absorbers of U.S.-technology.

Conclusion

The findings of this paper have some bearing on the debate on U.S.-competitiveness in high-technology industries. The study confirms the widely held belief that foreign firms use U.S.-based plants to source technology. However, this is only one side of the story. The benefits of innovation by foreign firms are also enjoyed to a significant extent by other foreign and domestic firms in the same region. A more informed debate on the effects of foreign investment on national technological competitiveness should take into account both the learning and contributory aspects of multinational innovative activity. Foreign firms should not be viewed primarily as absorbers of technological knowledge, but rather, along with domestic firms, as players in the continuous process of interfirm knowledge exchange and regional technological development.

Acknowledgements

I thank Tom Brewer, Rob Grant, Bruce Kogut, Stan Nollen, Pietra Rivoli, Mike Ryan, and Dennis Quinn for their comments and suggestions. I am grateful to Patrick Abouchalache, Sarah Abramson, Maria Lopez, and Jerome Romano for their research assistance. Financial support for this project was provided by the Reginald H. Jones Center at the Wharton School of the University of Pennsylvania.

REFERENCES

Albert, M. B., D. Avery, F. Narin and P. McAllister (1991). 'Direct validation of citation counts as indicators of industrially important patents', *Research Policy*, **20**, pp. 251–259.

Almeida, P. and B. Kogut (1994). 'Technology and geography: The localization of knowledge and the mobility of patent holders', working paper, Huntsman Center for Global Competition and Innovation, Wharton School, Philadelphia, PA.

Bartlett, C. A. and S. Ghoshal (1989). *Managing Across Borders: The Transnational Solution*. Harvard Business School Press, Boston, MA.

Braun, E. and S. MacDonald (1982). *Revolution in Miniature* (2nd ed.). Cambridge University Press, New York.

Cantwell, J. (1989). *Technological Innovation and Multinational Corporations*. Basil Blackwell, Oxford.

Cantwell, J. (1993). 'The internationalization of technological activity and its implication for competitiveness'. In O. Granstrand, H. Hakanson and S. Sjolander (eds.). *Technological Management and Internatioanl Business*. John Wiley, Chichester, pp. 137–162.

Caves, R. (1971). 'International corporations: The industrial economics of foreign investment', *Economica*, **38**(141), pp. 1–27.

Cohen, W. and D. Levinthal (1989). 'Innovation and learning: the two faces of R&D', *Economic Journal*, **99**, pp. 569–596.

Griffin, A. J. (1989). 'Semiconductor technology flow through formal transfer mechanisms', *IEEE Transactions on Engineering Management*, **36**(2), pp. 148–154.

Hedlund, G. (1986). 'The hypermodern MNC: A heter-

archy?', *Human Resource Management*, **25**, pp. 9–25.
Hymer, S. (1957). *The International Operations of National firms: A Study of Direct Investment*. MIT Press, Boston, MA.
Jaffe, A., M. Trajtenberg and R. Henderson (1993). 'Geographic localization of knowledge spillovers as evidenced by patent citations', *Quarterly Journal of Economics*, **108**(3), pp. 577–598.
Kindleberger, C. (1969). *American Business Abroad*. Yale University Press, New Haven, CT.
Kogut, B. (1989). 'A note on global strategies', *Strategic Management Journal*, **10**(4), pp. 383–389.
Kogut, B. and S. J. Chang (1991). 'Technological capabilities and Japanese foreign direct investment in the United States', *Review of Economic and Statistics*, **73**(3), pp. 401–413.
Kogut, B. and M. Gittleman (1994). 'Foreign direct investment in the United States', working paper, Wharton School, Philadelphia, PA.
Kokko, A. (1992). *Foreign Direct Investment, Host Country Characteristics, and Spillovers*. Economics Research Institute, Stockholm, Sweden.
Krugman, P. (1991) *Geography and Trade*. MIT Press, Cambridge, MA.
Malecki, E. (1991). *Technology and Economic Development*. Wiley, New York.
Mariotti, S. and L. Piscitello (1995). 'Information costs and location of FDIs within the host country: Empirical evidence from Italy', *Journal of International Business Studies*, **26**(4), pp. 815–842.
Marshall, A. (1920). *Industry and Trade*. Macmillan, London.
Narin, F., E. Noma and R. Perry. (1987). 'Patents as indicators of technological strength', *Research Policy*, **16**, pp. 143–155.
Rogers, E. and J. Larson (1984). *Silicon Valley Fever*. Basic Books, New York.
Saxenian, A. (1990). 'Regional networks and the Resurgence of Silicon Valley', *California Management Review*, Fall, pp. 39–112.
Saxenian, A. (1991). 'The origins and dynamics of production networks in Silicon Valley', *Research Policy*, **20**, pp. 423–437.
Shan, W. and J. Song (1995). 'Foreign direct investment and the sources of technological advantage', working paper, Wharton School, Philadelphia, PA.
Teece, D. J. (1992). 'Foreign investment and technological development in Silicon Valley', *California Management Review*, **34**(2), pp. 88–106.
U.S. Department of Commerce (1992). *Technology Profile Report: Semiconductors*. Patent and Trademark Office, Department of Commerce, Washington, DC.
Yoffie, D. (1987). 'The global semiconductor industry, 1987', Case #9-388-052, Harvard Business School, Boston, MA.

APPENDIX 1: PATENT DATA AND THE SEMICONDUCTOR INDUSTRY

This study uses semiconductor patent data to analyze the extent of localization of knowledge in this industry. Though some firms may choose not to patent innovations, patenting in the semiconductor industry is commonly practiced and is indeed a vital part of maintaining technological competitiveness.

Firms may patent for a variety of reasons. (a) Patents establish intellectual property rights which often become an important revenue source (during the late 1980s Texas Instruments earned several hundred million dollars from patent royalties). (b) Patent portfolios are used by firms as bargaining chips to prevent or to fight legal battles. (c) Patents facilitate licensing and cross-licensing which often serves as an important source of learning for the firm. (d) Interviews with 'headhunting' firms and engineers revealed that patents are valuable to individual engineers and researchers as indicators of personal technological expertise.

Recent government actions have increased the protection for intellectual capital supplied by the patent system and thereby increased the incentive to patent. In 1982, Congress created the US Court of Appeals for the Federal Circuit in Washington, DC, specifically for the purpose of hearing patent cases. This action significantly increased the protection provided by patents. Partly as a result of the above governmental actions, to protect intellectual property rights, patenting in the semiconductor industry doubled in the 5-year period between 1986 and 1990. The total number of semiconductor patents granted in 1991 numbered over 4500.

APPENDIX 2: U.S. PATENTS AND INTERNATIONAL FIRMS

The use of U.S. patent data (patents filed and granted by the U.S. Patent Office) to compare domestic and international firms may raise questions of possible biases introduced by the use of U.S. data. To address the issue of bias we make two points.

First, the U.S.A. is both a major market and manufacturing location for semiconductors, making U.S. intellectual property rights critical for any firm in the semiconductor industry. The U.S. patent system is widely used by both foreign and domestic firms. In fact since 1988 over 50 percent of the semiconductor patents granted are to international firms (U.S. Department of Commerce, 1992). This study analyzes the patenting activities of several multinationals in the U.S.A., all of which patent extensively in the U.S.A. The importance of U.S. patents to international firms was confirmed through field interviews. Several international firms awarded employees prizes and other benefits for successful patenting activity.

Secondly, this study does not compare firms by the magnitude of their patenting output but rather analyzes the geographic patterns of citations to patents belonging to foreign and domestic firms. The list of citations for each patent (which forms the basis of our study) is finalized by the patent examiner applying the same rigorous process regardless of the national identity of the applicant. The citation information of patents belonging to foreign or domestic firms is thus comparable and the analysis based on citation information (as carried out in this paper) is not biased.

[21]

The geographical sources of the competitiveness of firms: some results of a new survey

John H. Dunning*

This article presents some results of a new field survey, designed to assess the extent to which the executives of some of the largest industrial corporations in the world perceived how they had augmented their global competitive advantages as a result of their foreign direct investments. Among other interesting findings, the survey revealed that the more transnational a firm was, the more it derived its competitive advantages from its foreign affiliates; while, over the last decade or so, an increasing proportion of these competitive advantages were obtained from this source.

1. Introduction

This article presents some of the results of a new field survey of the geographical sources of firm-specific competitiveness as perceived by executives of 144 of the world's 500 largest industrial firms,[1] all of which are transnational corporations (TNCs). In 1993, the sample firms were responsible for 39 per cent of the global sales and nearly two fifths of the foreign direct investment (FDI) of the largest 500 industrial corporations. For some industries (e.g. motor vehicles, petroleum refining, tobacco, electronics and computers) and for some countries (e.g. Japan, Sweden and the United Kingdom), the coverage of the survey is particularly good; for others, it is less so; but given the country and industry groups used here, it is entirely acceptable. Full details of the sample and the way in which the survey was organized and carried out are presented in the appendix.

* State of New Jersey Professor of International Business, Rutgers University, Newark, New Jersey, United States. The author is much indebted to Sarianna Lundan of Rutgers University for research assistance and to the anonymous referees for helpful suggestions.

[1] As identified by *Fortune* magazine, 25 July 1994.

The purpose of the field survey, conducted in the second half of 1994, was to obtain the opinions of senior business executives[2] on the geographical origin of the kind of firm-specific competitive advantages or core competencies—sometimes called ownership-specific advantages of firms—identified by the literature.[3] There has been much debate in recent years over whether or not the competitive advantages of firms—some of which may be internalized by the firms themselves—stem from the location-bound characteristics of their home countries, or whether or not, as FDI and strategic alliances have become aimed increasingly at acquiring created assets[4] and as firms have become more transnational,[5] firms' competitive advantages—particularly knowledge of one kind or another—are being sourced increasingly from outside their home countries. In other words, to put it in Michael Porter's terminology, whether or not firms source from "multiple home bases" (Porter, 1997).

The concept of the "double" diamond of competitive advantage, as developed by Alan Rugman and his colleagues over the past few years,[6] is used here as the analytical basis. Essentially, the Rugman approach extends the concept of Porter's "single" diamond of competitive advantage (first set out in Porter, 1990) that argues that a firm's competitive advantages are essentially a function of the domestic economic environment in which it operates. Rugman and others, including the present author Dunning (1993b), have argued that the deepening structural integration of the world economy and the burgeoning of alliance capitalism (Dunning, 1995) are widening the geographical scope for creating or augmenting firm-specific competencies and learning experiences. Any attempt to identify the geographical sources of such advantages must embrace the diamonds of other countries, particu-

[2] Mainly Vice-Presidents or Directors of Foreign Operations.

[3] For a summary of the competitive advantages identified by economists, see, for example, Dunning (1993a, chap. 4), and for those identified by business strategists, see, for example, Porter (1990) and Peteraf (1993). For recent analyses of knowledge sourcing by TNCs outside their national boundaries, see Almeida (1996), Kuemmerle (1995 and 1996) and Pearce and Singh (1991 and 1992).

[4] The aim of strategic asset-acquiring FDI is to gain access to assets that protect or augment the acquiring firm's competitive advantages and/or lessen those of its rivals. In the language of the eclectic paradigm, firms engage in FDI not to exploit existing ownership (O) advantages, but to gain new advantages, which, when deployed with their existing O advantages, help sustain or further their global competitive competencies.

[5] As documented, for example, in various *World Investment Reports* (UNCTAD-DTCI, 1995 and 1996). The most recent United States data (Mataloni, 1995) show that the foreign affiliate sales of non-bank United States parent firms were 45 per cent of their parent sales in 1993. This figure compares with a corresponding 1983 figure of 37 per cent.

[6] See the special issue of *Management International Review* edited by Alan Rugman (1993), and Rugman, Van den Broeck and Verbeke (1995).

larly those with which the home-country firms have the most dealings by way of trade, FDI and non-equity cooperative ventures.

This field study aims at offering some new evidence to that recorded by a number of country case-studies[7] which, in the main, have confirmed the Rugman/Dunning hypothesis. The uniqueness of the present contribution is sixfold:

- The information is obtained directly from firms.
- The competitive advantages identified in the literature are classified into groups using as the framework the four components of Porter's diamond.
- Its findings are related to a number of critical firm-specific variables, viz. size of firm, degree of transnationality, country (or region) of origin and industry.
- Those foreign countries that provide the principal access to non-domestic competitive advantages that complement those offered by the home country are identified.
- The sample firms are ranked by the significance of the three main modes of foreign involvement used by these firms in acquiring competitive advantages from a foreign location.
- The perceptions of the sample firms about the positive or negative influences of different components of home-government policy on their global competitiveness are set out.

To keep this article within manageable dimensions, a threefold classification of industries is adopted, based upon their degree of research and development (R & D) intensity in the United States in 1992. Thus, high-technology (HT) industries are defined as those in which the average R & D expenditure as a percentage of sales was at least 4 per cent, and in which scientists and engineers employed in R & D as a percentage of total employment were 2 per cent or more.[8] Medium-technology (MT) industries are

[7] Including Australia, Canada, Denmark, Republic of Korea, Mexico and New Zealand. See Rugman (1993).

[8] More specifically, HT industries include pharmaceuticals; MT sectors include industrial equipment, motor vehicles, petroleum refining, soap and cosmetics; and LT sectors include beverages, building materials, food, metal products, paper, publishing and printing, rubber and plastics, textiles and tobacco.

defined as those in which the corresponding ratios vary between 2 per cent and 3.9 per cent, and 1 per cent and 1.9 per cent, respectively; and low-technology (LT) industries are defined as those that embrace those ratios under 2 per cent and under 1 per cent, respectively. As to home country or region of origin, five main groups, grouped according to their size and economic profiles, are considered here. These are, respectively, "larger" European countries,[9] "smaller" European countries,[10] the United States, Japan, developing economies,[11] and other countries.[12]

The firms in the sample are also grouped by their degrees of transnationality. This was obtained by averaging out (for each firm) the percentages of their global assets and global employment[13] accounted for by their foreign affiliates,[14] and then classifying firms into four groups on the basis of the degree of transnationality, viz. under 15 per cent (low), 15-29 per cent (medium-low), 30-59 per cent (medium-high) and 60 per cent and over (high).

Finally, although all of the sample firms were large companies, their global sales in 1993 varied from US$ 0.22 billion to US$ 138.2 billion. For the purposes of the analysis, firms were categorized into four groups according to size; medium-size (M) firms, defined as those with sales of under US$ 5 billion in 1993; large (L) firms, with sales between US$ 5 billion and US$ 24.9 billion in that year; very large (VL) firms, with sales between US$ 25 billion and US$ 49 billion; and mega (Mg) firms, with US$ 50 billion and over in sales.

In future research, an econometric analysis of the relationship between these four contextual variables and the competitive advantages of firms will be undertaken, since, with few exceptions, data exist for each of the 144 firms. For now however, the analysis is confined to identifying and exploring a number of possible bi-variant (and occasionally tri-variant) relation-

[9] These include France, Germany, Italy, Spain and the United Kingdom. Two United Kingdom/Netherlands-based TNCs are also included as originating from this group of countries.

[10] These include Belgium, Denmark, Finland, the Netherlands, Sweden and Switzerland.

[11] These include Brazil, Chile, Hong Kong, the Republic of Korea, Mexico and Taiwan Province of China.

[12] These include Canada, Australia and New Zealand.

[13] Or one of these when data on either employment or assets were not available. Additionally, for some firms, sales data from Worldscope/Disclosure were used.

[14] As defined by the TNCs themselves.

ships between the sources of competitiveness and firm- and/or country-specific characteristics, which—at the very least—may help point the way to more rigorous statistical evaluation.

2. The sample firms: some details

Tables 1 to 3 set out some details about the characteristics of the sample firms. In general, they mirror the existing knowledge about the industrial and geographical composition of TNCs, as well as the size and degree of transnationality as it has been set out in various publications, for example, *World Investment Reports 1991-1996* (UNCTC, 1991 and 1992; UNCTAD-DTCI, 1993, 1994, 1995, 1996) and *World Investment Directory* (4 vols.,

Table 1. The industrial distribution and extent of transnationality of the sample firms, 1993

(Billions of dollars and percentage)

Industry	Number of Firms	Sales (*Billions of dollars*)	Sales (*Percentage*)	Extent of transnationality[a] (*Percentage*)		
				Assets	Employment	R&D
High technology (HT), of which:	53	892.0	36.0	38.2	39.8	22.7
Aerospace	5	73.1	2.9	6.5	6.2	2.0
Chemicals and pharmaceuticals	19	215.9	8.7	51.3	54.2	26.6
Computers	9	213.9	8.6	28.9	28.7	9.6
Electronics	20	389.1	15.7	38.8	39.2	27.5
Medium technology (MT), of which	34	1,152.2	46.5	39.6	39.4	15.4
Industrial equipment	8	73.8	3.0	28.0	39.1	12.7
Motor vehicles	12	580.0	23.4	32.2	35.5	7.5
Petroleum refining	14	498.4	20.1	49.7	42.9	24.3
Low technology (LT), of which	57	434.5	17.5	41.7	45.5	28.5
Food, drink and tobacco	20	244.3	9.9	50.4	58.1	32.5
Paper	10	39.0	1.6	22.4	20.9	8.1
Building materials	5	16.4	0.7	49.6	46.4	39.7
Metals and metal products	14	107.3	4.3	35.0	38.0	24.6
Other industries	8	27.6	1.1	53.5	59.6	45.6
Total	144	434.5	100.0	40.0	41.8	25.3

Source: Author's estimates, based on the survey of sample firms.

[a] Measured as the percentage share of total assets, employment and R & D of the respondent firms accounted for by their foreign affiliates.

Table 2. The distribution of the sample firms, by region or country of origin and extent of transnationality, 1993

(*Billions of dollars and percentage*)

Region/country	Number of firms	Sales (*Billions of dollars*)	Sales (*Percentage*)	Extent of transnationality (*Percentage*) Assets	Employment	R & D
"Large" European countries, of which	44	922.8	77.2	49.1	49.9	32.0
Germany	13	359.5	14.5	39.7	46.0	23.1
United Kingdom and United Kingdom/Netherlands	23	400.3	16.1	57.3	57.9	38.6
Other countries	8	163.0	6.6	28.8	32.1	27.8
"Small" European countries of which	21	164.5	6.6	55.2	56.8	32.7
Sweden	7	32.0	1.3	70.0	70.3	53.0
Switzerland	5	68.9	2.8	58.0	61.8	33.2
Other countries	9	63.6	2.6	46.1	43.6	19.7
United States	31	710.2	28.6	36.1	34.5	14.7
Japan	26	606.3	24.5	19.5	28.7	6.0
Developing countries	14	42.5	1.7	27.2	25.4	15.6
Other countries	8	32.4	1.3	30.5	30.6	18.3
Total	144	2,478	100.0	40.0	41.8	23.3

Source: Author's estimates, based on the survey of sample firms.

UNCTAD-DTCI, 1992-1994). For the most part, too, the data set out in the tables are self-explanatory. However, three points should be highlighted:

- Although 82 per cent of the sales of the sample firms were within the high- or medium-technology industries, the degree of technological intensity did not appear to affect significantly the extent of their transnationality. There were, indeed, as wide differences among industries within the three broad technological intensity groupings identified in table 1 (compare, for example, aerospace with chemicals) as between these groupings.

- The extent to which TNCs undertake R & D activities outside their home countries is, on average, about one half of the share of their foreign assets and employment in their respective totals, although this

Table 3. The distribution of the sample firms, by size and extent of transnationality, 1993

(Billions of dollars and percentage)

Size by value of sales	Number of firms	Sales *(Billions of dollars)*	Sales *(Percentage)*	Extent of transnationality *(Percentage)* Assets	Employment	R & D
Medium (<5 billion dollars)	56	141.6	5.7	40.6	39.7	20.1
Large (5–24.9 billion dollars)	58	706.2	28.5	37.3	41.1	26.5
Very large (25–49.9 billion dollars)	16	561.5	22.6	46.1	45.0	29.0
Mega (>50 billion dollars)	14	1,069.4	43.1	42.3	49.5	12.8
Total	144	2,478.8	100.0	40.0	41.8	23.3

Source: Author's estimates, based on the survey of sample firms.

does vary between groupings (table 1). Table 2, however, shows that the R & D transnationality ratio ranges from 6 per cent in the case of Japanese TNCs to 53 per cent in the case of Sweden-based TNCs.

- Table 3 shows that the size of a firm does not appear to be a significant influence on the degree of transnationality.[15] Somewhat surprisingly, however, the mega-firms—although the most transnational in terms of their sales and assets—were the least transnational in terms of their R & D expenditures.

3. The geographical sources of competitiveness

Tables 4 to 7 present the evaluation of the respondents of the sample firms of the extent to which they believe they currently derive a number of identified specific competitive advantages as a direct result of their FDI and/or strategic alliances with foreign firms. The respondents were asked to use a scale of 1 to 7 in their evaluations, with a rank of 1 indicating that their

[15] However, it should be remembered that all the firms in this sample are large, compared with the great majority of the 38,000 parent firms identified by UNCTAD in the *World Investment Reports 1995* and *1996*.

competitive advantages were *entirely* derived from the resources, capabilities, markets and inter-firm rivalries specific to their home countries. (Conversely, a rank of 7 would indicate that such advantages were derived entirely from the location-specific attributes of foreign countries.) A rank of 4 would suggest that, in the respondent's opinion, the origin of the competitive advantages stemmed equally from the home country and foreign locations.[16]

(a) *Sector-specific differences*

Table 4 considers four groups of competitive advantages that correspond broadly to Porter's fourfold diamond of a country's competitive advantage.[17] Three of these four groups have been subdivided further into a number of more specific advantages, identified by other scholars. In table 4 the sample firms are also classified into high, medium- and low-intensity groupings. Economic theory would suggest that firms should seek to acquire competitive assets from those locations that provide these assets at the lowest cost and with the greatest security.

Essentially, the Porter hypothesis is that, independently of their size, industry, country of origin and degree of transnationality, firms will create, or obtain, their competitive advantages (e.g. innovatory capacity) in, or from, their home countries. The contrasting hypothesis, set out by Rugman, Dunning and other scholars researching into the behaviour of TNCs, is that, as firms become more transnational and globally integrate as regards their value-added activities, they are likely to derive an increasing proportion of their competencies from outside national boundaries. And, indeed, firms may deliberately seek out foreign assets that they perceive to augment, or be complementary to, these competencies.

The data in table 4 reveal a mixed picture, although, taken as a whole, they do suggest that the sample firms derive an important part of their competitive advantages—on average between 40 per cent and 50 per cent[18]—from their presence in foreign countries, either by way of FDI or strategic

[16] The comparison in this exercise is between advantages thought to emanate from the portfolio of assets located in the home country and that in all other countries.

[17] Viz. factor conditions, demand conditions, firm strategy, structure and rivalry, and related and supporting industries (Porter, 1990).

[18] Each ranking is being expressed as a percentage of 7 (a ranking that would suggest that all firms' advantages are derived from outside the national boundaries).

Table 4. The sourcing of competitive advantages, classified by the technological intensity of the sample firms, 1993

(Mean value)

	All	High technology	Medium technology	Low technology
Group 1	3.54	3.46	3.48	3.65
Access to resources and assets	(1.39)	(1.20)	(1.03)	(1.71
(i) Natural resources	4.24	3.67	4.44	4.62
	(1.70)	(1.36)	(1.67)	(1.88)
(ii) Unskilled labour	3.98	4.10	3.79	3.99
	(1.70)	(1.55)	(1.65)	(1.89)
(iii) Skilled and professional labour	2.98	2.98	2.76	3.12
	(1.27)	(1.06)	(1.17)	(1.50)
(iv) Innovatory capacity	2.88	2.75	2.71	3.11
	(1.40)	(1.16)	(1.17)	(1.70)
(v) Organizational capacity	3.12	3.21	2.88	3.18
	(1.32)	(1.13)	(1.04)	(1.61)
(vi) Managerial expertise	3.24	3.19	3.12	3.38
	(1.42)	(1.32)	(1.37)	(1.56)
(vii) Relational skills	3.75	4.00	3.41	3.71
	(1.62)	(1.64)	(1.33)	(1.73)
Group 2	3.94	4.06	3.37	4.15
Consumer demand	(1.59)	(1.36)	(1.50)	(1.81)
(i) Upgrading of product quality	3.31	3.40	2.94	3.44
	(1.37)	(1.36)	(1.14)	(1.49)
(ii) Making for more product innovation	3.44	3.40	3.06	3.71
	(1.45)	(1.26)	(1.50)	(1.55)
Group 3	4.60	4.68	4.56	4.55
Inter-firm competition/rivalry	(1.67)	(1.61)	(1.65)	(1.75)
Group 4	4.10	4.19	3.68	4.29
Linkages with foreign or domestic firms and institutions	(1.37)	(1.32)	(1.12)	(1.52)
(i) Related firms (agglomerative economies				
(ii) Universities and other research institutions	3.29	3.27	3.21	3.38
	(1.30)	(1.25)	(1.17)	(1.43)

Source: Author's estimates, based on the survey of sample firms.

NOTE: Unless otherwise indicated, in this and all subsequent tables the figures reported are mean responses with standard deviations in parentheses.

alliances. The access provided by a foreign location (compared with a domestic location) to natural resources (including unskilled labour), linkages with suppliers, industrial competitors and other foreign producers, and the benefits of larger markets and more stringent consumer demands, were all ranked particularly high (3.44 or above). By contrast, and in line with received theory and empirical evidence on innovatory activity, the sample firms perceived that their domestic operations and/or the indigenous resources and capabilities of their home countries continued to provide the main source of competitiveness—especially in terms of technological capacity and skilled professional manpower.

Predictably, there were some differences in the perceptions of firms depending on the technological intensity of their main activities. Most noticeably, LT firms claimed to obtain a higher proportion of their created assets (group 1 [iii-vii]) from foreign sources than did MT or HT firms, particularly so in the case of innovatory capacity. *Inter alia*, this finding tends to support the proposition that at least the part of the competitive advantages of firms that is obtained from foreign sources is likely to be different from (and complementary to) the part that is obtained from domestic sources. Nevertheless, the evidence suggests strongly that, for each of the advantages identified, a multiple (or at least a dual) location of value-added activities was perceived to yield positive gains.[19] Thus, it seems reasonable to conclude that, in the opinion of the leading industrial companies, FDI and/or cross-border strategic alliances *do* provide access to significant competitive advantages.[20]

(b) *Country-specific differences*

Table 5 reclassifies the data set out in table 4 by the source region or country of the sample firms. The data reveal few significant differences in the perceptions of the sample firms. As might be reasonably predicted, firms from high-wage economies, e.g. the United States and smaller European countries,[21] ranked the foreign sourcing of unskilled labour relatively higher than firms from low-wage economies, particularly developing countries. Likewise, and consistent with the principle of comparative advantage, was the above-average reliance of Japanese firms on foreign natural resources, and the below-average reliance of other developed-country firms on such

[19] A future article will attempt to relate the performances of the sample firms to the sources of their competencies.

[20] Of course, it is recognized that these data do not, in themselves, indicate whether the overall competitive position of the respondent firms would have been better or worse, if the FDI had *not* taken place, or the strategic alliances had *not* been formed.

[21] Notably Sweden and Switzerland.

Table 5. The sourcing of competitive advantages, classified by region or country of origin of the sample firms, 1993

(Mean value)

	All	"Large" European countries	"Small" European countries	United States	Japan	Other developed countries	Developing countries
Group 1	3.54	3.80	3.90	3.61	3.21	2.63	3.21
Access to resources and assets	(1.39)	(1.25)	(1.41)	(1.43)	(1.35	(1.30)	(1.63)
(i) Natural resources	4.24	4.26	3.65	4.03	4.87	3.29	4.86
	(1.70)	(1.58)	(1.81)	(1.61)	(1.63)	(2.14)	(1.66)
(ii) Unskilled labour	3.98	3.81	4.10	4.37	3.75	4.00	3.92
	(1.70)	(1.63)	(1.77)	(1.50)	(1.80)	(1.91)	(2.10)
(iii) Skilled and professional labour	2.98	3.18	3.30	3.27	2.46	3.00	2.21
	(1.27)	(1.22)	(1.45)	(1.15)	(1.26)	(1.26)	(1.01)
(iv) Innovatory capacity	2.88	2.95	3.38	2.94	2.23	3.29	2.79
	(1.40)	(1.31)	(1.56)	(1.39)	(1.34)	(1.25)	(1.37)
(v) Organizational capacity	3.12	3.30	3.43	3.35	2.85	2.43	2.43
	(1.32)	(1.37)	(1.12)	(1.05)	(1.54)	(1.27)	(1.28)
(vi) Managerial expertise	3.24	3.30	3.86	3.29	3.12	3.14	2.36
	(1.42)	(1.49)	(1.39)	(1.10)	(1.70)	(0.90)	(1.22)
(vii) relational skills	3.75	3.98	4.43	3.90	3.46	3.29	2.43
	(1.62)	(1.56)	(1.69)	(1.35)	(1.70)	(1.70)	(1.28)
Group 2	3.94	4.39	4.22	3.78	2.96	5.20	3.83
Consumer demand	(1.59)	(1.64)	(1.73)	(1.53)	(1.19)	(1.10)	(1.40)
(i) upgrading of product quality	3.31	3.65	3.67	3.20	2.36	4.29	3.14
	(1.37)	(1.46)	(1.53)	(1.16)	(1.04)	(0.95)	(1.17)
(ii) Making for more product innovation	3.44	3.58	3.90	3.33	2.76	4.14	3.43
	(1.45)	(1.48)	(1.30)	(1.37)	(1.48)	(1.07)	(1.50)
Group 3	4.60	5.43	5.38	4.23	3.23	4.29	4.36
Inter-firm competition/ rivalry	(1.67)	(1.34)	(1.72)	(1.23)	(1.45)	(0.49)	(2.17)
Group 4	4.10	4.41	4.95	3.84	3.65	3.86	3.43
Linkages with foreign or Domestic firms and Institutions	(1.37)	(1.39)	(1.36)	(1.07)	(1.29)	(1.46)	(1.40)
(i) Related firms (agglomerative economies)							
(ii) Universities and other research institutions	3.29	3.40	3.52	3.32	3.17	2.83	3.00
	(1.30)	(1.14)	(1.29)	(1.40)	(1.46)	(1.47)	(1.30)

Source: Author's estimates, based on the survey of sample firms.

resources (in the survey, these firms were all based in resource-rich countries).[22] Relative to Japanese firms, too, European and United States firms appeared to value the access to foreign-based created assets, e.g.

[22] Viz. Australia, Canada and New Zealand.

organizational capacity, management, expertise and relational skills, more highly than firms in other developed countries, although, in general, the competitive advantages of Japanese firms were perceived to be more home-specific than those of their major competitors. (This also reflects their lower degree of transnationality (see sect. 3*c*).)

In respect of each of the other three components of Porter's diamond, European-owned firms appeared to rely more on foreign (than on domestic), sources of competitiveness compared with their United States and Japanese counterparts. This was especially the case with respect to inter-firm competition and, for firms from smaller European countries, with respect to linkages with related firms.

Notwithstanding the below-average significance attached by the sample firms to the acquisition of, or access to, foreign-created assets, the rankings—even for TNCs from large and advanced economies, such as Germany and the United States, which record a comparative patenting advantage in high-technology products (Cantwell and Hodson, 1991)—suggest that large TNCs in the high and medium-technology industries are increasingly seeking a technology-related presence[23] in each of the main Triad countries or regions (e.g. United States firms in the European Union (EU) and Japan; Japanese firms in EU and the United States, and EU firms in Japan and the United States).

(c) *Degree of transnationality*

Perhaps the most popular hypothesis about the propensity of TNCs to derive competitive advantages from their foreign operations is that the competitive advantages will be positively related with the extent and depth of a firm's foreign operations, relative to domestic operations. In this section, two exercises are attempted. The first exercise is to group the data presented earlier on the basis of the four bands of transnationality set out in the introduction. The second exercise is to relate the extent of the two measures of transnationality identified earlier—viz. the average of the percentage of global assets and employment outside the home country and the average

[23] By this is meant either the pursuance of R & D-related activities by the foreign affiliates and/or strategic alliances with foreign firms, or the monitoring of the innovatory activities of foreign firms.

Table 6. The sourcing of competitive advantages, classified by extent of transnationality[a] of the sample firms, 1993

(Mean value)

	All	Low	Medium Low	Medium High	High
Group 1	3.54	2.33	2.83	3.76	4.50
Access to rersources and assets	(1.39)	(1.28)	(1.13)	(1.00)	(1.31)
(i) Natural resources	4.24	3.82	3.90	4.21	4.62
	(1.70)	(2.01)	(1.84)	(1.34)	(1.84)
(ii) Unskilled labour	3.98	2.26	3.57	4.38	4.52
	(1.70)	(1.58)	(1.80)	(1.44)	(1.60)
(iii) Skilled and professional	2.98	2.06	2.48	3.33	3.48
labour	(1.27)	(1.18)	(0.90)	(1.19)	(1.20)
(iv) Innovatory capacity	2.88	2.82	2.67	2.80	3.43
	(1.40)	(1.81)	(1.20)	(1.32)	(1.22)
(v) Organizational capacity	3.12	1.94	2.88	3.37	3.57
	(1.32)	(1.25)	(1.26)	(1.09)	(1.19)
(vi) Managerial expertise	3.24	2.35	2.71	3.55	3.73
	(1.42)	(1.27)	(1.30)	(1.29)	(1.28)
(vii) Relational skills	3.75	2.35	3.21	4.13	4.43
	(1.62)	(1.84)	(1.38)	(1.29)	(1.48)
Group 2	3.94	3.27	3.32	3.98	4.96
Consumer demand	(1.59)	(1.98)	(1.39)	(1.43)	(1.49)
(i) Upgrading of product	3.31	3.00	2.79	3.34	4.18
quality	(1.37)	(1.80)	(1.28)	(1.11)	(1.33)
(ii) Making for more	3.44	2.94	3.17	3.49	4.04
product innovation	(1.45)	(1.48)	(1.46)	(1.36)	(1.32)
Group 3	4.60	4.29	4.08	4.68	5.37
Inter-firm competition/rivalry	(1.67)	(1.79)	(1.67)	(1.56)	(1.43)
Group 4	4.10	3.06	3.50	4.27	4.87
Linkages with foreign or	(1.37)	(1.48)	(1.25)	(1.25)	(1.14)
domestic firms and institutions					
(i) Related firms					
agglomerative economies)					
(ii) Universities and other	3.29	2.12	2.87	3.54	3.90
research institutions	(1.30)	(0.93)	(1.55)	(1.15)	(1.11)

Source: Author's estimates, based on the survey of sample firms.

[a] For a more precise definition of the extent of transnationality, see the introduction.

percentage of global R & D expenditures undertaken outside the home country—to the values placed on Porter's four facets of competitive advantages (groups 1 to 4), in a series of bi-variate relationships.

Table 7. Bi-variate correlation coefficients between perceived competitive advantages derived from foreign activities and the extent of transnationality of the sample firms, 1993

(*Correlation coefficients*)

	A	B
Natural assets (group 1 [i–ii])	0.31982	0.18998
Created assets (technological) (group 1 [iii–v])	0.36478	0.37895
Created assets (managerial) (group 1 [vi–viii])	0.43228	0.39619
Consumer demand (group 2)	0.36339	0.35007
Inter-firm rivalry (group 3)	0.27146	0.28016
Linkages with related firms (group 4 [i])	0.46452	0.39511

Source: Author's estimates, based on the survey of sample firms.

NOTE: A = Extent of transnationality = average foreign employment and assets.
B = Extent of transnationality = average foreign R & D.

Table 6 shows quite clearly that a greater degree of transnationality of firms is likely to be associated with the perception that an increasing proportion of global competitive advantages is derived from foreign sources. This is the most striking finding presented here. It would also appear that, for created assets (group 1 [iii-vii]), the biggest rise in the significance of foreign operations for global competitiveness occurs when the average degree of transnationality (in terms of assets and employment) is 30 per cent or above. However, for innovatory capacity, it is only when the degree of transnationality exceeds 60 per cent that the (average) ranking exceeds 3.00; while, as far as the influence of consumer demand and inter-firm rivalry is concerned, it exceeds 5.00. The distinction between the sources of competitive advantages perceived to be derived from the foreign activities of firms, or from cross-border strategic alliances is particularly noticeable in the case of firms with low and high degrees of transnationality.

Table 7 shows some bi-variate correlation coefficients between the two measures of transnationality (mentioned above) and six indices of competitive advantage. It suggests that, for the main measure of transnationality (A in table 7), the coefficients are positive and significant at the 99 per cent level or above, and that the most likely benefits of increased transnationality are likely to arise from access to foreign organizational capacity and managerial expertise, or linkages forged with foreign firms.

The results for the second measure of transnationality, the foreign R & D ratio (B in table 7), are virtually identical to the results obtained from the first measure. Except for the correlation with natural assets (group 1 [i-ii]), all of the coefficients are positive and significant at the 99 per cent level or above. Not surprisingly, the two measures of transnationality are also significantly positively correlated with each other (0.69, $p<0.0001$), confirming that they essentially measure aspects of the same phenomenon.[24]

(d) *Size of firm*

The final contextual variable which, to some extent, is correlated with the degree of transnationality,[25] is the size of the firm. While it is not obvious why size itself should be related to the geographical sourcing of competitiveness-enhancing assets, the literature (Dunning, 1993, chap. 6) suggests that large firms are more likely to engage in FDI than small firms, and that transnationality itself may help a firm to preserve (or increase) its share of global markets.

At the same time, it might be hypothesized that medium-size firms are likely to be more specialized in their portfolio of global assets and, hence, more reliant on foreign sources to enhance or complement that portfolio. However, in recent years, the evidence suggests that very large firms are just as likely to engage in merger-and-acquisition (M & As) activities as are their smaller competitors. Finally, it might be predicted that the portfolio of foreign competitiveness-enhancing assets sought by medium-size firms,[26] relative to larger firms, may have more to do with gaining access to specialized resources, capabilities and markets, and/or establishing linkages with local foreign firms. (Medium-size firms are typically expected to engage in less competition with foreign-owned firms than their larger counterparts.)

The results of the table 8 survey show that the size of firms is only of marginal importance in affecting the sourcing of most categories of competitive advantage.

---

[24] Since the group 1-4 variables are not continuous, the significance levels reported here should be taken as indicators of a possible relationship (to be investigated further in a future article), rather than a form of final analysis.

[25] Although the Pearson correlation coefficient between size and transnationality is not significantly different from zero.

[26] It should be remembered that all the firms in the sample are large, relative to the universe of firms, but that some are much larger than others.

Table 8. The sourcing of competitive advantages, classified by the size of the sample firms, 1993

(Mean value)

	All	Medium (<5 Billion dollars)	Large (5–24.9 Billion dollars)	Very Large (25–49.9 Billion dollars)	Mega (>50 Billion dollars)
Group 1	3.54	3.58	3.53	3.57	3.43
Access to resources and assets	(1.39)	(1.54)	(1.44)	(1.09)	(0.94)
(i) Natural resources	4.24	3.96	4.41	4.20	4.69
	(1.70)	(1.79)	(1.76)	(1.61)	(1.03)
(ii) Unskilled labour	3.98	4.03	3.89	4.25	3.86
	(1.70)	(1.68)	(1.92)	(1.39)	(1.17)
(iii) Skilled and professional	2.98	2.96	3.06	2.88	2.86
labour	(1.27)	(1.28)	(1.41)	(1.12)	(0.77)
(iv) Innovatory capacity	2.88	2.71	2.98	3.25	2.71
	(1.40)	(1.41)	(1.49)	(1.13)	(1.20)
(v) Organizational capacity	3.12	3.13	3.22	3.00	2.79
	(1.32)	(1.39)	(1.39)	(1.15)	(0.89)
(vi) Managerial expertise	3.24	3.45	3.09	3.19	3.14
	(1.42)	(1.56)	(1.38)	(1.42)	(0.95)
(vii) Relational skills	3.75	3.82	3.78	3.50	3.64
	(1.65)	(1.66)	(1.72)	(1.59)	(1.08)
Group 2	3.94	4.07	3.90	3.93	3.67
Consumer demand	(1.59)	(1.68)	(1.63)	(1.58)	(1.15)
(i) Upgrading of product	3.31	3.30	3.40	3.13	3.14
quality	(1.37)	(1.38)	(1.46)	(1.25)	(1.17)
(ii) Making for more	3.44	3.57	3.33	3.53	3.29
product innovation	(1.45)	(1.46)	(1.48)	(1.46)	(1.33)
Group 3	4.60	4.58	4.72	4.50	4.29
Inter-firm competition/rivalry	(1.67)	(1.85)	(1.60)	(1.26)	(1.68)
Group 4	4.10	4.24	4.07	4.25	3.57
Linkages with foreign or	(1.37)	(1.36)	(1.45)	(1.29)	(1.16)
domestic firms and institutions					
(i) Related firms					
(agglomerative economies)					
(ii) Universities and other	3.29	3.28	3.30	3.20	3.43
research institutions	(1.30)	(1.29)	(1.41)	(1.08)	(1.16)

Source: Author's estimates, based on the survey of sample firms.

4. Which foreign countries provide the most valuable assets to upgrading competitiveness?

The next section gives some details of the foreign countries most frequently identified by the sample firms as sources of competitiveness-enhancing assets. Here, it is not possible to evaluate the relative importance of one source country *vis-à-vis* another country; rather, the kinds of competitive advantages that are likely to be most closely associated with each country are identified.

What do the received theories of trade and investment predict in this respect? The answer is that much will depend on the type of FDI undertaken (i.e. market seeking, resource seeking, efficiency seeking, or strategic-asset seeking) and on the static and dynamic competitive advantages of the home and host countries. Also, like trade, FDI and strategic alliances might be either between or within industries. In the case of the former, one might expect the FDI recipient countries to provide assets in which the home country is deficient. In the case of the latter, FDI is likely to be more of the efficiency-enhancing and strategic-asset seeking variety, in which case the foreign country will tend to provide assets similar to those in which the investing firm (or country) has a competitive edge.

In carrying out the survey, each respondent was asked to name up to three foreign countries which were hosts to the firm's foreign affiliates and which, in their perception, have had the most positive impact on their own competitiveness. Table 9 presents some evidence derived by relating each of the competitive advantages identified to the number of times a particular host country was mentioned.

The data in table 9 confirm most of the predictions made here. Thus, developing countries clearly have a competitive advantage in affording access to natural resources, while Germany and Japan provided more than the average share of technological assets, and the United States more than the average share of managerial assets and consumer-demand conditions favouring competitiveness (e.g. large markets). United States firms were also perceived to offer a more competitive stimulus than other firms. Relative to other countries, Germany, Japan and the United States were perceived as good locations for establishing linkages with other firms.

Table 9. The sourcing of competitive advantages, classified by region or country, 1993

(Number of times countries are mentioned among three most important sources of advantages)

	Europe			North America		Rest of the World			
	France	Germany	United Kingdom	Other	United States	Canada	Japan	Developed countries	Developing countries	Sum
Group 1 **Access to resources** Natural assets	0.50	2.00	5.00	4.00	6.00	2.00	0.50	2.50	30.00	52.50
Created assets (technological)	1.33	12.33	6.67	9.00	18.00	1.33	7.33	1.67	7.67	65.33
Created assets (managerial)	1.00	4.33	6.00	3.67	10.00	2.67	3.33	1.33	6.33	38.66
Group 2 **Consumer demand**	0.00	4.00	4.00	0.00	8.00	0.00	0.00	0.00	3.00	19.00
Group 3 **Inter-firm rivalry**	2.00	4.00	4.00	5.00	16.00	1.00	2.00	1.00	4.00	39.00
Group 4 **Linkages with related firms**	6.00	8.00	7.00	4.00	17.00	1.00	9.00	0.00	4.00	56.00
Totals	10.83	34.66	32.67	25.67	75.00	8.00	22.16	6.50	55.00	270.40
Number of firms	(139)	(131)	(125)	(116)	(113)	(138)	(118)	(142)	(130)	9

Source: Author's estimate, based on the survey of sample firms.
NOTE: The number of firms reflects the sample firms excluding firms from the host country.

One surprising finding of the survey was the frequency with which developing countries were identified as providing access to created assets and to competitive advantages (groups 3 and 4). Although mostly this result is the outcome of pooling a number of countries at different stages of development into the same group, China and Brazil were both mentioned a number of times as providing a locale for more efficient production via economies of scale.[27] As might be expected, the host countries named varied on the basis of both the industry of the investing TNCs and their home countries. For example, firms in HT sectors accounted for 45-57 per cent of all mentions in the high-technology countries, i.e. the United States, Germany, Japan and the United Kingdom. Taking European, United States and Japanese firms as a group, 25 per cent of the countries mentioned by the respondents as sources of competitive advantages were located elsewhere in the Triad (or, alternatively, nearly three quarters of their advantages were derived from within the Triad).

5. Modes of foreign entry most likely to advance competitive advantages

The respondents of the sample firms were asked to rank (on a scale of 1 to 7) the importance of each of three modes of acquiring and/or tapping into the resources and capabilities of foreign countries, i.e. the "foreign" diamonds of competitive advantage. The modes are FDI, non-equity cooperative agreements (e.g. strategic alliances, management contracts, licensing and franchising agreements etc.) and arm's length transactions (in both intermediate and final goods and services).

The hypothesis here is that deeper forms of cross-border structural integration, such as FDI and non-equity cooperative agreements, are more likely to result in an addition of competitive advantages to the home company (and country) than shallower forms of transactions, such as arm's length trade. It might also be predicted that firms are more likely to internalize their assets in industries that are technologically intensive than in those that are not. A related hypothesis is that non-equity cooperative ventures are likely to be ranked higher as a modality for competitive advantages in less

[27] The figures in table 9 somewhat understate the importance of scale economies, as the figure for group 2 indicates the overall importance of consumer demand conditions and not scale economies specifically.

Table 10. Perceptions of the importance of the mode of foreign involvement of the sample firms, by technological intensity, extent of transnationality and size, 1993

(Mean value)

	Foreign direct investment	Non-equity arrangements	Arm's length transactions
All firms	5.22 (1.63)	4.66 (1.46)	3.96 (1.45)
High technology	5.23 (1.63)	5.31 (1.08)	4.27 (1.59)
Medium technology	5.53 (1.42)	4.62 (1.41)	3.55 (1.28)
Low technology	5.02 (1.74)	4.07 (1.57)	3.92 (1.36)
Low transnationality	4.06 (2.19)	4.71 (1.69)	3.59 (1.62)
Medium-low transnationality	5.46 (1.67)	5.25 (1.07)	4.25 (1.26)
Medium-high transnationality	5.44 (1.38)	4.83 (1.28)	4.12 (1.53)
High transnationality	5.28 (1.67)	3.93 (1.65)	3.55 (1.210
Medium size	4.83 (1.84)	4.48 (1.50)	4.00 (1.41)
Large	5.19 (1.61)	4.69 (1.31)	3.95 (1.46)
Very large	6.07 (0.80)	5.07 (1.79)	4.13 (1.46)
Mega	5.93 (0.92)	4.79 (1.58)	3.71 (1.64)

Source: Author's estimates, based on the survey of sample firms.

technological-intensive industries than other modes of foreign entry. Because of the need to integrate their global operations, it might be expected that these firms that are the most transnationalized may be expected to rank FDI higher than firms whose foreign operations are less significant to their overall prosperity. Finally, it might be supposed that firms from countries whose domestic institutions are organized along hierarchical lines (e.g. the United States) would be more likely to rank FDI as a foreign asset-acquiring modality than firms from countries whose institutions practice more cooperative modes of governance (e.g. Japan).

Tables 10 and 11 offer evidence to assess these hypotheses. Overall, it is clearly true that deeper forms of integration were perceived to offer the most benefits as regards the acquisition of competitive advantages abroad. In table 10, the average rankings of each of the three modalities were classified according to the technological intensity of the firm. While the LT and MT firms behaved as expected, in other words, FDI was perceived to offer the most advantages, and arm's length transactions were perceived to offer the least, HT firms claimed that they derived more advantages from cooperative alliances than from FDI.

Table 11. Perceptions of the importance of the mode of foreign involvement of the sample firms, by region or country of origin, 1993

(Mean value)

	Foreign direct investment	Non-equity arrangements	Arm's length transactions
All firms	5.22 (1.63)	4.66 (1.46)	3.96 (1.45)
"Large" European countries	5.47 (1.420	4.60 (1.45)	3.81 (1.24)
"Small" European countries	4.70 (1.95)	4.81 (1.63)	4.32 (1.34)
United States	4.97 (1.80)	4.58 (1.26)	3.29 (1.64)
Japan	5.96 (1.17)	5.36 (1.41)	4.88 (1.26)
Developing countries	4.79 (1.37)	4.29 (1.14)	4.21 (1.25)
Other countries	4.57 (2.15)	3.14 (1.57)	3.29 (1.38)

Source: Author's estimates, based on the survey of sample firms.

Also, somewhat contrary to what was expected, the medium-low and medium-high transnationalized firms actually derived more advantages from all three modalities of foreign involvement than did firms in the high-transnationalization group. Whether this is another manifestation of the law of diminishing returns, or of the differences between the first mover versus late mover experiences, remains to be seen. More in line with what was expected, the connection between internationalization and size of firm was confirmed. The figures in table 10 appear to indicate that the gains from FDI accrue predominantly to the largest firms, whereas the benefits from cooperative alliances and trade do not exhibit any obvious pattern.

Finally, table 11 sheds some light on the advantages derived from the different modalities classified by the nationality of the firm. The results here contradict the hypothesis that firms from countries with typically hierarchical organizational structures would prefer FDI over other modes of foreign entry. Japanese firms stand out not only in terms of their perceived advantages being derived from FDI, but also in terms of their perceived advantages being derived from cooperative ventures and arm's length trade. In summary, although FDI is most likely to be the preferred route by which the domestic and foreign diamonds of competitive advantages are linked, the relative significance of this route is the greatest in the case of (large) firms that have medium-to-high technology and are moderately-to-highly transnational.

Table 12. Perceptions of firms as to whether foreign sourcing of competitive advantages has become relatively more important in recent years, 1993

(Mean value)

All	1.50 (1.36)	All	1.50 (1.36)
High technology	1.69 (1.35)	Medium size	1.47 (1.32)
Medium technology	1.79 (0.95)	Large	1.31 (1.55)
Low technology	1.16 (1.51)	Very Large	1.87 (0.83)
"Large" European countries	1.64 (1.27)	Mega	2.00 (0.96)
"Small" European countries	1.90 (1.04)	Low transnationality	0.94 (1.55)
United States	1.26 (1.53)	Medium-low transnationality	1.29 (1.37)
Japan	1.62 (1.39)	Medium-high transnationality	1.62 (1.44)
Developing countries	1.43 (0.85)	High transnationality	1.96 (0.92)
Other countries	0.38 (2.00)		

Source: Author's estimates, based on the survey of sample firms.

6. The dynamics of the geographical sourcing of competitive advantage

To what extent do firms perceive that their access to foreign diamonds of competitive advantage is becoming a more important contribution to their overall competitive advantages? It might be hypothesized that, as firms increase their degrees of transnationality, they will augment increasingly—rather than exploit—their competitive advantages from their foreign operations.[28] In the course of the survey, firms were asked to rank, on a scale of –3 to +3, whether they perceived that their competitive advantages were becoming less or more dependent on their foreign operations (indicated by negative or positive signs).

The results presented in table 12 essentially confirm the impressions already gained from tables 10 and 11 concerning the benefits derived from internationalization. Although all the sample firms, overall, perceived that the foreign sourcing of competitive advantages has become more important

[28] For an interesting article that supports this contention in relation to the R & D activity of TNCs in the electronics and pharmaceuticals industries, see Kuemmerle (1996).

over time, medium- and high-technology firms had a more positive view of that importance, as did the very large and mega firms. When classified by their regions of origin, firms from small European countries were the most positive in their responses, a finding that may not only reflect their size, but also the aftermath of the wave of investment created by the perceived threat of a "fortress" Europe by 1992 for the countries outside the European Union. Not surprisingly, the firms that were the most transnationalized also perceived the influence of foreign sourcing of competitive advantages to be very important.

7. Home-country governments and competitiveness

The role of governments in affecting the competitiveness of firms is a subject which has long been of fascination to economists and other scholars (Dunning, 1997, forthcoming). In the present context, we are interested in the perceived influences of home governments on the ability and willingness of the sample firms to be competitive globally. In the questionnaire, eighteen possible ways in which governments might exert such an influence were identified. For each of these ways, firms were asked to assign a figure from a range of –3 to +3, according to whether or not, in their opinion, home governments had influenced their *global* competitiveness negatively or positively between 1988 and 1993 (0 would indicate no influence at all). Table 13 presents the findings classified by the three groups of industries used here. Table 14 does the same for firms classified on the basis of nationality and table 15 does the same for firms classified on the basis of the degree of transnationality.

Table 13 shows that, on average, the sample firms thought that actions by their home governments either had a marginally beneficial effect on their global competitiveness (e.g. the provision of infrastructure, trade, or industrial and technology policies), or a marginally adverse effect (e.g. in terms of market facilitating, social and environmental policies). There were few consistent differences between industry groupings, except that LT firms thought more favourably of most government actions (notably, trade, industrial, technology, monetary and exchange rate policies) than either HT or MT firms.

More interesting differences are revealed in table 14. While United States and European firms perceived that their home governments pursued

Table 13. Perceived influence of home governments on competitive advantages of the sample firms, by technological intensity, 1993

(*Mean value*)

		All	High technology	Medium technology	Low technology
1	Provision and upgrading of infrastructure	0.52 (1.00)	0.48 (1.08)	0.60 (1.00)	0.51 (0.93)
2	Social policies	–0.13 (1.44)	–0.10 (1.47)	–0.31 (1.42)	–0.05 (1.43)
3	Monetary and exchange-rate policies	–0.11 (2.16)	–0.04 (1.94)	–0.73 (2.00)	0.20 (2.40)
4	Trade policies	0.35 (1.45)	0.26 (1.40)	–0.12 (1.52)	0.70 (1.39)
5	Industrial and technology policies	0.36 (2.08)	0.35 (2.11)	0.06 (2.46)	0.54 (1.80)
6	Education and training policies	0.09 (1.25)	0.15 (1.43)	0.03 (1.29)	0.07 (1.04)
7	Environmental policies	–0.06 (1.24)	–0.34 (1.22)	0.09 (1.33)	0.13 (1.16)
8	Market-facilitating policies	–0.37 (1.99)	–0.38 (2.05)	–0.55 (2.15)	–0.25 (1.85)
9	Promoting an ethos of competitiveness	0.28 (2.91)	0.53 (3.57)	0.13 (2.72)	0.13 (2.30)
10	Promoting a culture of investment and saving	0.18 (1.91)	0.51 (2.12)	–0.09 (2.07)	0.04 (1.56)

Source: Author's estimates, based on the survey of sample firms.

macroeconomic, i.e. monetary and exchange rate and trade policies marginally favourable to their competitiveness, Japanese firms considered the same policies of their home government to work to their disadvantage. By contrast, Japanese firms believed that most macro-organizational policies of their Government (e.g. environmental, industrial and technology policies), together with the promotion of a culture of investment and saving, aided their global competitiveness, while United States respondents generally thought their Government worked to the detriment of their competitiveness in terms of these policies. Respondents of firms from countries with small and/or relatively unsophisticated domestic markets were generally more favourably disposed to the actions of their home governments than those of

Table 14. Perceived influence of home governments on competitive advantages of the sample firms, by region or country of origin, 1993

(Mean value)

	All countries	"Large" European countries	"Small" European countries	United States	Japan	Other developed countries	Devl-oping countries
1 Provision and upgrading of infrastructure	0.52 (1.00)	0.68 (1.03)	0.81 (0.96)	–0.08 (0.85)	0.44 (0.96)	0.67 (0.67)	1.00 (0.99)
2 Social policies	–0.13 (1.44)	–0.45 (1.50)	0.38 (1.50)	–0.74 (1.37)	0.20 (1.00)	–0.13 (1.55)	0.92 (1.12)
3 Monetary and exchange-rate policies	–0.11 (2.16)	–0.21 (2.33)	0.52 (2.46)	0.19 (1.49)	–1.62 (1.77)	–0.13 (1.81)	–0.08 (2.43)
4 Trade policies	0.35 (1.45)	0.30 (1.15)	0.29 (1.95)	0.19 (1.47)	–0.19 (1.47)	1.13 (0.64)	1.54 (0.88)
5 Industrial and technology policies	0.36 (2.08)	–0.12 (1.87)	1.19 (2.40)	–0.53 (1.96)	0.72 (2.03)	0.13 (1.46)	2.08 (1.38)
6 Education and training policies	0.09 (1.25)	0.07 (1.32)	0.81 (1.47)	–0.58 (0.81)	0.12 (0.95)	0.00 (0.93)	0.62 (1.50)
7 Environmental policies	–0.06 (1.24)	–0.14 (1.32)	0.00 (1.00)	–0.55 (1.36)	0.50 (0.86)	–0.25 (1.39)	0.31 (1.18)
8 Market-facilitating policies	–0.37 (1.99)	–0.37 (2.02)	–0.14 (2.31)	–1.13 (1.87)	–0.54 (1.79)	–0.25 (1.39)	1.31 (1.38)
9 Promoting an ethos of competitiveness	0.28 (2.91)	0.21 (3.00)	0.57 (3.65)	0.10 (3.28)	0.19 (1.30)	–0.50 (2.20)	1.17 (3.41)
10 Promoting a culture of investment and saving	0.18 (1.91)	0.29 (2.00)	0.38 (1.72)	–0.77 (1.81)	0.84 (1.70)	–0.50 (1.07)	0.85 (2.19)

Source: Author's estimates, based on the survey of sample firms.

firms from other countries. Rather surprisingly, while firms from all the major investing countries believed that their governments did little to facilitate the efficient workings of markets, firms from developing countries were noticeably more appreciative in this respect.

Table 15 suggests that the degree of transnationality of firms seems to be unrelated to their perception of the competitiveness-enhancing actions by their home-country governments. Thus, while the least transnationalized firms regarded industrial and technology policies and the promotion of an investment and savings culture more favourably than did the most transna-

Table 15. Perceived influence of home governments on competitive advantages of the sample firms, by extent of transnationality, 1993

(Mean value)

	All firms	Low firms	Medium-low firms	Medium-high firms	High firms
1 Provision and upgrading of infrastructure	0.52 (1.00)	0.67 (1.12)	0.67 (0.87)	0.37 (1.03)	0.61 (0.87)
2 Social policies	–0.13 (1.44)	–0.39 (1.20)	0.10 (1.64)	–0.27 (1.39	0.03 (1.59)
3 Monetary and exchange-rate policies	–0.11 (2.16)	–0.39 (2.35)	–0.68 (1.89)	–0.03 (2.09)	0.83 (2.19)
4 Trade policies	0.35 (1.45)	0.50 (1.62)	0.32 (1.59)	0.27 (1.41)	0.60 (1.38)
5 Industrial and technology policies	0.36 (2.08)	1.33 (1.97)	0.59 (2.17)	0.17 (1.96)	–0.13 (2.13)
6 Education and training policies	0.09 (1.25)	0.28 (1.49)	–0.09 (1.06)	0.17 (1.15)	–0.10 (1.45)
7 Environmental policies	–0.06 (1.24)	–0.06 (1.43)	0.00 (0.98)	–0.08 (1.39)	–0.07 (1.08)
8 Market-facilitating policies	–0.37 (1.99)	–0.06 (1.66)	–0.41 (1.44)	–0.53 (2.18)	–0.40 (2.03)
9 Promoting an ethos of competitiveness	0.28 (2.91)	0.24 (2.25)	0.32 (2.93)	0.33 (2.79)	0.13 (3.73)
10 Promoting a culture of investment and saving	0.18 (1.91)	0.72 (1.84)	0.23 (2.43)	–0.02 (1.59)	0.03 (2.18)

Source: Author's estimates, based on the survey of sample firms.

tionalized firms, the reverse appeared to be the case for macroeconomic and social policies. Here, however, bi-variant comparisons tend to break down. It so happens that Japanese firms are the least transnationalized, and it could be (and most likely is) that the country of the respondent firms is a more important determinant of the role of government than the degree of their transnationality. It is precisely for this and related reasons why a multi-variant approach is necessary to complement the descriptive interpretation of the data set out here.

8. Summary and some policy conclusions

The findings of the survey show that a not insignificant part of the competitive advantages of some of the world's leading industrial TNCs are derived from their foreign-based activities. It has also been revealed here that a firm's competitive advantages are dependent on the types of FDI and/or cross-border alliances in which the firm engages; the technology intensity of the industries in which the firm is involved; the country from which the firm emanates; and, particularly, the degree of the firm's transnationality. The findings of the survey also reveal that an overwhelming majority of firms from all industries and countries believed that the importance of the foreign sourcing of their competitive advantages had increased in recent years, and that FDI (followed by inter-firm cooperative agreements) is the favoured modality for acquiring these advantages.

The role of home-country governments in influencing the global competitive advantages of the sample firms is also discussed briefly. In general, that role was not perceived to be of critical significance. However, there were noticeable differences observed among firms of different nationalities, and between large developed countries (particularly the United States and Asian developing countries, especially Japan, and smaller European countries, such as Sweden) that *inter alia* reflect some very distinctive country-specific philosophies of the role of governments, hierarchies and markets.

The policy implications of the findings are straightforward. Governments need to recognize that firms engage in foreign activities both to exploit their existing competitive advantages and to protect, or augment, these advantages. There is also a strong suggestion that, as the foreign operations of firms become a more significant component of the world economy, technology, organizational skills and other tangible assets are likely to be transferred across national boundaries, not only from the investing firms to their affiliates abroad, but also from the foreign affiliates back to the investing firms. This is particularly likely to be the case for TNCs (e.g. Asea Brown Boveri, Unilever, IBM) that operate from multiple home bases. This being so, it behoves governments to pursue as liberal policies towards both inward and outward FDI as their macro-organizational strategies would allow; at the same time, governments can pursue policies that help create the kinds of domestic economic environments that enable indigenous firms to become strong contestants in the global marketplace, and the country to attract high value-added FDI into its domestic arena. ■

Appendix: Characteristics of the sample

The sample consists of a total of 144 responses from 131 firms. For those firms that provided multiple responses, the statistics on the size and degree of transnationality, as well as industry classification, follow those of the largest corporate unit. Apart from the size and degree of transnationality, all other multiple responses are treated as unique individual responses in the analyses. Of the 144 responses to the survey, 110 came from firms that are ranked in the *Fortune* 500 (based on 1993 sales). The remaining firms were contacted to improve the industrial and/or geographical representation of the data. These firms are all among the largest TNCs in their respective home countries. The identity of one firm in the sample was concealed and therefore that firm could not be classified by industry. However, values for the other variables used in the various classifications were obtained.

References

Almeida, P. (1996). ''Knowledge sourcing by foreign multinationals: pattern citation in the US semi-conductor industry'', Washington, D.C.: Georgetown University (mimeo).

Cantwell, J. and C. Hodson (1991). ''Global R & D and UK competitiveness'', in M. Casson, ed., *Global Research Strategy and International Competitiveness* (Oxford: Basil Blackwell), pp. 133-182.

Dunning, J. H. (1993a). *Multinational Enterprises and the Global Economy* (Wokingham, Berkshire: Addison Wesley).

________ (1993b). ''Internationalizing Porter's diamond'', *Management International Review*, 33, 2 (Special Edition), pp. 7-15.

________ (1995). ''Reappraising the eclectic paradigm in the age of alliance capitalism'', *Journal of International Business Studies*, 26, 3, pp. 461-491.

________ , ed. (1997). *Governments, Globalization and International Business* (Oxford: Oxford University Press) (forthcoming).

Kuemmerle, W. (1995). ''Home base and investment into research and development abroad'', Boston, Mass., Harvard Business School Working Paper No. 2685.

________ (1996). ''The drivers of foreign direct investment into research and development: An empirical investigation'', Boston, Mass., Harvard Business School Working Paper No. 96:062.

Mataloni, R. (1995). ''U.S. multinational companies: Operations in 1993'', *Survey of Current Business*, 6 (June), pp. 31-51.

Pearce, R. D. and S. Singh (1991). ''The overseas laboratory'', in M. Casson, ed., *Global Research Strategy and International Competitiveness* (Oxford: Basil Blackwell), pp. 183-212.

———— (1992). *Globalizing Research and Development* (London: Macmillan).

Peteraf, M. (1993). "The cornerstones of competitive advantage: a resource based view", *Strategic Management Journal*, 14, pp. 179-191.

Porter, M. E. (1990). *The Competitive Advantage of Nations* (New York: The Free Press).

———— (1997). "Globalization, governments and international business: A business analytic approach", in John H. Dunning, ed., *Governments, Globalization and International Business* (Oxford: Oxford University Press) (forthcoming).

Rugman, A. M., ed. (1993). *Management International Review*, 33, 2 (spring) (Special Edition on Michael Porter's Diamond of Competitive Advantage).

————, J. Van den Broeck and A. Verbeke, eds. (1995). *Beyond the Diamond: Research in Global Management*, vol. 5 (Greenwich, Conn.: JAI Press).

UNCTAD, Division on Transnational Corporations and Investment (UNCTAD-DTCI) (1992-1994). *World Investment Directory* (4 vols.). Sales No. E.94.II.A.10; E.93.II.A.9; E.93.II.A.1 and E.92.II.A.11.

———— (1993). *World Investment Report 1993: Transnational Corporations and Integrated International Production*. Sales No. E.93.II.A.14.

———— (1994). *World Investment Report 1994: Transnational Corporations, Employment and the Workplace*. Sales No. E.94.II.A.14.

———— (1995). *World Investment Report 1995: Transnational Corporations and Competitiveness*. Sales No. E.95.II.A.9.

———— (1996). *World Investment Report 1996: Transnational Corporations: Investment Trade and International Policy Arrangements*. Sales No. E.96.II.A.14.

United Nations Centre on Transnational Corporations (UNCTC) (1991). *World Investment Report 1991: The Triad in Foreign Direct Investment*. Sales No. E.91.II.A.12.

———— (1992). *World Investment Report 1992: Transnational Corporations as Engines of Growth*. Sales No. E.92.II.A.19.

[22]

J Evol Econ (1997) 7: 169–192

Journal of
Evolutionary
Economics

Multinational enterprises and technological spillovers: An evolutionary model

Tommaso Perez*

Ufficio Italiano dei Cambi, Servizio Operazioni in Cambi, Via delle Quattro Fontane, 123, I-00184 Roma, Italy (Fax: +39-6-466 342 76)

Abstract. This paper analyses the determinants of the impact of foreign multinational enterprises on the technological development of domestic firms. It argues that the consequences of the foreign presence change according to market and technological conditions. An evolutionary model of technological competition between foreign and domestic firms is developed which is able to generate both vicious and virtuous circles of development in locations affected by foreign MNE activities. Theoretical analysis is tested against the empirical evidence for the UK economy (1983–1989).

Key words: Multinational enterprises – Technology transfer – Technological spillovers

JEL-classification: F23, O19, O33

1 Introduction

Economic analysis has devoted a considerable effort to the analysis of the impact of international production on the competitiveness of indigenous firms. The debate has centred on the existence of technological spillovers from foreign multinational enterprises (MNEs) onto domestic firms, which may spur their pace of technological development by means of technology diffusion, imitation or increasing competition.

* I am grateful to John Cantwell, Aldo Femia, Pier Carlo Padoan, Paolo Ramazzotti and Alessandro Sterlacchini, the participants to the 21st Annual Conference of the European International Business Academy and to two anonymous referees for helpful comments and suggestions. Any opinions expressed in this paper are those of the author and do not reflect those of the Ufficio Italiano dei Cambi. Usual disclaimers also apply.

Although all authors seem to agree on the importance of the local activity of foreign MNE affiliates for a country's development, opinions differ widely concerning its impact. While most scholars argue for a positive effect of the presence of foreign multinationals on the technological development of domestic firms (e.g., Caves, 1974; Findlay, 1978; Blomstrom, 1989), the opinion of some (e.g., Hymer, 1970) is entirely different, stressing the relevance of the negative consequences arising from the exportation of big firms' market power to foreign countries. Others, such as Cantwell and Dunning (Dunning and Cantwell, 1986; Cantwell, 1987, 1989), believe that the sign of the impact of the foreign presence depends on a set of conditions, and that it may turn into both virtuous and vicious circles of technological development in the locations affected by foreign MNE activity.

Just as there is no agreement at the theoretical level, so there are evident contrasts in the scant empirical evidence available. Using cross-sector industry data, Caves (1974), Globerman (1979), Blomstrom (1989), Blomstrom and Wolff (1994) and Kokko (1992) find that a foreign presence helps domestic firms to achieve higher levels of productivity.[1] Haddad and Harrison (1993) do not detect any spillover effect in a sample of Moroccan firms. Cantwell (1989) finds that the ability of European firms to react to competition by US MNEs during the postwar period has changed according to their initial level of technological development *vis-à-vis* US competitors. Drawing on this and other evidence, Cantwell highlights three stylised cases which emerge from extensive empirical research regarding the industrialised countries (Cantwell, 1993). First, in industries where the technological tradition of indigenous firms is well-established, domestic firms are generally able to react promptly to foreign threats, to assimilate foreign technologies and to mobilise resources to compete with foreign MNEs in their own home markets. In the centres of excellence for development of a particular technology, the process of technological competition between foreign and domestic firms may generate technological spillovers in both directions through the concentration in these locations of high-tech phases of production and/or R&D activities by foreign and domestic firms. Thus a virtuous circle of technological development may propel the indigenous industry towards new technological frontiers. Second, where technological and industrial development of indigenous firms is very weak, the activity of foreign affiliates may benefit the host country by increasing employment and, if it is the case, by stimulating the development of indigenous vertically related firms. Third, in the intermediate case, where the domestic technological tradition is moderately developed and the prosperity of indigenous firms depends on a protected local market, the presence of foreign MNEs is likely to damage their positions heavily. Drawing on their superior technological capabilities and on the competitive advantages deriving from exploitation of an international network of production, marketing and innovation, foreign MNEs easily displace domestic firms, which

[1] It is worth noting that all the above studies, with the exception of that by Caves (Australia) and Globerman (Canada), examine the Mexican manufacturing industry.

"fall further and further behind and are gradually driven out of world markets in a process of cumulative decline."[2]

In contrast to Cantwell's analysis, existing formal models of technological spillovers from foreign MNEs generally reflect the optimistic view of MNE activity referred to above. Like the spread of a disease, the larger the foreign presence in the national industry, the more foreign technologies spill over to domestic firms (Koizumi and Kopecky, 1977; Findlay, 1978; Das, 1987; Wang, 1990). Moreover, the wider the 'distance' between backlog indigenous firms and foreign MNE affiliates, the greater the room for improving the technological capability of domestic firms by the imitation of foreign technologies (Findlay, 1978; Wang and Blomstrom, 1992).[3]

The model to be formulated here differs from the previous ones with respect to two main points. Firstly, it explains the interaction between foreign and indigenous firms at the technological level in a way which allows the generation of two different outcomes: one corresponding to a virtuous circle of technological development and the other to a process of cumulative decline in the domestic industry. Secondly, unlike catching-up models, it is formulated to include cases in which domestic firms are ahead of foreign MNEs and spillovers are multidirectional. Thirdly, by assuming the existence of different levels of technological development among domestic firms, the model is able to explain the impact of the foreign presence onto the average productivity of domestic firms as the sum of two different effects: (i) the technological spillovers and (ii) the crowding-out of the less efficient domestic enterprises.

In Sec. 2, attention is focused on a set of circumstances that may affect the impact of international production on the technological development of domestic firms. More specifically, it is argued that both technological conditions (technological disparities, ease of learning, the appropriability regime, etc.) and market conditions (the extent of foreign penetration and its rate of increase, etc.) play a crucial role. A relatively wide technological gap can be easily reduced when a foreign presence is modest and slowly growing, or when industrial policies support R&D efforts by indigenous firms. Conversely, left alone to face market competition and a rapidly increasing foreign penetration, even indigenous firms characterised by a relatively narrow technological gap may be driven out of the market by foreign competitors. Drawing on this analysis, a model is developed in which different outcomes can be generated according to market and technological conditions (Sec. 3). The empirical evidence for the UK economy during the period 1983–1989 is reported in Sec. 4. Conclusions follow as usual.

---

[2] Cantwell (1987), p. 28.

[3] Of the latter models, Wang and Blomstrom's is the richest of implications. It assumes that the majority of spillovers do not arise 'automatically' from the foreign presence, but instead result from technological competition between foreign and indigenous firms. I will return to this point throughout the text.

2 Foreign presence, technological gap and host country's technological development: some critical issues

Even if a large number of interrelated circumstances[4] are likely to play an important role in shaping the impact of international production on the technological development of indigenous firms, technological and market conditions are probably of major importance. Among the latter, the analysis in this section will focus on three main factors which may affect spillovers, namely, the initial technology gap of domestic firms, the level of foreign penetration and its rate of increase, and the implications of the existence of market selection processes driven by technological disparities.

Technological gap and technological spillovers

An important point to be discussed in the search for a formal representation of the impact of foreign MNEs activity on the technological development of indigenous firms is the relationship between technological spillovers and technological disparities. Two very different perspectives exist in the literature on this matter. On the one hand, it is argued, according to the 'advantages of backwardness hypothesis', that the wider the disparities, the greater the opportunities for indigenous firms to achieve higher levels of efficiency by imitating foreign technologies (Findlay, 1978). On the other hand, it is asserted that a wide technological gap impairs indigenous firms' ability to catch up with foreign competitors (Cantwell, 1989, 1993).

Clearly, these two assumptions have very different implications. In the first case, international production would show a tendency to reduce inequalities among nations. In the other case, MNE activity would be a reinforcing factor in diverging development patterns. Even if it might be accepted that a minimum level of technological disparity must exist in order to create room for the possibility that the "leaking out" of technological knowledge from MNE affiliates will benefit indigenous firms' technological capabilities,[5] there exist several reasons for giving priority to the other point of view. I shall deal with them briefly.

In the first place, recent empirical and theoretical literature about technological activity and technology diffusion demonstrates that a firm's ability both to follow and to adapt to technological developments by other firms and organisations – viz., the firm's *absorptive capacity* – and also to pursue independent lines of technological development, depends on its existing technological capability (Cohen and Levinthal, 1989; Cantwell, 1993). Thus, the hypothesis of the existence of 'advantages of backwardness'

[4] For instance, Dunning (1988) argues for the existence of a set of Economic-Social and Policy conditions (the ESP paradigm) which influences the impact of international production on a country's development.

[5] However, if technological change is 'localised' (Atkinson and Stiglitz, 1969), in the sense that firms from different countries tend to follow different technological trajectories, it is argued that a sufficient condition for spillovers might be the existence of 'diversities', instead of 'disparities', in technological development between indigenous and foreign firms.

ensuring to backlog firms faster rates of productivity growth has been partly replaced by the idea that 'success breeds success', so that the most competitive firms are likely to consolidate their positions cumulatively. In the case of the host country's effects of international production, this implies that the ability of indigenous firms to benefit from the dissemination of foreign technology increases with their initial level of technological competence. Indigenous firms that lag far behind foreign competitors may be unable to sustain – by imitating foreign technologies or by pursuing indigenous lines of technological development – sufficient rates of productivity growth and product innovations to defend their relative competitiveness *vis-à-vis* other firms (which include foreign MNE affiliates) operating in the market.

Secondly, it is widely recognised that the existence and the amount of technological spillovers depend on a heterogeneous set of conditions, which include the linkages that foreign MNE affiliates establish with vertically-related indigenous firms and the quality of the technologies imported by foreign MNE affiliates. While linkages affect the size and the frequency of technological spillovers, and technology imports the amount of technological knowledge that may leak out from foreign MNE affiliates to indigenous firms, both factors are likely to be positively related with the level of technological development of the domestic industry. In fact, on the one hand, a low level of technological competence will render indigenous firms unable to meet the technological standards required by foreign firms and, hence, will reduce local sourcing by foreign enterprises with perverse effects both on the balance of payments and technological spillovers (Dunning, 1986). On the other hand, since technology imports are expensive (Teece, 1977), foreign firms will have no reason to import more and newer technologies from their parent companies, if competition from domestic firms does not threaten their market shares and profits. Thus, in such cases "... *a large foreign presence may coexist with a slow technology transfer and the transfer of old technologies*" [6] and, hence, it may result in a low amount of technology 'available' for imitation by indigenous firms.

The main problem arising from the above discussion is to find a way to take into account both the positive relationship between the technological disparities and the opportunities of improving the technological level of domestic firms by the imitation of foreign technologies suggested by the 'advantages of backwardness hypothesis', and the negative relationship that links technological disparities with the ability of indigenous firms to appropriate technological spillovers from foreign MNE affiliates. One reasonable way to deal with this problem is to assume that spillovers increase with the technological gap up to a certain critical level. Beyond this point, technological competence by indigenous firms will be so low that they will generally not be able to exploit fully the technological opportunities arising from a foreign MNE presence. Thus, as technological disparities increase

[6] Wang and Blomstrom (1992), p. 138.

further, spillovers diminish as does the ability of domestic firms to catch up with foreign competitors.[7]

Obviously, various factors determine the critical size of the gap compatible with the competitive upgrading of indigenous firms. The appropriability regime, the level of complexity of the technologies in use by foreign firms, and other technological factors influence the "leaking out" of technological knowledge from foreign plants. Market conditions, which includes the level of foreign penetration and its rate of increase, affect, as we shall soon discuss, a firm's profits and, hence, the resources that are available to monitor, decode and imitate foreign technologies, and/or to pursue indigenous lines of technological development. Government intervention influences spillovers by modifying competitive environment, by helping the competitive upgrading by domestic firms and/or by stimulating technology import by foreign MNE affiliates.

Foreign presence and technological spillovers

Another valuable point of discussion is the relationship between technological spillovers and the level of the foreign presence within a national industry, as well as its rate of increase. Both factors may be the outcome of technological disparities between foreign and indigenous firms, so that "*...widespread foreign investment usually indicates that the process of competitive upgrading in an economy is not entirely healthy because domestic firms in many industries lack the capabilities to defend their market position against foreign firms.*"[8]

On the other hand, it is also likely that the same factors will affect to some extent domestic firms' ability to imitate foreign technologies or to exploit indigenous lines of technological development.

Wang and Blomstrom (1992) hypothesise that the majority of spillovers do not arise 'automatically' from the foreign presence. Instead, they assume that the imitation of foreign technologies requires specific investment in learning activities by indigenous firms. Hence, since learning investment depends on a firm's profits, which are negatively correlated with foreign firms' output and technological disparities, it follows that the greater the foreign presence, the lower, *ceteris paribus*, spillovers will be. Similarly, Cantwell (1989, 1993) suggests that, where the domestic technological tradition is weak, the ability of foreign MNEs to capture market shares may force indigenous firms to cut back on research or to concentrate efforts on narrower fields of specialisation. Empirical evidence collected by Globerman and Meredith (1984) for Canada and by Veugelers and Vanden Houte (1990) for Belgium suggests that a foreign presence reduces indigenous firms' R&D expenditure, while Kokko (1992) does not detect any

[7] Similarly, in a north-south framework, Verspagen (1991) assumes that the wider the distance between backlog and developed countries, the greater 'potential' spillovers will be. However, a wide technological gap might also impair the ability of southern firms to imitate northern technologies and, hence, cause a limited amount of 'actual' spillovers.

[8] Porter (1990). p. 671.

positive effect of foreign presence on the productivity level of Mexican firms in sectors characterised by a high (over 50%) foreign share and/or wide technological gap.[9]

Combining these pieces of evidence and analytical reflections with the discussion of the preceding section, it follows that the reaction of indigenous firms is likely to be determined jointly by market and technological conditions. More specifically, it is argued that the critical level of technological disparities, beyond which indigenous firms' absorptive capacity declines, does not depend on technological factors alone, but also on other circumstances affecting indigenous firms' R&D efforts and profitability. In industries where a relatively wide technological gap is associated with a relatively low and slowly growing foreign share, indigenous firms may be able to catch up gradually. Conversely, where technological disparities are relatively modest but indigenous firms have to cope with a foreign presence that is 'too high' or that grows 'too fast', the same firms may be unable to compete with foreign rivals and may enter into a process of cumulative decline.

Indigenous firms' productivity growth and market selectiveness

In the preceding discussion, it has been pointed out that both technological and market conditions contribute to determine the impact of foreign direct investment over the technological development and the market performance of domestic firms. What is missing is an explicit analysis of the process explaining the interaction between these two sets of conditions.

In what follows, I suggest that market selection processes driven by technological disparities may constitute the interface by which technological and market conditions interact with one another. A tight selective environment will imply a strong relationship between a firm's competitiveness and its level of profits. Hence, market selectiveness will ensure the survival of only few firms very homogenous in terms of technological competence. *Vice versa*, a loose selective environment will permit even firms that lag far behind the technological frontier to earn sufficient profits to finance the investment required in order not to fall further behind competitors (Teece, Pisano and Shuen, 1990; Cantwell, 1993). As it is represented in Fig. 1a, technological and market conditions constitute the inputs of the selective process in determining a firm's relative competitiveness (which depends on each firm's market share and absolute competitiveness, as it will be more formally set out in the following section) and its absorptive capacity. The selection mechanism processes this information and sets the resultant change in a firm's market share and its actual level of profits. Consequently (Fig. 1b), the output of the selective process influences the technological

[9] It should be noted, as Kokko does, that the latter results may be due to a possible selection bias in the grouping of industry. In fact, by splitting the sample of industries according to the level of the technological gap at a certain point of time, we can automatically place in the group of sectors characterised by a high technology gap those industries where the current level of the gap is the result of a lack of technological spillovers.

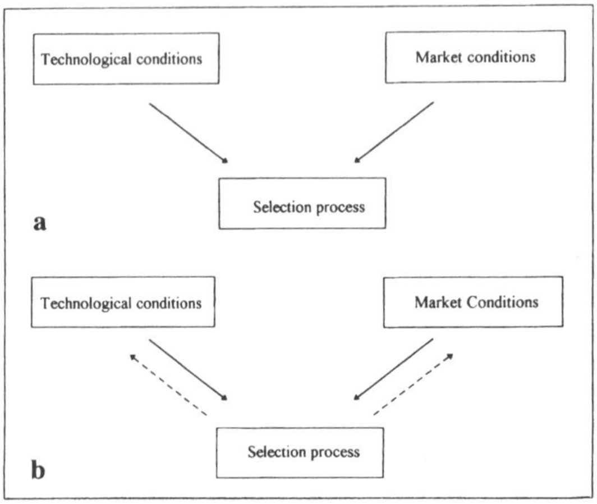

Fig. 1a,b. The interaction between technological and market conditions: a simplified schema

and market conditions of the subsequent period by determining resources available for learning investment, new machinery and capital equipment, each firm's market share, and, hence, each firm's productivity growth and relative competitiveness.

Clearly, additional circumstances might alter or reinforce the causation mechanism at work by affecting one or more elements of this stylised representation. For instance, the growth of demand associated with new foreign direct investment (FDI) and average productivity growth might partly offset the tightness of the selective environment. Government intervention by financing R&D expenditure, favouring technological co-operation between foreign and indigenous firms and/or regulating FDI flows can alter the competitive game and the spread of the leaking out of technological knowledge from foreign plants.

3 The model

The preceding section has been dedicated to a presentation of the principal factors affecting the interrelationship between FDI and the technological development of indigenous firms. Accordingly, the following pages will be devoted to developing a model that embodies the following properties:

(i) technological spillovers depend on the absorptive capacity of firms;
(ii) technology imports by foreign firms are inversely related to the existing technological gap;
(iii) a firm's market share dynamic depends on its technological level relative to the other firms operating in the market;
(iv) the dimension of the market grows with the national income according to a Keynesian-type demand formation mechanism.

Some possible extensions of the model will be discussed in the final part of this section.

In formulation of the model, I have drawn extensively on the work by Dosi and Freeman (1992)[10] aimed at explaining the process of forging ahead and falling behind among countries according to an evolutionary perspective. However, substantial changes have been made with respect to the original formulation. Firstly, given the scope of this paper, equations describing the interaction between foreign MNEs and indigenous firms at the technological level have been incorporated into the model. This has been done to take account of some of the more interesting lines of development in the recent literature concerning technological spillovers from foreign MNEs. As in the Wang and Blomstrom (1992) model, the decision by foreign firms to import new technology depends on the existing technological gap between foreign and indigenous firms, while the imitation of foreign technology requires specific investment. Secondly, for the sake of simplicity, the number of parameters has been reduced by eliminating stochastic processes and foreign trade. Thirdly, the model refers to a single-sector economy.

Although comprising these substantial modifications with respect to Dosi and Freeman's work, the model illustrated below still belongs within the realm of evolutionary economics. In fact, it displays two of the main characteristics of that tradition.[11] The first is the existence of a strong path-dependency: the absorptive capacity of indigenous firms depends on their past process of technological accumulation, this being a major cause of virtuous or vicious circles of technological development. Secondly, heterogeneity among individuals (e.g., between foreign and domestic enterprises or within each group of firms) is explicitly recognised as a cause of evolutionary paths, driven by market selectiveness.[12]

Market selection process

Following Dosi, Silverberg and Orsenigo (1988), it is assumed that the evolution of the market share of indigenous firms in terms of total output (f) depends on their relative competitiveness (RC) *vis-à-vis* the other firms operating in the market (the symbol $\dot{f}(t-1,t)$ stands for the change occurred in the value of f between t−1 and t):

$$\dot{f}_i(t-1,t) = \alpha\,[RC_i(t-1) - 1]\,f_i(t-1)$$

Taking into account that $f_i(t) = \frac{Q_i(t)}{Q(t)}$, where Q_i and Q are a firm's and total output, and assuming, for the sake of simplicity, the existence of only two firms – one domestic (firm 1) and one foreign (firm 2) representative of the

[10] See also Dosi and Fabiani (1994).

[11] For a wider discussion of these issues see, for instance, Hall (1994).

[12] Conversely, a third characteristic of the evolutionary approach, namely its probabilistic nature, is not present in my model, so as not to complicate my formulation and numeric simulations with elements that do not affect significantly the issues analysed. However, probabilistic processes may be easily reintroduced without changing the main framework of the model.

two group of firms[13]– the previous equation has been reformulated as regards the domestic firm into:

$$\dot{Q}_1(t-1,t) = \alpha\,[RC_1(t-1)-1]\frac{Q_1(t-1)}{Q(t-1)}Q(t) + \frac{Q_1(t-1)}{Q(t-1)}\dot{Q}(t-1,t) \quad (1)$$

with

$$RC_i(t) = \frac{E_i(t)}{\overline{E}(t)} \quad (2)$$

$$E_i(t) = \frac{1}{P_i(t)} \quad (3)$$

$$\overline{E}(t) = \sum_i f_i(t)E_i(t) \quad (4)$$

where P_i is the price charged by the i-th firm and E_i its absolute competitiveness.

Similarly, the output dynamic of the foreign firm depends on its relative competitiveness and on the flows of new foreign direct investment (FDI), measured in terms of the additional investment available to employ new workers (N^{FDI}):

$$\dot{Q}_2(t-1,t) = \alpha\,[RC_2(t-1)-1]\frac{Q_2(t-1)}{Q(t-1)}Q(t) + \frac{Q_2(t-1)}{Q(t-1)}\dot{Q}(t-1,t) + N^{FDI}(t)\pi_2(t)P_2(t) \quad (5)$$

with

$$N^{FDI}(t) = \frac{FDI(t)}{w_2(t)} \quad (6)$$

$$FDI(t) = \tau Y(t-1) \quad (7)$$

where π is labour productivity and Y the real national product,

$$Y(t) = \sum_i \frac{Q_i(t)}{P_i(t)} \quad (8)$$

Each firm fixes prices according to a simple mark-up rule over labour costs:

$$P_i(t) = (1+\rho)\frac{w_i(t)}{\pi_i(t)} \quad (9)$$

where w is the nominal wage and ρ the mark-up.

Workers use all the wages received at time $t-1$ to buy goods in the following period. Thus, total demand, which is equal to total output, as we shall see later, is equal to:

$$Q(t) = \sum_i w_i(t-1)N_i^P(t-1) + w_2N^{FDI}(t-1) + \sum_i w_iN_i^{|M|}(t-1) \quad (10)$$

[13] This assumption will be removed later in this section.

where N^P and $N^{|M|}$ are the number of workers engaged in, respectively, productive and imitating activities.

Technology import and imitation

As has been discussed above, it is assumed that a firm's capacity to absorb foreign technologies depends:

(a) on its existing level of technological competence; and
(b) on its learning efforts.

Thus, the indigenous firm's labour productivity growth is related to the technological gap *vis-à-vis* the foreign MNE (proxy for domestic firm's level of technological competence), which is defined as the ratio between labour productivity in foreign and domestic plants (i.e., $GAP(t) = \pi_2(t)/\pi_1(t)$), and to the number of workers engaged in the learning activity. Furthermore, technological spillovers are greater, the wider the gap up to a certain critical level: thereafter absorptive capacity declines:

$$\frac{\dot{\pi}_1(t-1,t)}{\pi_1(t-1)} = \mu_1 + \lambda_0 \left[1 - \exp\left\{-\eta N_1^{|M|}(t-1)\right\}\right] \frac{GAP(t-1) - 1}{\exp\{\lambda_1 GAP(t-1)\}} \tag{11}$$

In the numerical simulations reported below, parameter λ_1 has been set equal to 1. Therefore, the maximum of the function lies at the point where the distance between foreign and domestic firms is equal to its minimum value (i.e., where GAP = 1), as suggested by John Cantwell's approach to technological competition. Lower values of the parameter would instead be more close to the catching-up approach.

As regards the foreign firm, it is assumed that productivity growth depends on two different factors: (i) the decision to import new technologies from its mother company and (ii) the imitation of domestic firms' technologies. For what concerns the technology import decision, it has been simply assumed that technology import is inversely related to the existing productivity gap between foreign and domestic firms. As concerns point (ii), technological spillovers from indigenous to foreign firms generate alike spillovers from foreign to indigenous enterprises. They are maximised for small gaps and decrease as the gap move away in both directions.[14]

[14] The formulation of equation 12 has given rise to many problems. In fact, in the case of foreign firms it is difficult to identify an upper limit of resources that can be devoted to technology import and imitation. In some circumstances, it is possible that such resources would be higher than profits, because of the easier access to capital with respect to domestic counterparts and/or because international sourcing strategies by MNEs might suggest the need to tap into foreign lines of technological development independently from the profitability of the foreign subsidiary. However, it should be noted that, if either the second or the third element of the rhs of equation 12 are eliminated, the dynamics of the model do not substantially change.

$$\frac{\dot{\pi}_2(t-1,t)}{\pi_2(t-1)} = \mu_2 + \delta_0\left[1 - \exp\left\{-\eta N_2^{|M|}(t-1)\right\}\right] \times \frac{[1/GAP(t-1)] - 1}{\exp\{\delta_1[1/GAP(t-1)]\}} + \delta_2 GAP \exp\{-GAP(t-1)\} \quad (12)$$

Labour market dynamic

The last part of the model describes labour market dynamics.

All profits are reinvested by firms in the imitating activity. Thus, we have:

$$N_1^{|M|}(t) = \frac{RD_i(t)}{w_i(t)} \quad (13)$$

with

$$RD_i(t) = P_i(t)Q_i(t) - w_i(t)N_i^P(t) \quad (14)$$

As regards workers employed in the productive activity, firms instantaneously adjust employment to the level of demand:

$$N_i^P(t) = \frac{Q_i(t)}{P_i(t)\pi_i(t)} \quad (15)$$

Nominal wages change with inflation, average productivity growth and total employment:

$$\frac{\dot{w}_i(t-1,t)}{w_i(t)} = \vartheta_1 \frac{\dot{\overline{\pi}}(t-2,t-1)}{\overline{\pi}(t-2)} + \vartheta_2 \frac{\dot{\overline{P}}(t-2,t-1)}{\overline{P}(t-2)} + \vartheta_3 \frac{\dot{N}(t-2,t-1)}{N(t-2)} \quad (16)$$

with

$$\overline{\pi} = \sum_i f_i \pi_i \quad (17)$$

$$\overline{P} = \sum_i f_i P_i \quad (18)$$

$$N = \sum_i N_i^P + \sum_i N_i^{|M|} \quad (19)$$

In the numerical simulations set out below, it is hypothesised that the wage dynamic is the same for both firms, but that wages paid by foreign firms are equal to $w_2(t) = (1+\varepsilon)w_1(t)$ with $\varepsilon > 0$, in order to take into account the additional costs that foreign firms have to sustain to penetrate the host country's market.

Comments

Before discussing simulation results, a few comments on the working of the model are in order.

Firstly, the expansion of foreign MNE's output produces two opposite effects. Increasing the income produced and distributed tends to expand domestic firms' profits and, hence, their R&D efforts. However, the increasing foreign share reduces the relative competitiveness of indigenous firms and raises wages, eroding their profits. The algebraic sum of these two effects is the principal engine driving the dynamics of the model.

Secondly, the major deficiencies of the model are the closed-economy assumption and the exogeneity of FDI flows. The incorporation of foreign trade dynamics will probably reinforce the causation mechanisms at work. Endogenising FDI would be an important extension of my analysis.

Simulation results

A large number of different parametrisations were performed, which showed the model's robustness to variations in the parameters. In the numerical simulations reported, parameters were set up as follows: $\alpha = 0.1$ (relatively weak market selection process), $\lambda_0 = 0.1$, $\lambda_1 = 1$, $\delta_0 = 0.1$, $\delta_1 = 1$, $\delta_2 = 0.04$, $\mu_1 = \mu_2 = 0$, $\vartheta_1 = 1$, $\vartheta_2 = 1$ (wages adjust completely to productivity and price growth), $\vartheta_3 = 0$ (labour supply elasticity at a given wage rate equal to infinity), $\eta = 0.06$, $\varepsilon = 0.2$ (foreign firms face labour costs 20% higher than domestic firms), $\rho = 0.1$ (mark up is 10% over production costs). The results reported refer to 30 iterations given the initial conditions $f_1 = 0.99$ and $f_2 = 0.01$.

In a first group of simulations, where $\tau = 0$ (no new FDI after time $t = 0$), the results obtained highlight the influence of the initial technological gap on the 'success' or the 'failure' of domestic firms faced with foreign competition. Three different types of reaction emerge.

When the initial technological gap is wide, domestic firms do not have the necessary absorptive capacity to imitate successfully foreign technology (Fig. 2). The market selection process drives indigenous competitors out of the market, eroding domestic firms' profitability and, hence, their investment in learning. Foreign firms permeate the economy. Moreover, since indigenous firms exert a very moderate competitive pressure over foreign

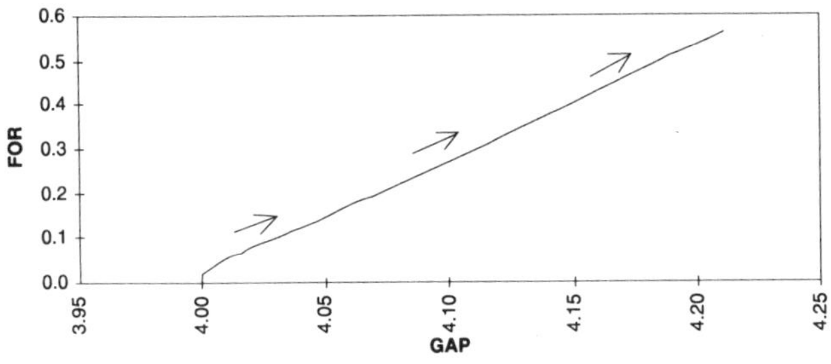

Fig. 2. The vertical axis measures foreign penetration (FOR) in terms of total output, the horizontal axis the productivity gap (GAP)

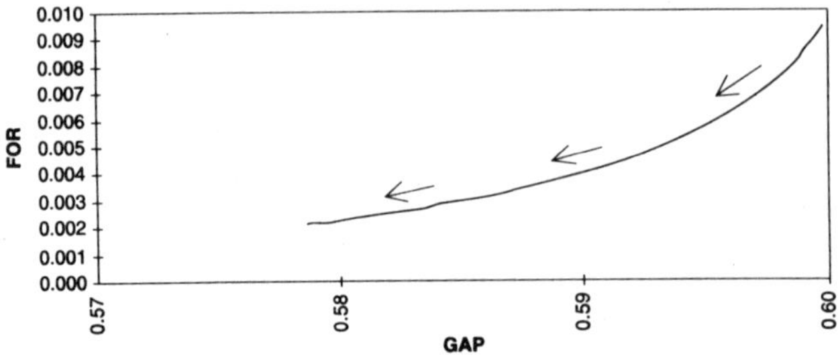

Fig. 3. Axes as in Fig. 2

firms, technology import is modest and the location concerned becomes the basis for low value-added activities by foreign MNEs.

Conversely, when the initial technological competence of domestic firms is high, they are able to reduce the original gap (Fig. 3) and to contrast successfully competition from foreign firms, while the location attracts high-tech foreign MNE productive activities. Technological competition and spillovers interact with one another in a virtuous circle of technological development.

Of particular interest is an intermediate case characterised by an initial apparent process of catching up by domestic firms, which subsequently changes into a rapid process of falling behind (Fig. 4). In this case, the initial conditions in terms of technological gap and foreign penetration ensure sufficient profits for domestic firms to finance R&D activities for the successful imitation of foreign technologies. However, as the foreign presence increases, market conditions erode the domestic firms' profitability, causing the cumulative decline of the national industry.

A second set of simulations was performed in order to detect the influence of FDI flows on the technological development of indigenous firms. Figures 5 to 7 report the effects of a change in the parameter τ from 0 to 0.1 on the technological gap, domestic firms' productivity and real output. The results obtained seem consistent with Cantwell's analysis. Locations characterised by a high level of technological development benefit from an increase in the FDI flows. The technological gap diminishes, while productivity

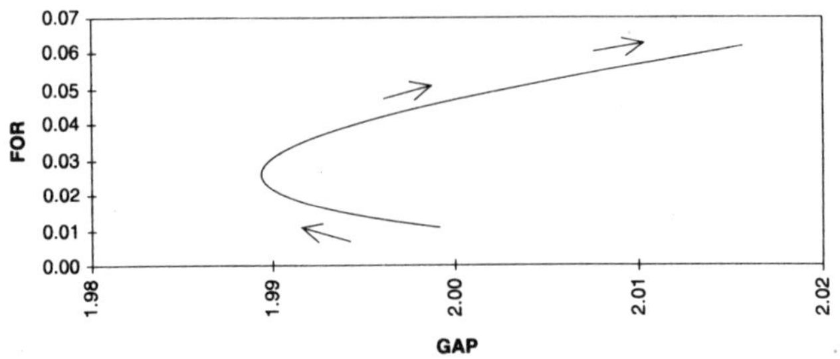

Fig. 4. Axes as in Fig. 2

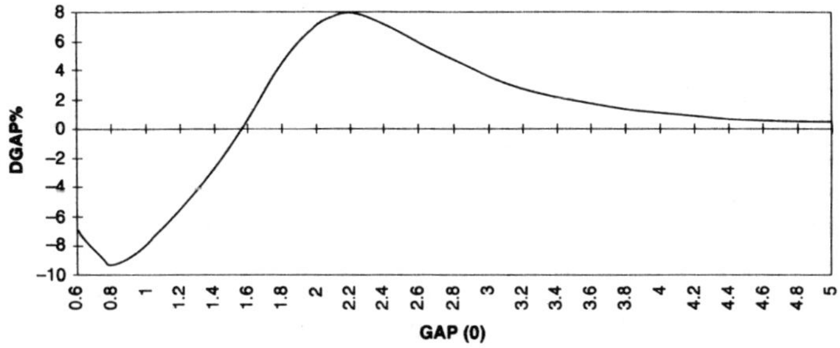

Fig. 5. The vertical axis measures the percentage change in the productivity gap (DGAP%) which occurs when τ increases from 0 to 0.1, the horizontal axis the initial productivity gap (GAP 0)

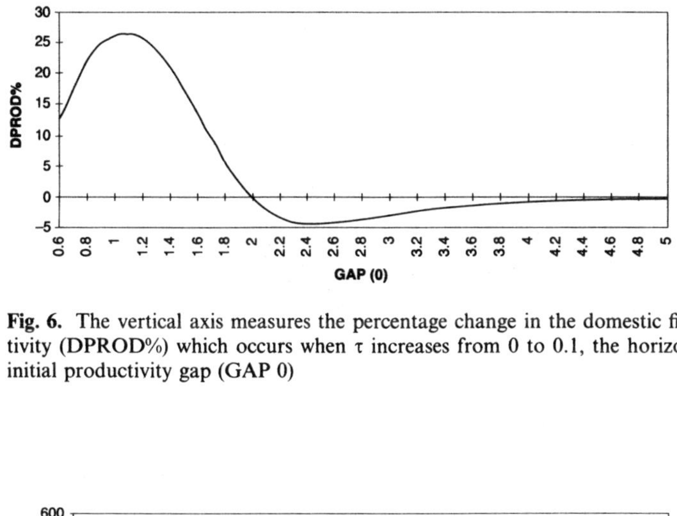

Fig. 6. The vertical axis measures the percentage change in the domestic firm's productivity (DPROD%) which occurs when τ increases from 0 to 0.1, the horizontal axis the initial productivity gap (GAP 0)

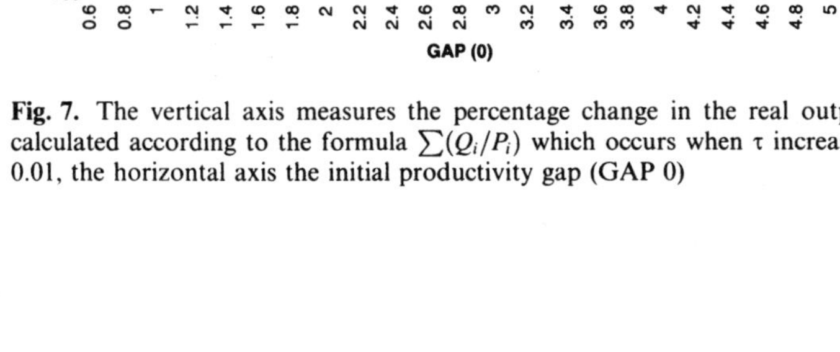

Fig. 7. The vertical axis measures the percentage change in the real output (DQ/P%) calculated according to the formula $\sum(Q_i/P_i)$ which occurs when τ increases from 0 to 0.01, the horizontal axis the initial productivity gap (GAP 0)

and real output increase. A completely different dynamic prevails in production sites characterised by a mid/low level of technological development. Here, rapidly increasing foreign penetration erodes domestic firms' profits and, hence, their investment in learning. The gap increases, productivity and real output decrease. Finally, locations characterised by a very low level of technological development are not harmed by an increase in FDI flows. Here, the lack of any technological and productive capability by indigenous firms entails that the market expansion effects of FDI outweigh the crowding-out effects on domestic firms. Technological gap, domestic firms' productivity and real output are almost unchanged.

Some possible extensions of the model

Up to now it has been assumed that there exist only two firms, one domestic and one foreign. The latter hypothesis hides one important implication of selection mechanisms described in the preceding section. In fact, average productivity growth by a group of firms depends not only on the productivity growth by individual enterprise, but also on the shrinking market shares of the least efficient ones (e.g., Nelson, 1968).

Thus, when domestic firms are heterogeneous with respect to their level of technological development, foreign entry might cause higher rates of average productivity growth by domestic firms because of two distinct effects: the technological spillovers and the faster elimination of the most backward indigenous firms. While the spillover effects are clearly beneficial for the host country's economy, the crowding-out of laggard domestic competitors is not.

Figures 8 and 9 confirm these intuitive results. They have been obtained after having modified the original model in order to allow the existence of two domestic firms at a different level of initial technological development. The horizontal axis in both figures reports the initial market share of the most backward ones (firm 1). The vertical axis measures in Fig. 8 the change from time 0 to 30 of domestic firms' average productivity gap and in Fig. 9 the correspondent variation in their average productivity.

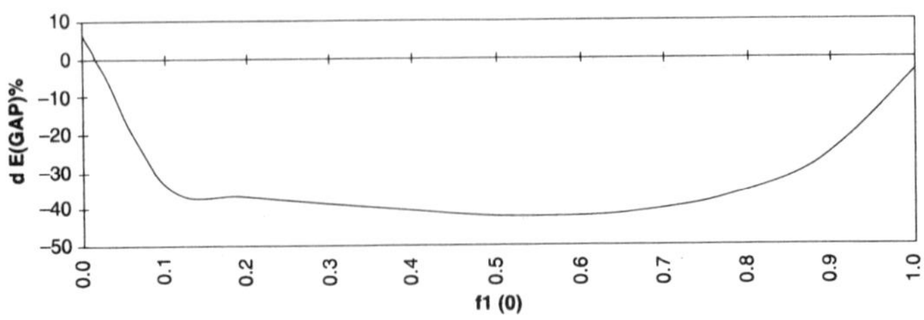

Fig. 8. The vertical axis measures the percentage change in the average productivity gap of domestic firms. The horizontal axis the initial market share of firm 1

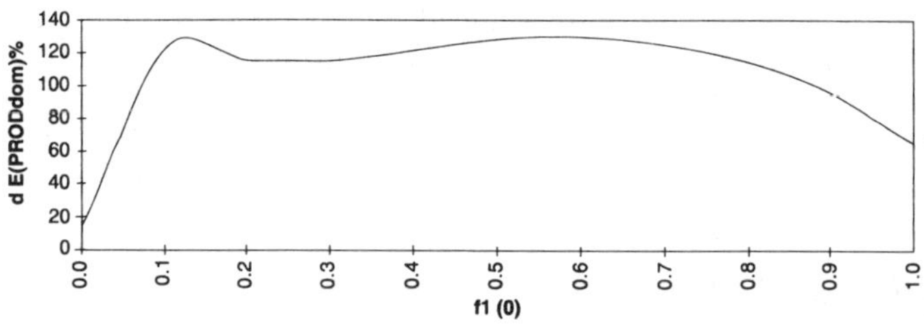

Fig. 9. The vertical axis measures the percentage change in the average productivity by domestic firms. The horizontal axis the initial market share of firm 1

What clearly emerges from these simulations is that the relationship between initial average technological disparities between foreign and domestic firms and average productivity performance by domestic enterprises is non-linear. In most of the cases a given rate of average productivity growth by domestic firms might be associated with very different initial conditions: a large number of technologically backward indigenous firms or a predominant presence of technologically developed domestic competitors. However, while in the first case the productivity growth by domestic firms is mainly due to the crowding-out process of domestic enterprises, in the second case it is principally associated with the competitive upgrading of the indigenous industry.

To sum up, the analysis carried out in this section concerns two different aspects of the impact of inward FDI on the host countries' technological development. Firstly, the technological development of domestic firms is a crucial factor in shaping the ability of indigenous competitors to benefit from spillovers and to generate the conditions for a virtuous interplay between foreign and indigenous firms. Secondly, the productivity performance of individual enterprises is likely to be different from the performance of the "average" firm. In fact, the combination of technological spillovers with crowding-out effects might be the cause of non-linear relationships between observed rates of average productivity growth by indigenous firms and spillovers.

4 Empirical evidence

The scarcity of available data on international production does not permit in-depth assessment of the goodness-of-fit of my theoretical analysis to empirical evidence. However, some indications can be drawn from the data available on foreign MNE operations for some countries. Hereafter, I shall refer to the experience of the UK economy during the 1980s; this case is of particular relevance to my analysis given the particular strategy pursued by UK governments over the period under investigation. In fact, industrial policies were characterised by measures designed to attract a larger share of FDI and by close reliance on market forces to generate industrial development.

For empirical purposes, use has been made of data drawn from the UK Census of Production, which refers to 18 two-digit manufacturing sectors in 1983 and 1989. The data set, which is described in further detail in the appendix, allows the figures for domestic and foreign enterprises to be separated.

Test 1. Technological competence and productivity growth

A main point of theoretical analysis has been the dependence of indigenous firms' ability to catch up with foreign competitors on their initial level of technological competence. Given the lack of data on the technological development of indigenous and foreign firms with the same classification as my industry-level data, I have used the initial technological gap as a proxy for this variable. The latter is defined as the ratio between average productivity in foreign firms and average productivity in domestic firms (respectively, GAPVAE when referring to value-added per employee and GAPNOE when using net output per employee).

A first attempt to regress domestic firm's productivity growth between 1983 and 1989 (DLVAED and DLNOED) on the initial technological gap (GAPVAE83 and GAPNOE83) for the whole sample of 18 industries yielded very unsatisfactory results (results not reported). This finding is not surprising, bearing in mind the analysis of previous sections. On the one hand, industries at a very low level of development may have benefited from in-coming FDI because the expansive effects of the presence of foreign firms may have outweighed the crowding-out effects on domestic firms. The setting up of new plants by foreign firms may have enhanced the development of vertically-related indigenous firms, and this phenomenon may have offset the negative effects produced on indigenous firms in direct competition with foreign MNE plants.[15] On the other hand, differences in the conditions possibly affecting technological spillovers (such as technological complexity, the appropriability regime, etc.) may result in non-linear relationships between the initial technological gap and productivity growth when referring to cross-sector data. Both points of view seem to find some empirical support, as we shall see.

To investigate the first issue, a non-linear OLS regression was performed on the same variables employed in the preceding estimations. Indigenous firms' productivity growth was regressed against a quadratic function of the initial technological gap, with very good results obtained in terms of fit, especially when value-added per employee was employed as a performance indicator (eqs. a and b in table 1). In both cases, as expected, an upward-oriented parabola was obtained: productivity growth is higher when indigenous firms are characterised by high or very low levels of the initial technological gap.

Even if apparently very satisfactory, the estimates reported above may be heavily affected, given the small number of observations, by the existence

[15] Note that given the high level of aggregation of the data it is likely that, in many cases, vertically-related firms belong to the same industrial sector.

Table 1. Domestic firms' productivity growth and technological gap, OLS regressions, 18 UK manufacturing industries. 1983–1989

(a) DLVAED = 2.609 −	3.730 GAPVAE83 +	1.439 GAPVAE83^2	$\bar{R}^2 = 0.56$	$n = 18$
(4.028)**	(4.138)**	(4.917)**		
(b) DLNOED = 4.511 −	6.620 GAPNOE83 +	2.538 GAPNOE83^2	$\bar{R}^2 = 0.29$	$n = 18$
(3.199)**	(3.002)**	(2.989)**		
(c) DLVAED = 0.562 −	0.285 GAPVAE83		$\bar{R}^2 = 0.15$	$n = 15$
(2.526)*	(−1.550)			
(d) DLNOED = 0.689 −	0.363 GAPNOE83		$\bar{R}^2 = 0.30$	$n = 15$
(3.773)**	(2.471)*			

Notes: absolute values of the t-statistics in brackets. *,** and denote respectively significance at the 5 and 1 percent level of confidence (two-tailed test). Standards errors have been corrected for heteroschedasticity. n is the number of observations

of a few outliers. Inspection of the plot of the data suggests the existence of only 3 industries in which an above-average initial productivity gap is associated with a catching-up process by domestic firms, while for the remaining industries there seemingly prevails a linear relationship between the initial productivity gap and its rate of increase. These three industries are: manufacturing of metal goods, man-made fibres and other manufacturing industries. Metal manufacturing is a sector which probably uses relatively simple technologies, which can easily be imitated even by firms lagging far behind foreign MNEs. Man-made fibres is a very small sector compared with the other industries at a 2-digit level.[16] Thus, the close-down of a few domestic enterprises or their acquisition by foreign MNEs may have heavily affected the performance of the two groups of firms. As regards the "other manufacturing industries", the heterogeneity of the firms belonging to it may by itself be a cause of abnormal behaviour. It is accordingly argued that the non-linear relationship between domestic firms' productivity growth and initial technological gap depends on the specific characteristics of a few sectors.

To test this hypothesis, a linear OLS was re-run after the data referring to these three sectors were excluded (eqs. c and d). The explanatory power of the estimated regressions is significantly reduced when we use value-added per employee as a performance indicator. Conversely, using net output per employee the goodness-of-fit of our estimation is almost unchanged.

Test 2. Technological competence and catching up processes

More than domestic firms' productivity growth *tout court*, it may be relevant to analyse their performance relative to foreign competitors. To this

[16] In 1983, it accounted for roughly 0.2% of total manufacturing employment and net output. In the same year, there were only 4 foreign enterprises operating in the sector.

Table 2. Catching up by domestic firms and technological gap, OLS regressions, 18 UK manufacturing industries, 1983–1989

(e) DLGAPVAE =	−2.011 + (2.488)*	3.090 GAPVAE83 − (2.754)**	1.125 GAPVAE83^2 (3.158)**	$\bar{R}^2 = 0.20$	$n = 17$
(f) DLGAPNOE =	−3.457 + (2.542)*	5.305 GAPNOE83 − (2.609)**	1.963 GAPNOE83^2 (2.634)**	$\bar{R}^2 = 0.15$	$n = 17$
(g) DLGAPVAE =	−0.553 + (3.100)**	0.517 GAPVAE83 (3.381)**		$\bar{R}^2 = 0.36$	$n = 15$
(h) DLGAPNOE =	−0.616 + (4.001)**	0.562 GAPNOE83 (4.535)**		$\bar{R}^2 = 0.51$	$n = 15$

Notes: as Table 1

end, I turn now to changes occurring in the technological gap over the period concerned (DLGAPVAE and DLGAPNOE) and to their relationships with the initial technological disparities.

Table 2 (eqs. e and f) sets out results for the non-linear specification of the model, after excluding observations for the manufacturing of metal goods sector (the full-sample gives a poor fit). Both equations give satisfactory results in terms of R-squared. All coefficients are statistically different from zero and show the expected signs. In fact, the downward oriented parabolas obtained from the OLS regressions suggest that the gap has decreased (or increased less) in sectors characterised by a low initial productivity gap or a very high one.

After excluding the three sectors identified above, the OLS linear regressions were re-run (eqs. g and h). Using changes in the technological GAP as the dependent variable, linear regressions fit much better than non-linear ones. Again, industries characterised by a high initial technological gap seemed to be falling further behind foreign MNEs. Conversely, where the technological gap was modest, indigenous firms seem to have been able to catch up with foreign firms.

Test 3. Market selection and foreign presence

The last issue investigated here is the relationship between the initial technological gap and output growth by domestic firms. In previous sections it was supposed that a high technological gap is reflected in a relatively slow-growing market share by domestic firms because of the existence of market selection mechanisms driven by technological disparities.

If technological capabilities are not a major factor affecting a firm's competitiveness, we should find no statistically significant relationship between the technological gap and output growth by domestic firms. *Vice versa*, when technological factors are important, a negative relationship between the two variables should be registered.

The results reported in Table 3 seem to confirm the latter hypothesis. Using the gap in terms of either value-added or output per employee, indigenous firms' output growth (DLNOD) seems to have been negatively affected by the initial technological disparities. The larger the technological gap between domestic and foreign firms, the smaller the output growth by domestic firms.

Table 3. Domestic firms' output growth and technological gap, OLS regressions, 18 UK manufacturing industries, 1983–1989

(i) DLNOD = 0.655 − 0.349 GAPVAE83 (4.152)** (2.850)**	$\bar{R}^2 = 0.22$	$n = 15$
(j) DLNOD = 0.750 − 0.417 GAPNOE83 (5.087)** (3.711)**	$\bar{R}^2 = 0.34$	$n = 15$

Notes: as Table 1

5 Concluding remarks and policy implication

This paper suggests that the effect of foreign MNE presence on the technological development of indigenous firms is likely to vary according to a number of circumstances. In particular, I have focused attention on two factors affecting the sign of the impact of in-flowing FDI on domestic firms' competitiveness, namely, the original technological gap between foreign and domestic firms and the pace of FDI flows. In this respect, a large technological gap may impair the capacity of domestic firms to absorb foreign technology, while a foreign presence in rapid expansion might dramatically reduce their profitability and hence their investment in R&D activity, new machinery and capital equipment.

On the basis of a theoretical model that adopts an evolutionary perspective, I have argued that when domestic firms excel in the development of a particular technology, they benefit from in-coming FDI. In this case, technological competition between foreign and domestic firms creates room for a large amount of technological spillovers. By the rapid absorption of foreign technologies, indigenous firms reduce the original gap and spur foreign MNEs to import more advanced technology. Conversely, indigenous firms lagging far behind foreign competitors are penalised by in-flowing FDI. The greater competitive pressure generated by the foreign presence erodes their profits and hence reduces their investment in learning activities, new machinery and capital equipment. This gives rise to a cumulative process of technological decline in the locations concerned, which eventually leads to a large presence of foreign firms specialised in low-tech productions.

Similarly, I have suggested that an increase in the pace of FDI inflows produces different effects according to the level of technological development of indigenous industries. Highly technologically developed indigenous competitors may easily offset the negative influences of greater competitive pressure by exploiting the expansive effects associated with FDI. Backward domestic firms are instead likely to be almost completely crowded out by a rapidly growing foreign presence.

The empirical evidence for the UK seems to confirm these predictions. In all regressions, indigenous firm's productivity growth and catching-up *vis-à-vis* foreign MNEs are inversely correlated with the initial productivity gap. Similarly, output growth by domestic firms is negatively affected by technological disparities.

Both the theoretical and the empirical analyses of the preceding pages challenge the *laissez-faire* view about in-flowing FDI as the optimal policy prescription. Instead, it is suggested that the positive (or negative) effects of the foreign MNE entry might be enhanced (offset) by regulating FDI flows and by taking actions to speed up the dissemination of foreign technologies among domestic firms, either by encouraging technological co-operation between foreign and indigenous enterprises or by financing R&D efforts by indigenous firms.

The scant empirical evidence available does not allow the measurement of costs and benefits of the available policy instruments, and to assess the best combination of them. However, it is able to show that international production is not a process which unambiguously leads to the competitive upgrading of the indigenous industry. Instead, it entails losers and winners which may correspond, alternatively, with foreign or indigenous firms, or with a group of enterprises independently from their nationality. Governments may use their power to change significantly the result of the game in order to maximise economic and social gains and to minimise loses.

This does not imply that they should try to stop the increasing movements towards the internationalisation of productive activities. Instead, as Dunning said, governments should "*...be aware both of the role MNEs may play in restructuring their economies, and of their own responsibilities in ensuring that MNEs, in achieving their corporate objectives, do not drive out indigenous firms (suppliers as well as competitors) ...*"[17] To this end, I believe, particular attention should be given to measures able to restrain competitive pressure by regulating FDI flows for the time necessary for the competitive upgrading of the domestic industry, while governments do their best to rapidly improve the technological competence of the indigenous firms.

Appendix. Data sources and description of variables

All the data are from UK Census of Production PA1002. Data for domestic-owned firms have been obtained by subtracting the figures for majority-owned foreign enterprises (Table 19)[18] from the corresponding figures for all UK enterprises (Table 13). Percentage changes have been obtained by log differentiating the original variables after dividing them for the producer price index (source OECD).

The 18 two-digit sectors used in the empirical analysis are:

[17] Dunning (1985), p. 429.

[18] Table numbers refer to UK Census of Production PA1002.

Table A.1. SIC classification of British manufacturing sectors

SIC(REV. 2)	Sector
22	Metal manufacturing
24	Manufacturing of non metallic mineral products
25	Chemical industry
26	Production of man made fibres
31	Manufacture of metal goods not elsewhere specified
32	Mechanical engineering
33	Manufacture of office machinery and data processing equipment
34	Electrical and electronic engineering
35	Manufacture of motor vehicles and other parts thereof
36	Manufacture of other transport eqiupment
37	Instrument engineering
41/42	Food, drink and tobacco
43	Textiles
45	Footwear and clothing
46	Timber and wooden furniture
47	Paper and paper products; printing and publishing
48	Processing of rubber and plastic
49	Other manufacturing industries

References

Atkinson AB, Stiglitz JE (1969) A new view of technological change. Economic Journal 79: 573–78

Blomstrom M (1989) Foreign investment and spillovers: a study of technology transfer to Mexico. Routledge, London

Blomstrom M, Wolff EN (1994) Multinational corporations and productivity convergence in Mexico. In: Baumol W, Nelson R, Wolff E (eds) International convergence of productivity: cross-national studies and historical perspectives. Oxford University Press, Oxford

Cantwell J (1987) The reorganisation of European industries after integration: selected evidence on the role of multinational enterprise activities. Journal of Common Market Studies 26: 127–151. Reprinted in: Dunning J, Robson P (eds) Multinationals and the European Community. Basil Blackwell, Oxford

Cantwell J (1989) Technological innovation and multinational corporations. Basil Blackwell, Oxford

Cantwell J (1993) Technological competence and evolving patterns of international production. In: Cox H, Clegg J, Ietto-Giles G (eds) The growth of global business. Routledge, London

Caves RE (1974) Multinational firms, competition and productivity in host-country markets. Economica 41: 176–193

Cohen WM, Levinthal DA (1989) Innovation and learning: the two faces of R&D. Economic Journal 99: 569–96

Das S (1987) Externalities, and technology transfer through multinational corporations. Journal of International Economics 22: 171–182

Dosi G, Freeman C et al. (1992) The diversity of development patterns: on the process of catching up, forging ahead and falling behind. Paper presented at the International Economic Association Meeting, Varenna (Italy), October 1–3

Dosi G, Fabiani S (1994) Convergence and divergence in the long-term growth of open economies. In: Silverberg G, Soete L (eds) The economics of growth and technological change. Edward Elgar, Aldershot

Dunning J (1985) Multinational enterprises, economic structure and international competitiveness. Wiley/IRM, Chichester
Dunning J (1986) Japanese participation in the British industry. Croom Helm, London
Dunning J (1988) Multinationals, technology and competitiveness. Allen and Unwin, London
Dunning J, Cantwell J (1986) The changing role of multinational enterprises in the international creation, transfer and diffusion of technology. In: Arcangeli F, David PA, Dosi G (eds) Technology diffusion and economic growth: international and national policy perspectives. Oxford University Press, Oxford
Findlay R (1978) Relative backwardness, direct foreign investment, and the transfer of technology: a simple dynamic model. Quarterly Journal of Economics 92: 1–16
Globerman S (1979) Foreign direct investment and 'spillover' efficiency benefits in Canadian manufacturing industries. Canadian Journal of Economics 12: 42–56
Globerman S, Meredith L (1984) The foreign ownership-innovation nexus in Canada. Columbia Journal of World Business 19: 53–62
Haddad M, Harrison A (1993) Are there positive spillovers from direct foreign investment? Evidence from panel data for Morocco. Journal of Economic Development 42: 51–74
Hall P (1994) Innovation, economics and evolution. Harvester Wheatsheaf, Hemel Hempstead
Hymer S (1970)The efficiency (contradictions) of multinational corporations. American Economic Review 60: 441–448
Koizumi T, Kopecky KJ (1977) Economic growth, capital movements and the international transfer of technical knowledge. Journal of International Economics 7: 45–65
Kokko A (1992) Foreign direct investment, host country characteristics and spillovers. Stockholm School of Economics, Stockholm
Mansfield EM, Romeo A (1980) Technology transfer to overseas subsidiaries by US-based firms'. Quarterly Journal of Economics 95: 737–750
Nelson RR (1968) A "diffusion" model of international productivity differences in manufacturing industries. American Economic Review 58: 1219–48
Porter ME (1990) The competitive advantages of nations. Free Press, New York
Silverberg G, Dosi G, Orsenigo L (1988) Innovation, diversity and diffusion: A self-organisation model. Economic Journal 98: 1032–54
Teece DJ (1977) Technology transfer by multinational firms: the resource costs of transferring technological know-how. Economic Journal 87: 242–261
Teece DJ, Pisano G, Shuen A (1990) Firm capabilities, resources, and the concept of strategy. Mimeo, University of California, Berkeley
Verspagen B (1991) Uneven growth between interdependent economies. Avebury, Aldershot
Veugelers R, Vanden Houte R (1990) Domestic R&D in the presence of multinational enterprises. International Journal of Industrial Organisation 8: 1–15
Wang Y (1990) Growth, technology transfer, and the long-run theory of international capital movements. Journal of International Economics 29: 255–271
Wang Y, Blomstrom M (1992) Foreign investment and technology transfer: a simple model. European Economic Review 36: 137–155

Name Index